ESO ASTROPHYSICS SYMPOSIA
European Southern Observatory

Series Editor: Bruno Leibundgut

Springer
Berlin
Heidelberg
New York
Barcelona
Hong Kong
London
Milan
Paris
Tokyo

Physics and Astronomy ONLINE LIBRARY

http://www.springer.de/phys/

ESO ASTROPHYSICS SYMPOSIA
European Southern Observatory

Series Editor: Bruno Leibundgut

G. Meylan (Ed.), **QSO Absorption Lines**
Proceedings, 1994. XXIII, 471 pages. 1995.

D. Minniti, H.-W. Rix (Eds.), **Spiral Galaxies in the Near-IR**
Proceedings, 1995. X, 350 pages. 1996.

H. U. Käufl, R. Siebenmorgen (Eds.),
The Role of Dust in the Formation of Stars
Proceedings, 1995. XXII, 461 pages. 1996.

P. A. Shaver (Ed.), **Science with Large Millimetre Arrays**
Proceedings, 1995. XVII, 408 pages. 1996.

J. Bergeron (Ed.), **The Early Universe with the VLT**
Proceedings, 1996. XXII, 438 pages. 1997.

F. Paresce (Ed.), **Science with the VLT Interferometer**
Proceedings, 1996. XXII, 406 pages. 1997.

D. L. Clements, I. Pérez-Fournon (Eds.),
Quasar Hosts
Proceedings, 1996. XVII, 336 pages. 1997.

L. N. da Costa, A. Renzini (Eds.), **Galaxy Scaling Relations: Origins, Evolution and Applications**
Proceedings, 1996. XX, 404 pages. 1997.

L. Kaper, A. W. Fullerton (Eds.), **Cyclical Variability in Stellar Winds**
Proceedings, 1997. XXII, 415 pages. 1998.

R. Morganti, W. J. Couch (Eds.), **Looking Deep in the Southern Sky**
Proceedings, 1997. XXIII, 336 pages. 1999.

J. R. Walsh, M. R. Rosa (Eds.), **Chemical Evolution from Zero to High Redshift**
Proceedings, 1998. XVIII, 312 pages. 1999.

J. Bergeron, A. Renzini (Eds.), **From Extrasolar Planets to Cosmology: The VLT Opening Symposium**
Proceedings, 1999. XXVIII, 575 pages. 2000.

A. Weiss, T. G. Abel, V. Hill (Eds.),
The First Stars
Proceedings, 1999. XIII, 355 pages. 2000.

A. Fitzsimmons, D. Jewitt, R. M. West (Eds.),
Minor Bodies in the Outer Solar System
Proceedings, 1998. XV, 192 pages. 2000.

L. Kaper, E. P. J. van den Heuvel, P. A. Woudt (Eds.), **Black Holes in Binaries and Galactic Nuclei: Diagnostics, Demography and Formation**
Proceedings, 1999. XXIII, 378 pages. 2001.

G. Setti, J.-P. Swings (Eds.), **Quasars, AGNs and Related Research Across 2000**
Proceedings, 2000. XVII, 220 pages. 2001.

A. J. Banday, S. Zaroubi, M. Bartelmann (Eds.), **Mining the Sky**
Proceedings, 2000. XV, 705 pages. 2001.

E. Costa, F. Frontera, J. Hjorth (Eds.),
Gamma-Ray Bursts in the Afterglow Era
Proceedings, 2000. XIX, 459 pages. 2001.

S. Cristiani, A. Renzini, R. E. Williams (Eds.),
Deep Fields
Proceedings, 2000. XXVI, 379 pages. 2001.

Series homepage – http://www.springer.de/phys/books/eso/

E. Costa F. Frontera J. Hjorth (Eds.)

Gamma-Ray Bursts in the Afterglow Era

Proceedings of the International Workshop
Held in Rome, Italy, 17-20 October 2000

 Springer

Volume Editors

Enrico Costa
IAS/CNR, Via Fosso del Cavaliere
00133 Roma, Italy

Filippo Frontera
Physics Department, University of Ferrara
Via Paradiso 12, 44100 Ferrara, Italy

Jens Hjorth
Astronomical Observatory, Juliane Maries Vej 30
2100 Copenhagen, Denmark

Series Editor

Bruno Leibundgut
European Southern Observatory
Karl-Schwarzschild-Strasse 2
85748 Garching, Germany

Library of Congress Cataloging-in-Publication Data applied for.

Die Deutsche Bibliothek - CIP-Einheitsaufnahme

Gamma ray bursts in the afterglow era : proceedings of the international workshop,
held in Rome, Italy, 17 - 20 October 2000 / E. Costa ... (ed.). –
Berlin ;Heidelberg ; New York ; Barcelona ; Hong Kong ; London ; Milan ; Paris ;
Tokyo : Springer, 2001
 (ESO astrophysics symposia) (Physics and astronomy online library)

ISBN 978-3-642-07668-8   e-ISBN 978-3-540-45505-9

Springer-Verlag Berlin Heidelberg New York
a member of BertelsmannSpringer Science+Business Media GmbH

http://www.springer.de

Cover design: Erich Kirchner, Heidelberg

Preface

This book contains the Proceedings of the 2nd Workshop on 'Gamma-Ray Bursts in the Afterglow Era' held from 17 to 20 October 2000 at the Consiglio Nazionale Ricerche (CNR) Headquarters in Rome, Italy.

The success of the first conference on the same topic held at the same location two years before (3–6 November 1998) prompted us to organize another workshop. Following the increased engagement of ESO in GRB science, Alvio Renzini joined the organizers of the previous workshop (Enrico Costa, Filippo Frontera and Luigi Piro).

Two years of Gamma-Ray Burst (GRB) observations since the last workshop have more than doubled the sample of GRBs localized and followed up for afterglow search. Most of this information came from BeppoSAX, with important contributions from other satellites, in particular from RXTE, Chandra and the InterPlanetary Network (IPN). The number of cases monitored with multiwavelength observations was increased and several new events were followed up. An important discovery of the last two years was the simultaneous observation of gamma- and optical radiation from the prompt emission of GRB 990123, a nice confirmation of a prediction firmly stated by Re'em Sari at the first workshop three months before the occurrence of GRB 990123! Another relevant observational result was the definite discovery of breaks in the afterglow light curves of some GRBs and the detection, with BeppoSAX and Chandra, of other emission features in the X-ray afterglow of a few GRBs. All these and other important discoveries have increased the interest in the GRB phenomenon from both the observational and the theoretical points of view. The Rome Workshop was just the meeting needed to report and discuss, among the wide scientific community interested in the GRB phenomenon, these discoveries, their possible interpretations, and the perspectives on the field.

The bad news from the two-year interval from the first to the second workshop was the premature death of Jan van Paradijs, one of the leaders of the GRB afterglow era, who led the team who discovered the first optical afterglow of a GRB. Another piece of bad news was the similarly premature death of Daniele Dal Fiume, one of the key people in the BeppoSAX team, who provided a relevant contribution to the development of the BeppoSAX data analysis software, especially the high-energy instruments PDS and GRBM. At the workshop, Jan van Paradijs was remembered by Ed van den Heuvel, and Daniele Dal Fiume was remembered by Filippo Frontera.

The workshop also marked an important evolution in the development of space- and ground-based instrumentation performing episodic or systematic observations of GRBs and associated phenomena. It was the first dedicated conference after the unexpected and unfortunate deorbitation of the Compton Gamma-Ray Observatory, which corresponded to the termination of the real-time alerts from the BACODINE system. An important development, on the other hand, was the rejuvenated capability of the IPN, deriving from the activation of the burst detector aboard NEAR. One of the highlights of the workshop was the first presentation, by George Ricker, of the HETE-2 mission status shortly after its launch. High expectations of a new step forward in GRB science rely on this new dedicated mission.

Most of the data presented in the workshop were based on BeppoSAX observations, by itself or as part of a network. A central role in the domain of X-ray measurements was played by the reports concerning Fe features in GRB spectra. After the two marginal line detections from GRB 970508 (Piro et al. 1999) and GRB 970828 (Yoshida et al. 1999), two much more statistically significant detections of afterglow X-ray lines were reported at the workshop and are contained in this book: a line from the GRB 000214 afterglow detected using BeppoSAX (Antonelli et al., this book) and a line from the afterglow of GRB 991216 observed using Chandra (Piro et al., this book). The reported line properties confirm the importance of X-ray spectroscopy as a tool to probe the circumburst matter distribution. Moreover, the report (Frontera et al., this book) of the first detection of a transient absorption feature in the earliest emission of GRB 990705 confirms the scenario of a medium that is completely photoionized by the burst radiation. All these data, although deriving from different bursts, indicate a significant amount of Fe in the GRB proximity, providing important hints on the GRB progenitor.

Another topic on the afterglow physics reported at the workshop is the study of deviations of the optical light curves from the straight simple power law, that steer the evolution of the most widely accepted synchrotron shock model toward unavoidable complications. The major role of the ESO instruments, in particular of the Very Large Telescope (VLT), apart from providing a better coverage of the Southern sky, is producing important data for photometry, spectroscopy and polarimetry, and this was reported on during the workshop. The most brilliant result is, without any question, the detection of the thus far highest redshift, 4.5, for GRB 000131 (Andersen et al., this book). Also, ample space was devoted to future satellites able to accurately localize GRBs (e.g. INTEGRAL, AGILE, Swift, GLAST). Most of the observational and theoretical results from the workshop are included in this book. The different parts of the book trace the various sessions of the conference.

The workshop organization was managed by two CNR Institutes (Istituto Tecnologie e Studio Radiazioni Extraterrestri, Bologna, and Istituto Astrofisica Spaziale, Roma), with a contribution from ESO. Many people contributed to the success of the 2nd Workshop on GRBs. We wish to thank all the members of the Scientific Advisory (see Table 1) and Local Organizing (see Table 2) Committees;

Table 1. Scientific Advisory Committee

E. Costa (Co-chair)	IAS/CNR, Rome, Italy
G. Fishman	MSFC/NASA, Huntsville, USA
D. Frail	NRAO, Socorro - USA
F. Frontera (Co-chair)	University of Ferrara and ITESRE/CNR, Bologna, Italy
J. Heise	SRON, Utrecht, The Netherlands
E. van den Heuvel	University of Amsterdam, The Netherlands
J. Hjorth	Astronomical Observatory, Copenhagen, Denmark
K. Hurley	University of California, Berkeley, USA
C. Kouveliotou	Univ. Space Res. Assoc., MSFC/NASA, Huntsville, USA
S. Kulkarni	Caltech, Pasadena, USA
M. Matsuoka	RIKEN, Wako-shi, Japan
P. Mészáros	Penn State University, University Park, USA
H. Pedersen	Astronomical Observatory, Copenhagen, Denmark
L. Piro (Co-chair)	IAS/CNR, Rome, Italy
M. Rees	IoA, Cambridge University, United Kingdom
A. Renzini (Co-chair)	ESO, Garching, Germany
G. Ricker	MIT, Boston, USA
M. Ruderman	Columbia University, New York, USA
L. Scarsi	IFCAI/CNR, Palermo, Italy
M. Tavani	IFC/CNR, Milan, Italy
G. Vedrenne	CESR/CNRS, Toulouse, France
M. Vietri	University of Rome III, Rome, Italy
S. Woosley	UCSC, Santa Cruz, USA

Table 2. Local Organizing Committee

L. Barbanera	IAS/CNR, Rome, Italy
M. Feroci (Chair)	IAS/CNR, Rome, Italy
G. Gandolfi	IAS/CNR, Rome, Italy
M. Orlandini	ITESRE/CNR, Bologna, Italy
J. Hjorth	Astronomical Observatory, Copenhagen, Denmark

in particular we thank Marco Feroci for the hard work of coordinating the many operative and scientific issues of the workshop, Lidia Barbanera for her skill and nice solutions for many logistical and administrative problems, Mauro Orlandini and Giangiacomo Gandolfi for managing the webpage of the workshop, Giuseppe Di Persio and Giorgio Patria for managing the computer system, Rossana Morani for secretarial support, Isabella Vannutelli for the organization of the press-conference and the press-release, and Pamela Bristow from ESO for taking charge of the contacts with Springer-Verlag for the publication of these proceedings.

We also wish to thank Aldo Spizzichino for his original design and skillful realization of the workshop poster. Finally a very warm thanks goes to Nicola Masetti for taking charge of the contact with the authors, paper collection, and

VIII   Preface

for helping us a lot, along with Lorenzo Amati, in the editing of this book. We are also grateful to our sponsors, CNR, ESO, Alenia Aerospazio, Laben and Nuova Telespazio for their support.

Bologna,
October 2001

Enrico Costa
Filippo Frontera
Jens Hjorth

Contents

Part I   Global Properties of GRBs

Some Recent, Interesting Observations of Gamma-Ray Bursts
K. Hurley   3

On the Fast Spectral Variability of GRBs
E.P. Mazets, R.L. Aptekar, P.S. Butterworth, T.L. Cline, D.D. Frederiks,
S.V. Golenetskii, V.N. Il'inskii, and V.D. Pal'shin   9

Spectral Properties of Short Gamma-Ray Bursts
W.S. Paciesas, R.D. Preece, M.S. Briggs, and R.S. Mallozzi 13

X-Ray Flashes and X-Ray Rich Gamma Ray Bursts
J. Heise, J. in 't Zand, R.M. Kippen, and P.M. Woods 16

BATSE Observations of Fast X-Ray Transients
Detected by BeppoSAX-WFC
R.M. Kippen, P.M. Woods, J. Heise, J. in't Zand, R.D. Preece,
and M.S. Briggs 22

Observations of Gamma-Ray Bursts
with the Rossi X-Ray Timing Explorer
H. Bradt, A.M. Levine, F.E. Marshall, R.A. Remillard, D.A. Smith,
and T. Takeshima 26

Testing the Optically Thin Synchrotron Shock Model
for Gamma-Ray Bursts Spectra from 2 to 700 keV
with BeppoSAX
L. Amati, F. Frontera, M. Tavani, J.J.M. in 't Zand, E. Costa,
C. Guidorzi, E. Montanari, and P. Soffitta 34

On the Spectra of the Gamma-Ray Bursts
Z. Bagoly, I. Csabai, A. Mészáros, I. Horváth, R. Vavrek,
and L.G. Balázs 37

Short Gamma-Ray Bursts Are Different
J.P. Norris, J.D. Scargle, and J.T. Bonnell 40

New Possibilities Offered by BeppoSAX:
Automatic GRB Alerts Using GRBM
*C. Guidorzi, F. Frontera, E. Montanari, F. Calura, L. Amati, E. Costa,
and M. Feroci* 43

Probing the Isotropy in the Sky Distribution
of the Gamma-Ray Bursts
A. Mészáros, Z. Bagoly, L.G. Balázs, I. Horváth, and R. Vavrek, ... 47

Analysis of the BATSE GRB Light Curves
B.M. Belli 50

Gamma-Ray Burst Follow Up Observations with BOOTES
in 1998–2000
*J.M. Castro Cerón, A.J. Castro-Tirado, R. Hudec, J. Soldán, M. Bernas,
P. Páta, T.J. Mateo Sanguino, A. de Ugarte Postigo, J.Á. Berná,
M. Nekola, J. Gorosabel, B.A. de la Morena, J.M. Más-Hesse,
Á. Giménez, and J. Torres Riera* 53

Non-isotropic Angular Distribution
for Very Short-Time Gamma-Ray Bursts?
D.B. Cline, C. Matthey, and S. Otwinowski 56

Tools for Gamma-Ray Burst Data Mining
*J. Hakkila, R.S. Mallozzi, R.J. Roiger, D.J. Haglin, G.N. Pendleton,
and C.A. Meegan* 60

Broadband Spectral Deconvolution of GRBs
*L. Hanlon, D. Kinsella, N. Murphy, B. McBreen, K. Bennett,
O.R. Williams, C. Winkler, and R. Preece* 63

A Gamma-Ray Bursts' Fluence-Duration Correlation
I. Horváth, L.G. Balázs, P. Mészáros, Z. Bagoly, and A. Mészáros 66

Estimation of Emission Time Parameter for APEX Experiment
A. Kozyrev, I. Mitrofanov, D. Anfimov, and C. Barat 69

Quiescent Times in Gamma-Ray Bursts
A. Merloni and E. Ramirez-Ruiz 72

Neutrino Astrophysics with the MACRO Detector
*T. Montaruli, F. Cei, R. Pazzi, and F. Ronga,
for the MACRO Collaboration* 75

The Fingerprints of the GRB Process
*F. Quilligan, B. McBreen, K. Hurley, L. Hanlon, D. Watson,
and S. McBreen* 78

A Variety of Decays of Gamma-Ray Burst Pulses
F. Ryde and R. Svensson 81

The GRBs at Rest Frames of Emitters
A. Sanin, I. Mitrofanov, D. Anfimov, M. Litvak, M. Briggs, W. Paciesas,
G. Pendleton, R. Preece, G. Fishman, and C. Meegan 84

The Unique Signature of Shell Curvature in Gamma-Ray Bursts
A.M. Soderberg and E.E. Fenimore.................................. 87

Final Results of the Off-Line Scan of the BATSE Daily Records
B.E. Stern, Ya. Tikhomirova, D. Kompaneets, and R. Svensson.......... 91

Part II   GRB Afterglows

X-Ray Afterglows and Features of Gamma-Ray Bursts
L. Piro 97

Transient Spectral Features in the Prompt Emission
of Gamma-Ray Bursts with BeppoSAX
F. Frontera, L. Amati, E. Costa, C. Guidorzi, M. Vietri,
and J.J.M. in 't Zand 106

Discovery of a Redshifted Iron K-Line
in the X-Ray Afterglow of GRB 000214
L.A. Antonelli, M. Vietri, L. Piro, E. Costa, P. Soffitta, M. Feroci,
L. Amati, F. Frontera, E. Pian, J. in 't Zand, L. Stella,
and G.C. Perola 112

Observations of Iron Features with ASCA
T. Murakami, D. Yonetoku, and A. Yoshida......................... 115

Temporal and Spectral Analysis of X-Ray Afterglows of GRBs
Observed by BeppoSAX
G. Stratta, L. Piro, P. Soffitta, A. Antonelli, E. Costa, M. Feroci,
F. Frontera, G. Gandolfi, J. Heise, J. in 't Zand, L. Nicastro,
and E. Pian... 118

Optical/Near-IR Observations of Gamma-Ray Bursts
in the Afterglow Era
A.J. Castro-Tirado... 121

The GRB Followup Euro–US Consortium:
Results from the ESO Telescopes
N. Masetti, on behalf of a large collaboration 127

GRB 000301C: A Possible Short/Intermediate Duration Burst
Connected to a DLA System
J. Gorosabel, J.U. Fynbo, B.L. Jensen, P. Møller, H. Pedersen, J. Hjorth,
M.I. Andersen, and K. Hurley 130

Hunting Gamma-Ray Bursts in the Lyman Forest;
GRB 000131 at $z = 4.50$
M.I. Andersen, J. Hjorth, H. Pedersen, B.L. Jensen, L.K. Hunt,
J. Gorosabel, P. Møller, J. Fynbo, and B. Thomsen 133

Evidence for a Supernova in the Ic Band Light Curve
of the Optical Transient of GRB 970508
V.V. Sokolov 136

Hypernovae and Gamma-Ray Bursts
P.A. Mazzali, K. Nomoto, K. Maeda, and T. Nakamura 139

Theoretical Implications
of the Gamma-Ray Burst–Supernova Connection
R.A. Chevalier 142

Properties of GRB Optical Afterglows: Colors and Luminosities
V. Šimon, N. Masetti, R. Hudec, and G. Pizzichini 148

GRB Optical Afterglows:
Correlation between Pair of Parameters
C. Bartolini, G. Beskin, G. Cosentino, A. Guarnieri, A. Piccioni,
and A. Pozanenko 151

Broad-Band Modelling of GRB Afterglows
E. Berger, R. Sari, D. Frail, and S. Kulkarni 154

The Jet and the Supernova in GRB 990712
G. Björnsson, J. Hjorth, P. Jakobsson, L. Christensen, E.J. Lindfors,
and S. Holland 157

On the Transient Fe K-Edge
in the Prompt Emission of GRB 990705
M. Böttcher, C.D. Dermer, L. Amati, and F. Frontera 160

On the Detectability of Delayed X-Ray Flashes from GRBs
M. Böttcher, C.L. Fryer, E.P. Liang, and I.A. Smith 163

Wolf-Rayet Stars and GRB Connection
A. Cherepashchuk and K. Postnov 166

Follow-Up Observations from Observatories Based in Spain
J. Gorosabel, A.J. Castro-Tirado, J. Greiner, J.M. Castro Cerón, S. Klose,
and N. Lund 169

Optical Observations of the Dark Gamma-Ray Burst GRB 000210
J. Gorosabel, J. Hjorth, H. Pedersen, B.L. Jensen, L.F. Olsen,
L. Christensen, E. Mediavilla, R. Barrena, J.U. Fynbo, M.I. Andersen,
A.O. Jaunsen, S. Holland, and N. Lund 172

The Light Curves of GRB 990123 and GRB 990510
G. Björnsson, S. Holland, J. Hjorth, and B. Thomsen 175

GRB/SNe Correlations:
Search for Unrecognized OAs of GRBs among Detected SNe
V. Hudcová, R. Hudec, N. Masetti, G. Pizzichini, and E. Palazzi 178

Optical GRB Analyses: Results from Archival Plates
R. Hudec and W. Wenzel 181

EN: Simultaneous Optical Data for GRBs
R. Hudec, J. Florian, R. Smida, I. Stoklasova, J. Pálek, N. Masetti,
E. Palazzi, and G. Pizzichini 185

Near-Infrared Polarimetric Observations of GRB Afterglows
S. Klose, B. Stecklum, and O. Fischer 188

Iron Line Emission in X-Ray Afterglows
D. Lazzati, G. Ghisellini, M. Vietri, F. Fiore, and L. Stella 191

BeppoSAX Observation of GRB990806:
From the Prompt Emission to the X-Ray Afterglow
E. Montanari, L. Amati, F. Frontera, C. Guidorzi, M. Capalbi, E. Costa,
M. Feroci, L. Piro, J. Heise, J.J.M. in't Zand, E. Pian, E. Palazzi,
N. Masetti, and L. Nicastro 195

GRB000615 in X-Rays
L. Nicastro, G. Cusumano, A. Antonelli, L. Amati, F. Frontera,
E. Palazzi, E. Pian, E. Costa, M. Feroci, L. Piro, J. in 't Zand,
and J. Heise 198

GRB980613 a Very Faint Burst with a Not So Faint Afterglow
Detected by BeppoSAX
P. Soffitta, L. Amati, L.A. Antonelli, E. Costa, M. Feroci, F. Frontera,
J. Heise, L. Nicastro, L. Piro, G. Stratta, and J.J.M. in't Zand 201

Physical Constraints from Broadband Afterglow Fits:
GRB000926 as an Example
S.A. Yost, R. Sari, F.A. Harrison, E. Berger, A. Diercks, T. Galama,
D. Reichart, D. Frail, and P.A. Price................................. 204

Part III   Host Galaxies and Cosmology with GRBs

The Observed Offset Distribution of GRBs about Their Hosts
J.S. Bloom and S.R. Kulkarni 209

A Deep, High-Resolution Imaging Survey of GRB Host Galaxies
*N.R. Tanvir, S. Holland, M.I. Andersen, G. Björnsson, J.U. Fynbo,
J. Gorosabel, J. Hjorth, A. Jaunsen, P. Møller, P. Natarajan,
and B. Thomsen* 212

Host Galaxies as Gamma-Ray Burst Distance Indicators
D. Band, R. Jimenez, and T. Piran 215

The GRB Host Galaxies and Redshifts
*S.G. Djorgovski, S.R. Kulkarni, J.S. Bloom, D.A. Frail, F.A. Harrison,
T.J. Galama, D. Reichart, S.M. Castro, D. Fox, R. Sari, E. Berger,
P. Price, S. Yost, R. Goodrich, and F. Chaffee* 218

Gamma-Ray Bursts as a Probe of Cosmology
D.Q. Lamb and D.E. Reichart 226

Construction of the Variability $\rightarrow$ Luminosity Estimator
D.E. Reichart and D.Q. Lamb 233

Early Afterglows as Probes for the Reionization Epoch
D. Lazzati, G. Ghisellini, F. Haardt, and A. Fernández-Soto 236

γ-Ray Burst Remnants: How Can We Find Them?
R. Perna 239

Progenitors of GRBs Originated in the Dense Star Clusters
Yu.N. Efremov 243

The Multiband Photometry of GRB Host Galaxies:
Comparison with the Spectral Energy Distributions
of Nearby Galaxies and Theoretical Modeling
*V.V. Sokolov, T.A. Fatkhullin, V.N. Komarova, E.R. Kasimova,
and V.I. Korchagin* 246

Multiscale Statistical Methods and the Angular Distribution
of Gamma-Ray Bursts
R. Vavrek, L.G. Balázs, A. Mészáros, I. Horváth, and Z. Bagoly 249

Determining the Gamma-Ray Burst Rate
as a Function of Redshift
N. Weinberg, C. Graziani, D.Q. Lamb, and D.E. Reichart 252

Part IV   Theories for GRBs and Their Afterglows

Gamma-Ray Burst Models: The Central Engine
S.E. Woosley 257

High-Energy Particles from γ-Ray Bursts
E. Waxman 263

**Ultra-high Energy Cosmic Rays and Neutron-Decay Halos
from Gamma Ray Bursts**
C.D. Dermer 269

On the Neutrino Flux from Gamma-Ray Bursts
D. Guetta, M. Spada, and E. Waxman 272

Efficiency and Spectrum of Internal γ-Ray Burst Shocks
M. Spada, D. Guetta, and E. Waxman 275

Observational Consequences of $e^{\pm}$ Pair Creation in γ-Ray Bursts
E. Ramirez-Ruiz, P. Madau, M.J. Rees, and C. Thompson 278

The Close Environment of GRB
P. Mészáros 281

Failed Optical Afterglows
G. Ghisellini, D. Lazzati, and S. Covino 288

Winds from Massive Stars and the Afterglows of γ-Ray Bursts
E. Ramirez-Ruiz, L.M. Dray, P. Madau, and C.A. Tout 291

**The Effects of a Gamma-Ray Burst
on Nearby Preplanetary Systems**
*P. Duggan, B. McBreen, L. Hanlon, L. Metcalfe, A. Kvick,
and G. Vaughan* 294

The Role of Dust in GRB Afterglows
D.Q. Lamb 297

Theory of GRB Afterglow
T. Piran and J. Granot 300

**Strange Afterglows from Embedded GRBs:
Reconciling Hypernovae with Slow Decays**
R.A.M.J. Wijers 306

Light Curves from an Expanding Relativistic Jet
J. Granot, M. Miller, T. Piran, W.M. Suen, and P.A. Hughes 312

GRBs and Fireball versus Precessing Gamma Jets
D. Fargion 315

**Neutrino Pair Annihilation above a Kerr Black Hole
with the Accretion Disk**
K. Asano 318

Dissipation Efficiency of Internal Shocks in GRB
A.M. Beloborodov 321

**Interaction between Internal Shocks and the Reverse Shock
for a GRB in a Dense Stellar Wind**
F. Daigne and R. Mochkovitch 324

**GRB Synchrotron-Self-Compton Emission
Generated by Self-Consistent Electron Distribution**
E.V. Derishev, V.V. Kocharovsky, Vl.V. Kocharovsky, and P. Mészáros ... 327

Effects of Free Neutrons on Gamma-Ray Bursts
E.V. Derishev, V.V. Kocharovsky, and Vl.V. Kocharovsky 330

Bursts and Black Holes
A. Gomboc and A. Čadež 333

Light Curves of Optical Afterglows
F. Hroch 336

Ultra Efficient Internal Shocks
S. Kobayashi and R. Sari 339

Light Curves of GRB Optical Flashes
S. Kobayashi 342

100 GeV Photons from Gamma-Ray Burst Fireballs
Vl.V. Kocharovsky, E.V. Derishev, and V.V. Kocharovsky 345

New Results on the Temporal Structure of GRBs
E. Nakar and T. Piran 348

**Signature of a Highly Magnetized Millisecond Pulsar
in GRB Afterglows**
B. Zhang and P. Mészáros 351

Part V   Experiments: Present and Future

The Swift Panchromatic GRB Mission
N. Gehrels 357

The Italian Contribution to the Swift Mission
G. Chincarini, G. Tagliaferri, and F.M. Zerbi 360

The INTEGRAL Burst Alert System
S. Mereghetti, D.I. Cremonesi, and J. Borkowski 363

AGILE and Gamma-Ray Bursts
M. Tavani, G. Barbiellini, A. Argan, N. Auricchio, P. Caraveo, A. Chen,
V. Cocco, E. Costa, G. Di Cocco, G. Fedel, M. Feroci, M. Fiorini, M. Galli,
A. Giuliani, C. Labanti, I. Lapshov, P. Lipari, F. Longo, S. Mereghetti,
E. Morelli, A. Morselli, A. Pellizzoni, F. Perotti, P. Picozza, C. Pittori,
C. Pontoni, M. Prest, M. Rapisarda, E. Rossi, A. Rubini, P. Soffitta,
M. Trifoglio, E. Vallazza, S. Vercellone, and D. Zanello 366

Gamma-Ray Bursts with SuperAGILE
M. Feroci, E. Costa, E. Del Monte, I. Lapshov, M. Rapisarda, P. Soffitta,
G. Barbiellini, M. Prest, E. Vallazza, S. Mereghetti, M. Tavani,
S. Vercellone, A. Morselli, and F. Longo 368

The GLAST Burst Monitor (GBM)
G.G. Lichti, M.S. Briggs, R. Diehl, G. Fishman, R. Georgii, R.M. Kippen,
C. Kouveliotou, C. Meegan, W. Paciesas, R. Preece, V. Schönfelder,
and A. von Kienlin 371

The IPN I: From the Past to the Future
T.L. Cline, K.C. Hurley, S. Barthelmy, P. Butterworth, M. Feroci,
F. Frontera, S. Golenetskii, E. Mazets, and J. Trombka 375

One Year of Rapid, Precise Gamma-Ray Burst Localizations
by the Interplanetary Network
K. Hurley, T.L. Cline, S. Barthelmy, P. Butterworth, M. Feroci,
F. Frontera, E. Montanari, C. Guidorzi, S. Golenetskii, E. Mazets,
and J. Trombka 378

GRBs of Energy E >10 GeV with ARGO-YBJ
S. Vernetto, for the ARGO-YBJ Collaboration 381

Progress on MARGIE, a Gamma-Ray Burst Ultra-long Duration
Balloon Mission
D. Band, M. Cherry, J. Stacy, T. Guzik, S. Kappadath, J. Buckley,
P. Hink, J. Macri, M. McConnell, J. Ryan, and J. Matteson, ... 384

Simultaneous Detection of the High Energy
and Optical Transients by Čerenkov Telescopes
G. Beskin,, C. Bartolini, A. Guarnieri, A. Piccioni, S. Biryukov,
D. Eichler, and D. Faiman 387

PREPROCESS: A Fast Image Processing Software Tool
A. Di Paola 390

The HETE Triggering Algorithm
E.E. Fenimore and M. Galassi 393

The Status of the Ondřejov BART Experiment
R. Hudec, M. Nekola, P. Kubánek, C. Polášek, and A.J. Castro-Tirado . . . 396

Progress in Lobster Eye X-Ray Telescope Development
R. Hudec, A. Inneman, and L. Pina 399

**On the Feasibility of Independent Detections
of Optical Afterglows of GRBs**
R. Hudec 402

High Precision Space Astrometry of Cosmic GRBs
S.M. Kopeikin, V.G. Kurt, O.S. Ougolnikov, and A.A. Sukhanov 405

**New Version of Optical Transient Monitor
for BOOTES Project**
P. Páta, M. Bernas, A.J. Castro-Tirado, and R. Hudec 412

**High Resolution Spectroscopy of the X-Ray Emission of GRBs
by IMXS-BOSS on the ISS**
*L. Piro, L. Colasanti, E. Costa, G. Gandolfi, P. Soffitta, F. Gatti,
M. Razeti, D. Pergolesi, R. Vaccarone, G. Testera, M. Pallavicini,
A. Ferrari, E. Trussoni, M. Orio, D. Mc Cammon, T. Sanders,
M. Galeazzi, A. Szymkowiak, and S. Porter* 415

Statistics of Faint Variable Sources for GRB OA Analyses
J. Polcar, R. Hudec, and H. Meusinger 418

Event Rates in SuperAGILE and HETE
B. Preger, E.E. Fenimore, and E. Costa 421

IR and Optical Observations of GRB from Campo Imperatore
*R. Speziali, F. D'Alessio, L. A. Antonelli, A. Di Paola, L. Burderi,
F. Fiore, G. Israel, D. Lorenzetti, F. Pedichini, L. Stella, and F. Vitali* . . . 424

**A GRB Detection System
Using the BGO-Shield of the INTEGRAL-Sectrometer SPI**
A. von Kienlin, N. Arend, and G.G. Lichti 427

**A Preview of the Swift UVOT Capabilities:
Imaging and Lightcurves from *XMM-Newton* OM**
*S. Zane, K.O. Mason, M.S. Cropper, T.E. Kennedy, J. Nousek,
P. Roming, and M. McLelland* 431

**REM – Rapid Eye Mount. A Fast Slewing Robotized Telescope
to Monitor the Prompt Infra-red Afterglow of GRBs**
*F.M. Zerbi, G. Chincarini, M. Rodonó, A. Antonelli, L. Burderi,
S. Campana, P. Conconi, S. Covino, G. Cutispoto, G. Ghisellini,
D. Lazzati, E. Martinetti, E. Molinari, S. Sardone, L. Stella,
and F. Vitali* 434

List of Participants 437

Part I

Global Properties of GRBs

Some Recent, Interesting Observations of Gamma-Ray Bursts

Kevin Hurley

UC Berkeley Space Sciences Laboratory, Berkeley CA 94720-7450

Abstract. I review some of the observations of time histories, energy spectra, and the statistical properties of gamma ray bursts (GRBs) which have been made over the past several years. The emphasis is on a) the more surprising or unusual results which have not been discussed in the BATSE review talk at this meeting, and b) the energy range above about 25 keV.

1 Time Histories

Over the years, there has been considerable discussion about just what can be learned by studying GRB light curves, with many ideas, but no clear concensus. Recently, several studies have pointed to some new answers to this question.

The shortest GRB observed, 910711, had a duration of only 8 ms [1], and evidence was found for a 0.2 ms feature in it, the shortest time structure ever observed in a burst. For years, this appeared to be an isolated example. Now another case has been found [2], GRB930229, in which there is a comparably short feature, a spike with rise time 0.22 ms and decay time 0.4 ms (see figure 1). (In a related study, it was established that millisecond variability is a common feature in GRB time histories[3].) Schaefer has studied the differences in the arrival times of GRB gamma-rays of different energies [4], and for an assumed distance to GRB930229, has concluded that the variation of the speed of light with frequency, $\delta c/c < 6.3 \times 10^{-21}$. Thus suitable gamma-ray burst time histories can set limits on the mass of the photon and the energy scale of quantum gravity.

At the "long" end of the GRB duration scale, it has been assumed for years that most bursts lasted less than about 100 seconds. Only one event, GRB840304, was found to have a duration around 1000 s [5]. Now, however, evidence is emerging that such durations may be relatively common, constituting a "gamma-ray afterglow". By summing the time histories of BATSE bursts, emission can be observed out to $\approx$ 1000 s[6]. Two PHEBUS events with durations of 1000 s have revealed that the time histories at late times could be described by a power law with index $\approx$-0.6, and that the spectral evolution at these times could be described by hardening in one case and softening in another(figure 2) [7]. The energy in these afterglows constitutes only $\approx$ 10% of the total burst energy, and accordingly these features are observed most clearly only in the very strongest bursts, which accounts for their apparent rarity. In retrospect, it seems possible that GRB840204 was another example of a gamma-ray afterglow, and that GRB940217, for which EGRET observed an 18 GeV photon $\approx$ 5000 s after

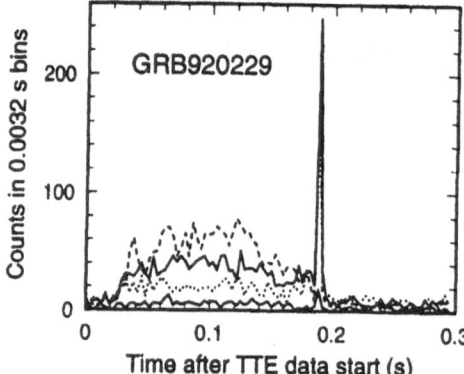

Fig. 1. From Schaefer and Walker [2]. The time history of this burst displays a spike with a 0.22 ms rise time, and has been used to set limits on the variation of the speed of light with frequency

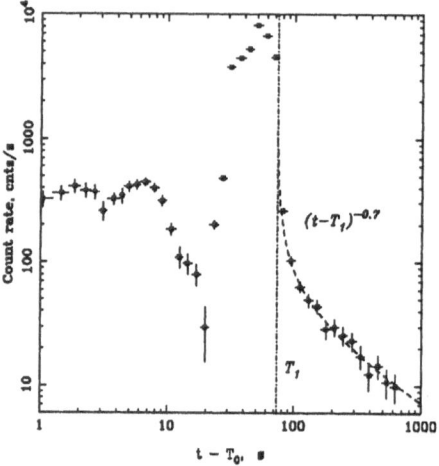

Fig. 2. From Tkachenko et al. [7]. Time history of GRB910402, an ≈ 700 s long gamma-ray burst which displays spectral hardening in the gamma-ray afterglow

the gamma-ray emission appeared to end [8], might be viewed as an extreme case of a gamma-ray afterglow with spectral hardening.

In those cases where the spectra of gamma-ray afterglows soften, there is clear evidence that they evolve into soft X-ray afterglows [9]. In fact, one example was recently found for a SIGMA burst which occurred long before afterglows of any kind were known, GRB920703[10]. Spectral softening began ≈ 6 s into the burst. Although the 2-10 keV soft X-ray afterglow was not observed in this case, a study of BeppoSAX WFC and GRBM data on bursts clearly displays evolution to this emission component[11]. Thus a second answer to the question of what we can learn from GRB light curves is that we may be able to predict the intensity of the X-ray afterglow, in order to decide whether to call a target-of-opportunity observation with an X-ray satellite such as XMM-Newton or *Chandra* .

Now that redshifts have been measured for 14 bursts, it is clear that GRBs are not standard candles, and that their fluxes or fluences alone cannot be used to determine their distances (figure 3). But, since time dilation is observed in the light curves of Type Ia supernovae, it is reasonable to ask whether a similar signature can be found in GRB time histories. The possible use of GRB light curves as distance indicators was considered earlier [12], and it was found that dim bursts have longer durations than bright ones, by a factor of about 2. Initially, this did not appear to be confirmed by independent studies. Now, however, two different approaches have revealed similar effects. GRB light curves may be co-added to produce an average curve of emissivity (ACE)[13]. Comparing the ACEs of dim and bright bursts, it is found that although the rise times of the two groups are similar, the decay times of the dim bursts are about a factor of two longer than those of the bright bursts. Similarly, the peak-to-peak timescales of bursts have been analyzed [14] , and it is found that the time between the peaks in dim bursts is 1.9 times greater on the average than that of bright bursts.

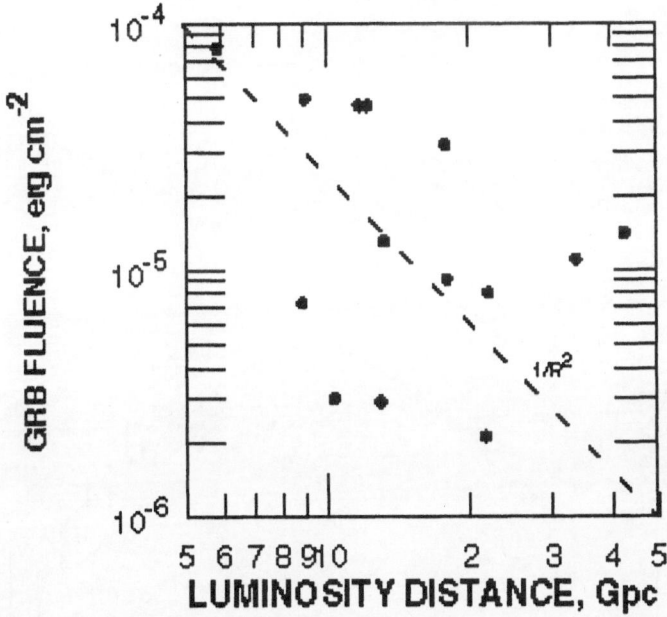

Fig. 3. Why gamma ray burst fluences cannot be used as distance indicators. The fluences are plotted as a function of luminosity distance for fourteen GRBs with well determined redshifts

Two more studies have concentrated on GRB time histories as distance indicators. In one [15] a variability parameter is defined which is correlated with the intrinsic luminosity of a burst. The correlation appears to hold over a wide dynamic range, starting with GRB980425, which may be associated with SN1998bw [16] and would therefore be one of the least luminous bursts observed to date. In the other [17]the lag, defined by cross-correlating the light curves of a given GRB in two spectral channels, is inversely correlated with the luminosity. However, this correlation does not hold for GRB980425.

2   Energy Spectra

The spectra of most gamma-ray bursts above 25 keV are smooth [18] and well-fit by the Band function [19]. Compelling evidence for line features has not been found in the BATSE data [20]. However, evidence for a broad, transient emission feature has recently been found in the PHEBUS data for GRB910402 (figure 4) [21]. Perhaps coincidentally, this is one of the bursts studied for its gamma-ray afterglow [7] . The feature occurred for about 7 seconds, and could be described by a Gaussian centered around 2.4 MeV. Its origin is unclear, but it indicates that high resolution GRB spectroscopy, e.g. by INTEGRAL, may reveal more surprises.

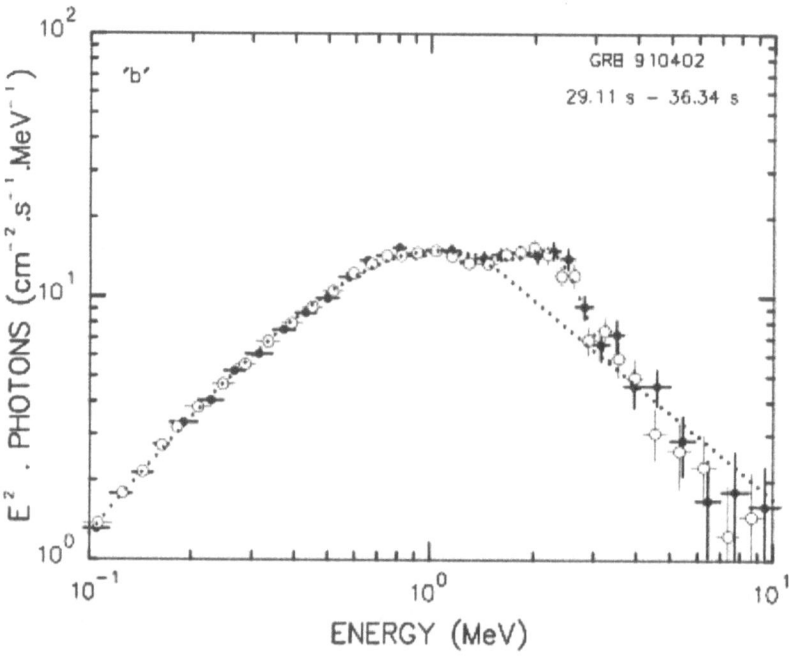

Fig. 4. From Barat et al. [21]. A time-resolved spectrum of GRB910402, displaying a 2.4 MeV excess with respect to the Band function fit

Practically every time a new experiment has come on line, something new and unexpected has been discovered about gamma-ray bursts. In particular, the record for the highest energy photons in GRBs has been broken by a significant amount about every 5 years on the average since 1980. The latest result comes from the ground-based Milagrito experiment [22] which has detected emission above 200 GeV possibly related to GRB970417. What makes this interesting is not just that it is more than an order of magnitude greater than the previous record [8], but also the fact that these energies are close to or greater than the opacity limit for sources at redshifts > 0.3 due to pair production on extragalactic starlight. This suggests that if all GRBs are cosmological and at redshifts of about one or more, then this record may stand for some time. This idea will be tested by experiments such as Milagro in the near future.

3   Statistical Properties

Out of many recent and interesting studies of GRB statistical properties, one on the number-intensity relation, or log N-log P relation, stands out. It was previously shown [23] that down to a peak flux of 0.3 photons/cm^2 s, the curve increases; but for lower fluxes, it is practically unconstrained, because different assumptions about the incident spectra lead to very different behaviors in this region. Two studies have now been carried out on the BATSE untriggered bursts; one indicates that the curve flattens at low fluxes, while the other indicates that it continues to rise. In Kommers' study [24] almost 2000 untriggered bursts were found down to peak fluxes of 0.18 photons/cm^2 s, and they defined a flattening log N-log P curve. This implies that more sensitive missions, such as INTEGRAL and Swift, will not detect more bursts per year and per steradian than BATSE. However, an independent but similar study by Stern [25] revealed almost 400 more bursts than Kommers et al. had found, down to peak fluxes of about 0.1 photons/cm^2, and they define a *rising* log N-log P curve. This implies a full-sky rate of at least 1200 GRBs/year. If these weak bursts are detected by future missions, as they should be, they promise to force revisions in many models.

4   Conclusions

Space did not allow a discussion of many interesting topics, but they should at least be mentioned for completeness. Evidence has been presented by Mukherjee et al. [26] for a third class of GRBs, based on duration, fluence, peak flux, and hardness ratio. This appears to confirm an earlier suggestion by Horvath [27] . In a study of 9 years of SMM data to search for a possible class of hard bursts which BATSE might have missed [28] it was concluded that, at least for bursts longer than 2 seconds with peak fluxes above 0.04 photons/cm^2 s, no such class exists. Fenimore [29] analyzed the time histories and peak energies in the $\nu - F\nu$ spectra of 98 BATSE bursts. By averaging the time histories and E_{peak} values as a function of time, he showed that the time histories could be described by a nearly linear rise and decay, with a nearly linear E_{peak} vs. time evolution, another confirmation of the general hard-to-soft trend in GRB energy spectra. Finally, Lloyd et al. [30] revisited Mallozzi's [31] hardness-intensity relation, which has been interpreted in the past as another piece of evidence of a cosmological origin for bursts. They confirm that a strong correlation exists between E_{peak} and the fluence, but they also conclude that, if bursts are not standard candles, as we now believe, then part of the correlation must be intrinsic rather than cosmological.

References

1. P. Bhat et al.: Nature **359**, 217 (1992)
2. B. Schaefer, K. Walker, K.: Ap. J. **511**, L89 (1999)
3. K. Walker, B. Schaefer: Ap. J. **537**, 264 (2000)
4. B. Schaefer: Phys. Rev. Lett. **82(25)**, 4964 (1999)
5. R. Klebesadel, R.: 'The Durations of Gamma-Ray Bursts'. In *Gamma Ray Bursts: Observations, Analyses, and Theories, Proc. 1990 Los Alamos Workshop at Taos, NM, July 29–August 3, 1990*, ed. by. C. Ho, R. Epstein, and E. Fenimore (Cambridge University Press, Cambridge 1992) pp. 161–168
6. V. Connaughton: 'BATSE Observations of Gamma-Ray Burst Tails'. In *Gamma-Ray Bursts, 5th Huntsville at Huntsville, AL, October 18–22, 1999*, ed. by R. M. Kippen, R. S. Mallozzi, and G. J. Fishman (AIP Conf. Proc. 526, AIP Press, New York 2000) pp. 385–389
7. A. Tkachenko et al.: Astron. Astrophys. **358**, L41 (2000)
8. K. Hurley et al.: Nature **372**, 652 (1994)
9. L. Piro: These Proceedings (2001)
10. R. Burenin et al: Astron. Astrophys. **344**, L53 (1999)
11. F. Frontera et al.: Ap. J. Supp. **127**, 59 (2000)
12. J. Norris et al.: Ap. J. **424**, 540 (1994)
13. I. Mitrofanov et al.: Ap. J. **523**, 610 (1999)
14. M. Deng and B. Schaefer, B.: Ap. J. **502**, L109 (1998)
15. D. Reichart et al.: Ap. J., in press (2001)
16. T. Galama et al.: Nature **395**, 670 (1998)
17. J. Norris et al.: Ap. J. **534**, 248 (2000)
18. B. Schaefer et al.: Ap. J. **492**, 696 (1998)
19. D. Band et al.: Ap. J. **413**, 281 (1993)
20. M. Briggs et al.: Astrophys. Lett. & Communications, **39(1-6)**, 237 (1999)
21. C. Barat et al.: Ap. J. **538**, 152 (2000)
22. R. Atkins et al.: Ap. J. **533**, L119 (2000)
23. W. Paciesas et al.: Ap. J. Supp. Ser. **122**, 465 (1999)
24. J. Kommers et al.: Ap. J. **533**, 696 (2000)
25. B. Stern et al.: Ap. J. **540**, L21 (2000)
26. S. Mukherjee et al.: Ap. J. **508**, 314 (1998)
27. I. Horvath: Ap. J. **508**, 757 (1998)
28. M. Harris and G. Share: Ap. J. **494**, 724 (1998)
29. E. Fenimore: Ap. J. **518**, 375 (1999)
30. N. Lloyd et al.: Ap. J. **534**, 227 (2000)
31. R. Mallozzi et al: Ap. J. **454**, 597 (1995)

On the Fast Spectral Variability of GRBs

E.P. Mazets[1], R.L. Aptekar[1], P.S. Butterworth[2], T.L. Cline[2], D.D. Frederiks[1],
S.V. Golenetskii[1], V.N. Il'inskii[1], and V.D. Pal'shin[1]

[1] Ioffe Physico-Technical Institute, St.Petersburg, 194021, Russia
[2] Goddard Space Flight Center, Greenbelt, MD 20771, USA

Abstract. Fast spectral variability of gamma-ray burst emission is considered for a number of events seen by the Konus-Wind experiment. The variability manifests itself as a strong correlation between instantaneous energy flux F and peak energy E_p. In the (F, E_p) plane, the correlation produces distinct tracks in the form of branches and loops representing the different parts of a burst time history. Despite the variety of features seen in different events, the main characteristics of the spectral evolution produce a quite consistent pattern.

The temporal evolution of GRB energy spectra has been clearly seen in first experiments with sufficient spectral and time resolution. Some characteristics of spectral evolution were pointed out in the Konus experiment aboard Venera 11 and Venera 12 [1]. Observations on Venera 13 and 14 revealed a strong correlation between instantaneous intensity and hardness of a radiation [2]. Then, many authors considered different aspects of spectral evolution of gamma-ray bursts [3,4]. The BATSE experiment aboard the CGRO offers a wealth of spectroscopic information permitting a rich study of the variability and related correlations in GRBs [5,6]. The Konus-Wind gamma-ray burst experiment flown onboard the GGS WIND spacecraft [7] has sampled about 900 bursts since lunch in 1994. A sufficient part of this sample consists of events which are strong enough that we can further extend studies of spectral evolution.

The Konus instrument records a time history and some prehistory of a burst in three energy windows: G1=10–50 keV, G2=50–200 keV, G3=200–750 keV (nominal values) with a time resolution of 2–256 ms, and also up to 64 multichannel energy spectra in the 10 keV–10 MeV range with accumulation times of 64 ms–8 s. Energy loss spectra obtained are deconvolved to incident photon spectra using the detector response function for known angle of incidence. For fitting the spectra, we used the empirical model by Band with spectral parameters α, β, E_0 [12]. Thus, for each energy spectrum we obtain values of an energy flux F and a peak energy $E_p = E_0(2-\alpha)$. These data allows the consideration of a spectral evolution with moderate time resolution. Finer time resolution can be obtained by considering time history measurements. The three time profiles G1, G2, G3 give two independent sets of hardness ratio measurements, for example G2/G1, and G3/G2. Each pair of these values depends on spectral parameters corresponding to a short time interval. We can calculate expected hardness ratios using the response function matrix and the incident photon spectrum with parameters α, β, E_0. In each time bin, the best fit parameters are determined by minimizing the quadratic sum of the two weighted differences between the

GRB 950822

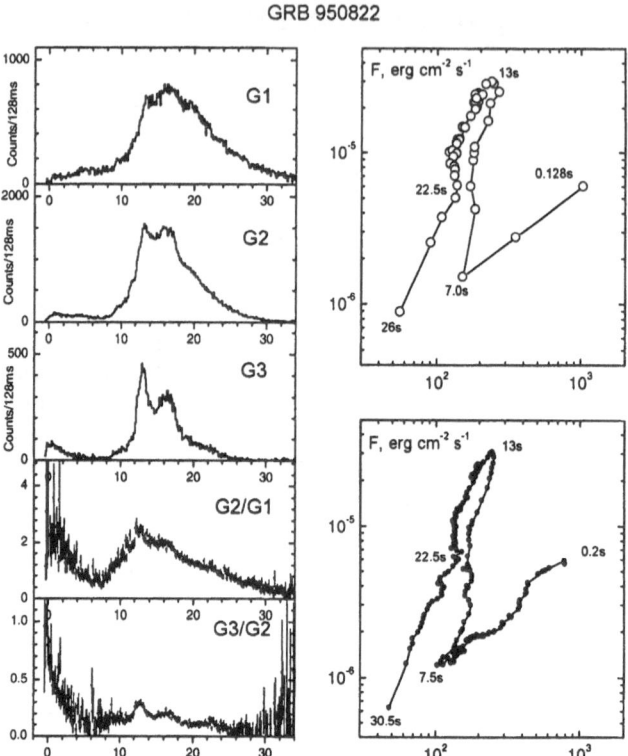

Fig. 1. The spectral evolution in GRB 950822. Left panel: time history and hardness ratio time profiles. Right panel (top): the correlation between intensity and hardness. Data are obtained from energy spectra measurement. Right panel (bottom): Intensity-hardness correlation. Data are obtained using the hardness ratio profiles. Numbers placed along tracks correspond to the times $T - T_0$ on the time history profile.

calculated and observed hardness ratio values. This procedure is illustrated with spectral evolution data for GRB 950822 in Fig. 1. This figure displays a lot information. The mutual dependence of the instantaneous values of the energy flux F and the hardness of a spectrum E_p forms a distinct track in the (F, E_p). The track consists of three main branches. The first branch, 0.2 to 7.5 s, represents a weak but very hard emission in initial stage of the burst. The second branch, 7.5 to 13 s, corresponds to the rising of the main pulse of the burst. Finally the long branch for burst decay, 13 to 30 s, ends with very soft emission. Each branch can be approximated by a power law relation $F \propto E_p^\gamma$ where indices γ are correspondingly ~ 0.7, ~ 6, and ~ 2.5. Such spectral behaviour is typical of a majority of gamma-ray bursts, especially bursts with well separated pulses. Similar branches and loops are seen in many events. A number of further examples are presented in Fig. 2.

Only count rate statistics constrain the time resolution with which correlation can be observed. This is seen in Fig. 2j which presents the correlation tracks of

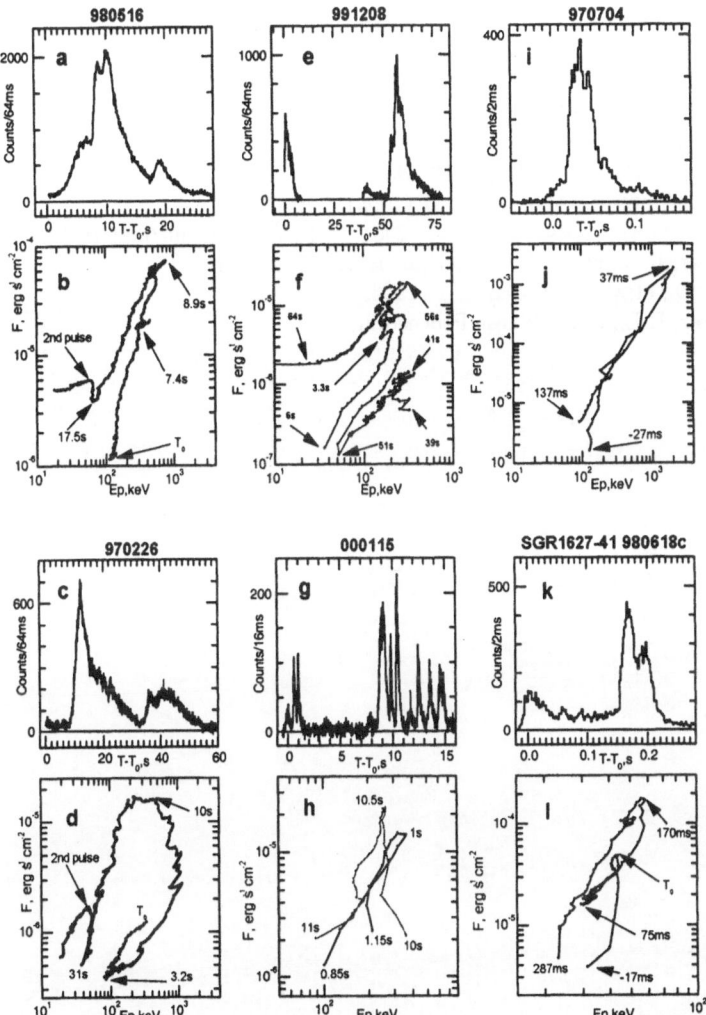

Fig. 2. Time histories and correlation tracks of five gamma-ray bursts and one recurrent burst from SGR 1627-41. Panel **h**: Only two tracks for two pulses are shown for clarity.

very short ($\sim$100 ms) and hard ($E_\gamma > 2$ MeV) GRB 970704 obtained with a time resolution of 2 ms.

Finally, it should be noted, that the SGR 1627-41 also exhibits a strong spectral variability [9]. Figure 2l demonstrates that the correlation tracks of SGR if shifted along the hardness axis E_p by one order of magnitude look like the tracks for GRBs.

A more complete description of the Konus-Wind data on the spectral evolution in GRBs will be published elsewhere.

This work was supported by Russian Aviation and Space Agency Contract, RFBR grant N 99-02-17031 and CRDF grant RP1-2260.

References

1. Mazets, E.P., et al.: Ap&SS, **75**, 47 (1981)
2. Golenetskii, S.V.,et al.: Nature, **306**, 451 (1983)
3. Norris, J.P., et al.: ApJ, **301**, 213 (1986)
4. Kargatis, V.E., et al.: ApJ, **422**, 260 (1994)
5. Ford, L.A., et al.: ApJ, **439**, 307 (1995)
6. Borgonovo, L., & Ryde, F.: ApJ (2001) in press, astro-ph/0009164
7. Aptekar, R.L., et al.: Space Science Rev., **71**, 265 (1995)
8. Band, D., et al.: ApJ, **413**, 281 (1993)
9. Aptekar, R.L., et al.: These Proceedings.

Spectral Properties of Short Gamma-Ray Bursts

W.S. Paciesas[1], R.D. Preece[1], M.S. Briggs[1], and R.S. Mallozzi[1,2]

[1] University of Alabama in Huntsville, AL 35899, USA
[2] deceased

Abstract. The distribution of GRB durations is bimodal, but there is little additional evidence to support the division of GRBs into short and long classes. Based on simple hardness ratios, several studies have shown a tendency for longer GRBs to have softer energy spectra. Using a database of standard model fits to BATSE GRBs, we compare the distributions of spectral parameters for short and long bursts. Our preliminary results show that the average spectral break energy differs discontinuously between short and long burst classes, but within each class shows only a weak dependence on burst duration.

Various studies have shown that short and long GRBs are statistically different classes [4,7,1,9,6]. Recently, additional evidence has come from a study by Norris et al [11], who found no measurable energy-dependent pulse lag in the time histories of short events. This is in contrast to long GRBs, which clearly show such a lag [10], even for short subpulses. Moreover, in bursts with measured redshifts (which thus far are all long events), the energy-dependent pulse lag appears to be anti-correlated with burst luminosity [10]. Thus, if the mechanism producing the lag works for short bursts, they must be intrinsically more luminous than long bursts, and therefore more distant. Alternatively, a different mechanism may operate in short events. Either way, the evidence seems to support separate classification of short and long GRBs.

In the currently favored fireball model, the prompt burst emission is thought to be optically thin synchrotron or synchrotron self-Compton emission from internal shocks, as external shocks are unable to produce the observed temporal structure [5,16]. Detailed studies of the spectra of a number of bright GRBs, including both long and short events, have shown good consistency with the synchrotron shock model [17,18,2]. However, more comprehensive analyses have uncovered problems with this interpretation [3,13,15]. In particular, some GRB spectra are harder at low energies than the synchrotron limit [3,13]. These conclusions are based mostly on spectroscopy of long bursts, but the problem may be most acute for short bursts because their spectra are on average harder.

Recent work [12,14] has characterized the range of spectral behavior in bright, long bursts in some detail, but the spectral properties of the class of short bursts have only been characterized using hardness ratios. Phebus data showed that short GRBs are harder than long ones [4] (confirmed by many succeeding analyses of BATSE data), but detailed study of the spectral differences between short and long bursts has not been done. In particular, the consistency of short burst spectra with the synchrotron shock model predictions has not been properly tested. It is clear that a better characterization of the spectral differences between short and long bursts is warranted. For the foreseeable future, the BATSE data base will provide the best sample of bursts for this purpose.

The BATSE *CONT* datatype is derived from the large area detectors and is independent of the BATSE trigger. However, the *CONT* data have 16-channel energy resolution and 2 s time resolution, so they are not optimal for the analysis of the spectra of short events because the 2 s integration degrades the signal-to-noise ratio. Nevertheless, a database of *CONT* fits was conveniently available [8], so we used these data to perform a preliminary study of spectral differences between short and long GRBs. The *CONT* fit database contains spectral fits for ∼1200 BATSE GRBs. Fit results for two spectra per burst (peak flux interval and total fluence interval) are available, generally from four different spectral models. For a given event, fit results may not be available for all models due to poor statistics and/or lack of fit convergence.

We extracted spectral parameters for all GRBs in the *CONT* database for three of the models (cut-off power law, broken power laws, and the Band GRB function). For short events, there is little difference between the peak flux and fluence intervals because of the 2 s *CONT* time resolution, whereas long GRBs typically have harder spectra at the time of peak flux. We binned the results for each spectral fit parameter according to the burst duration. Since the distribution of spectral parameters within each duration bin is broad and approximately Gaussian, we computed the centroid and width of the best-fitting Gaussian for each duration bin. Figure 1 shows an example of the parameters for the cut-off power law model fit to the peak flux intervals, plotted as a function of burst duration. The left panel shows the power law spectral index and the right panel shows the cut-off energy. Within a given duration interval, the thin vertical bars show the width of a Gaussian fit to the parameter distribution, and the thick vertical bars show the error in the mean of the distribution. Although the distributions are broad, there appear to be differences in the trend of the parameters with GRB duration. The hardening trend in the power law index is roughly continuous throughout, whereas the trend in the cut-off energy appears more like a step-function, with a discontinuity around a duration of 2 s, consistent with the minimum in the T_{90} duration distribution.

Although no quantitative analysis of the statistical significance of these results has yet been done, distributions of fit parameters for the other models show essentially the same trends for both peak-flux and fluence intervals. (The broken power law and Band GRB function fits provide a third parameter, the high energy power law index, but the statistics of this parameter are not yet good enough to define clearly its trend with duration.)

It would appear from the right panel of Figure 1 that the energy spectra of short and long GRBs have different characteristic break energies that otherwise depend only weakly on duration. Since the break energy is affected by the Lorentz factor of the expanding fireball as well as by the redshift of the emitting source, this places interesting limits on the nature of the sources. Either the short GRBs have higher bulk Lorentz factors or they are located closer to us than long GRBs,

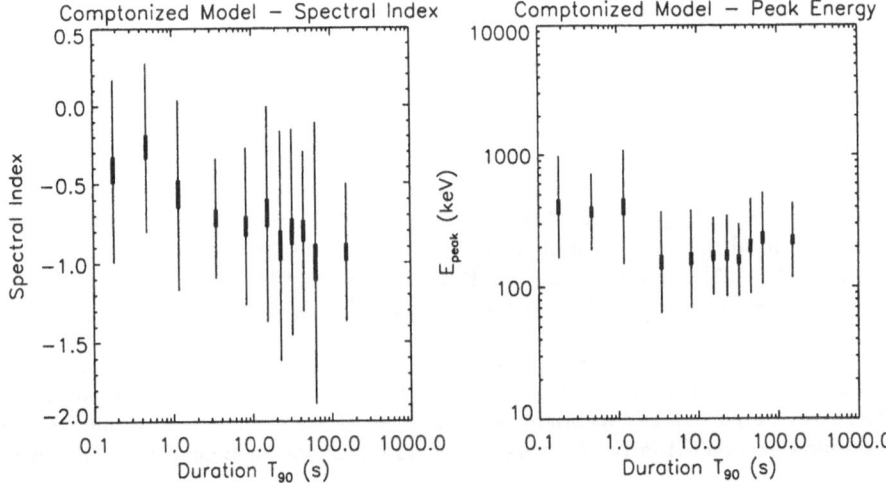

Fig. 1. Spectral parameters as a function of burst duration for fits of a Comptonized model (power law with exponential high-energy cut-off) using *CONT* data (see text). Thin vertical bars show the width of a Gaussian fit to the parameter distribution within a duration bin and thick vertical bars show the error in the mean of each distribution.

or both. Discovering optical counterparts for short GRBs and measuring their redshifts would clearly help resolve the nature of these sources.

References

1. B.M. Belli: Ap&SS **231**, 43 (1995)
2. E. Cohen, J.I. Katz et al.: ApJ **488**, 330 (1997)
3. A. Crider, E.P. Liang et al.: ApJ **479**, L39 (1997)
4. J.-P. Dezalay, C. Barat et al.: In: *Gamma-Ray Bursts: Huntsville, AL 1991*, ed. by W.S. Paciesas, G.J. Fishman (AIP Conf. Proc. 265, New York 1992) pp. 304–309
5. E.E. Fenimore, C. Madras et al.: ApJ **473**, 998 (1996)
6. I. Horváth: ApJ **508**, 757 (1998)
7. C. Kouveliotou, C.A. Meegan et al.: ApJ **413**, L101 (1993)
8. R.S. Mallozzi, G.N. Pendleton et al.: In: *Gamma-Ray Bursts: 4th Huntsville Symposium*, ed. by C.A. Meegan, R.D. Preece, T.M. Koshut (AIP Conf. Proc. 428, New York 1998) pp. 273–277
9. S. Mukherjee, E.D. Feigelson et al.: ApJ **508**, 314 (1998)
10. J.P. Norris, G.F. Marani et al.: ApJ **534**, 248 (2000)
11. J.P. Norris, J.D. Scargle et al.: *these proceedings*
12. R.D. Preece, G.N. Pendleton et al.: ApJ **496**, 849 (1998)
13. R.D. Preece, M.S. Briggs et al.: ApJ **506**, L23 (1998)
14. R.D. Preece, M.S. Briggs et al.: ApJS **126**, 19 (2000)
15. R.D. Preece, M.S. Briggs et al.: *in preparation* (2001)
16. R. Sari, T. Piran: ApJ **485**, 270 (1997)
17. M. Tavani: Phys. Rev. Lett. **76**, 3478 (1996)
18. M. Tavani: ApJ **466**, 768 (1996)
19. M. Tavani: ApJ **497**, L21 (1998)

X-Ray Flashes
and X-Ray Rich Gamma Ray Bursts

John Heise[1,2], Jean in 't Zand[2,1], R. Marc Kippen[3], and Peter M. Woods[4]

[1] Space Research Organization Netherlands, Utrecht, NL 3484CA Netherlands
[2] Astronomical Institute, Utrecht University, Utrecht, NL-3508 TA Netherlands
[3] University of Alabama in Huntsville, Huntsville, AL 35899, USA
[4] Universities Space Research Association, Huntsville, AL 35812, USA

Abstract. X-ray flashes are detected in the Wide Field Cameras on BeppoSAX in the energy range 2-25 keV as bright X-ray sources lasting of the order of minutes, but remaining undetected in the Gamma Ray Bursts Monitor on BeppoSAX. They have properties very similar to the x-ray counterparts of GRBs and account for some of the Fast X-ray Transient events seen in almost every x-ray satellite. We review their X-ray properties and show that x-ray flashes are in fact very soft, x-ray rich, untriggered gamma ray bursts, in which the peak energy in 2-10 keV x-rays could be up to a factor of 100 larger than the peak energy in the 50-300 keV gamma ray range. The frequency is ~ 100 yr^{-1}.

1 Fast X-Ray Transients/High-Latitude X-Ray Transients

Fast X-ray Transients have been observed with many x-ray satellites. In partic-ular they are seen with x-ray instruments that scan the entire sky on a regular basis. Such events are detected in one sky scan and disappeared in the next, typically limiting the duration to be longer than a minute and shorter than a few hours. For this reason they are called Fast Transients. The first transients of this type were seen with UHURU (Forman *et al.* [6]). We review some of the other observations.

Ariel V. Ariel V scanned the sky for 5.5 years with a time resolution of one satellite orbit (~ 100 min) and in 1983 Pye and McHardy [11] reported 27 events. In contrast to LMXBs, Fast Transients are also seen at high galactic latitude and are, therefore, also called High Latitude Transient X-ray Sources. About 20% of the sources seen with Ariel V are identified with RS CVn systems and have a duration of order hours. The authors conclude that all Ariel V observations are consistent with as yet unknown coronal sources, but remark that two of the transients are time coincident with gamma ray burst sources. One of them also was spatially coincident with a GRB to within $\sim 1^\circ$. The frequency is estimated as one Fast Transient every ~ 3 days above 4×10^{-10} erg/s/cm^2 (2-10 keV).

HEAO-1 also had complete sky coverage. Ambruster *et al.* [1] report the anal-ysis of the first 6 month of HEAO-1 scanning data. They observe 10 Fast X-ray Transients with the A1 instrument (0.5-20 keV) above $\sim 7 \times 10^{-11}$ erg/s/cm^2, of which 4 are identified with flare stars. The duration is > 10 s and < 1.5 hr. They estimate an all sky rate of ~ 1500 per year. In the A2 instrument (2-60 keV), 5 more Fast Transients have been detected [2] with peak fluxes between $\sim 10^{-10}$ and $\sim 10^{-9}$ erg s^{-1}cm^{-2} (2-10 keV). They suggest that most of the events are hard coronal flares from dMe-dKe stars, with a flare rate of 2×10^4 per year above $\sim 10^{-10}$ erg s^{-1}cm^{-2} and they rule out the identification with GRBs.

Einstein Observatory. Gotthelf *et al.* [7], searched for X-ray counterparts to GRBs in the data from the Imaging Proportional Counter (IPC) on-board the Einstein Observatory to a limiting sensitivity of 10^{-11} erg s^{-1}cm^{-2} in the 0.2-3.5 keV band. On a time scale of up to ~ 10 s they find 42 events of which 18 have spectra consistent with an extragalactic origin and light curves similar to x-ray counterparts of GRBs, although many events are much shorter than 10 s. The events are not identified on a one arc-minute spatial scale. The implied rate of 2×10^6 yr^{-1} is far more numerous than known GRBs.

After the identification of X-ray afterglows in GRBs (Costa *et al.* [3]) Grindlay [9], suggested that a fraction of the Fast Transients might be X-ray afterglows of GRBs. The approximate agreement in rates, and derived $\log N - \log S$ between fast Transients and GRB afterglows would rule out strong beaming differences between prompt γ-rays of GRB and X-ray afterglows.

ROSAT. Greiner *et al.* [8] have searched for GRB X-ray afterglows in the ROSAT all-sky survey. They find 22 afterglow candidates, where about 4 are predicted. Follow-up spectroscopy strongly suggested a flare star origin in many, if not all, cases.

Ginga. X-ray (1-8 keV) counterparts of GRBs were first detected in 1973 and 1974 (reviewed in [12]). Strohmayer *et al.* [12] summarize the results observed with the *Ginga* satellite in the range 2-400 keV. Out of 120 GRBs in the operational period of 4.5 years between 1987 March and 1991 October, 22 events were studied. The average flux ratio of the X-ray energy (2-10 keV) to the gamma ray energy (50-300 keV) is 0.24 with a wide distribution from 0.01 to more than unity. Photon spectra are well described by a low-energy slope, a bend energy, and a high-energy slope. The distribution of the bend energy extends to below 10 keV, suggesting that GRBs might have two break energies, one in the 50-500 keV range and the other near 5 keV.

2   The Wide Field Cameras WFC on Board BeppoSAX

The WFCs comprise 2 identically designed coded aperture cameras on the BeppoSAX satellite which were launched in April 1996. During every pointing of the Narrow Field Telescopes on BeppoSAX the two WFCs observe fields perpendicular to the main target. The two WFCs look at oppositie location in the sky. Each camera covers $40° \times 40°$ (full width to zero response), covering 2-25 keV

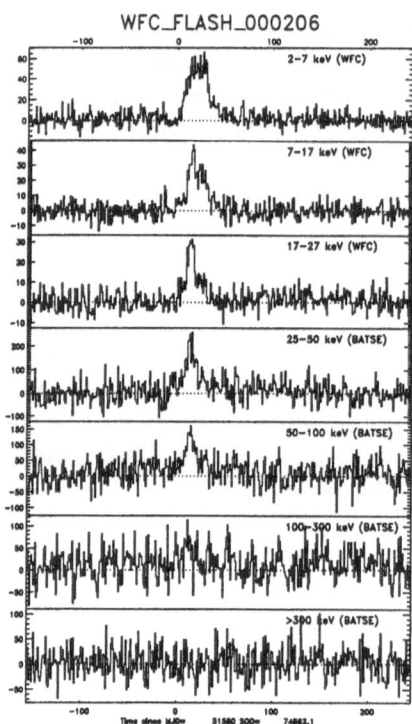

Fig. 1. Example of a light curve in different energy bands of an X-ray flash (Fast Transient X-ray source on a time scale of minutes), seen also in the lowest channels of untriggered BATSE data

photons. The WFCs combines a large field of view with a resolving power of 5 arcmin and allow for a fast position determination.

Since 1996 the X-ray counterparts of GRBs have been localized and studied with Wide Field Cameras (WFCs) on BeppoSAX. Typically error circles around the localization are given with 99% confidence levels between 2 and 3 arcmin. These positions have been established after an average delay of 4 to 5 hours. The error regions fall within the field of view of most optical, radio and x-ray telescopes and have triggered the discovery of afterglows in these wavelength bands. The average deviation between the WFC position and the optical transients found is within the error circle radius and consistent with statistics.

A total of 49 GRB counterparts have been observed at a rate of about 9 per year. The X-ray counterparts can be very bright and range between 10^{-8} and 10^{-7} erg/s/cm^2 (see Fig. 2). The average spectra are characterized by a power law shape, with photon indices between 0.5 and 3 (see Fig. 2). The T90 durations (the duration of the interval above 90% of the peak flux) range between 10 and 200 sec, a histogram is plotted in Fig. 2.

2.1 Two Types of Fast Transients

Apart from X-ray transients associated with LMXBs (including x-ray bursts) and transient phenomena in known x-ray sources, the X-ray transients observed in the WFC are of the type Fast X-ray Transients. A total of about 39 sources have been detected in about 5 years of operations of BeppoSAX. They are seen at positions including high galactic latitude. Although the sky coverage if far from isotropic and favors two regions perpendicular to the Galactic Center, the sky distribution is consistent with being isotropic. A histogram of durations shows that the WFC-Fast Transient have a bi-modal distribution. 17 out of 39 last typically between 10 and 200 s, which we call X-ray flashes, and a class of 22 sources which last typically of the order of an hour, between 2×10^3 and 2×10^5 s. In the latter class 9 sources have been identified with galactic coronal sources (6 flare stars and 3 RS-CVn variables). We do not know the identity of the remaining sources on times scales of an hour, but it is suggestive that all hour-long Fast X-ray transients are coronal sources. We now concentrate on the properties of the Fast Transients with durations of order minutes.

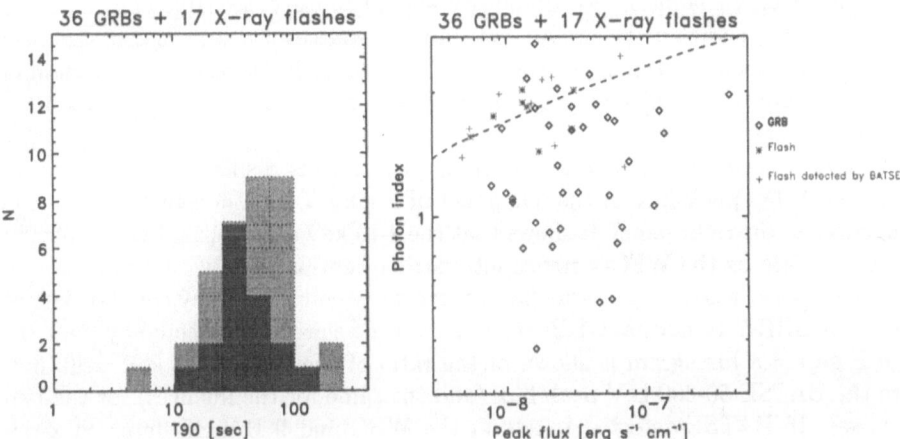

Fig. 2. left: Histogram of T90 durations for the x-ray counterpart of GRBs (light shaded) and x-ray flashes (dark shaded); right: Photon power law index in 2-25 keV versus peak flux in the same range

2.2 Properties of X-Ray Flashes

An operational definition of an X-ray flash in the WFC is a Fast Transient x-ray source with duration less than 1000 s which is not triggered and not detected by the Gamma Ray Burst Monitor (GRBM) in the gamma ray range 40-700 keV. This definition excludes the x-ray counterparts of the typical Gamma Ray Bursts as observed with BeppoSAX, which we will refer to with the term classical GRBs. 17 x-ray flashes have been observed in the WFC in about 5 years of BeppoSAX operations. They are bright x-ray sources, in the range 10^{-8} and 10^{-7} erg/s/cm^2. An example of the light curve of an x-ray flash in different energy bands is given in Figure 2. The durations range between 10 s and 200 s (see Figure 2). The energy spectra in the range 2-25 keV fit with a single power law photon spectrum adding an absorption column consistent with galactic absorption.

The photon index as a function of peak flux (shown in Figure 2 right panel) range between very soft spectra with photon index 3 to hard spectra with index 1.2. We extrapolated such power law spectra into in the 40-700 keV range and present a rough estimate of the sensitivity of the GRBM as the dashed line in Figure 2. It shows that the soft x-ray flashes indeed are not observable with the GRBM, assuming an extension of the power law spectrum. The GRBM upper limit, however, is not consistent for two hard x-ray flashes, indicating that a spectral break must occur in the energy range between the WFC and GRBM, typically between 30 to 50 keV.

The BATSE energy range extends to lower energies than the GRBM. Using the times and positions of the x-ray flashes, we checked for the observability of these sources with the BATSE. 10 out of 17 were potentially observable and 9 out of these 10 actually are detected in either the lowest or the lowest two BATSE energy channels, resp. 25-50 keV and 50-100 keV, see [10]. These BATSE events did not trigger the instrument, but the sources are seen in the standard accumulations in 1 s timebins. The 50-300 keV peak flux is near the threshold of detectability in BATSE, whereas the 2-10 keV flux is bright: between 5×10^{-9} and 10^{-7} erg s^{-1} cm^{-2}.

The ratio of the 30-500 keV BATSE peak flux is displayed against the 2-10 keV WFC peak flux in the left panel of Figure 3 and the same ratio of the fluences in the right panel. It shows that the 2-10 keV x-ray peak fluxes typically are as bright as the WFC-x-ray counterparts of normal GRBs,

How x-ray rich are the x-ray flashes as compared to the x-ray counter part of normal GRBs? 16 normal GRBs seen in the WFCs are also detected by BATSE. In Figure 3 a histogram is shown of the ratio of the WFC 2-10 keV peak flux to the BATSE 50-300 keV peak flux (and the same for the fluences) for the two classes: 16 BATSE-detected bursts in the WFC and 9 BATSE-detected x-ray flashes. These ratios for the x-ray flashes typically extend the range observed in normal GRBs by a large factor. Peak flux ratios of x-ray flashes extend up to a factor of 100, and fluence ratios extend up to a factor of 20.

2.3   Origin of X-Ray Flashes

In principle X-ray flashes could be GRBs at large redshift $z > 5$, when gamma rays would be shifted into the x-ray range and the typical spectral break energy at 100 keV shows at 20 keV. However, in all aspect the x-ray flashes have the same properties as the x-ray counterpart of normal GRBs. In particular the T90 duration histogram would not be expected to be the same for a sample at high redshift GRBs, because of time dilation.

The statistical properties of X-ray flashes display in all aspects a natural extension to the properties of GRBs. X-ray flashes therefore probably show an extension of the physical circumstances which lead to relativistic expansion and the formation of gamma ray bursts, the process called the cosmic fireball scenario. In almost all progenitor models for GRBs, the gamma burst is produced by the final collapse of an accretion torus around a recently formed collapsed object (black hole). In many cases the stellar debris around the birthgrounds of

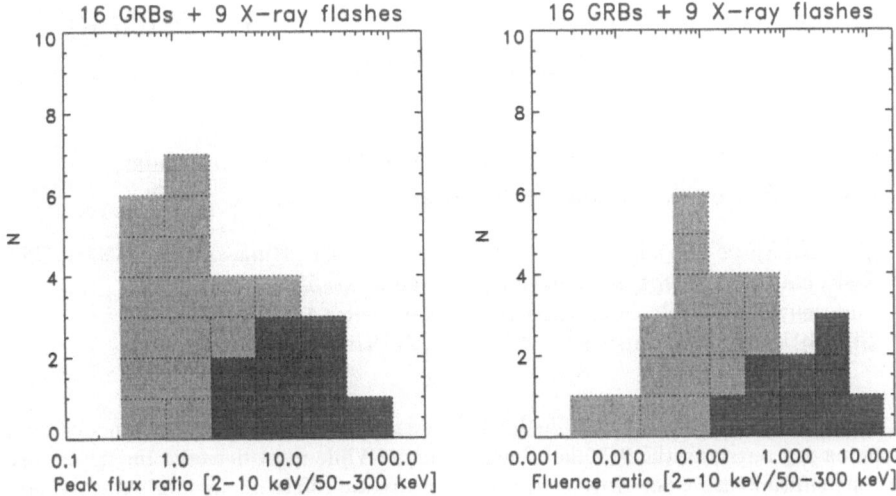

Fig. 3. Ratio of fluxes and fluences between X-ray and γ-ray range for the BATSE detected subsample of WFC flashes and x-ray counterpart of GRBs.

gamma bursts prevents one from observing any prompt high energy radiation at all, making GRBs a rare phenomenon as compared to supernovae. It seems plausible that x-ray flashes bridge the stellar collapses in a relatively clean direct circumburst environment observed as normal GRBs and stellar collapses unobserved in x- and γ radiation. An important factor in the cosmic fireball scenario is the initial relativistic Lorentz factor Γ of the bulk motion producing the gamma ray burst. Γ depends on the fraction of restmass energy (the baryon load) to the total energy of the burst. A low baryon load leads to high Lorentz factors and probably high energy gamma ray bursts (see e.g. Dermer [4,5]). A high baryon load (also called a "dirty fireball") leads to a smaller Lorentz factor and presumably a softer gamma ray burst: the x-ray flash.

References

1. C.W. Ambruster and K.S. Wood, Ap.J. **311**, 258 (1986)
2. A. Connors, P.J. Serlemitsos and J.H. Swank, Ap.J. **303**, 769 (1986)
3. E. Costa, F. Frontera, J. Heise *et al.*, Nature, **387**, 783 (1999)
4. C.D. Dermer, K.E. Mitman, Ap.J.Lett. **513**, L5 (1999)
5. C.D. Dermer, J. Chaing, M. Böttcher, Ap.J.Lett. **513**, 656 (1999)
6. W. Forman, C. Jones, L. Cominsky, *et al.*, Ap.J.Suppl. **38**, 357 (1978)
7. E.V. Gotthelf, T.T. Hamilton, D.J. Helfand, Ap.J. **466**, 779 (1996)
8. J. Greiner, D.H. Hartmann, W. Voges, *et al.*, A&A **353**, 998 (2000)
9. J.E. Grindlay, Ap.J. **510**, 710 (1999)
10. R.M. Kippen, P.M. Woods, *et al.*, these proceedings
11. J.P. Pye and I.M. McHardy, Mon.Not.R.Astr.Soc. **205**, 875 (1983)
12. T.E. Strohmayer, E.E. Fenimore, T. Murakami, A. Yoshida, Ap.J. **500**, 873 875 (1983)

BATSE Observations of Fast X-Ray Transients Detected by BeppoSAX-WFC

R. Marc Kippen[1,2], Peter M. Woods[1,3], John Heise[4], Jean in't Zand[4], Robert D. Preece[1,2], and Michael S. Briggs[1,2]

[1] National Space Sci. and Tech. Ctr., 320 Sparkman Dr., Huntsville, AL 35805, USA
[2] University of Alabama in Huntsville, Huntsville, AL 35899, USA
[3] Universities Space Research Association, Huntsville, AL 35805, USA
[4] SRON-Utrecht, Sorbonnelaan 2, NL-3584 CA Utrecht, The Netherlands

Abstract. The *Beppo*SAX Wide Field Cameras have been successful in detecting gamma-ray bursts in the 2–26 keV energy range. While most detected bursts are also strong emitters at higher energies, a significant fraction have anomalously low gamma-ray flux. The nature of these "Fast X-ray Transients" (FXTs), and their relation to gamma-ray bursts (GRBs), is unknown. We use BATSE untriggered continuous data to examine the >20 keV gamma-ray properties of the events detected in common with *Beppo*SAX. Temporal and spectral characteristics, such as peak flux, fluence, duration, and spectrum are compared to the full population of triggered BATSE GRBs. We find that FXTs have softer spectra than most triggered bursts, but that they are consistent with the extrapolated hardness expected for low-intensity GRBs.

1 Introduction

Several papers in these proceedings detail how measurements made with the *Beppo*SAX Wide Field Cameras (WFCs) have revolutionized gamma-ray burst (GRB) astronomy in two key areas: First, the timely, arc-minute burst locations provided by the WFCs led directly to ground-breaking discoveries of multi-wavelength afterglow emission and host galaxies at cosmological distances. Second, the 2–26 keV WFC burst measurements extend GRB spectroscopy into the x-ray band, where few previous observations exist. This latter capability has led to the exciting discovery of several "Fast X-ray Transients" (FXTs), which resemble GRBs in their x-ray properties and spatial distribution, but lack the strong gamma-ray emission typical of "classical" GRBs [1]. Some bursts with a strong x-ray spectral component were observed previously with *Ginga* [2], hinting that the WFC FXTs could represent a class of x-ray rich GRBs. However, a completely unrelated and unknown phenomenon cannot be ruled out.

The key to understanding the relation between FXTs and GRBs lies in the gamma-ray domain, where we have the most knowledge of classical GRB behavior to compare with. Unfortunately, observational data are scarce due to the weakness of the gamma-ray emission from these events. In this paper, we present preliminary results on the gamma-ray properties of WFC FXTs compared to those of classical GRBs using data from the *Compton*-BATSE Large Area Detectors (LADs, >20 keV). BATSE is the only instrument to detect significant gamma-ray emission from WFC FXTs, and thus provides the best means of understanding their true nature.

2   BATSE Observations

BATSE and *Beppo*SAX-WFC operated simultaneously for 3.8 years, ending with the termination of the *Compton* Observatory on 4 June, 2000. In this interval, ~53 GRB-like transient events were observed by the WFCs – 17 of which were classified as FXTs due to their lack of detectable gamma-ray emission in the SAX GRB Monitor (40–400 keV). Based on the WFC source locations, we know that 12 of these FXTs were observable (unocculted by Earth) by BATSE, but none activated BATSE's on-board transient event trigger system. This indicates that the 50–300 keV peak fluxes were near, or below ~0.2 ph $\cdot$ cm^{-2} $\cdot$ s^{-1} (1024 ms timescale).

To investigate with higher sensitivity than the on-board trigger allows, we performed a *post facto* search for the 10 FXTs where BATSE continuous data are available. Nine of the 10 candidates were detected with $\geq 5\sigma$ significance in the 20–100 keV energy range. These detections are strengthened by the fact that the times and independent BATSE locations of the events agree (within the uncertainties) with those measured with the WFCs (see Figure 1).

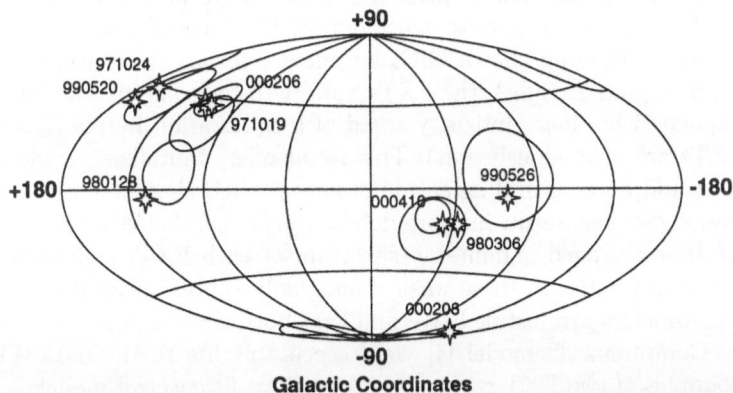

Fig. 1. Locations (and dates) of the nine WFC FXTs detected with BATSE in this study. The BATSE 1σ error circles (solid lines) are compared to the independent WFC locations (stars whose size is not representative of the ~10 arc-minute uncertainties).

3   Gamma-Ray Properties of FXTs Compared to GRBs

Qualitatively, the nine FXTs detected in the BATSE search are very similar to GRBs in their temporal and spectral characteristics. For instance, they have rapidly varying lightcurves, durations from ~1–50 s, and strong spectral evolution. To quantify the comparison, 4-channel, 1024-ms data were used to compute several standard GRB parameters, including peak flux, fluence, duration, and

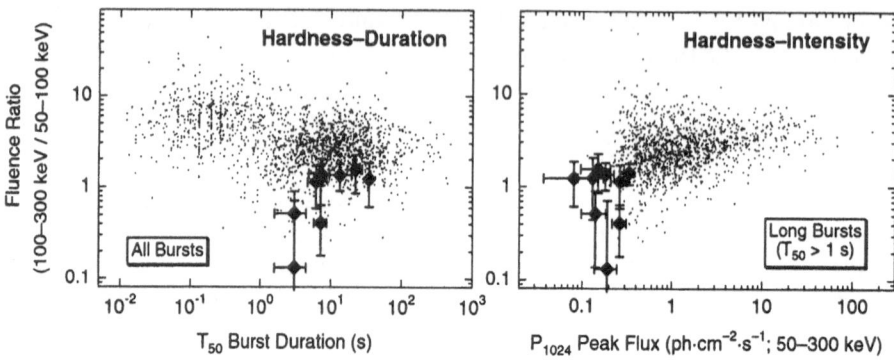

Fig. 2. Gamma-ray parameters of nine FXTs (diamonds) compared to GRBs from the BATSE trigger catalog (dots). The plot at right shows only "long" GRBs ($T_{50} > 1$ s).

hardness ratios. These quantities were produced using the same software and procedures employed for all BATSE GRBs, so we can make direct comparisons to the large BATSE GRB trigger catalog [3].

As can be seen in Figure 2 (left), the FXTs have durations comparable to the long class ($T_{50} > 1$ s) of GRBs, but are noticeably softer than the average long GRB based on their fluence hardness ratios. This spectral difference is also apparent in the distribution of peak flux hardness ratios. However, since the FXTs are less intense than most triggered GRBs, spectral differences may be expected due to the well-known GRB hardness–intensity correlation (e.g., [4]). As shown in Figure 2 (right), the FXTs appear to be generally consistent with the extrapolated hardness–intensity trend of long-duration bursts (albeit three of the FXTs are anomalously soft). This result offers tantalizing evidence that the FXTs could be a natural extension of known GRB characteristics.

To investigate this result in more detail, 16-channel, 2.048-s data were used to fit the time-averaged gamma-ray spectrum of each FXT. Due to the large statistical uncertainties in these weak events, only simple spectral models with few free parameters are justified. We find that both the single power law model and the "Comptonized" model [4] yield acceptable fits to the data. Figure 3 shows examples of two FXT spectra and their best-fit spectral models.

The Comptonized model, which includes curvature parameterized by the $\nu\mathcal{F}_\nu$ peak energy E_{peak}, was used by Mallozzi et al. [5] to fit all time averaged 16-channel spectra of 1023 bursts from the BATSE 4B catalog. In Figure 4, the fitted values of E_{peak} for 802 long GRBs from this collection are compared to those of the FXTs. Apart from the three soft outliers mentioned above (note also the large errors), the FXTs appear to be consistent with E_{peak}–intensity trend of the GRB sample. For further clarity, this trend was modeled by fitting a power-law function to the distribution of E_{peak} in eight intensity bins (each containing ~100 long GRBs). The 68% confidence region for this fit is indicated by the two dashed lines in Figure 4. The extrapolated fit is clearly consistent with most of the FXT spectra, in agreement with the previous result based on hardness ratios.

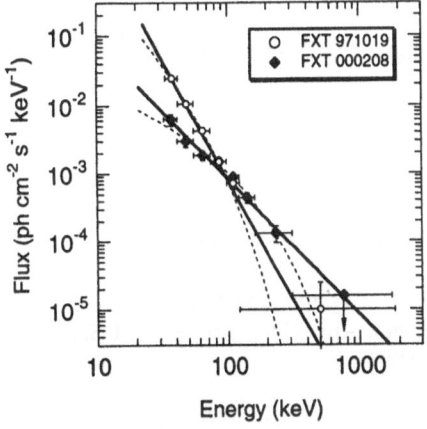

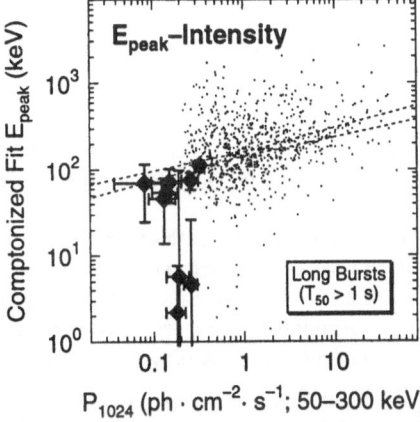

Fig. 3. Time-averaged energy spectra of two FXTs measured with BATSE. Solid and dashed lines indicate the best-fit power-law and Comptonized models, respectively.

Fig. 4. Best-fit time-averaged E_{peak} vs. peak flux for FXTs (diamonds) and long GRBs (dots). Dashed lines indicate the 68% confidence region of a power-law fit to the GRB E_{peak}–intensity correlation.

4 Conclusions and Future Work

Thanks to the high sensitivity of BATSE, we have been able to directly compare the gamma-ray properties of FXTs to those of the full GRB population. The preliminary result is that FXTs appear to be consistent with an extrapolation of known GRB behavior – indicating that they represent a previously unexplored sector of the GRB population. If confirmed, this result implies that there are a large number of undetected bursts with peak energies in the x-ray regime. In fact, the untriggered BATSE burst catalog of Stern et al. [6] suggests that the all-sky rate of FXT-like events with $P_{1024} \gtrsim 0.1$ ph $\cdot$ cm$^{-2} \cdot$ s^{-1} is ~400/yr. This represents nearly half of all long GRBs in their catalog! It is tempting to speculate that these x-ray rich events could represent a large population high-redshift GRBs. In future work, we will attempt to confirm the findings presented here by jointly analyzing the x-ray and gamma-ray spectral data from WFC and BATSE. We will also investigate in more detail the similarities between the WFC FXTs and the large sample of untriggered BATSE events.

References

1. J. Heise, J. in't Zand, R. M. Kippen, et al.: these proceedings
2. T. E. Strohmayer, E. E. Fenimore, T. Murakami, et al.: ApJ **500**, 873 (1998)
3. http://gammaray.msfc.nasa.gov/batse/grb/catalog/current/
4. R. S. Mallozzi, W. S. Paciesas, G. N. Pendleton, et al.: ApJ **454**, 597 (1995)
5. R. S. Mallozzi, G. N. Pendleton, W. S. Paciesas, et al.: In *Gamma-Ray Bursts: 4th Huntsville Symp.* eds. C. Meegan et al. (AIP CP428, New York 1998) p. 273
6. B. E. Stern, Y. Tikhomirova, et al.: ApJ, in press (astro-ph/0009447)

Observations of Gamma-Ray Bursts with the Rossi X-Ray Timing Explorer

Hale Bradt[1], Alan M. Levine[1], Francis E. Marshall[2], Ronald A. Remillard[1], Donald A. Smith[3], and Toshi Takeshima[2,4]

[1] Massachusetts Institute of Technology, Cambridge MA 02139-4307, USA,
   Room 37-587; bradt@mit.edu
[2] Code 662, Goddard Space Flight Center, NASA, Greenbelt MD 20771, USA
[3] 2477 Randall Laboratory, University of Michigan, Ann Arbor MI 48109, USA
[4] Present address: NASDA, Tokyo, Japan 105-8060

Abstract. The role of the Rossi X-ray Timing Explorer (RXTE) in the study of Gamma-ray Bursts (GRBs) is reviewed. Through April 2001, the All-Sky Monitor (ASM) and the Proportional Counter Array (PCA) instruments have detected 30 GRBs. In 16 cases, an early celestial position was released to the community, sometimes in conjunction with IPN results. The subsequent optical and radio searches led to the detection of 5 x-ray afterglows, to at least 6 optical or radio afterglows, to 3 of the 17 secure redshifts known at this writing, and to 2 other likely redshifts. The decay curves of early x-ray afterglows have been measured. The rapid determination of the location of GRB 970828 and the absence of optical afterglow at that position gave one of the first indications that GRBs occur in star-forming regions [6]. The location of GRB 000301C led to the determination of a break in the optical decay rate [16] which is evidence for a jet, and to variability in the optical light curve that could represent gravitational lensing [7]. X-ray light curves of GRB from the ASM in conjunction with gamma-ray light curves exhibit striking differences in different bands and may reveal the commencement of the x-ray afterglow [18].

1 Introduction

The Rossi X-ray Timing Explorer (RXTE) [2] has played a highly significant role in the explosive growth of GRB studies in the afterglow era. The RXTE contributions have been possible because of its ability to rapidly point to a new target, the sensitivity and high-energy response ($2 - 60$ keV) of the PCA [9], and the wide-field position-determining capability of the ASM [11]. This overview encompasses work through April 2001.

2 Observations of Afterglows with the PCA

Searches for early x-ray afterglows with the PCA have been accomplished mostly by means of rapid pointings (few hours) of RXTE that direct the PCA FOV toward the near real time BATSE "Locburst" position, when the position has uncertainty less than a few degrees. In addition early GRB positions from the ASM, BeppoSAX, and the Interplanetary Network (IPN) have been used as

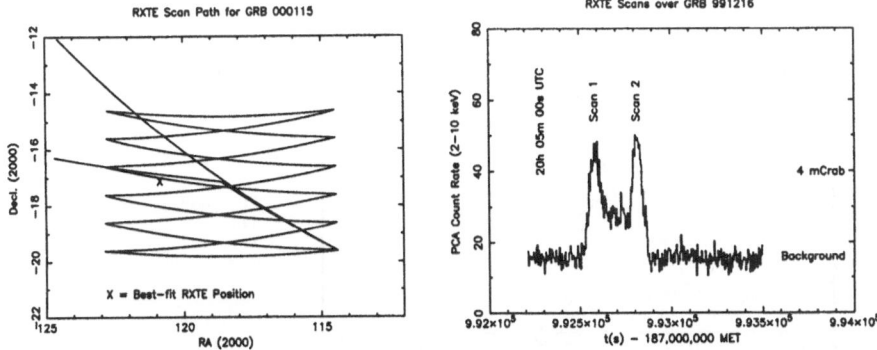

Fig. 1. Typical raster scans by RXTE PCA. Left: the scan track used for GRB 000115. Right: PCA counting rate vs. time for the raster scan of GRB 991216

targets. Raster scans of the region with the 1.0 deg (FWHM) PCA field of view are used to carry out the search; a detection usually can be localized to a few arc minutes; Fig. 1. Such searches by the RXTE/PCA are dependent on the schedule of upcoming command contacts, and upon the burst location relative to the current celestial pointing direction of RXTE. The PCA typically reached its target within a few hours after notification of the burst location.

There have been 31 attempts to view x-ray afterglows with the PCA. Of these 24 were triggered by BATSE Locburst notices and 7 by one or more of the RXTE/ASM, BeppoSAX, and the IPN. Five x-ray afterglows were detected, four derived from Locburst and one from the ASM; Table 1. In each case celestial positions were promptly reported to the community. Highlights of PCA detections and follow-on measurements are:

Table 1. PCA detections of x-ray afterglows

GRB	BATSE peak flux cm^{-2} s^{-1} (50 − 300 keV)	PCA flux @ 3 h (mCrab)	X-ray decay index[a]	Comments[b]
970616*		0.5		3σ PCA detection
970828*		1.0	1.6	See Table 2
990506*	19	3	0.9	Radio; no opt afterglow; host at $z = 1.3$
991216*	68	8	1.8	Opt. and radio afterglow; jet in ISM, $z > 1.02$?
000115*	57	2	1.9	No opt. afterglow

[a] Spectral indices are from PCA data alone.
[b] See references elsewhere in this report and also on J. Greiner web page: http://www.aip.de/~jcg/grbgen.html
* Prompt position notice provided to community

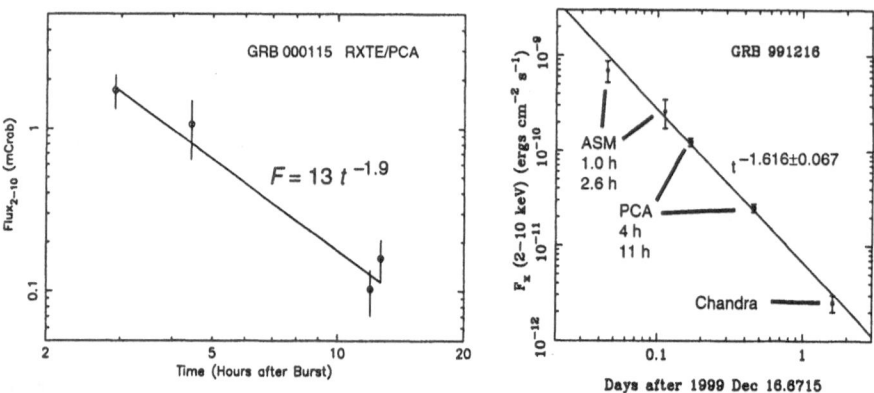

Fig. 2. X-ray decay light curves from RXTE. Left: PCA determination of index for GRB 000115. Right: ASM/PCA/Chandra decay of GRB 991216, reported in Halpern et al. [8]

- Radio afterglow in GRB 990506, but with no optical [19]; host galaxy has $z = 1.3$ [5].
- Optical afterglow in GRB 991216 with a break in the decay curve indicating a jet in interstellar medium [8], radio emission [4], and a reported lower-limit to the redshift, $z > 1.02$ [20].
- X-ray decay indices of 4 bursts; see Table 1 and Fig. 2.

3   Observation of GRBs with the ASM

The ASM, with an instantaneous field of view of ~ 3000 sq. deg. and $\sim 40\%$ duty cycle, has upon occasion serendipitously observed a burst in the FOV of one of its three Scanning Shadow Cameras (SSCs). A detection in a single camera yields a line of position of a few arc minutes by a few degrees. Such an error region can be reduced with IPN lines of position or with PCA scans. Less frequently, a GRB is detected in both of the two azimuthal cameras that have overlapping FOVs (~ 300 sq. deg.). This yields crossed lines of position and an error region of order 10 sq arc min. This is sufficient for radio and optical searches, though PCA or IPN results can further refine the position.

The rate of GRB detection in the overlapping FOVs was initially expected to be ~ 1 per year. The actual rate (see "2-SSC" events in Tables 2 and 3) turns out to be about twice this because (1) the x-ray portion of the bursts tends to be longer than the gamma-ray portion and (2) the ASM rotates every 100 s which shifts the FOVs 6° on the sky. This shift can bring a FOV onto a long-duration burst. It can also bring a camera onto a burst that had just been observed with the other azimuthal camera, thus providing crossed lines of position.

In 5.2 years of operation, up to the present (April 2001) a total of 27 probable GRBs have been detected in the ASM. Of these 24 are confirmed as GRBs by gamma-ray detectors on other satellites; Table 2. Ten of these were detected in

2 SSCs and three were located in a recent global reprocessing of the entire data base. The ASM also recorded at least three other events for which there has been no reported gamma burst from other satellites; Table 3. Each of the listed events has a hard x-ray spectrum, is at high galactic latitude, and is a one-time detection. Thus they could be x-ray bright GRBs. Two of these appear to be afterglow detections as no fast variability is apparent during the 100-s ASM exposures.

The ASM has been providing rapid celestial positions to the community since the commencement of the afterglow era in 1997; to this date 12 such positions have been reported, sometimes in conjunction with IPN reports. Of these, four were observed in two cameras and hence yielded small uncertainties in two dimensions.

The nearly continuous telemetry stream from RXTE, via the TDRSS satellites, makes possible rapid position determinations, in principle less than an hour and in practice usually within a few hours of the burst occurrence time in the RXTE detectors. The events were typically confirmed as due to GRBs in near real time by a BATSE event and thus could be released rapidly with high confidence, well before IPN positions became available. In one case, GRB 970828, the ASM position determination made possible acquisition of the burst by the PCA only 3.6 h after the burst occurrence at RXTE.

The ASM positions have led to 5 optical afterglows and 2 to 4 redshifts. The range in the latter number depends upon the confidence in the determined redshift and the role the ASM played in instigating the search. As noted, the one-camera detections required other information (IPN, SAX, or PCA) to permit fruitful optical and radio searches. In the case of GRB 991216, the ASM did not catch the burst itself, but did capture the early afterglow at 1.0 h and 2.6 h after the burst. The PCA measured the afterglow of this same burst at 4 h and 11 h. Together with a Chandra observation, these provided an excellent measure of the very early x-ray decay of a GRB [8]; Fig. 2. Notable accomplishments that stemmed from ASM results are:

- No optical afterglow was detected in GRB 970828 possibly due to its occurring in a star forming region [6]. Variability in the x-ray afterglow decay was found with ASCA [21].
- Redshifts were obtained for the host galaxy of GRB 980703 ($z = 0.966$) [3] and for the afterglow of GRB 000301C ($z = 2.04$) [10]
- The optical afterglow decay of GRB 000301C showed a clear break [16] and remarkable rapid optical variability on time scales of $\sim$1/2 day [17] [12] suggestive of a microlensing event [7].
- The x-ray afterglows of GRB 960524 and GRB 991216 were measured by the ASM only 25 min and 1.0 h after the burst, respectively (see above).
- Coincident x-ray (ASM) and gamma-ray light curves of 15 GRBs show marked differences between low and high energies; Fig. 3.

Table 2. ASM detections of confirmed GRBs

GRB[a]	Confirming satellite[b]	Comments[c,d]
960228	k	2 SSCs; found in reprocessing
960416	b k u	2 SSCs; soft intermediate x-ray peak; extended x-ray tail
960524	b	2 SSCs; afterglow in ASM 25 min. after burst, fnd. in reproc.
960529	k	2 SSCs
960610	b	2 SSCs; found in reproc.
960727	k u	1 peak; extended x-ray tail
961002	k u	1 peak; extended x-ray tail
961019	b k u	Delayed x-ray peak; x-ray tail
961029	k	Limited x-ray data
961216	b k	Poor ASM position
961230	u	2 SSCs; weak x ray
970815*	b k s u	2 SSCs; strong tertiary x-ray peak
970828*	b k o u	2 SSCs; no opt. afterglow; $z = 0.958$?, 0.33? (x-ray line); rad.?
971024*	b k	Weak; position uncertain
971214*	b k n s u	Single longer x-ray peak; OT; $z = 3.42$ from BSAX position
980703*	b k u	2 SSCs; OT; $z = 0.966$; long γ/x tails; radio
981220*	k s u	8 Crab (5−12 keV); γ peaks smeared in x rays
990308*	b k u	OT; host galaxy underluminous or distant
991216	b n u	OT; x-ray afterglow at opt. pos. at 1.0 h, 2.6 h (ASM)
000301C*	k n u	3 Crab (5−12 keV); OT; $z = 2.04$; rapid opt. var. (lensing?)
000508*	b u	2 Crab; emerged from earth occultation
001025*	n u	∼4 Crab (5−12 keV); no optical
010126*	k n u	∼5 Crab (5−12 keV); no optical
010324*	s u	2 SSCs; det. 360 s after GRB onset; no det. rapid var. in ASM, but decays factor of 4 in 100 s (afterglow?); no opt.

[a] The GRBs detected in 1996 were located in archival searches.
[b] b=BATSE; k=KONUS; n=NEAR; o=SROSS-C; s=BSAX; u=Ulysses
[c] See refs. herein and on J. Greiner: http://www.aip.de/∼jcg/grbgen.html
[d] OT = optical transient
* Prompt position notice provided to community

4   Coincident X/Gamma Light Curves

The ASM provides light curves for GRB detections in three x-ray energy bands: 1.5 − 3 keV, 3 − 5 keV, and 5 − 12 keV. In addition, the ASM time series data mode may reveal temporal variability in each channel down to a time resolution of 0.125 s if the GRB is the dominant x-ray source in a camera's FOV. These data provide valuable information in a band not typically monitored by gamma-

Table 3. ASM GRB not confirmed by gamma detectors

Date	α, δ[a]	Start (MJD)	I_{max}	Gal.	Comments
	(J2000)	(interval)	(mC)	lat.	
961225	154.97, 64.04	50442.2620 (>90s)	360	+46°	2 SSCs; flat l.c.; afterglow?
970123	184.68, −21.20	50471.0707 (>190 s)	62	+41°	2 SSCs; ramping down; afterglow?
000913	16.355, −16.008	51800.6711	700	−78°	

[a] ~90% errors: 0.2° for the first two bursts. For the third: line length = 2.50°, line width = 0.090°, pos. angle = −36.930°

ray experiments. (But note, e.g., the x-ray detections of GRBs (< 10 keV) with Ginga [14] and with the currently operational WFC on BeppoSAX [5].)

The ASM in conjunction with gamma-ray detectors on BATSE, BeppoSAX, Ulysses, Konus, and NEAR have yielded multifrequency light curves over the range 1.5 to >300 keV. Fifteen of these have been collected and analyzed by Smith et al. [18]. They show diverse morphologies with striking differences between the x-ray and gamma-ray bands; Fig. 3. For example, the pronounced 3rd ASM peak of GRB 970815 is not detected in the upper BATSE energy bands. This peak could represent the beginning of the afterglow in the external shock model [13]. This supposition is supported by (1) the $\sim$ 8−s delay of the peak time in the highest ASM energy channel ($\sim$7 keV) relative to that of the lowest channel ($\sim$2.25 keV) and (2) the achromatic decay of this peak, with decay indices in each of the 3 ASM energy channels consistent with $\alpha = 1.3 \pm 0.1$.

Smith et al. [18] also studied the durations of gamma bursts as a function of energy and compared them with the $E^{-0.5}$ power law expected from an origin in synchrotron radiation [15]. Examples that match well are the two distinct peaks of GRB 960416 taken separately (Figs. 3, 4a,b) and the single peak (see [18]) of GRB 000301C (Fig. 4h). The more complex bursts (see light curves in [18]) are not consistent with this prediction. Some have flatter curves which widen more slowly than expected with decreasing energy; Fig. 4c,d. Other complex bursts show a flat curve at high energies but an excess at low energies; Fig. 4e,f. These inconsistencies are probably indicative of complex interactions that violate the assumption of a single infusion of energy followed by cooling through radiation.

5 RXTE Burst Studies in the HETE Era

The RXTE will continue to contribute occasional additional positions, light curves, and decay indices of GRBs with increased reliance on ASM self-triggered events and on IPN crossing lines of positions and confirmations, now that BATSE is no longer in orbit. Most important, the RXTE/PCA has the unique potential to acquire a HETE burst extremely rapidly, say within 1/2 hour of the burst and

32   H. Bradt et al.

Early Evolution of Gamma Ray Bursts

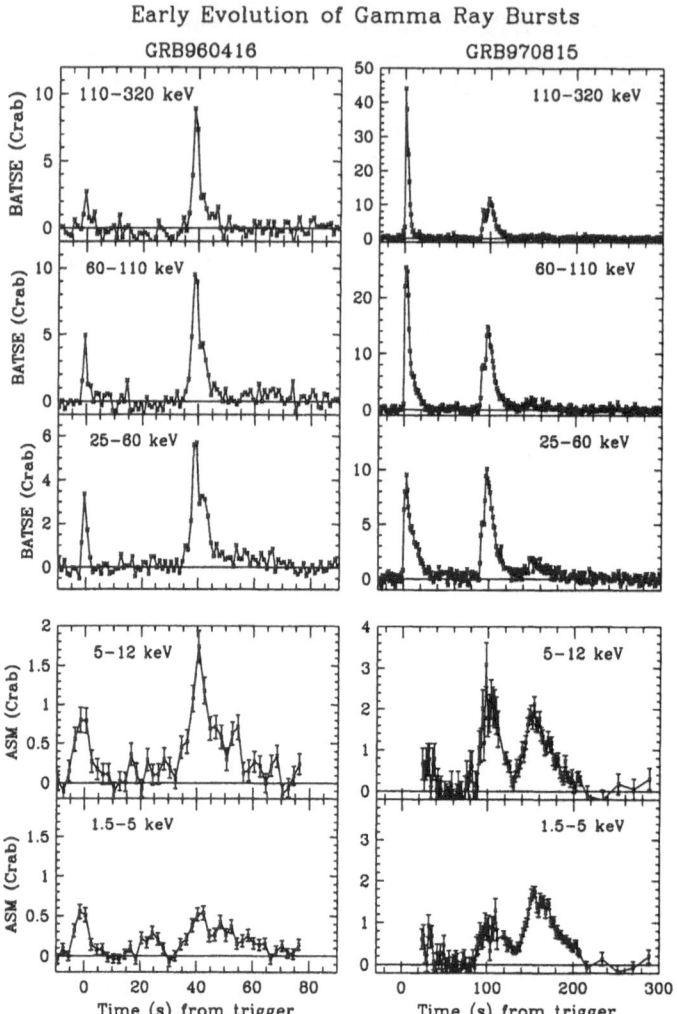

Fig. 3. Light curves in 3 BATSE energy bands and 2 ASM bands of GRB 960416 and GRB 970815, from Smith et al. [18]

to study the temporal activity in the afterglow with high statistics, a domain not heretofore explored.

Acknowledgments

We acknowledge the contributions of the many individuals on the ASM and PCA teams, the observers whose data are included in Fig. 3, and NASA for the continued operation of RXTE.

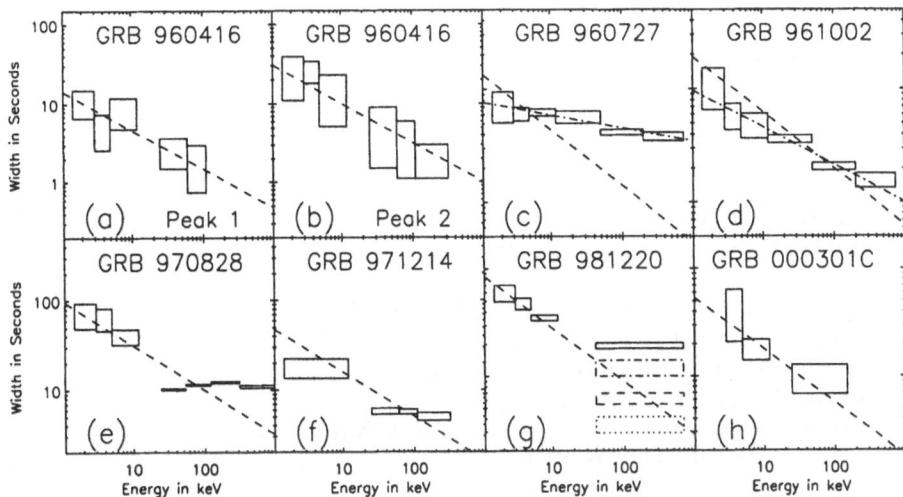

Fig. 4. Peak width vs. energy for seven GRBs. The dashed lines indicate a logarithmic slope of −0.5, from Smith et al. [18]

References

1. J. S. Bloom, D. A. Frail, R. Sari: AJ, in press (2001, astro-ph 0102371; see Table 1)
2. H. V. Bradt, R. E. Rothschild, J. H. Swank: A&AS **97**, 355 (1993)
3. S. G. Djorgovski, S. R. Kulkarni, J. S. Bloom, R. Goodrich, D. A. Frail, L. Piro, E. Palazzi: ApJ **508**, L17 (1998)
4. D. A. Frail, et al.: ApJ **538**, L129 (2000)
5. F. Frontera, et al.: ApJS **127**, 59 (2000)
6. P. J. Groot et al.: ApJ **493**, L27 (1998)
7. P. M. Garnavich, A. Loeb, K. Z. Stanek: ApJ **544**, L11 (2000)
8. J. P. Halpern, et al.: ApJ **543**, 697 (2000)
9. K. Jahoda, J. H. Swank, A. B. Giles, M. J. Stark, T. Strohmayer, W. Zhang, E. H. Morgan: Proc. SPIE, **2808**, 59 (1996)
10. B. L. Jensen et al.: A&A, in press (astro-ph 0005609)
11. A. M. Levine, H. Bradt, W. Cui, J. G. Jernigan, E. H. Morgan, R. Remillard, R E. Shirey, D. A. Smith: ApJ **469**, L33 (1996)
12. N. Masetti et al.: A&A **359**, L23 (2000)
13. P. Meszaros, M. Rees: ApJ, **476**, 232 (1997)
14. Y. Ogasaka, T. Murakami, J. Nishimura, A. Yoshida, E. E. Fenimore: ApJ **383**, L61 (1991)
15. T. Piran: Phys. Rep., 314(6), 575 (1999, astro-ph 9810256):
16. J. E. Rhoads, A. S. Fruchter: ApJ **546**, 117 (2001)
17. R. Sagar, V. Mohan, S. B. Pandey, A. K. Pandey, C. S. Stalin, A. J. Castro Tirado: Bull. Astr. Soc. India **28**, 499 (2000)
18. D. A. Smith et al. 2001: ApJ, submitted (astro-ph 0103357)
19. G. B. Taylor, J. S. Bloom, D. A. Frail, S. R. Kulkarni, S. G. Djorgovski, B. A. Jacoby: ApJ, **537**, L17 (2000)
20. P. M. Vreeswijk, et al.: GCN Circ. 496 (1999)
21. A. Yoshida, M. Namiki, C. Otani, N. Kawai, T. Murakami, Y. Ueda, R. Shibata, S. Uno: AdSpR **25**, 761 (2000)

Testing the Optically Thin Synchrotron Shock Model for Gamma-Ray Bursts Spectra from 2 to 700 keV with BeppoSAX

L. Amati[1], F. Frontera[1,2], M. Tavani[3], J.J.M. in 't Zand[4], E. Costa[5], C. Guidorzi[2], E. Montanari[2], and P. Soffitta[5]

[1] Istituto Tecnologie e Studio Radiazioni Extraterrestri, CNR,
   Via Gobetti 101, 40129 Bologna, Italy
[2] Dipartimento di Fisica, Università di Ferrara,
   Via Paradiso 12, 44100 Ferrara, Italy
[3] Istituto Fisica Cosmica e Tecnologie Relative, C.N.R.,
   Via Bassini 15, 20133 Milano, Italy
[4] Space Research Organization in the Netherlands,
   Sorbonnelaan 2, 3584 CA Utrecht, The Netherlands
[5] Istituto Astrofisica Spaziale, C.N.R.,
   Via Fosso del Cavaliere 100, I-00133 Roma, Italy

Abstract. We present preliminary results of the fits with the Optically Thin Synchrotron Shock Model of GRB time-averaged spectra from 2 to 700 keV basing on BeppoSAX data. This model gives a satisfactory description of the majority of the data. Nevertheless, ~30% of the spectra cannot be fitted by the model, indicating the need of accounting for substantial radiative cooling of the emitting electrons during the event.

1 Introduction

The joint study of the X- and gamma-ray emission of Gamma-Ray Bursts (GRBs) is a crucial diagnostic of theoretical models for the main mechanisms responsible of the emission.

Through a combination of the data from the Wide Field Cameras (WFC) [1] and Gamma-Ray Burst Monitor (GRBM) [2] on board the BeppoSAX satellite [3] it is possible to study the spectra of γ-ray bursts (GRBs) down to few keV X-rays. Frontera et al. [4] investigated the spectra of 8 events simultaneously detected by the WFC and GRBM. Now, more than 40 GRBs have been spectrally measured by both instruments.

In this contribution we present the preliminary results of the fits with the Optically Thin Synchrotron Shock Model by M. Tavani [5] of 19 BeppoSAX time averaged spectra in the 2–700 keV energy band. With respect to previous studies, this work has the advantage of being based on fits with a physical model and of extending GRB spectral analysis down to X-rays with high accuracy. In addition, by using time-integrated spectra we take also advantage of a much better statistical quality of the data and can perform the fits using the GRBM 240 channels spectra (128 s integration time) [6].

2   The Spectral Model

The Optically Thin Synchrotron Shock Model (OTSSM) by M. Tavani [5] is based on synchrotron emission by a combination of relativistic 3D Maxwellian distribution plus a power-law tail for the electrons that are efficiently and fastly accelerated by a relativistic shock in an optically thin environment. This combination of electrons distributions gives a synchrotron photon spectrum with a shape similar to the smoothed broken power-law (the Band model [1]) generally used to fit GRBs spectra. The model is characterized by only two parameters (in addition to spectral normalization): E_c, the relativistic synchrotron energy of the Maxwellian distribution which sets the scale of the peak of the $\nu F\nu$ spectrum, and δ, the index of the power law distribution for supra-Maxwellian particles. The spectral index below E_p asymptotically approaches $-2/3$.

3   Results and Discussion

The sample on which we performed our analysis is composed by 19 GRB events detected from 11 January 1997 to 14 October 1999. The gamma-ray (40–700 keV) fluences span over 2 orders of magnitude, the ratio between X-ray (2–28 keV) fluence and gamma-ray fluence ranges from 0.03 to 0.64 with the exception of GRB990704 (2.84). Gamma-ray durations vary from 3 to 150 s, X-ray durations vary from 10 to 260 s, and very different kinds of morphology are included

For this analysis we used XSPEC package ver. 10 [8]. The OTSSM description software was added as a model in XSPEC. The fits were performed accounting for galactic photo-electric absorption in the direction of the events. The GRBM response function, calibration and spectral data analysis techniques are discussed in [9,6]. For the WFC, details on response function, data reduction and background subtraction can be found in [1].

The fits with the OTSSM give acceptable values of reduced χ_ν^2 ($= \chi^2/\nu$, where ν is the number of d.o.f.) for 14 events out of 19. Values of Ec range from few keV to $\sim$500 keV, values of δ range from 3 to 6.6 (corresponding to high energy spectral index ranging from -2.0 to -3.8). Among the events better described by the OTSSM there are the brightest ones (GRB970111, GRB980329, GRB990123, GRB990705).

Thus, the OTSSM model is successful in describing average spectra of most GRBs. Nevertheless, in $\sim$30 % of the cases the measured spectral shape deviates substantially from this model. In Fig. 1 we show the spectra of four GRBs of our sample fitted with the OTSSM. It can be seen that for GRB970111 and GRB980329 the fits are very good ($\chi_\nu^2 = 39.8/44$ and $25.5/24$, respectively), whereas the spectra of GRB970508 and GRB990217 cannot be described satisfactorily by this model ($\chi_\nu^2 = 54.3/9$ and $34.7/9$, respectively).

By fitting the spectra with the standard Band model [1] we find that the low-energy spectral index, when Ep is inside our energy passband, is always between -0.67 and -1.5, the two extremes being in agreement with the synchrotron shock model when the effect of electron cooling is taken into account. However, the fits

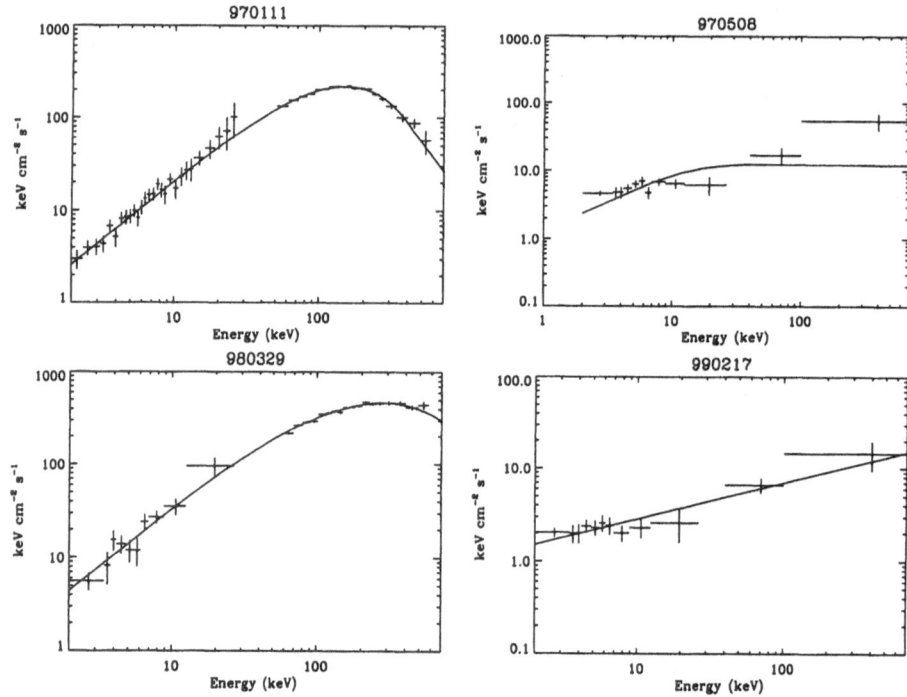

Fig. 1. $\nu F\nu$ spectra of four GRBs of our sample fitted with the OTSSM

with the OTSSM model that does not take into account the electron cooling are clearly bad when the index α is softer than $-2/3$.

The evidence for substantial radiative cooling of the electron energies during some events requires an improvement of the OTSSM model that includes the time dependence of the electron distribution parameters. This will be the next step of our work.

References

1. Jager, R., et al. 1997, A&AS, 125, 557
2. Frontera, F. et al. 1997, A&AS, 122, 357
3. Boella, G. et al. 1997, A&AS, 122, 299
4. Frontera, F. et al. 2000, ApJS, 127, 59
5. Tavani, M. 1996, ApJ, 466, 768
6. Amati, L. et al. 1999, A&AS, 138, 403
7. Band, D. et al. 1993, ApJ, 413, 281
8. Arnaud, K.A. 1996, Astronomical Data Analysis Software and Systems V, eds. Jacoby, J. and Barnes, J., ASP Conf. Series 101, 17.
9. Amati, L. et al. 1997, SPIE Proceedings, 3114, 176

On the Spectra of the Gamma-Ray Bursts

Zsolt Bagoly[1], István Csabai[2], Attila Mészáros[3], István Horváth[4],
Roland Vavrek[5], and Lajos G. Balázs[5]

[1] Lab. for Information Technology, Eötvös University, H-1518 Budapest, Hungary
[2] Dept. of Physics for Complex Systems, Eötvös University,
   H-1518 Budapest, Hungary
[3] Astron. Inst. Charles Univ. V Holešovičkách 2, CZ-180 00 Prague 8, Czech Republic
[4] Dept. of Physics, Bolyai Military University, Box 12, H-1456 Budapest, Hungary
[5] Konkoly Observatory, Box 67, H-1505 Budapest, Hungary

Abstract. The soft tails of gamma-ray bursts spectra obtained by different instruments are discussed. It is known that some of them have excesses from the exact power law dependence. It is conjectured that this soft-excess may estimate the redshift of a long gamma-ray burst.

1 Introduction

The long subgroup [9] of the gamma-ray bursts detected by BATSE [4] are at high redshifts; the highest directly confirmed redshift is $z = 4.5$ [1]. This conclusion holds only in statistical sense; it can be said that long GRBs have redshifts up to $z \simeq 4.5$ (or probably even up to $z \simeq 20$; [13], [7], [8]). Only for a few cases are the spectroscopic redshifts known from the host galaxies being detected by the BeppoSAX satellite [1] or other instruments [8].

In [6] and [17] a linear relation between the intrinsic peak luminosities of GRBs and their so called "variabilities" is found from a few cases of GRBs, when the bursts were observed *both* by BATSE and other instruments. Similarly, [12] found a relation between the so called spectral lag and the peak-luminosity allowing to determine the redshifts of long GRBs, too. The physical meaning of the two methods is unclear yet.

In this letter we present a third method of determination of the redshifts for the given long GRBs. This has a well defined physical meaning following from the spectra of long GRBs. Details will be published elsewhere [2].

2 The Method Based on the Soft-Excess

The spectra of GRBs usually may well be approximated by broken power-law spectra [3]. Nevertheless, it is also well-known that there may be an essential departure from the power-law at low energies (around $\sim (10 - 50)\ keV$). This is the so called soft-excess, which is confirmed for $\simeq 15\%$ of GRBs on the high confidence level [14],[15],[16]. Spectral evolution and soft excess were clearly observed by [4] around the peak emission in the case of GRB 990123.

Let us define the following quantity:

$$SER = \frac{P(E_3, E_2) - P(E_2, E_1)}{P(E_3, E_2) + P(E_2, E_1)} \tag{1}$$

where $P(E_j, E_i)$ defines the observed flux between energies E_j and E_i ($i, j = 1, 2, 3$). For a pure power-law spectrum any flux ratio will not change with redshift. The SER *will* change with the redshift for a given Band spectrum.

3   Results and Conclusion

First, we consider GRBs having well defined redshifts and HER and DISCSC data[4]. We analyzed the HER spectra around the peak emission and found a definite correlation between SER values based on the HER fluxes and z. The deconvolved HER spectrum has high energy resolution but its time resolution is strongly non-uniform. The $64\mu s$ time resolution DISCSC data avoid this problem, so to calculate the SER values ±2 bins around the peak emission in the lower two DISCSC channels ($E_1 = 25\ keV$, $E_2 = 50\ keV$ and $E_3 = 100\ keV$) were used. It is clearly visible on Fig. 1 that there is an remarkable correlation between SER and z. This also means that SER seems to be a good indicator of z.

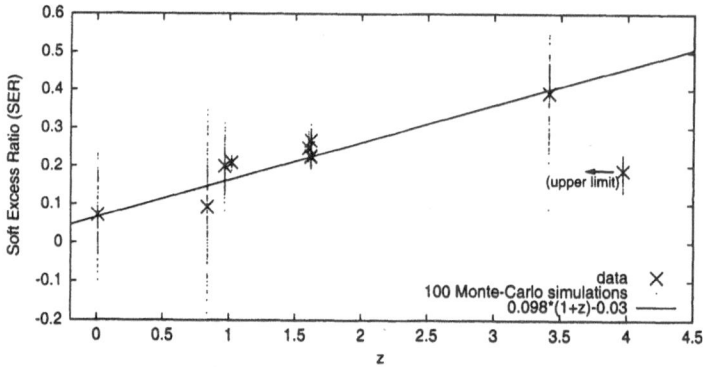

Fig. 1. The dependence of the measured SER on the observed z (GRBs 980425, 970508, 970828, 980703, 991216, 990123, 990510, 971214, and 980329). For every burst further 100 Monte-Carlo simulations show the effect of the Poisson noise. The best linear fit is also shown: the uncertainties of *obtained* z-s are tipically around $dz \sim 0.3$.

Second, we consider all such GRBs from BATSE Catalog, for which DISCSC data given, and hence SER can be calculated. We deduce their redshifts from the previous linear approximation (because of $|SER| \leq 1$, it's clearly an approximation only). We restricted ourselves, in this stage of our study, for the bright ($F_{256} \geq 2$ photon/(cm^{2}s)) long ($T_{90} \geq 10$ s) GRBs only. The results are shown on Fig. 2. One can observe the long tail with redshifts up to $z \sim 5$.

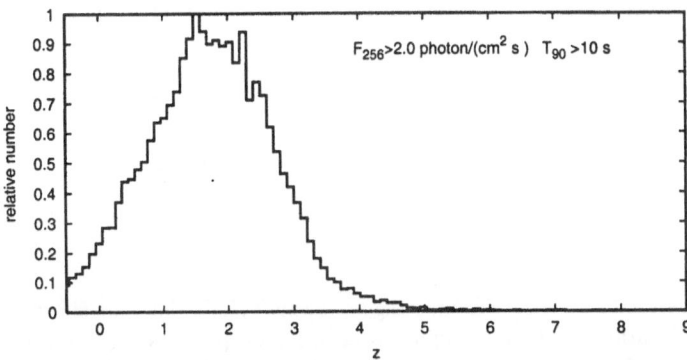

Fig. 2. The redshift distribution of 402 bright long GRBs having DISCSC data.

As the conclusion, we emphasis that the Soft Excess Ratio method gives for every GRB its redshift using the plausible physical explanation of this Cepheid-like behaviour.

This research was supported in part through Domus Hungarica Scientiarum et Artium grant (A.M.); OTKA grants T024027 (L.G.B.), F029461 (I.H.) and T034549.

References

1. M.I. Andersen, et al., A&A, **364L** p. 54. (2000)
2. Z. Bagoly et al.: *in preparation* (2001)
3. D.L. Band, et al., ApJ **413** 281 (1994)
4. M.S. Briggs et al., ApJ **524** 83 (1999)
5. E. Costa, et al. , IAU Circ. 6572 (1997)
6. E.E. Fenimore, E. Ramirez-Ruiz, ApJ, *submitted* (astro-ph/0004176) (2000)
7. I. Horváth, P. Meszáros, A. Mészáros, 1996, ApJ **470** 56 (1996)
8. S. Klose: In: *Reviews in Modern Astronomy*, **13**, Astronomische Gesellschaft, Hamburg, p. 129 (2000)
9. C. Kouveliotou, et al.. ApJ **413**, L101 (1993)
10. D.Q. Lamb, D.E. Reichart, ApJ **536** 1 (2000)
11. C.A. Meegan, et al.: Current BATSE Gamma-Ray Bursts Catalog, http://gammaray.msfc.nasa.gov/batse (2000)
12. J.P. Norris, G.F. Marani, J.T. Bonnell: In: *Gamma-Ray Bursts: 5th Huntsville Symposium*, eds. R.M. Kippen et al. (AIP Conference Proceedings, Vol. 526, Melville, New York p. 78. (2000)
13. A. Mészáros, P. & Mészáros, ApJ **466** 29 (1996)
14. R.D. Preece, et al., ApJ **473** 310 (1996)
15. R.D. Preece, et al., ApJS **126**, 19 (2000)
16. R.D. Preece, J.R. Espley, M.S. Briggs: In: *Gamma-Ray Bursts: 5th Huntsville Symposium*, eds. R.M. Kippen et al. (AIP Conference Proceedings, Vol. 526, Melville, New York p. 175. (2000)
17. D.E. Reichart, et al., ApJ, *submitted* (astro-ph/0004302) (2000)

Short Gamma-Ray Bursts Are Different

J.P. Norris[1], J.D. Scargle[2], and J.T. Bonnell[1]

[1] NASA/Goddard Space Flight Center, Greenbelt, MD 20771, USA
[2] NASA/Ames Research Center, Moffett Field, CA 94035-1000

Abstract. We analyze BATSE time-tagged event (TTE) data for short gamma-ray bursts (T_{90} duration < 2.6 s), studying spectral lag vs. peak flux and duration, as well as the number of distinct pulse structures per burst. Performing the cross-correlation between two energy bands, we measure an average lag $\sim$ 20–40 $\times$ shorter than for long bursts, and a lag distribution close to symmetric about zero – unlike long bursts. Using a "Bayesian Block" method to identify significantly distinct pulse peaks, we find an order of magnitude fewer pulses than found in studies of long bursts. The disparity in lag magnitude is discontinuous across the $\sim$ 2-s valley between long and short bursts. Thus, short bursts do not appear to be representable as a continuation of long bursts' temporal characteristics.

1 Introduction

Our understanding of short bursts is limited to their gamma-ray characteristics, whereas for long bursts we have multi-wavelength afterglows, redshifts, host galaxies, evidence of progenitors arising in star-forming regions, plus lots of theory. For short bursts, we have the "Smaller Bump" in the GRB duration distribution. The following references serve as a summary of the sparse research history of short bursts: their identification as a subclass by duration [5]; early indications that their fraction is $\sim$ 1/4 of the total [6]; their harder spectra [3]; and truncation effects that could be hiding even more bursts at shorter durations [13]. Most relevant to the present work, was the suggestion by two groups that a continuous deformation of some combination of long bursts' temporal characteristics – the distributions of number of pulses per burst, pulse width, and intervals between pulses – could explain short bursts' durations and the 2-s minimum [7,16]. Here, we further quantify the temporal differences between long and short GRBs, finding the properties of the two classes to be disjoint.

2 Spectral Lag and Pulse Analysis

The sample comprises all 261 short bursts (durations measured via technique similar to that in reference [1]) with "good" TTE data (the burst was contained within the TTE 32k event buffer), and peak flux (PF) > 2 photons cm^{-2} s^{-1} (50–300 keV). We used a "Bayesian Block" cell coalescence method [14,15] to determine the burst region and significant peaks and valleys, yielding the number of pulses per burst. We cross-correlated the 25–50 keV and 100–300 keV energy channels, bootstrapping the TTE data to make 51 profiles per burst. We fitted a cubic near the peak of the cross correlation function (CCF), requiring 51 consecutive fits to be concave down; else, we lengthened the fitted region, repeating the procedure for a new set of 51 profiles. The mode of the CCF peaks per burst was taken as the spectral lag measure, and error bars were generated.

3 Lag Results for Short Bursts

Whereas CCF lags for long bursts [9,10] extend up to ~ 2 s, with the core near 50 ms, the maximum lags for short bursts are ~ 30 ms, or $\sim 60 \times$ shorter (Figure 1a). Virtually all long bursts analyzed (save for two apparently anomalous cases) have positive lags, or insignificantly negative determinations [9,10]. Figure 1a shows that short burst lags are distributed approximately symmetrically about zero lag with the mode near 0–2 ms. The better determined lags at higher peak flux are also concentrated near zero. In fact, for PF > 10 photons cm^{-2} s^{-1}, the absolute value of the CCF lag is less than 2.5 ms for 25 of 35 bursts. When divided into four duration ranges as shown in Figure 1b, short bursts do not evidence a trend towards longer lags for longer short bursts: The core of the lag distribution is contained within $\pm$ 15 ms with modes of less than $\pm$ 5 ms, essentially independent of duration. Similarly, a duration-independent discontinuity in lag is evident for long bursts as well [11].

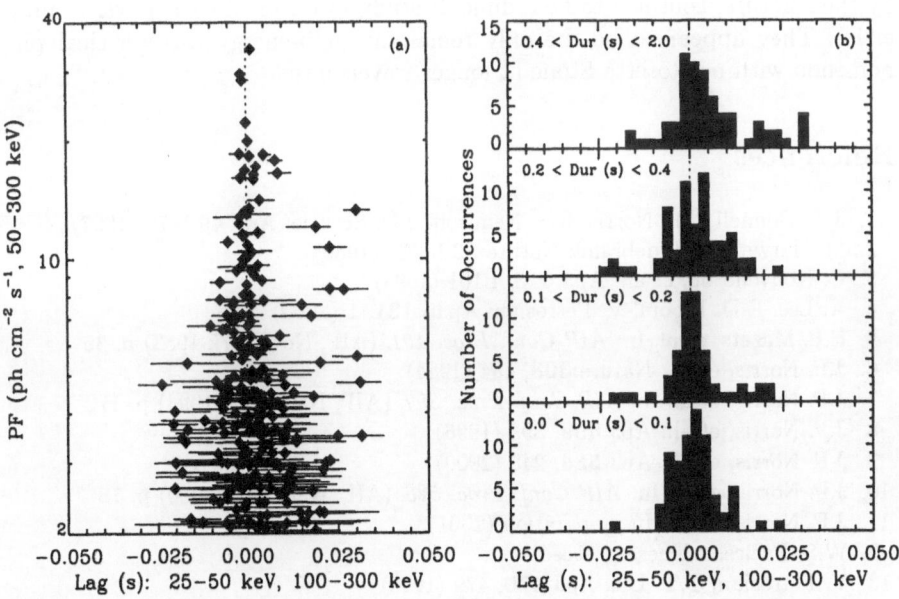

Fig. 1. (a) CCF lag vs. peak flux for short bursts. (b) Histograms of CCF lags for four duration ranges, illustrating that the mode is nearly independent of duration.

From our Bayesian Block coalescence analysis, we determine that bright short bursts usually have $\sim$ 1–2 pulses per burst [11]. In comparison, bright long bursts have a broad range in number of pulses – a few to tens – as can be appreciated from Figure 5 of Lee et al. [4] (see also [8]). Fewer pulses are discernible in weaker short bursts [11], a brightness bias also evident in long bursts [4]. Thus long bursts average an order of magnitude more pulses per burst than do short bursts. Combined with the discontinuity in lag, the results suggest that the temporal properties of the two classes are disjoint, in the vicinity of $\sim$ 2 s.

4   Summary and Conclusions

Short bursts appear to be a "different" phenomenon (by Harwit criteria [2]): Their lags are of order $\sim$ 20–40 $\times$ shorter than lags for long bursts, with the discontinuity near $\sim$ 2 seconds (T_{90} duration). The lag distribution appears close to symmetric, about zero. They have an order of magnitude fewer pulses than do long bursts. In the duration distribution the two modes are separated by a factor of order 100, $\sim$ 250 ms compared to $\sim$ 25 s. Strengthening this picture is the analysis of Paciesas et al. [12] which finds a discontinuity in the distribution of E_{peak} values between short and long bursts.

Thus, based on our knowledge of their gamma-ray characteristics alone, we conclude that short bursts cannot be represented as a continuation of long bursts. The two classes are disjoint, temporally and spectrally. Moreover, short bursts, by their nature, continue to be a difficult study in terms of their physical properties: They appear to be the only remaining high-energy astrophysical phenomenon with no Rosetta Stone at longer wavelengths.

References

1. J.T. Bonnell, J.P. Norris, R.J. Nemiroff, J.D. Scargle: ApJ **490**, 79 (1997)
2. M. Harwit, R. Hildebrand: Nature **320**, 724 (1986)
3. C. Kouveliotou, et al.: ApJ **413**, L101 (1993)
4. A. Lee, E.D. Bloom, V. Petrosian: ApJS **131**, 1 (2000)
5. E.P. Mazets, et al. In: *AIP Conf. Proc. 101*, (AIP, New York 1983) p. 36
6. J.P. Norris, et al.: Nature **308**, 434 (1984)
7. J.P. Norris, et al. In: *AIP Conf. Proc. 307*, (AIP, New York 1994) p. 172
8. J.P. Norris, et al.: ApJ **459**, 393 (1996)
9. J.P. Norris, et al.: ApJ **534**, 248 (2000)
10. J.P. Norris, et al. In: *AIP Conf. Proc. 526*, (AIP, New York 2000) p. 78
11. J.P. Norris, et al.: In preparation (2001)
12. W.S. Paciesas: These proceedings
13. V. Petrosian, T.T. Lee: ApJ **470**, 479 (1996)
14. J.D. Scargle: ApJ **504**, 405 (1998)
15. J.D. Scargle: To appear in MaxEnt99 (2000)
16. V. Wang, In: *AIP Conf. Proc. 384*, (AIP, New York 1996) p. 106

New Possibilities Offered by BeppoSAX: Automatic GRB Alerts Using GRBM

Cristiano Guidorzi[1], Filippo Frontera[1,2], Enrico Montanari[1],
Francesco Calura[1], Lorenzo Amati[2], Enrico Costa[3], and Marco Feroci[3]

[1] Universitá di Ferrara, via Paradiso 12, 44100 Ferrara, Italy
[2] ITeSRE, CNR, Via Gobetti 101, 40129 Bologna, Italy
[3] IAS, CNR, Via Fosso del Cavaliere, 00133 Roma, Italy

Abstract. A new, fully automatic gamma-ray burst alert system, analysing the data from the Gamma-Ray Burst Monitor on board BeppoSAX, has been developed and it is operating since April, 2000. When a true celestial burst is recognized, this S/W yields some relevant data, like peak counts, total counts, duration. In the near future, the possible incoming direction will be included as well. This info is promptly and automatically distributed, in a fixed format, to a list of potential observers. Also a message for mobile phones, containing the same data and reported in a compact format, is automatically sent. Potential users can apply to receive these alerts.

1 The Gamma-Ray Burst Monitor (GRBM/BSAX)

The Gamma-Ray Burst Monitor (GRBM) [1], [2], aboard the BeppoSAX satellite [3] consists of four CsI(Na) detector units, each of which has a geometrical area of 1136 cm^2 and continuously records ratemeters in two energy bands, namely 40-700 keV and > 100 keV. When the on-board elecronics trigger a transient event in the 40-700 keV band, High Time Resolution (HTR) ratemeters are available from 8 seconds before the trigger time and they last 106 seconds, with an integration time ranging from 0.48828125 ms (for the first 10 s after the trigger time) to 7.8125 ms.

2 An Automatic On-Line Quest of GRBs

In order to recognize all the Gamma-Ray Bursts (GRBs) detected by the GRBM, both triggered and not on board, a systematic analysis of the 1 s ratemeters in both energy bands is automatically applied to the GRBM data in a real-time mode. The fully automatic operations can be summarized in the following steps:

1) *Periodic Data Download*: the GRBM raw data are grouped into single orbits and once per orbit (about 103 min.) they become available for the automatic analysis performed with the S/W developed in Ferrara.

2) *GRB Quest*: the light curves recorded by the four detector units in both 40-700 keV and > 100 keV bands are cross-checked by applying the *S/W Trigger Conditions* (SWTC, see below), in order to identify possible GRBs.

3) *Alert Procedure and Data Distribution*: when a likely GRB is recognized, its peak counts (counts/s), fluences (counts), durations and other useful parameters characterizing the burst (one for each unit that detected it) are distributed in a fixed format to a mailing list, including the BeppoSAX Science Operation Center (for cross-checking with the WFC data), the GRBM team, the Ulysses team (for IPN triangulations, when possible) and some machines devoted to automatic cross-checks with the NEAR data and with the XTE data (since October 2000, some HETE II team members have been receiving these alerts, as well).

This automatic procedure performs the following useful tasks:

1) the recognition of true GRBs (that are only about 4%) among the transient events triggered on board, mostly due either to high energy particles crossing the detectors or ionospheric events or solar flares;

2) the identification of GRBs that are ignored by the on-board logic (about 20%);

3) a fast identification of the GRBs that can be localized, when possible, with an IPN triangulation or by means of the Wide Field Cameras (WFCs) aboard BeppoSAX.

3   GRB Quest: Trigger Conditions

Because the constraints requested on board for triggering GRBs are different from the SWTC, we called the formers *On-Board Trigger Conditions* (OBTC).

OBTC. Whenever the $40 - 700$ keV counting rate, accumulated over an interval that can be preset in the range 7.8125 to 4000 ms, exceeds the background level (mean count rate on a moving time interval from 8 to 128 s long) by $n\sigma$ ($n = 4$, 8 or 16) in at least two units, OBTC are satisfied and high time resolution mode starts and lasts 106 s. The current accumulation time is 1s.

SWTC. One second ratemeters in both (40–700) and (> 100) keV energy bands are considered. The background level in each energy band is continuously estimated using two intervals: 80 and 30 s long, respectively (the first preceeding the time bin to be analysed by 5 s and the second following it by 50 s). The SWTC consist of different combinations of several thresholds for both energy bands of the different units, together with a so called Hardness Ratio (HR) condition ($HR_w > 0.3$), where HR_w is the weighted average of the different HR_i ($i = 1, ..., 4$, unit number), defined as the ratio between the total counts in the harder band (> 100 keV) and the softer one ($40 - 700$ keV). Only units that detected the event are taken into account to calculate HR_w. Given that GRBs show a greater HR_w than other transient events, the HR condition turned out to be very effective. For a detailed description of the SWTC, refer to [4].

4   GRB Mean Rate

During the period April 20, 2000–October 11, 2000 ($175 - 32 = 143$ days during which the system was active: in fact, it went off for 32 days due to technical problems), the alert procedure automatically recognized 131 likely GRBs; among them, 96 were confirmed as GRBs after visual inspection, the other 35 being mainly due to hard spikes and hard X-ray solar flares. We define the system reliability as the ratio between the number of true GRBs automatically identified over the number of all transients events detected: this turns out to be 0.73 ± 0.07. The GRB mean rate comes out to be: 0.67 ± 0.07 GRB/day. The value obtained for the GRB rate is quite sensible.

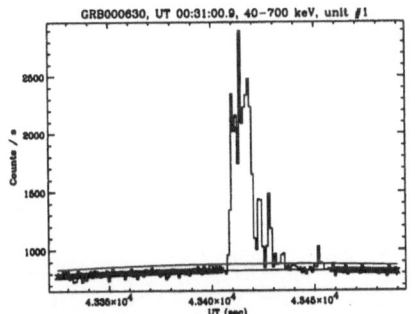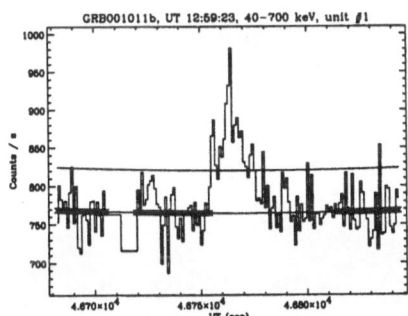

Fig. 1. Examples of the light curves of two on-line detected GRBs. The background fit automatically calculated and the $+2\sigma$ level are shown: the 2 intervals used for fitting the background are highlighted. GRB000630 (left panel): it was promptly localized with an IPN triangulation (Ulysses-BeppoSAX-KonusWind-NEAR, GCN 736), allowing the discovery of the optical afterglow (GCN 737-748). GRB001011b (right panel): it did not trigger the on-board electronics, because under the on-board threshold sensitivity: nevertheless, it triggered SWTC # 4 (before the burst a small data gap is visible).

5   Work in Progress

The final purpose of these efforts is the automatic localization of a burst. To this aim, we will use the GRB total counts automatically calculated to estimate the GRB incoming direction, by means of a response matrix with a refined grid, developed on the basis of a Monte Carlo model of the BeppoSAX payload. A preliminary work performing the same task had been done with the on-ground calibrations [5]. The best direction estimate is determined by means of χ^2 techniques. For a complete description of the Monte Carlo model and of the localization techniques, see [6]. Currently, the typical statistical error radius in locating the GRB is about $10°$, sufficiently small to resolve the ambiguity in the case of two possible IPN directions for the same GRB, when these are many degrees far from each other.

References

1. Frontera F., Costa E., Dal Fiume D., et al., *A&AS* **122**, 357 (1997).
2. Feroci M., et al. 1997, *Proc. SPIE* **3114**, 186 (1997).
3. Boella G., Butler R.C., Perola G.C., et al., *A&AS* **122**, 299 (1997).
4. Guidorzi C. et al., *Conf. Proc. of the It. Soc. of Phys.*, Vol. **68**, 261, SIF, (2000).
5. Preger, B., et. al., *A&AS Suppl. Ser.*, **138**, 559 (1999).
6. Calura F. et al., in AIP Conf. Proc. **526** (2000).

Probing the Isotropy in the Sky Distribution of the Gamma-Ray Bursts

Attila Mészáros[1], Zsolt Bagoly[2], Lajos G. Balázs[3],
István Horváth[4], and Roland Vavrek[3]

[1] Astronomical Institute of the Charles University,
   V Holešovičkách 2, CZ-180 00 Prague 8, Czech Republic
[2] Laboratory for Information Technology, Eötvös University,
   Pázmány Péter sétány 1/A, H-1518 Budapest, Hungary
[3] Konkoly Observatory, Box 67, H-1505 Budapest, Hungary
[4] Department of Physics, Bolyai Military University,
   Box 12, H-1456 Budapest, Hungary

Abstract. The statistical tests – done by the authors – are surveyed, which verify the null-hypothesis of the intrinsic randomness in the angular distribution of gamma-ray bursts collected at BATSE Catalog. The tests use the counts-in-cells method, an analysis of spherical harmonics, a test based on the two-point correlation function and a method based on multiscale methods. The tests suggest that the intermediate subclass of gamma-ray bursts are distributed anisotropically.

1 Introduction

At the last years the authors carried out several statistical tests in order to verify the isotropy of the angular distribution of the gamma-ray bursts (GRBs) collected at BATSE catalog ([4]). In this contribution we collect the results of them; these results were partly published in several articles ([1], [2], [5], [6], [7]).

2 Tests

Spherical Harmonics

The key idea of this test is based on the fact that the sky-exposure function of BATSE instrument is not depending on right ascension. Therefore in equatorial coordinates the theoretically expected values of spherical harmonics of the distribution of GRBs are zeros for any $m \neq 0$ term. Then these expectations are tested.

Counts-in-Cells

This is a simple statistical test. The idea is the following: The sky is separated into equal areas, and then, e.g., χ^2 test is used to test the null hypothesis of isotropy. The sky-exposure function can be eliminated by the use of equatorial coordinates; then "effective" equal areas are taken. For example, if the sky is separated into 8 equal areas, then the boundaries are $\alpha = 0, 90, 180, 270$ degrees, and $\delta = -30.8, +1.5, +33.6$ degrees (instead of $\delta = -30, 0, +30$ degrees).

Two-Point Angular Correlation Function

The key idea of this method is the following. Having N GRBs on sky we have $N(N-1)/$ angular distances among them. If N GRBs are distributed randomly, then these distances should be distributed randomly, too. Then the observed distances are compared with the pseudo-randomly generated $N(N-1)/2$ distances coming from Monte Carlo simulations, which are provided in accordance with the sky-exposure function. Hence, the sky-exposure function is eliminated by Monte Carlo simulations.

Multifractal Analysis, Minimal Spanning Tree, Voronoi Tesselation

For the detailed description of these three methods see the contribution [10] in these Proceedings.

3 Results

The results of done tests for the three subclasses ([5], [3]) of GRBs separately are collected at Table 1.

The done tests of isotropy suggest the existence of **anisotropy for the intermediate subclass** on the confidence level > 95%.

For the remaining two subclasses the situation is unclear; there is no unambiguous rejection of isotropy for them yet on the higher than 95% confidence level. It can only be said that the short subgroup is highly "suspicious".

save

Table 1. Survey of the results of the isotropy tests. The question "Is the null hypothesis of isotropy rejected?" is answered. When the answer is "Yes", then the significance level of rejection is also given. We required a higher than 95% level.

short $T_{90} < 2s$	intermediate $2s < T_{90} < 10s$	long $T_{90} < 10s$	
No	Yes	No	Spherical
	> 97%		harmonics
No	Yes	No	Counts-
	> 96.4%		in-cells
Yes	Yes	Yes	Two-Point
> 99.2%	> 99.8%	> 99.8%	Correlation
No	Not	Not	Voronoi
	done	done	tesselation
No	Not	Not	Minimal
	done	done	spanning tree
Yes	Not	Not	Multifractal
> 99.9%	done	done	analysis

4   Conclusions

The long GRBs seems to be distributed isotropically – the positive result from two-point angular correlation function is probably an unknown instrumental effect.

For the short GRBs the isotropy is not rejected yet on a satisfactorily high confidence level, but there are indications for the anisotropy both from the the multifractal analysis and also from the two-point angular correlation function. Add also that the statistical comparison of the short and the intermediate + long subgroups also suggests anisotropy here [1], [2]. Simply the situation is highly "suspicious" here. Note still that the shortest "tail" $T_{90} < 0.1$ s, which is doubtlessly anisotropic [3], was not considered separately.

The intermediate subclass [3] **is anisotropic**; only the concrete value of confidence level is a question – it "fluctuates" between 96.4–99.9 %. The character of anisotropy of intermediate subclass is incomprehensible, because the "dimmer" half of this subsection is more anisotropic [6]. In addition, there is no concentration toward the Galactic or Supergalactic planes.

This research was supported by Research Grant J13/98: 113200004 (A.M.), by OTKA grants T024027 (L.G.B.), F029461 (I.H.) and T034549.

References

1. L.G. Balázs, A. Mészáros, I. Horváth: A&A **339** 1 (1998)
2. L.G. Balázs, A. Mészáros, I. Horváth, R. Vavrek: A&A Suppl. **138** 417 (1999)
3. D.B. Cline, C. Matthey, S. Otwikowski: In: *Gamma-Ray Bursts: 5th Huntsville Symposium*, eds. R.M. Kippen et al. (AIP Conference Proceedings, Vol. 526, Melville, New York 2000) pp. 97-101
4. I. Horváth: ApJ **508** 757 (1998)
5. C. Kouveliotou, et al.: ApJ **413** L101 (1993)
6. C.A. Meegan, et al.: Current BATSE Gamma-Ray Bursts Catalog, http://gammaray.msfc.nasa.gov/batse (2000)
7. A. Mészáros, Z. Bagoly, R. Vavrek: A&A **354** 1 (2000)
8. A. Mészáros, Z. Bagoly, I. Horváth, L.G. Balázs, R. Vavrek: ApJ **539** 98 (2000)
9. A. Mészáros, Z. Bagoly, I. Horváth, L.G. Balázs, R. Vavrek: In: *Gamma-Ray Bursts: 5th Huntsville Symposium*, eds. R.M. Kippen et al. (AIP Conference Proceedings, Vol. 526, Melville, New York, 2000) pp. 102-106
10. R. Vavrek, L.G. Balázs, A. Mészáros, I. Horváth, Z. Bagoly: these Proceedings

Analysis of the BATSE GRB Light Curves

B.M. Belli

Istituto di Astrofisica Spaziale, CNR, Via del Fosso del Cavaliere 00131 Roma, Italy

Abstract. Different progenitors and physical processes could be at the origin of GRBs. New information about the properties of the gamma-ray emission can help to select among the different proposed models and possibly to identify different classes with different origins. We have tried to study the gamma light-curve behaviour in relation to other physical parameters of the event as duration, peak counting rate, energy spectra hardness, et etc. There are hints that some of the mentioned correlations are present.

1 Introduction

Studying the gamma-ray burst (GRB) temporal histories it has been possible to identify different groups of GRBs [3] characterized by the morphology of their light curve. This division is obviously tentative. At the moment when this classification has been suggested any 'a priori' knowledge of the physical mechanisms and of the modalities occurring in the phenomenon were known. We report here the classification criteria that we have followed and the related marks used to distinguish the different groups in the figures (fig.1 on the left):

1 – temporal profiles with spiky structures with large fluctuations in all time scales (starlet),
2 – temporal profile with rather simple structures with few peaks (full circle),
3 – temporal profile with fast rise time and exponential dacay (FRED) (triangle),
4 – single or few pulses, spike events (square).

This classification is not always possible specially for weak events. In addition not all the light curves of the BATSE catalog are published or available at the moment. For these reasons in our analysis the classification is limited to those events which are strong enough and for which the light curve is known, i. e. the events of the first BATSE catalog, 260 events [2] and those for which the light curve has been provided to the scientific community by the GRB Coordinates Network (GCN), 179 events. This circumstance strongly reduces the number of events. For the spike events we have included in this class those events with one or few spikes and also those with a small bump in the photon count before or after the main feature. The group of events, provided by the GCN, have as temporal bin 1.024 s.

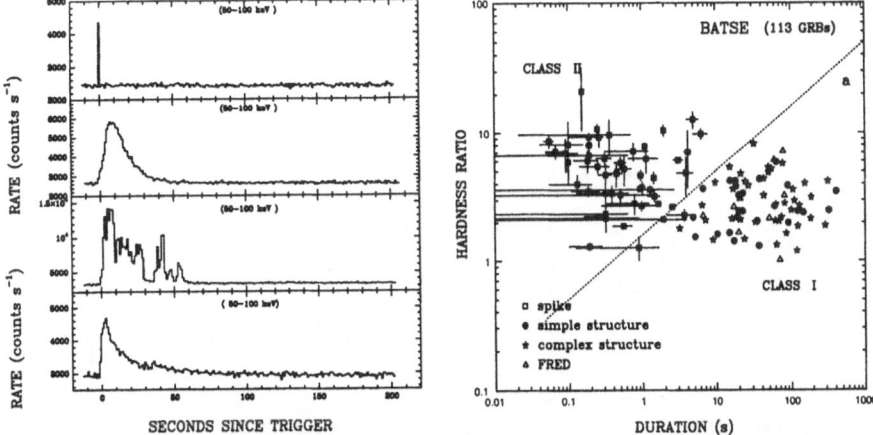

Fig. 1. On the left examples of the different shapes of light-curve from the available ones of the BATSE catalog; on the right hardness-ratio against duration plot for 113 events selected among the BATSE GRBs with available light-curve. The events are represented with different marks according to the light-curve shape. The error bars are those reported in the catalog.

2   Duration Spectral-Hardness Correlation

We first study a possible duration spectral-hardness correlation plotting the available events in the duration and spectral hardness plane (fig. 1, on the right). The spectral hardness HR is given by the hardness-ratio, the ratio of the fluences in the energy ranges 100-300 and 50-100 keV; and the duration T90 is the time during which the integral counting rate goes from the 5% to the 95% of the total. In a previous paper we divided GRBs in two classes I and II in the plane D-HR, duration hardness-ratio. The two classes are separated by the straight line HR = 2 $D^{0.5}$. Class I is essentially composed of the longest events and Class II of the shortest ones but not only [1]. It appears clearly in the figure that all the spike type events belong to the Class II on the left of the straight line a).

The short events (D<2 s) belonging to the other light-curve types, have been not studied in this analysis for lack of a sufficient time definition. For this reason we are not able to evaluate how numerous they are. On the right of the line a) i.e. in the Class I, the events of all the other light-curve types appear completely mixed. Different values of the duration or the hardness ratio do not give any information on the possible shape of the light-curve.

3   Other Possible Correlations for GRBs

We plot now the events in the plane duration-peak counting rate and in the plane duration-fluence, fig.2 on the left and on the right respectively. For Class I events we can see, by a qualitative evaluation, that the spiky events are presented over

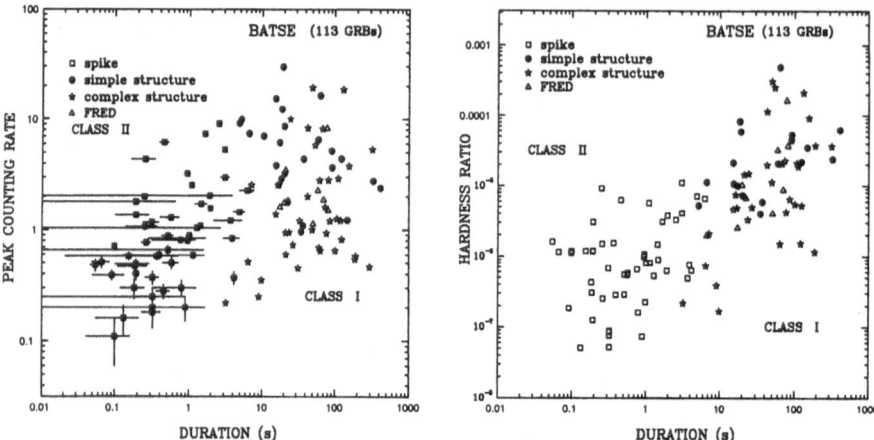

Fig. 2. On the left plot in the plane dutation-peak photon flux $(cm^{-2}, s^{-}1)$ of the selected BATSE GRBs, divided in different light-curve shape with different marks. On the right plot of fluence $(ergcm^{-2})$ versus duration (s) of the same eve nts.

the whole parameter value range while the simple structure events are grouped mostly on the high values of the peak counting rate and of the fluence, the FRED events in the center.

4   Discussion

As it is shown in the figures the spike events appear to belong to the Class II and could even be a separate class of GRBs. For the events of Class II no afterglow has been yet observed so far. We could say, if our hypotheses are right, that there is not a real experimental difficulty to observe them, because there are also events lasting up to ten second in this class. The simple structure events seem to emit on average a stronger energy than the complex structure events. It is impossible to say whether this circumstance is due to different physical processes or to different enviromental conditions. At the moment we cannot completely exclude that this effect could be also due to poor statistics or to selection effects. We hope that all the BATSE light-curves will be soon published and so a complete analysis will be possible.

References

1. Belli, B. M. Ap&SS, **231**, 43 (1995)
2. Fishman G. J. et al.: ApJS **1**, 229 (1994)
3. Fishman, G. J. and Meegan C. A. Ann. Rev. Astron. Astrophys. **33**, 415 (1995)

Gamma-Ray Burst Follow Up Observations with BOOTES in 1998–2000

J.M. Castro Cerón[1], A.J. Castro-Tirado[2,3], R. Hudec[4], J. Soldán[4],
M. Bernas[5], P. Páta[5], T.J. Mateo Sanguino[6], A. de Ugarte Postigo[7],
J.Á. Berná[8], M. Nekola[4], J. Gorosabel[9], B.A. de la Morena[6],
J.M. Más-Hesse[3], Á. Giménez[3], and J. Torres Riera[10]

[1] Real Instituto y Observatorio de la Armada, Sección de Astronomía
   11.110 San Fernando-Naval (Cádiz) Spain, josemari@roa.es
[2] IAA-CSIC, 18.080 Granada Spain
[3] LAEFF-INTA, 28.692 Villafranca del Castillo (Madrid) Spain
[4] ASÚ, AVČR, 251 65 Ondřejov Czech Republic
[5] ČVUT, FEL, Radioelectronics, 166 27 Prague 6 Czech Republic
[6] CEA-INTA, 21.130 Mazagón (Huelva) Spain
[7] Facultad de CC.FF., UCM, 28.040 Madrid Spain
[8] FISTS, UA, 03.690 San Vicente del Raspeig (Alicante) Spain
[9] DSRI, 2.100 Copenhagen Ø Denmark
[10] DCE-INTA, 28.850 Torrejón de Ardoz (Madrid) Spain

Abstract. The Burst Observer and Optical Transient Exploring System (**BOOTES**) provides an automated real time observing response to the detection of Gamma Ray Bursts (GRBs). Error box size depending, it uses wide field cameras attached to small robotic telescopes or the telescopes themselves. To date we have acquired photometry for about 30 events with the Ultra Wide (UWFC) and the Narrow Field Cameras (NFC) and about 50 events with the Wide Field Camera (WFC).

1 Introduction

GRBs were first reported in 1973 [1,2]. The first optical counterparts were detected in 1997, 3-20 hours after the onset of the burst. Their luminosity declines as a power law with $F \propto t^{\alpha}$ (-1.1 < α < -2.0). See [3] for a review.

2 Scientific Objectives

To study GRBs it is of the utmost importance to perform prompt optical follow up observations, to detect longer wavelength transient emission associated to them. BOOTES can perform such follow ups. Its scientific objectives include [4]:

- Simultaneous and quasi simultaneous observations of GRB error boxes.
- Detection of optical flashes of cosmic origin.
- Sky monitoring in the I, R^1 & V bands.

---

[1] BOOTES's CCD cameras have their peak response in the red, thus the reference to R band/magnitude for the unfiltered ones, which should not be taken as a standard filtre. See http://www.laeff.esa.es/~ajct/BOOTES/ for additional information.

- Monitoring of different types of objects in search of recurrent transient optical emission.
- Discovery of comets, meteors, asteroids, variable stars, novae, supernovae...

3   Instrumentation and Results

BOOTES is part, within the framework of a Spanish-Czech collaboration, of a wide ongoing effort to prepare for the ESA's satellite INTEGRAL. See [5,6] for an account of BOOTES's early and current set up and instrumentation respectively. BOOTES has been now performing rapid follow up observations of events detected by *BATSE*, *BeppoSAX*, *RossiXTE* and the *IPN* [7], for two years. Results include:

- Predetection images: they set up upper limits for any possible precursors. Some examples are available on line [8].
- Simultaneous images: it was achieved last 20 February 2001 [9], although no counterpart was detected (See Fig. 1).
- Quasi simultaneous images: GRB 000313. See [6,10,11].
- Follow up images: table 1 shows all the follow up observations performed with the BOOTES-1B UWFC to date. See [5,6] for a listing of the BOOTES-1A WFC and the BOOTES-1B NFC follow up observations respectively; also [12-14].

Table 1. Follow up observations of GRBs with the BOOTES-1B UWFC

GRB	$T_0 + (min.)$	% Error Box	R_{lim}	GRB	$T_0 + (min.)$	% Error Box	R_{lim}
991219	4	100	11	000302	8	100	11
000110	231	100	11	000302a	1,441	100	11
000111	17	100	11	000302b	740	100	11
000114	8	100	11	000303	10	100	11
000115	1,300	100	11	000306	90	100	11
000122	246	100	11	000312	93	100	11
000130	24	100	11	000313	4	100	11
000217	487	100	11	000315	281	100	11
000220	13	100	11	000323	643	100	11
000221	1,623	100	11	000324	3	100	11
000222	510	100	11	000326	883	100	11
000225	10	100	11	000520	617	100	11
000226	8	100	11	000730b	2,270	100	10
000227	10	100	11	000812	1,539	100	10
000229	17	100	11	010220	0	100	10
000301	24	100	11				

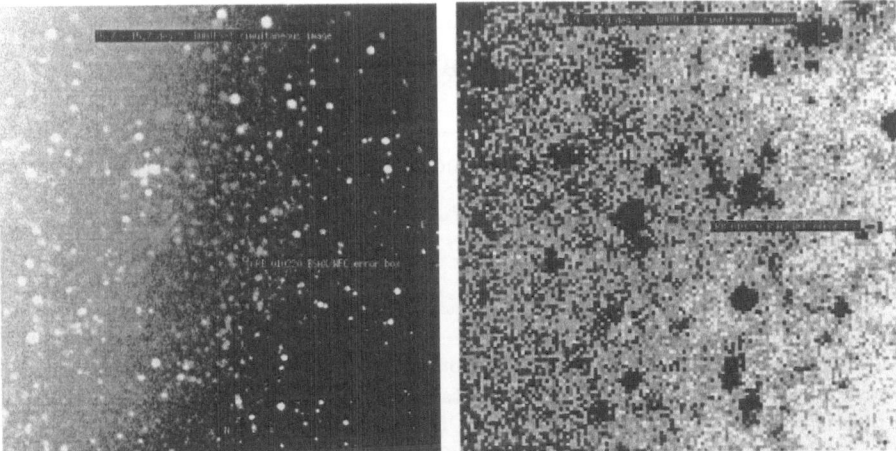

Fig. 1. GRB 010220 simultaneous optical observations. No optical transient was found within the BSAX/WFC error circle, probably due to high extinction ($A_V = 3.3$ mag). The error circle was located at a galactic latitude of $b = +1.38°$ with the galactic nebula IC 1805 lying along the line of sight (**left frame**) ending at 22:51:15 UT and covering the first 8 s of the event (**right frame**) starting at 22:52:15 UT and covering the time interval from $t_0 + 68$ s to $t_0 + 128$ s

Acknowledgments

We thank INTA's "División de Ciencias del Espacio" for their support through project IGE 4900506. JMCC, JÁB and JG acknowledge research grants from "Ministerio de Ciencia y Tecnología", "Generalidad Valenciana" and the European Comission respectively. This work is partially supported by Spain's CICYT Grant ESP95-0389-C02-02. The Czech contribution is supported by the Ministry of Education and Youth of the Czech Republic, projects ES02 and ES36.

References

1. R.W. Klebesadel, I. Strong, R.A. Olson: ApJ **182**, L85 (1973)
2. J.T. Bonnell, R.W. Klebesadel: ed. by C. Kouveliotou et al., AIP **384**, 977 (1996)
3. A.J. Castro-Tirado. In: 4th INTEGRAL Workshop, ESA-SP, in press (2001)
4. A.J. Castro-Tirado et al.: A&ASS **138**, 583 (1999)
5. J.M. Castro Cerón et al.: In: IV Scientific Meeting of SEA, ASSL, in press (2001)
6. J.M. Castro Cerón et al.: In: 4th INTEGRAL Workshop, ESA-SP, in press (2001)
7. S. Barthelmy et al.: ed. by C. Meegan et al., AIP 99 (1998)
8. J.M. Castro Cerón et al.: (2001) http://hug.phys.huji.ac.il/schools/WINTER01/
9. A. Castro-Tirado, J.M. Castro Cerón, T. Mateo et al.: GCN **957** (2001)
10. A. Castro-Tirado, J. Soldán, R. Hudec et al.: GCN **612** (2000)
11. A.J. Castro-Tirado et al.: A&A in preparation (2001)
12. A. Castro-Tirado, R. Hudec, J. Soldán et al.: GCN **528** (2000)
13. A. Henden, A.J. Castro-Tirado, J.M. Castro Cerón et al.: GCN **621** (2000)
14. N. Masetti, E. Palazzi, E. Pian et al.: GCN **720** (2000)

Non-isotropic Angular Distribution for Very Short-Time Gamma-Ray Bursts?

David B. Cline, Christina Matthey, and Stanislaw Otwinowski

University of California, Los Angeles
Department of Physics and Astronomy, Box 951457
Los Angeles, CA 90095-1547 USA

Abstract. We analyse the gamma-ray bursts (GRBs) **with time duration (T90) less than 100 ms** from the BATSE Catalogue. The study of the angular location of these GRBs shows **a strong deviation from isotropy**:
1. The angular distribution is very different from the long-duration GRBs,
2. $< V/V_{max} >= 0.5$, consistent with a quasi-Euclidean distribution of sources.
It indicates that the very-short GRBs likely form a separate class of GRBs, probably from Galactic sources or near solar origin.

1 Comparison of an Angular Distribution and V/V_{max} Distribution for S and M GRBs

The distribution of the duration time **T90** for GRBs from the BATSE detector (up to May 26, 2000) is shown on the Fig. 1. We divide the GRBs into three classes and fit the time distribution:
L ($\tau > 1$ s)long;
M (1 s $\geq \tau \geq 0.1$ s) medium ([2]) ; and
S ($\tau < 100$ ms) very short.
In this study we assume that the S GRBs constitute a separate class of GRBs.
We contrast the distribution of the S and M GRBs:
1. the angular distributions for S is shown on Fig. 2A (46 events) and for M on Fig. 3A. We break up the Galactic map into eight equal probability regions.

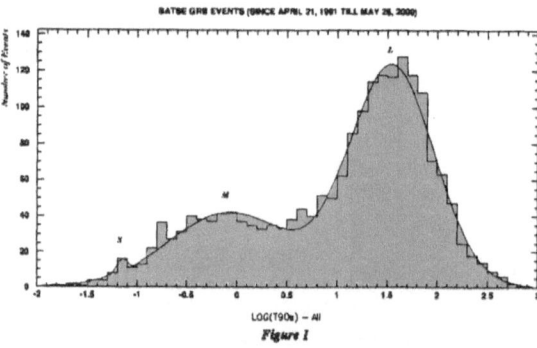

Fig. 1. Time distribution T90 for all GRB events with three Gaussian curve fit.

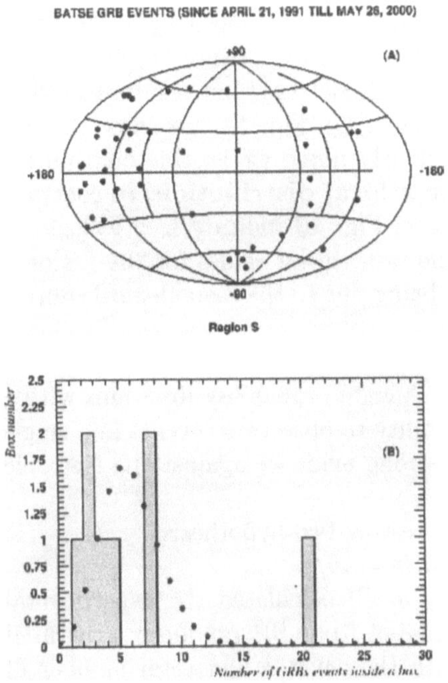

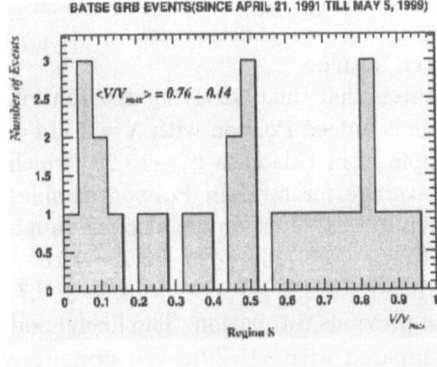

Figure 2

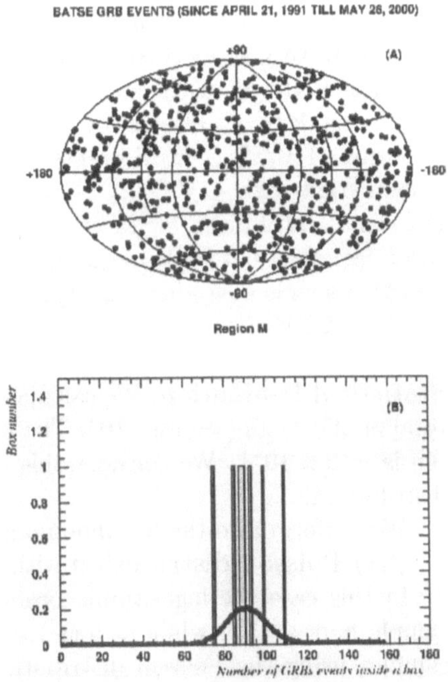

Figure 3

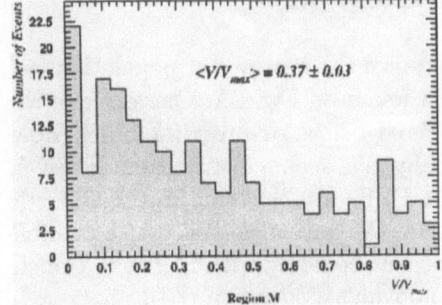

Fig. 2. Fig 3. Fig 4. **Comparison of angular distribution and V/V_{max} distribution for the GRB events (S) and (M).** The Galactic coordinate angular distribution of the GRB events: **FIG. 2A** – with very short time duration (S), **FIG. 3A** – with medium time duration (M). The **Poisson distribution** of GRB events in each of the eight regions: **(FIG. 2B)** – for **46** events; **(FIG. 3B)** – for **269** events. **FIG. 4** Distribution for V/V_{max}: **(4A)** – very short time duration,(S);**(4B)** – medium time duration,(M)

Figure 2B and Figure 3B show the Poisson distribution for those regions. On Fig. 2A, the 46 very short S GRBs, there are 20 in the excess region: from Fig. 3A, we see that the M distribution is consistent with isotropy(first results see([2])).

2. the V/V_{max} calculs for S and M events. On Fig. 4A, we show the distribution of V/V_{max} for S events. As previously noted ([8]) , this distribution is consistent with $< V/V_{max} > = 0.5$ for a local distribution. In contrast, the same distribution for M events shown on Fig. 4B indicate a $< V/V_{max} >$ much less than 0.5 ([3]) consistent with the same mean values for the L (long) events, which is now widely interpreted as being due to the cosmological sources for those GRBs ([4]).

Statistical Evaluation: We calculate the Poisson probability for 8 bins with a total of 46 events (see Fig. 2B). The probability to observe 20 events in a single bin is $\sim 2 \times 10^{-6}$. We consider this as a strong evidence against the isotropic distribution.

We perform also the likelihood analysis testing two hypothesis:

(h1) Poisson distribution with $\lambda = 5.75 = 46/8$

In this case the logarithmic probability $\ln(P)$ calculated the experimental sample is $\simeq -31.3$, while $< \ln(P) >$ estimated from 10^5 randomly generated samples using the Poisson distribution with the same λ and total number of events is $\simeq -17.76$ with the standard deviation $SD \simeq 1.74$. We conclude that the observed value is about 7.8 SD below the Poisson average. **This corresponds to the $\sim 1.4 \times 10^{-6}$ chance to observe such a configuration** and discards (h1).

(h2) Poisson distribution with extra source.

Testing (h2) we first estimate the hypothesis that the underlying distribution for 7 angular bins except the anomalous one is indeed Poisson with $\lambda = 3.714 = 26/7$. The $\ln(P)$ of the experimental sample is in this case $\simeq -15.29$, which is less than one SD (~ 1.532) from the average for random Poisson samples (~ -13.905). That conforms the Poisson hypothesis. One can get also the raugh estimate of the parameters characterising (h2): $\lambda \sim 3.7$, $X \sim 20 - 3.7 = \sim 16$.

The direct minimisation of the likelyhood for h2 gives $\lambda \sim 3.78 \pm 0.7$, $X \sim 15.7 \pm 1.5$ in good agreement with the previous estimation. The likelyhood minimum value is $\simeq -15.5$ (should be compared with –15.29). **We consider these results as a strong evidence to accept (h2).**

Sources of Short Bursts. We have studied the nearby star population and found no pattern that fits the distribution found on Fig. 2A. One explanation for the very short bursts that we have offered before **primordial black hole** (PBH evaporation ([5,6]) might not naturally give such a distribution; however, we cannot exclude the possibility that the PBHs are clustered in the Galactic plane in a manner such as is seen on Fig. 2A. We have also looked at a possible **Oort cloud explanation** (i.e., comets, comets colliding with PBHs, or comets colliding with each other), and there is no obvious association ([7]).

Conclusions. After a carefully studies of very short-duration GRBs (**T90** less than 100 ms) we conclude that:

1) have a harder energy spectrum than the bulk of GRBs,

2) have a $< V/V_{max} >$ consistent with 1/2, and

3) display a non-isotropic angular distribution.

We believe that this can indicate a separate class of GRBs that are most likely from Galactic sources.

References

1. C. Kouveliotou, et al. ApJ. **413**, L101 (1993)
2. D.B. Cline, C.Matthey, S.Otwinowski, UCLA Astronomy Preprint: UCLA APH-0124-04/00 (2000)
3. D.B. Cline, C.Matthey and S.Otwinowski, ApJ.**527** 827 (1999)
4. C. A. Meegan, et al. 1996, ApJ.**106**, 65 (1996)
5. D.B. Cline, W.P. Hong: ApJ. **L57** 401 (1992)
6. D.B.Cline, D.A.Sanders, and W.P.Hong, ApJ. **486**, 169 (1997)
7. E. Maoz, ApJ. **414**, 877 (1993)
8. D.B.Cline, C.Matthey and S.Otwinowski, 1998, Proc. 4th Huntsville Symp. 438 (New York: AIP)

Tools for Gamma-Ray Burst Data Mining

Jon Hakkila[1], Robert S. Mallozzi[2], Richard J. Roiger[3], David J. Haglin[3],
Geoffrey N. Pendleton[4], and Charles A. Meegan[5]

[1] College of Charleston, Charleston SC 29424, USA
[2] Science Communications Inc., Huntsville AL 35899, USA
[3] Minnesota State University, Mankato, MN 56001 USA
[4] University of Alabama in Huntsville, AL 35899, USA
[5] NASA/MSFC, Huntsville, AL 35812, USA

Abstract. Statistical evidence exists for two or more gamma-ray burst (GRB) subclasses. Pattern recognition algorithms also find this. However, not all statistical clus-
terings necessarily indicate separate source populations. Subclass identification can aid
our understanding of systematic observational and instrumental biases. We demonstrate several computational pattern-finding tools, and show how these can be applied
to the problems of GRB classification and subclass interpretation.

1 Introduction

Gamma-ray burst (GRB) observed characteristics (or *attributes*) overlap, making
subclass behaviors difficult to identify. Some "attribute dispersion" is caused by
subclass behaviors, some is intrinsic, some is caused by statistical measurement
error, and some is due to systematic (*e.g.* instrumental and sampling) errors.
Two GRB subclasses are known to exist [1] [4], but it has been difficult to
assign individual GRBs to a specific subclass because of attribute overlap. Class
assignment has been complicated by the identification of a third GRB subclass
[10] that overlaps the other two.

We are developing a web-based tool [2] for GRB classification. The tool
(http://grb.mnsu.edu/grb/) contains a preprocessed GRB database, data visualization software, and artificial intelligence pattern recognition algorithms (or
AI classifiers). AI classifiers can be *supervised* and/or *unsupervised*. Supervised
classifiers require training *instances* (data elements) in order to develop classification rules for unknown instances. Unsupervised classifiers try to classify a
data set by searching for clusters in multidimensional attribute spaces.

2 Supporting the Existence
of Three or More GRB Subclasses

We demonstrate the power of pattern recognition algorithms in GRB classification using the decision tree C4.5 [6] and the concept hierarchy ESX [7]. A
decision tree is a supervised classifier that develops rules by sorting through
training instances via a series of branching tests; the test results are turned into

IF THEN ELSE rules. A concept hierarchy is a supervised/unsupervised classifier in which lower-level entries inherit much of their meaning from the broader
category of which they are subsets; the basic link is a IS-A or A-KIND-OF.

Three GRB subclasses have been identified using statistical clustering analysis [10]. The attributes delineating these are fluence, T90 duration, and a hardness ratio. Subclass 1 (long) bursts are long, bright, and of intermediate hardness.
Subclass 2 (short) bursts are short, faint, and hard. Subclass 3 (intermediate)
bursts are of intermediate duration, intermediate fluence, and soft.

The presence of these three subclasses can be demonstrated with C4.5 in a
new data visualization technique we call "Fuzzy Controlled Learning" (or FCL).
FCL helps users to visualize the attribute space in which subclasses reside, while
recognizing that the subclass distributions overlap in this space. FCL is best used
when a principal attribute is available that serves as a performance indicator.
We assume for this analysis that T90 duration is the principle attribute, since
the longest and shortest GRBs have quite different characteristics.

We withhold 50 GRBs from the long and short ends of the T90 distribution as "comparison" GRBs. These GRBs are considered to have fluences and
hardnesses most indicative of the long and short subclasses. Fifty long and fifty
short GRBs from the remaining data are used as C4.5 training instances. C4.5
produces a rule set for classifying these GRBs. The rules are applied to the
comparison bursts; from this rule accuracy is determined. On each subsequent
application, training instances are selected farther from the ends of the T90 distribution; rule accuracies are determined for each training set. The accuracies
indicate how closely GRBs in that region of the attribute space compare to those
in the comparison set; a score near 100% indicates that the training set is in-
distinguishable from the comparison set, while a score near 50% indicates that
C4.5 could only guess at subclass characteristics.

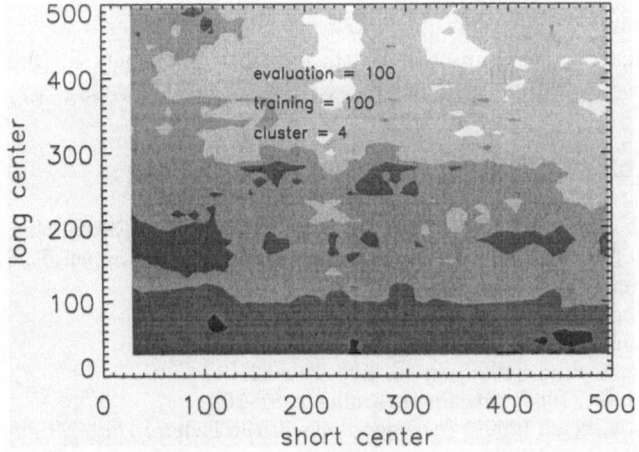

Fig. 1. FCL contour plot. Contours indicate the following agreements between training
and comparison data: 90% (dark), 80%, 70%, 50%, and 30% (light).

Figure 1 is a contour plot of these FCL rule accuracies. The vertical axis is the distance of the long training cluster center (in units of numbers of GRBs) from the longest GRB, while the horizontal axis is the distance of the short training cluster center from the shortest GRB. The darkest contours (accuracies $\geq 70\%$) near the x-axis indicate several hundred GRBs with long burst characteristics; the clearly defined short GRBs near the y-axis are fewer in number. The lightest contours (accuracies $\leq 30\%$) occur for GRBs with T90s between 4.5 seconds and 16.8 seconds because these GRBs are dissimilar to both long and short bursts; FCL has found the parameter space of the third (intermediate) subclass.

We have applied the concept hierarchy algorithm ESX in unsupervised mode to the same dataset examined by statistical clustering methods [10]. The results are fairly sensitive to the choice of dataset (number of bursts used, number of attributes required, and required quality of observation). Although evidence for three subclasses exists, ESX has found four or more subclasses when the dataset is modified [36]. The additional subclasses are formed as separate clusters upon rearrangement of subclass 1 and subclass 3 boundaries; this indicates that subclass 2 is the most clearly defined.

3 Conclusions

There is statistical evidence that three or more GRB subclasses exist. However, caution must be exercised in interpreting subclasses as separate source populations. For example, we have demonstrated [3] that subclasses 1 and 2 appear to be intrinsically different. We have also found that certain subclass 1 bursts can take on subclass 3 characteristics; faint, soft subclass 1 (presumably distant, high-redshift) bursts are most likely to be biased by instrumental and sampling biases [3]. Pattern recognition algorithms have played crucial roles in demonstrating these results. The results should not deter observers from searching for subclasses; understanding systematic biases capable of causing data clustering is as important as understanding intrinsic GRB properties.

We dedicate this manuscript to the memory of Robert S. Mallozzi (1965–2000), and we gratefully acknowledge NASA support (NAG5-8142).

References

1. Cline, T. L., Desai, U. D. (1974) Proc. 9th ESLAB Symp. ESRO, Noordwijk, 37–45
2. Haglin, D. J. et al. (2000) In Gamma-Ray Bursts, ed. M. Kippen, R. S. Mallozzi, & G. J. Fishman (AIP: New York) 877–881
3. Hakkila J. et al. (2000) Ap. J. **538**, 165–180
4. Kouveliotou, C. et al. (1993) Ap. J. **413**, L101–L104
5. Mukherjee, S. et al. (1998) Ap. J. **508**, 314–327
6. Quinlan, J. R. (1986) Machine Learning **1**, 81–106
7. Roiger, R. J. et al. (1999) In Proceedings of the Federal Data Mining Symposium & Exposition, ed. W. T. Price (AFCEA International: Fairfax, VA) 109-120
8. Roiger, R. J. et al. (2000) In Gamma-Ray Bursts, ed. M. Kippen, R. S. Mallozzi, & G. J. Fishman (AIP: New York) 38–42

Broadband Spectral Deconvolution of GRBs

L. Hanlon[1], D. Kinsella[1], N. Murphy[1], B. McBreen[1], K. Bennett[2],
O.R. Williams[2], C. Winkler[2], and R. Preece[3]

[1] Department of Experimental Physics, University College Dublin, Dublin 4, Ireland
[2] Astrophysics Division, Space Science Department of ESA,
Noordwijk, The Netherlands
[3] Department of Physics, University of Alabama, Huntsville, USA

Abstract. Data from GRBs which occurred in the field of view of COMPTEL have
been jointly analysed with the corresponding BATSE data using the Band spectral
form, yielding deconvolved spectra over the energy range from 30 keV to 10 MeV and
in some cases, from 5 keV to 10 MeV. In particularly bright bursts, time-resolved, jointly
deconvolved spectra were also obtained. The results from this study are compared with
those obtained using BATSE data alone and interpreted in the light of current models
of the physical processes taking place in GRBs.

1 Introduction

Spectral fitting of a large number of GRBs has provided distributions of parameters such as break energies and photon indices below and above the breaks (α and
β) which are then used to constrain models of the burst emission mechanism. For
example, it has been shown that the distribution of α values is inconsistent with
the most idealised version of the synchrotron emission model [1] although more
realistic implementations of this model can reproduce the observed distribution
[2], including positive values of α [3].

BATSE's energy range is well positioned to constrain the distribution of
α values. However, at the high energy end of the spectrum, the effective area
of BATSE decreases rapidly leading to poor constraints on β values for many
GRBs. Clearly, broadening the energy range used in spectral fitting may yield
improved constraints for β values and hence on physical parameters such as the
index of the electron distribution [2].

In this paper, joint BATSE-COMPTEL fits to about 50 intervals from 18
GRBs, detected between 1991 and 1995, are presented.

2 Data Analysis and Results

The detectors used for the joint deconvolution are: the COMPTEL D2-14 (0.3–
1.7 MeV) and D2-7 (0.6–10.6 MeV) modules; the illuminated BATSE Spectroscopy
Detector (SD) with the highest gain setting (0.03–(1.3-3.0) MeV) and the BATSE
DISCSP #1 data type, which is the lowest energy SD discriminator channel (5–
10 keV). The DISCSP #1 data are only available for a few events. The model

used for deconvolution in XSPEC is the Band 'GRB' model which is a functional form that provides a reasonable approximation to the broadband GRB spectral shape and has as free parameters E_b, the break energy; α, the asymptotic slope below the break energy and β, the slope above the break. In the minimisation procedure, a normalisation correction factor ('recor') is applied to the COMPTEL detectors' data to ensure that cross-calibration uncertainties between detectors do not introduce biases into the fits.

In order to make combined BATSE/COMPTEL analysis of GRB spectra useful, there must be sufficient counts in the COMPTEL high range burst module (D2-7) so that at least 2 bins remain in the spectrum after background subtraction and rebinning. Of the 28 GRBs which COMPTEL imaged between April 1991 and April 1995, 18 satisfy this requirement and form the sample presented here.

2.1   Distributions of Parameters

α *Values:* As expected, the distributions of α values from BATSE and combined BATSE+COMPTEL (Fig. 1(a)) fits are consistent with being drawn from the same parent population (K-S probability of 56%). The mean value of α is –0.7 for BATSE+COMPTEL fits and –0.8 for BATSE-only fits to the same events. However, a comparison with the larger ensemble of BATSE bursts analysed by Preece et al. (2000) illustrates significant differences between the two samples. There, the mean value of α was –1, steeper than our –0.7. Furthermore, the distribution obtained here is narrower than that of Preece's sample, which extends to $\alpha = -2$. The 'standard' synchrotron model predicts α in the range –3/2 to –2/3 [4]. Our jointly fitted sample, having a minimum α of –1.5 and a mean α of –0.7 may be representative of a set of hard, bright GRBs with 'cleaner' synchrotron signatures than the population as a whole.

β *Values:* Fig. 1(b) shows the distribution of β values obtained. In 14/42 intervals the BATSE-only fits were not constrained at all (pegged at the soft limit of

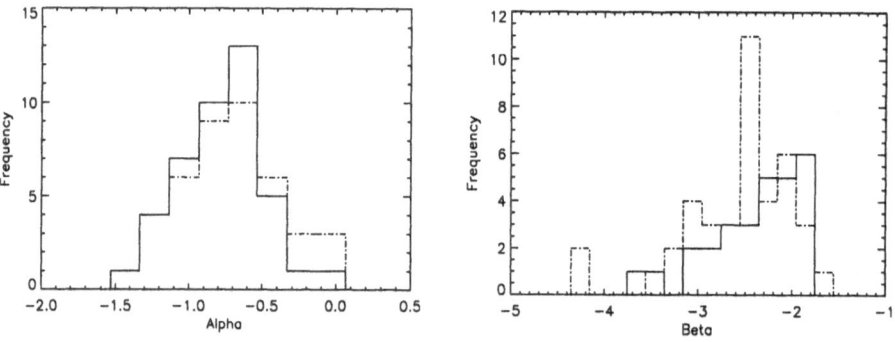

Fig. 1. Histogram of (left) α and (right) β values derived from fits of the Band model to BATSE plus COMPTEL burst sub-intervals (dashed line) and BATSE-only fits (solid line) to the same intervals.

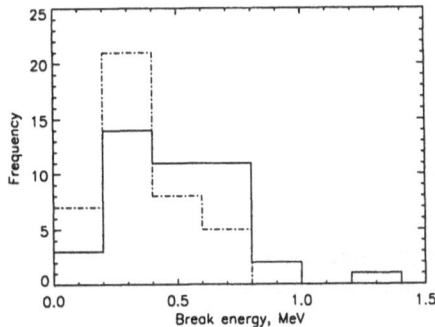

Fig. 2. Histogram of break energies for fits of the Band model to BATSE plus COMP-TEL burst sub-intervals (dashed line) and BATSE-only fits (solid line) to the same intervals.

−10) and are not included in the histogram. In only 2/42 intervals were the joint fits unconstrained. Therefore, as expected, the joint fitting method clearly constrains β in many cases where BATSE alone cannot. The median value of β is −2.43 for BATSE+COMPTEL and −2.26 for BATSE only, excluding $\beta <$ −5 intervals. The 2 distributions have a K-S probability of 5% of being drawn from the same parent population. More than 60% of the intervals have $\beta >$ −2.1 causing problems for both simple cooling (electron distribution index $p >$ 2.2) and instantaneous ($p > 3.2$) spectra if Fermi acceleration ($p = 2.2$) is the mechanism [2].

Break Energies: Fig. 2 shows the distribution of break energies derived from joint fits of the Band model to BATSE plus COMPTEL burst sub-intervals (dashed line) and BATSE-only fits (solid line) to the same intervals. The mean value of E_{break} is $\sim 390\,keV$ (joint fits) and $\sim 500\,keV$ (BATSE only) with a probability of $\sim 1\%$ that the distributions are drawn from the same parent population. These values are harder than the mean of 250 keV obtained for the full BATSE sample, but the distributions are broad. Further analysis, including the remaining sample of COMPTEL GRBs, will be presented in a future publication.

References

1. R.D. Preece, M.S. Briggs et al.: In: *4th Huntsville Symposium*, ed. by C.A. Meegan, R.D. Preece, T.M. Koshut (AIP Conference Proceedings 428, 1998), pp. 319-323
2. N. Lloyd, V. Petrosian: astro-ph/0007061, (2000)
3. A. Crider, E.P. Liang et al.: ApJ **479**, L39 (1997)
4. A. Crider, E.P. Liang: ApJS **127**, 283 (2000)
5. R.D. Preece, M.S. Briggs et al.: ApJS, **126**, 19 (2000)

A Gamma-Ray Bursts'
Fluence-Duration Correlation

István Horváth[1], Lajos G. Balázs[2], Peter Mészáros[3],
Zsolt Bagoly[4], and Attila Mészáros[5]

[1] Dept. of Physics, Bolyai Military University, Budapest, Box 12, H-1456, Hungary
[2] Konkoly Observatory, Budapest, Box 67, H-1525, Hungary
[3] Department of Astronomy and Astrophysics, Pennsylvania State University,
525 Davey Lab. University Park, PA 16802, USA
[4] Lab. for Information Technology, Eötvös Univ., Pázmány s. 1/A, H-1518, Hungary
[5] Astron. Inst. Charles Univ., 180 00 Prague 8, V Holešovičkách 2, Czech Republic

Abstract. We present an analysis indicating that there is a correlation between the fluences and the durations of gamma-ray bursts, and provide arguments that this reflects a correlation between the total emitted energies and the intrinsic durations. For the short (long) bursts the total emitted energies are roughly proportional to the first (second) power of the intrinsic duration. This difference in the energy-duration relationship is statistically significant, and may provide an interesting constraint on models aiming to explain the short and long gamma-ray bursts.

1 Introduction

The gamma-ray bursts (GRBs) measured with the BATSE instrument on the Compton Gamma-Ray Observatory are usually characterized by 9 observational quantities (2 durations, 4 fluences, 3 peak fluxes) [12], [7], [13]. In a previous paper [1] we have shown that these 9 quantities can be reduced to only two significant independent variables (principal components). Here we present a new statistical analysis of the correlation between these variables and show that there is a significant difference between the power law exponents of long and short bursts. The details of this analysis will be presented elsewhere [2].

2 Distributions of Durations and Total Emitted Energies

We consider here those GRBs from the current BATSE Gamma-Ray Burst Catalog [13] which have measured T_{90} durations and fluences (F_1, F_2, F_3, F_4). Therefore, we are left with $N = 1929$ GRBs, all of which have defined T_{90} and $F_{tot}(= F_1 + F_2 + F_3 + F_4)$, as well as peak fluxes P_{256}.

The distribution of the $\log T_{90}$ clearly displays two peaks reflecting the existence of two groups of GRBs [8]. This bimodal distribution can be fitted by two log-normal distributions [3]. The fact that the distribution of T_{90} within a subclass is log-normal has important consequences. Let us denote the observed duration of a GRB with T_{90} (which may be subject to cosmological time dilatation) and with t_{90} those measured by a comoving observer (intrinsic duration).

Then one has $T_{90} = t_{90}f(z)$ where z is the redshift, and $f(z)$ measures the time dilatation. For the concrete form of $f(z)$ one can take $f(z) = (1 + z)^k$, where $k = 1$ or $k = 0.6$, depending on whether energy stretching is included or not.

Taking the logarithms of both sides of this equality one obtains the logarithmic duration as a sum of two independent stochastic variables. According to a theorem of Cramér [3], if a variable ζ, which has a Gaussian distribution, is given by a sum of two independent variables, i.e. $\zeta = \xi + \eta$, then both ξ and η have Gaussian distributions. Therefore, from this theorem it follows that the Gaussian distributions of $\log T_{90}$, confirmed for the two subclasses separately [3], implies the same type of distribution for the variables of $\log t_{90}$ and of $\log f(z)$. However, unless the space-time geometry has a very particular structure, the distribution of $\log f(z)$ cannot be Gaussian. This means that the Gaussian nature of the distribution of $\log T_{90}$ must be dominated by the distribution of $\log t_{90}$, and the latter must then necessarily have a Gaussian distribution. This holds for both duration subgroups separately. (Note here that several other authors, e.g. [20], [15], [16], have already suggested, that the distribution of T_{90} reflects predominantly the distribution of t_{90}.)

One also has $F_{tot} = (1 + z)E_{tot}/(4\pi d_l^2(z)) = c(z)E_{tot}$, where d_l is the luminosity distance, and E_{tot} is the total emitted energy. Once there is a log-normal distribution for F_{tot} (for the two subgroups separately), then the previous application of Cramér theorem is also possible here. The existence of this log-normal distribution is not obvious, but may be shown as follows.

Assume both the short and the long groups have distributions of the variables T_{90} and F_{tot} which are log-normal. In this case, it is possible to fit *simultaneously* the values of $\log F_{tot}$ and $\log T_{90}$ by a single two-dimensional ("bivariate") normal distribution. This distribution has five parameters (two means, two dispersions, and the angle (α) between the axis $\log T_{90}$ and the semi-major axis of the "dispersion ellipse"). Its standard form can be seen in [19] (Chapt. 1.25). When the r-correlation coefficient differs from zero, then the semi-major axis of the dispersion ellipse represents a linear relationship between $\log T_{90}$ and $\log F_{tot}$ with a slope of $m = \tan \alpha$. This linear relationship between the logarithmic variables implies a power law relation of form $F_{tot} = (T_{90})^m$ between the fluence and the duration, where m may be different for the two subgroups. Then a similar relation will exist between t_{90} and E_{tot}.

We obtain the best fit through a maximum likelihood estimation (e.g., [7], Vol.2., p.57-58). ¿From this estimation we obtain the dependence of the total emitted on the intrinsic duration in form

$$E_{tot} \propto \begin{cases} (t_{90})^1 \; ; & \text{(short bursts)}; \\ (t_{90})^{2.3} \; ; & \text{(long bursts)}. \end{cases} \tag{1}$$

Several papers discuss the biases in the BATSE values of F_{tot} and T_{90} (cf. [4], [9], [10], [18], [11], [7], [5], [14]). All types of biases are particularly essential for faint GRBs. To discuss these effects we provide several different additional calculations (for more details see [2]), which give the same results.

3 Conclusion

The exponent in the power laws differ significantly for the two subclasses of short $(T_{90} < 2$ s$)$ and long $(T_{90} > 2$ s$)$ bursts. These new results may indicate that two different types of central engines are at work, or perhaps two different types of progenitor systems are involved. While the nature of the progenitors remains so far indeterminate, our results indicate strongly that the nature of the energy release process giving rise to the bursts is different between the two burst classes. In the short ones the total energy released is proportional to the duration, while in the long ones it is proportional roughly to the square of the duration. This result is completely model-independent, and provides an interesting constraint on the two types of bursts.

This research was supported in part through OTKA grants T024027 (L.G.B.), F029461 (I.H.) and T034549, NASA grant NAG5-2857, Guggenheim Foundation and Sackler Foundation (P.M.) and Research Grant J13/98: 113200004 (A.M.).

References

1. Z. Bagoly, A. Mészáros, I. Horváth, L.G. Balázs, P. Mészáros: ApJ **498**, 342 (1999)
2. L.G. Balázs, P. Mészáros, Z. Bagoly, I. Horváth, A. Mészáros: astro-ph/0007438
3. H. Cramér: *Random variables and probability distr.* Cambridge Tracts in Maths and Mathematical Phys. No.36 (Cambridge Univ. Press, Cambridge 1937)
4. B. Efron, V. Petrosian: ApJ **339**, 345 (1992)
5. J. Hakkila, C.A. Meegan, G.N. Pendleton, R.S. Mallozzi, D.J. Haglin, R.J. Roiger: In: *Gamma-Ray Bursts; 5th Huntsville Symp.* ed. by R.M. Kippen, R.S. Mallozzi, G.J. Fishman (AIP, Melville 2000) pp. 48-52
6. I. Horváth: ApJ **508**, 757 (1998)
7. M. Kendall, A. Stuart: *The Advanced Theory of Statistics* (Griffin, London 1976)
8. C. Kouveliotou et al.: ApJ **413**, L101 (1993)
9. D.Q. Lamb, C. Graziani, I.A. Smith: ApJ **413**, L11 (1993)
10. T. Lee, V. Petrosian: ApJ **470**, 479 (1996)
11. T. Lee, V. Petrosian: ApJ **474**, 37L (1997)
12. C.A. Meegan et al.: ApJS **106**, 65 (3B BATSE Catalog) (1996)
13. C.A. Meegan et al.: The BATSE Current Gamma-Ray Burst Catalog (2000) http://gammaray.msfc.nasa.gov/batse/grb/catalog/current/
14. C.A. Meegan, J. Hakkila, A. Johnson, G. Pendleton, R.S. Mallozzi: In: *Gamma-Ray Bursts; 5th Huntsville Symp.* ed. by R.M. Kippen, R.S. Mallozzi, G.J. Fishman (AIP, Melville 2000) pp. 43–47
15. J.P. Norris et al.: ApJ **424**, 540 (1994)
16. J.P. Norris et al.: ApJ **439**, 542 (1995)
17. W.S. Paciesas et al.: ApJS **122**, 465 (4B BATSE Catalog) (1999)
18. V. Petrosian, T. Lee: ApJ **467**, 29L (1996)
19. R.J. Trumpler, H.F. Weaver: *Statistical Astronomy* (University of California Press, Berkeley 1953)
20. R.A.M.J. Wijers, B. Paczyński: ApJ **437**, L107 (1994)

Estimation of Emission Time Parameter for APEX Experiment

Alexandr Kozyrev[1], Igor Mitrofanov[1], Dmitrij Anfimov[1], and Claude Barat[2]

[1] Space Research Institute, Profsojuznaya str. 84/32, 117810, Moscow, Russia
[2] Centre d'Etude Spatiale des Rayonnements, 4346-31029, Toulouse Cedex, France

Abstract. The data on gamma-ray bursts are presented collected by Russian-France APEX experiment on Phobos-2 interplanetary spacecraft. The values of emission time are estimated for 49 long APEX bursts detected with time resolution of 0.125 sec and 16 short event detected by time-to-spill mode with time resolution of fraction of ms. The APEX data is shown to confirm the bi-modal distribution of emission time values found for BATSE bursts.

The Soviet-France experiment APEX was operated during 8 months from July 1998 until March 1989 [1]. It used large 10×10 *cm* cylindrical *CsI(Tl)* gamma-ray detector with spectral resolution 11% at 662 keV. APEX have ability to measure the variable flux of gamma-rays both in fixed time scale of $0.125sec$ and in time-to-spill mode.

Emission time parameter technique has been implemented to measure the active phase of gamma-ray bursts (GRB) [2]. The emission time τ_{50} is determined as the sum of all time intervals of GRB, which contribute 50% of total fluence with the highest flux. Emission time parameter is complementary to parameters T_{50}, which determine the duration of a burst and include high pulses as well as interpulse valleys.

Statistics of emission time τ_{50} for 463 bright BATSE bursts manifests the clear bi-modal distribution [2]. There are distinct classes of short and long bursts which are separated by $\tau_{50} = 0.4sec$. This signature confirms the previously known findings that bursts are bi-modal in respect to T_{50} statistics, but in the case of emission time the bimodality cannot be associated with interference of time scales of pulses and inter-pulses. The later are excluded from the emission time intervals. Therefore, bi-modality of emission time is thought to manifest the existence of two physically distinct classes of short and long bursts.

Emission times of BATSE bursts were measured with time resolution of $64ms$. This limit is well seen at bi-modal distribution, which has the sharp cut at $64ms$ [2]. The number of short events is increasing with decreasing emission time, and one could suspect that there are a lot of them at time scale $< 64ms$. The question arises: How many short bursts are missing due to the limit of time scale at $64ms$?

To respond this question we study the data for APEX GRBs, which we report below.

The algorithm of emission time estimates for BATSE PREB and DISCSC data with time resolution of 64 *ms* is totally applicable to APEX M2 data with time resolution of $125ms$. We used this algorithm for 51 bursts from APEX data base [1]. Time profile of M2 corresponds to energy range 122–1420 keV. We selected all time intervals which contribute 50% of the fluence with highest flux, and we co-add them in the single value τ_{50}. This selection of emission time interval is shown in Fig.1 for particular case of APEX burst GRB 881024.

We found that distribution of emission times for these APEX bursts is similar to the distribution for class of long BATSE bursts [2] (Fig.2, *left*; solid line). However, it seems that APEX data base [1] does not contain a class of short gamma-ray bursts.

There are 43 short triggering events detected by APEX, which were not included into the GRBs data base [1]. They are distributed in the following groups of events (Tab. 1).

Table 1. Groups of 43 triggering events which were not included into previous APEX data base [1].

Group of events	Duration	Number
Triggering events from charged particles	$\leq 2ms$	11
Noise-like, background fluctuations	Several seconds	14
Possible ordinary bursts	Several seconds	2
Possible short bursts	$\geq 5ms$	16

We found the group of 16 events, which are longer then the events from charged particles ($\leq 2ms$) and could be associated with short GRBs. They all have duration longer than $5ms$ and one-pulse profile with several time-to-spill intervals (see example in Fig.2, *right*).

We estimated emission times for short APEX bursts by new procedure, which sum-up time-to spill intervals starting from the shortest one until they contribute 50% of the total fluence. To get exact number of 50%, we take the fraction of the last longest interval.

The distribution for emission time for 16 short APEX bursts is shown in the Fig.2, *left* by dashed line. The total distribution is in a very good agreement with similar data obtained from BATSE. For total sample of 463 bright BATSE bursts we selected two classes of 323 long bursts and 140 short events [2]. Using this ratio, we may predict that for the class of 51 long APEX bursts one should find about 22 short events. We found 16, which is in a good agreement with the predicted value. There are probably some more short events in our data, but they would be so short that we can not disentangle them from the charged particle events (Tab. 1). We may conclude from the joint analysis of emission time of BATSE and APEX bursts that two classes of long and short events are detected at 5 orders of time scale form 1 millisecond up to 100 seconds, and the number of long bursts is about 2–3 times larger than number of short events.

Acknowledgements

Co-authors from Space Research Institute thanks Local Organizing Committee for kindful hospitality.

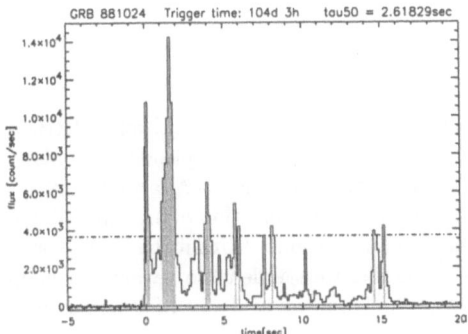

Fig. 1. Time profile of APEX GRB 881024 in $125ms$ time scale at 122-1420 keV. Time intervals are shown by which black which contribute 50% of total fluence with highest flux. The threshold level of flux is shown by dashed line.

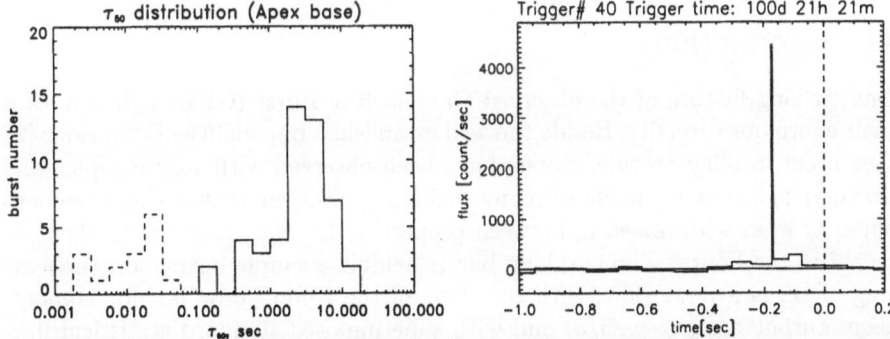

Fig. 2. *Left*: Emission time distribution for APEX bursts. Histogram for 51 long APEX bursts is shown by solid line, histogram for 16 short APEX bursts is shown by dashed line. *Right*: Time profile of APEX short burst GRB 881021 at energy range 122-1420 keV. Dashed line corresponds to triggering moment.

References

1. I. Mitrofanov et al.: Gamma-ray burst: observation, analyses and theories, Cambridge university press, Cambridge, p. 209, 1992.
2. I. Mitrofanov et al., ApJ, 522, 1069, 1999

Quiescent Times in Gamma-Ray Bursts

Andrea Merloni and Enrico Ramirez-Ruiz

Institute of Astronomy, Madingley Road, Cambridge, CB3 0HA, UK

Abstract. A number of time histories of bright, long GRBs exhibit *quiescent times*, defined as those intervals between adjacent episodes of emission during which the γ-rays count-rate drops to the background level. Analysing them, we found a quantitative proportionality relation between the duration of an emission episode and the quiescent time elapsed since the previous episode. Within the framework of the internal-external shocks model for γ-ray bursts, we studied the various mechanisms that can give rise to such quiescent times. Furthermore, we looked for signatures in the properties of the γ-rays, X-rays and optical afterglows that can help us to establish whether or not the central engine goes dormant for a period of time comparable to the duration of the gaps. In particular, we suggest that simultaneous broad band (e.g. from soft X-rays to γ-rays, as on board of the newly launched *HETE-2* satellite) observations of the prompt burst, as well as rapid observations of the prompt afterglow emission from the reverse shock will constrain the possible quiescent times production mechanisms.

1 Introduction

One striking feature of the observed Gamma Ray Burst (GRB) light-curves is their enormous diversity. Beside the well established bimodality of durations [2], they seem to obey no rule. Bursts have been observed with very complex and irregular light-curves, made of many different pulses, or with a single, smooth pulse, or even with mixed in-between properties. On the other hand, when observed in the Fourier domain, long bursts exhibit a simple behaviour: their average PDS is a power-law, with the slope of the Kolmogorov law for homogeneous turbulence ($\alpha \sim -5/3$) and with superimposed standard statistical fluctuation [1].

Here we focus our attention on another striking property of long bursts light-curves, namely their extremely large amplitude variation, such that different emission episodes in the same burst may appear separated by intervals during which the count-rate in the gamma-ray band drops to the background level. The very existence of such quiescent times in GRB light-curves poses severe restrictions on any emission model: it is well known, for example, that in the external shock scenario it is almost impossible to reproduce this property [3].

The work we present here has been done with the twofold aim of characterizing observationally such quiescent times and finding possible observational tests that could help us to discriminate between the two main possible explanations for this phenomenon: (1) a turning-off central engine or (2) a continuous relativistic outflow, modulated by the central engine or by the interaction with the ambient medium.

2   Observations

We have analysed all long ($T_{90} > 5s$), bright bursts from the 4B BATSE catalog, using the 64 ms resolution four-channel data. We found (see Ramirez-Ruiz & Merloni 2001) that about 15% of the analysed bursts contain at least one long (more than 5 per cent of the total burst duration) quiescent interval in their time history and that a quarter of these have more than one.

We have found a strong correlation between the duration of the quiescent interval and the duration of the following emission episode, while no correlation has been found between the quiescent time and the duration of the previous emission episode (see [6], Fig. 3).

3   Discussion

In order to extract as much information as possible from the correlation we found, and to gain more insight into the dynamics of GRB central sources, we need first to understand the physical processes responsible for the production of quiescent times. The fundamental issue to tackle is whether the gaps in the γ-rays light-curves are produced by a turning-off central engine or by a structure in the relativistic outflow velocity space, due possibly to its interactions with the ambient medium it propagates into. In the former case, the flow will be discretised in a number of thick shells, each one of size roughly comparable to the duration of the emission episodes observed at the detector. In the latter, instead, the flow could be regarded as a single shell whose thickness is related to the total burst duration.

If the gaps reflect a period of inactivity of the central source, the correlation found in [6] is generally indicative of an accumulation of energy in a *meta-stable* configuration. That is, a configuration in which any local instability can abruptly drain the system of all the stored energy; after such event it will tend to a more stable state, characterized by a threshold energy E_0. The source then becomes quiescent. If the lifetime of the GRB engine is long enough, and depending on the rate of energy extraction from the central source, the system can undergo another episode of emission. In this case, the longer the quiescent time, the higher the stored energy above the threshold available for the next emission episode, similarly to what is observed [1].

We do not speculate further on the physical nature of such meta-stable configuration, and on whether it is associated with some non-linear dynamo process in the accretion torus around the newly formed black hole, that many scenarios envisage as the central engine for long bursts, or to the complex interaction of the

---

[1] We stress that this is the opposite to what expected from a relaxation oscillator, in which the accumulation time depends on the amount of energy previously dissipated. This produces a correlation between the quiescent time and the duration of the *previous* emission episode (as, e.g., in the case of the Rapid Burster [5]) that is clearly ruled out by the observations.

relativistic jet with its environment (particularly for the Super/Hyper/Supra-nova scenarios; see e.g. [4]). Instead, in the framework of the internal shocks model for GRB, we simulated numerically [7] various different mechanisms that can give rise to the observed quiescent times, trying to assess their relative likelihood. We conclude that:

- A central engine that goes dormant for a long enough period will produce a quiescent time in the γ-ray light-curve of comparable duration.
- Internal shocks can produce significantly long periods of quiescence with an opportune modulation of a continuous relativistic wind: for example, when the wind ejecta have a constant (or progressively decreasing) Lorentz factor for a long enough time, the relativistic shells will not catch up with each other and the emission from the internal shocks may be delayed or shifted to lower energy bands.
- Consequently, a measurement in a single energy channel is inadequate for the correct identification of a complete turn-off of the central engine.

How is it possible to discriminate between a turning-off central engine and a continuous modulated outflow?

First of all, we stress the importance of simultaneous, high temporal resolution observations of GRBs. In particular, cross-correlating X/γ-rays light-curves will probe the velocity structure of the relativistic outflow responsible for the internal shocks [7]. The newly launched *HETE-2* satellite, with its large bandwidth (from $\sim$1 to 400 keV) and high temporal resolution (of the order of a millisecond) is the ideal instrument to accomplish this task.

Furthermore, we suggest that the prompt afterglow emission from the reverse shock, that strongly depends on the wind dynamics [8], is best suited to discriminate between these cases. In particular, the time delay of the prompt afterglow peak with respect to the main γ-ray event, will be much shorter in the case of a discontinuous wind. Thus, a very rapid optical follow-up of the long bursts exhibiting periods of quiescence would be crucial to distinguish between the two possibilities. This will help us to extract physical informations from the observed correlation and give us a hint of the properties of the central engine.

References

1. A. Beloborodov, B. Stern, R. Svensson: ApJ, **535**, 158 (2000)
2. C. Kouveliotou et al.: ApJ **413**, L101 (1993)
3. E.E. Fenimore, E. Ramirez-Ruiz, In: *PASP Conf. Proc. Gamma-Ray Bursts: The First Three Minutes*, astro-ph/9906125 (1999)
4. G. Ghisellini, D. Lazzati, A. Celotti, M. Rees: MNRAS, **316**, L45 (2000)
5. W.H.G. Lewin et al.: ApJ, **207**, L95 (1976)
6. E. Ramirez-Ruiz, A. Merloni: MNRAS, **320**, L25 (2001)
7. E. Ramirez-Ruiz, A. Merloni, M.J. Rees: MNRAS, submitted, astro-ph/0010219 (2001)
8. R. Sari, T. Piran: ApJ, **517**, 270 (1999)

Neutrino Astrophysics
with the MACRO Detector

Teresa Montaruli[1,3], Fabrizio Cei[2], Roberto Pazzi[2], and Francesco Ronga[3],
for the MACRO Collaboration

[1] Dipartimento di Fisica dell'Università di Bari and I.N.F.N., 70126, Bari
[2] Dipartimento di Fisica dell'Università di Pisa and I.N.F.N., 56010, Pisa
[3] Laboratori Nazionali di Frascati dell'I.N.F.N., 00044, Frascati (Roma)

Abstract. A sample of 1197 neutrino induced upward-going muons is used to look for point-like sources and for space-time correlations with BATSE and *Beppo*SAX gamma-ray bursts. We set an upper limit (90% c.l.) of 6.9×10^{-10} cm^{-2} upward-going muons per average burst. A search for high multiplicity events in coincidence with GRBs has been performed using the supernova trigger.

1 Point-Like Source Search and GRB-ν Coincidences

The MACRO detector at the Gran Sasso Laboratories [1], under an average rock coverage of ~ 3700 mwe, has overall dimensions of $12 \times 76.6 \times 9$ m^3. A streamer tube system of $\sim 20,000$ m^2 reconstructs tracks with angular resolution $\leq 1°$. Pointing capabilities have been checked using the Moon shadow [2]. The time-of-flight (T.o.F.) is measured by 600 tons of liquid scintillator (time resolution ~ 500 psec). Crossing vertical μs have an energy ≥ 1 GeV.

MACRO detects 3 ν topologies. Low energy neutrino ($< E_\nu > \sim 4$ GeV) results are reported in [3]. For the ν astronomy searches we consider higher energy events ($< E_\nu > \sim 50$ GeV) [4] which are detected through the T.o.F. measurement. Neutrinos are selected between atmospheric muons as upward-going muons. They were measured since Mar. 89 during construction, which was completed in Apr. 94. Here we show the results obtained until March 2000 with a sample of 1197 events. Of these, 100 are detected with 1/6 of the lower apparatus during 1.38 yrs and with the lower detector during 0.41 yrs of running; the others are detected in 4.80 yrs of the full detector running (the live-times include inefficiencies and the possibility that part of the apparatus has been off).

MACRO results on point-like ν sources are reported in [5]. We studied the response of MACRO to neutrinos produced by point-like sources including the effect of ν absorption in the Earth and the detector acceptance to the expected ν fluxes. Assuming a ν spectrum following a power law with spectral index 2.1, the median energy of neutrinos producing detectable muons is ~ 15 TeV; the corresponding mean energy for atmospheric neutrinos is ~ 50 GeV. The search for astrophysical point-like sources uses the direction information of upward-going muons. In order to evaluate the atmospheric neutrino background we mixed (with 100 independent extractions) the local angles and the arrival times of

Table 1. 90% c.l. ν induced μ-flux limits for some of the candidate sources. Corresponding limits on the ν flux are given in the last column for $E_{\nu min} = 1$ GeV. Limits are calculated for $\gamma = 2.1$ and for $E_\mu > 1$ GeV and include the effect of absorption in the Earth and the collection efficiency of the search half-cone of 3° for the expected signal. B indicates the results of Baksan [6]; I the ones of IMB [7].

Source	δ (°)	Events in 3°	Backg. in 3°	μ-flux limit 10^{-14}	Prev. limit 10^{-14}	ν-Flux l. $E_\nu > 1$GeV 10^{-6} cm^{-2} s^{-1}
SN1987A	−69.3	0	2.1	0.14	1.15 B	0.29
CenXR-3	−60.6	1	1.8	0.33	0.98 I	0.62
GX339-4	−48.8	6	1.7	1.51	−	2.79
VelaP	−45.2	1	1.7	0.45	0.78 I	0.84
SN1006	−41.7	1	1.5	0.50	−	0.92
Gal Cen	−28.9	0	1.0	0.30	0.95 B	0.57
Kep1604	−21.5	2	1.0	1.02	−	1.92
Sco XR-1	−15.6	1	1.0	0.77	1.5 B	1.45
Geminga	18.3	0	0.5	1.04	3.1 I	1.95
Crab	22.0	1	0.5	2.30	2.6 B	4.30
Her X-1	35.4	0	0.2	3.05	4.3 I	6.45
MRK 421	38.4	0	0.2	4.57	3.3 I	8.74

upward-going events. We considered the case of a possible detection of an unknown source represented by an excess of events clustered inside cones of half widths 1.5°, 3° and 5°. Hence we looked at the number of events falling inside these cones around the direction of each of the 1197 measured events. We find 77 clusters of ≥ 4 muons around a given muon (including the event itself), to be compared with 72.6 expected from the background of atmospheric neutrinos.

For our search among known point-sources, we considered several existing catalogues. We find no statistically significant excess from any of the considered sources with respect to the atmospheric neutrino background. Within these catalogues we selected 42 sources we consider interesting. We find 11 sources with ≥ 2 events in a search cone of 3° to be compared to 13.6 sources expected from the simulation. The 90% c.l. μ and ν flux limits are given in Tab. 1 for some of the considered sources. According to the models reported in [8], cold dark matter could be accreted by the presence of a black hole in the Galactic Center and if it is made of neutralinos they can annihilate into neutrinos. Most of these models are excluded by our experimental upper limit of $\sim 3 \cdot 10^{-15}$ cm^{-2} s^{-1} for a 3° cone.

MACRO has set limits on GRB ν emissions using the direction and time information. No excess has been observed in the direction of 2704 BATSE (Apr. 1991 to May. 2000) and of 67 *Beppo*SAX GRBs. Since we find no coincidence between 1178 ν events and BATSE GRBs in a search window of 10° and ±200 s, and 0.05 are expected from the background of atmospheric νs, the resulting

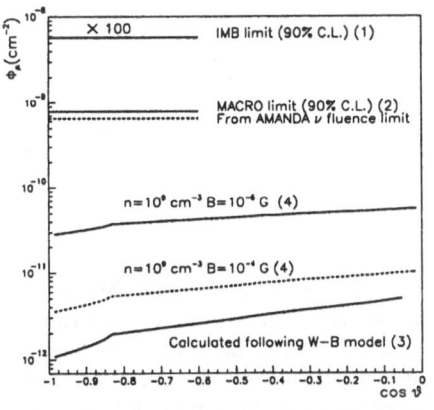 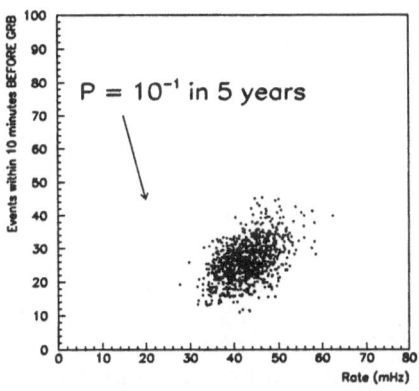

Fig. 1. On the left: MACRO, AMANDA and IMB (references in [9]) upper limits for GRBs vs cosine of the zenith angle (this dependence shows the effect of Earth absorption on the flux of upward-going μs). AMANDA limit is derived from ν limits using the calculated flux labeled W-B [10]. Other models presented in [9] are shown. On the right: events within 10 min before BATSE GRBs vs SN trigger rate (the plot of events after GRBs is similar). The line shown in the plot corresponds to a probability of 10^{-1} for observing a given number of events in a 10 min interval for a detector running time of 5 yrs vs background rate.

upper limit for average burst is $6.9 \cdot 10^{-10}$ cm^{-2} s^{-1}. In Fig. 1 (on the left) we show the comparison of MACRO, AMANDA and IMB μ upper limits and models [9]. The 2 experiments see complementary sky regions.

1453 BATSE GRBs occurred in $\sim$ 5 yrs of the SN trigger up-time. The rate of SN triggers before and after 1, 5 and 10 min the GRB occurrence has been compared to the background rate. No evidence is observed of a possible GRB-induced variation on the rate of SN triggers as shown in Fig. 1 (on the right).

References

1. S. P. Ahlen *et al.*, MACRO Collabor.: Nucl. Instr. Meth. A **324**, 337 (1993)
2. M. Ambrosio *et al.*, MACRO Collabor.: Phys. Rev. D **59**, 012003 (1999)
3. M. Ambrosio *et al.*, MACRO Collabor.: Phys. Lett. B **478**, 5 (2000)
4. S. P. Ahlen *et al.*, MACRO Collabor.: Phys. Lett. B **357**, 481 (1995) M. Ambrosio *et al.*, MACRO Collabor.: Phys. Lett. B **434**, 451 (1998)
5. M. Ambrosio *et al.*, MACRO Collabor.: ApJ **546** (2001) 1038
6. M.M., Boliev, Baksan Collabor.: in Proc. of the 24th Int. Cosmic Ray Conf., 28 Aug.-8 Sep. 1995, Roma, Vol. **1**, 722
7. R. Becker-Szendy *et al.*, IMB Collabor., ApJ **444**, 415 (1995)
8. P. Gondolo & J. Silk, Phys. Rev. Lett., **83**, 1719 (1999)
9. L. Perrone, presented at NOW2000, see http://www.ba.infn.it/%7enow2000
10. E. Waxman & J. N. Bahcall, Phys. Rev. Lett. **78**, 2292 (1997)

The Fingerprints of the GRB Process

F. Quilligan, B. McBreen, K. Hurley, L. Hanlon, D. Watson, and S. McBreen

Physics Department, University College Dublin, Dublin 4, Ireland

Abstract. A comprehensive temporal analysis has been performed on the 200 brightest GRBs with $T_{90} > 2\,$s from the BATSE 4B catalog. The rise times, fall times and full-widths at half maximum were measured and the frequency distributions are consistent with lognormal distributions provided the pulses are well separated. The distribution of time intervals between pulses is not random but compatible with a lognormal distribution when allowance is made for the BATSE time resolution and a small excess ($\sim 5\%$) of long duration intervals that is often referred to as a Pareto-Lévy tail. The time intervals between pulses are most important because they are an almost direct measure of the activity in the central engine. Lognormal distributions of time intervals also occur in pulsars and SGR sources and therefore provide indirect evidence that the time intervals between pulses in GRBs are also generated by rotation powered systems with super-strong magnetic fields.

1 Introduction

The light curves of GRBs are irregular and complex. Statistical studies are necessary to characterise their properties and hence to identify the physical properties of the emission mechanism. One of the first studies [3] revealed that lognormal distributions can adequately describe the pulse properties of GRBs.

2 Pulse Properties

The dataset used was taken from the BATSE 4B catalogue. A subset of the BATSE catalogue was selected based on the criteria that the GRB duration was greater than two seconds ($T_{90} > 2\,$s) and the peak flux $P_{256\mathrm{ms}} > 3.28$ photons $\mathrm{cm}^{-2}\mathrm{s}^{-1}$. In this way a sample of 200 bursts with good signal to noise and clearly resolved features was obtained [7]. The GRBs were denoised using wavelets and then passed through an automatic pulse selection algorithm as an objective way to identify pulses. The rise times, fall times and full widths at half maximum were measured and the frequency distributions are consistent with lognormal distributions (Fig. 1) for pulses that were reasonably well separated from each other. The lognormal distribution provides an elegant description of the pulses in GRBs. It is generated by statistical processes whose results depends on the product of probabilities arising from a combination of events. These complicated conditions must also apply to the generation of pulses in GRBs.

3   Time Intervals between Pulses

A parent lognormal distrbution ($\mu = 0.07$; $\sigma = 1.07$) can be shown to accurately represent the distribution of time intervals when allowance is made for (1) the 64 ms time resolution of BATSE (2) a small (5%) excess of time intervals longer than ~ 20 sec which is clearly visible in Fig. 1 (d).

It is often found that distributions that seem to be lognormal over a wide range change to an inverse fractional power distribution for the last few percent. An amplification model has been used to characterise the transition from a lognormal distribution to an inverse-power Pareto-Lévy tail [4]. The distribution of time intervals conform to the lognormal distribution over most of the range with the exception of about 5% of the time intervals longer than about 20s (Fig. 1 (d)).

The origin of the nonrandom distribution of time intervals between pulses is an important clue to the GRB process. In the internal shock model there is almost a one to one correspondence between the emission of shells and pulses resulting from the collisions of shells [3]. Hence the time intervals between pulses is an almost direct measure of the activity of the central engine. The temporal behaviour of soft γ-ray repeaters (SGR) and young pulsars provide an additional context in which to view the results of GRB time profiles. The time intervals between about 30 microglitches in the Vela Pulsar are consistent with a lognormal distribution with a mean of 50 days [1]. The amount of energy involved in the microglitches is about 10^{38} ergs. The macroglitches in the Vela Pulsar are about a thousand times more powerful but occur too infrequently to determine the distribution of time intervals but they have a wide range and do not seem inconsistent with the lognormal distribution. More energetic outbursts that have a lognormal distribution have been obtained from SGRs [1]. Hence lognormally distributed time intervals between outbursts and glitches are characteristic features of SGR sources and neutron star microglitches. It is widely accepted that these sources are rotating neutron stars with high magnetic fields. It is not unreasonable to argue that the coupled effects of rapid rotation and intense magnetic fields are also involved in powering GRBs since the time intervals between pulses are also consistent with a lognormal distribution. The magnetic fields and rotation rates in GRB models are at least 10^{15} gauss and milliseconds respectively [2]. The Pareto-Lévy tail of long time intervals have an amplification process that is not available to most time intervals. GRB models leave open many possibilities to account for the Pareto-Lévy tail such as the magnetic field getting wound up more than usual because of better coupling between the black hole and dense nuclear matter. The resulting extra magnetic stresses in the wind prevent the formation of shocks and pulses in the GRB [11]. Several other possibilities have been proposed to account for the long time intervals [8,5].

4   Conclusions

The distributions of rise times, fall times and FWHM in GRBs are consistent with lognormal distributions. The lognormal distribution is produced by a multiplicative process and this condition must apply to the temporal and spectral properties of GRBs. The time intervals between pulses are not random but are consistent with a lognormal distribution. This result implies that GRBs are generated by rotation powered systems with intense magnetic fields.

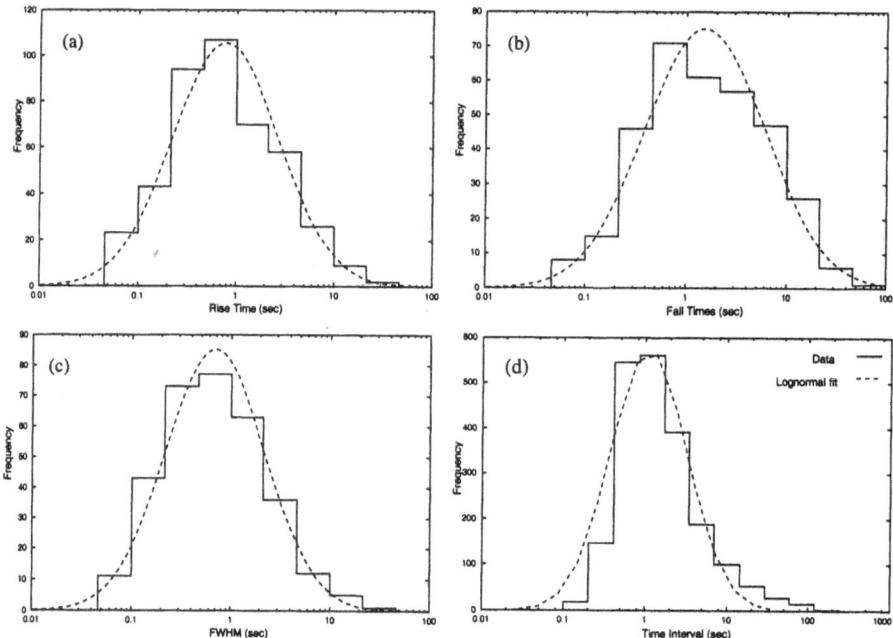

Fig. 1. The distributions of (a) rise time, (b) fall times and (c) FWHM of pulses and lognormal fits to the data. (d) The distribution of time intervals between pulses with the best lognormal fit. The Pareto-Lévy tail occurs at long time intervals and the deficit at short intervals is due to the 64 ms resolution limit of BATSE.

References

1. K.J. Hurley, B. McBreen, M. Rabbette, S. Steel: A&A **288**, L49 (1994)
2. W. Kluźniak, M. Ruderman: ApJ **508**, L113 (1998)
3. B. McBreen, K.J. Hurley, R. Long, L. Metcalfe: MNRAS **271**, 662 (1994)
4. E.W. Montroll, M.F. Shlesinger: Proc. Natl. Acad. Sci. **79**, 3380 (1982)
5. E. Nakar, T. Piran: (These Proceedings)
6. T. Piran: Phys. Report, pg. 575 (1999)
7. F. Quilligan, K.J. Hurley et al.: A&AS **138**, 419 (1999)
8. E. Ramirez-Ruiz, A. Merloni: MNRAS **320**, L25 (2001)
9. M.J. Rees: A&AS **138**, 491 (1999)

A Variety of Decays of Gamma-Ray Burst Pulses

Felix Ryde[1,2] and Roland Svensson[2]

[1] Center for Space Science and Astrophysics, Stanford University, Stanford CA 94305
[2] Stockholm Observatory, SE-133 36 Saltsjöbaden, Sweden

Introduction

The main target of this study is the GRB light curve during the decay phase of long, bright pulses. As shown by Ryde & Svensson (2000; hereafter RS00) approximately half of these decays can be described by a power law $\propto 1/(\text{time})$. This happens for cases when the hardness-fluence correlation (HFC) is an exponential function, $E_{pk}(\Phi) \propto e^{-\Phi/\Phi_0}$, and the hardness-intensity correlation (HIC) is a power law, $E_{pk}(N) \propto (N/N_0)^\delta$. Here, $N(t)$ is the instantaneous photon flux, $E_{pk}(t)$ is the corresponding photon energy, at which the $E^2 N_E$-spectrum peaks and is used as a measure of the spectral hardness, and the photon fluence is defined by $\Phi(t) = \int^t N(t')\,dt'$. These most commonly assumed correlations were found by Liang & Kargatis (1996; HFC) and Golenetskii et al. (1983; HIC).

There obviously exists a large group of GRB pulses which decay in a different way. In this paper, we search for alternative descriptions of the spectral/temporal evolution. We use the complete sample of long pulses in strong bursts presented in Ryde & Svensson (2001) consisting of 25 pulses within 23 bursts observed by BATSE on the *CGRO* during its entire mission (1991–2000). The spectral analysis of the LAD/HERB data ($\sim 25 - 1900$ keV) was performed with the WINGSPAN/MFIT package (Preece et al. 1996). For each time bin the photon spectrum with the background subtracted was determined using the Band et al. (1993) function with both its power law indices left free to vary. The instantaneous, integrated photon flux, $N(t)$, was found by integrating the modeled photon spectrum over the available energy band.

Other Types of Behaviors

There is no consensus on what shape the pulse decays have. Both power law and stretched exponential decays have been used. Guided by the findings of RS00 we study the following generalized power law decay:

$$N(t) = \frac{N_0}{(1 + t/\tau)^n},\tag{1}$$

where t is taken from the start of the decay, when $[N(t), E_{pk}(t)] = [N_0, E_{pk,0}]$ and the time constant $\tau \equiv \delta \Phi_0/N_0$, where Φ_0 is the exponential decay constant of the exponential HFC and δ is the index of the power law HIC. The photon fluence associated with equation (1) when n differs from 1 becomes

$$\Phi(t) = \frac{N_0\tau}{n-1}\left\{1 - (1+t/\tau)^{-(n-1)}\right\}, \quad n \neq 1, \tag{2}$$

which for n larger than 1, converges to the asymptotic value $f_0 \equiv N_0\tau/(n-1)$.

Now, we consider two different alternatives. First, for GRB pulse light curves whose decays follow equation (1), and for which the HFC $E_{pk}(\Phi)$ is an exponential, the HIC $E_{pk}(N)$ will follow

$$E_{pk}(N) = E_{pk,0}\exp\left\{\frac{f_0}{\Phi_0}\left[\left(\frac{N}{N_0}\right)^{(n-1)/n} - 1\right]\right\}, \quad n \neq 1. \tag{3}$$

When $\ln(N_0/N) << 2n/|n-1|$ the HIC approaches a power law with the exponent δ/n, which becomes identical to the original power law HIC, when n tends to 1. On the other hand, if the HIC $E_{pk}(N)$ actually is a power law then the HFC $E_{pk}(\Phi)$ will follow

$$E_{pk}(\Phi) = E_{pk,0}\left(1 - \frac{\Phi}{f_0}\right)^{n\delta/(n-1)}, \quad n \neq 1, \tag{4}$$

which behaves similarily to the exponential HFC as n tends to 1.

We also fitted the decays with a stretched exponential: $N \propto \exp(-(t/\tau_d)^\nu)$, where τ_d is the time constant for the decay phase and ν is the peakedness parameter. This function is the most commonly assumed pulse shape used so far (e.g. Norris et al. 1996).

Our study showed that, first, a power law gives a better description of the pulse decays than a stretched exponential and, second, the power law index (Eq. 1) has a bimodal distribution in that there are two preferred values $n = 1$ and $n \sim 3$ (See Fig.1). The sample is divided into approximately two equally large sets by $n \sim 2$. For the 11 pulse decays with n larger than 2, we found that for each case either the HIC $E_{pk}(N)$ or the HFC $E_{pk}(\Phi)$ is still valid, while the other corresponding correlation is different, and thus described by a new function. To be able to get constrained fits on all cases we had to freeze the values of N_0 and n to the values obtained from the fits of the light curve.

Six out of these eleven cases are, however, good enough for n to be constrained. Four out of these gave n-values that were the same to within the errors as the values obtained from fitting the light curve. In the last two cases the errors in the n-values were so large that no certain conclusion could be drawn. In all of these four cases the power law HIC $E_{pk}(N)$ is valid.

This suggests that the important relations for a GRB pulse decay are the power law HIC $E_{pk}(N)$ and the light curve, $N(t)$. The power law correlation between the hardness and the intensity is valid independent of the shape of the light curve. The HFC $E_{pk}(\Phi)$, on the other hand, is different for different light curve behaviors according to equation (4), since the fluence is the time integral of the instantaneous flux.

In Figure 2 one of these cases, GRB960807, is presented. The first panel shows the DISCSC data (all four energy channels) and indicates the time interval studied and the second panel shows the light curve with the LAD HERB data in the chosen time binning. The best fit is indicated with a solid curve. The two left-hand panels, show the correlations, the HIC $E_{pk}(N)$ in panel 3 and the HFC $E_{pk}(\Phi)$ in panel 4. The fit of an exponential HFC is shown by a dashed line.

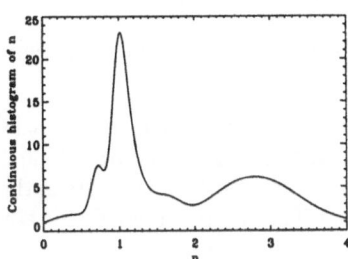

Fig. 1. Continuous histogram of the power law index n.

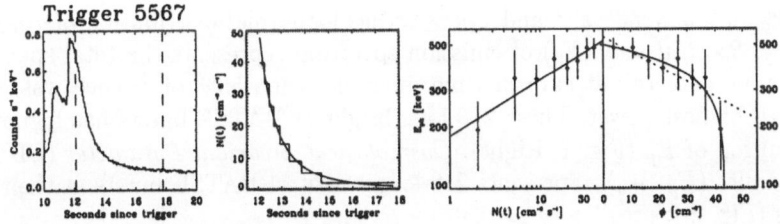

Fig. 2. Spectral and temporal behaviour of GRB960807 (BATSE trigger 5567).

Acknowledgments

We are grateful to the GROSSC at NASA/GSFC for providing the HEASARC Online Service. We acknowledge financial support from the Swedish Foundation for International Cooperation in Research and Higher Education (STINT) and the CF Liljevalch J:or Fund at Stockholm University.

References

1. Band, D., et al. 1993, ApJ, 413, 281
2. Golenetskii, S., Mazets, E., Aptekar, R., & Ilyinskii, V. 1983, Nature, 306, 451
3. Liang, E. P., & Kargatis, V. E. 1996, Nature, 381, 495
4. Norris, J. P., et al. 1996, ApJ, 459, 393
5. Preece, R., Briggs, M., Mallozzi, R., & Brock, M. 1996, WINGSPAN v 4.4
6. Ryde, F., & Svensson, R. 2000, ApJ, 529, L13
7. Ryde, F., & Svensson, R. 2001, ApJ, to be submitted

The GRBs at Rest Frames of Emitters

Anton Sanin[1], Igor Mitrofanov[1], Dmitrij Anfimov[1], Maxim Litvak[1],
Michael Briggs[2], William Paciesas[2], Geoffrey Pendleton[2], Robert Preece[2],
Gerald Fishman[3], and Charles Meegan[3]

[1] Institute for Space Research, Moscow 117810, Russia
[2] University of Alabama, Huntsville AL 35899, USA
[3] NASA/Marshall Space Flight Center, Huntsville AL 35812, USA

Abstract. The concept of the instrument-independent parametrization are presented for the statistical search of basic physical signatures of gamma-ray bursts. These new parameters are presented for several gamma-ray bursts in their rest frames with known red-shifts of optical afterglows. The group of these GRBs are compared with the group of several most bright BATSE bursts. Collective estimation of the average redshift factor of the comparison group are obtained.

Emission Time (τ_{50}) is defined, as a sum of time intervals when 50% of fluence was emitted with highest power (see [1]). 218 brightest BATSE bursts with $F_{peak} > 2 \; \gamma cm^{-1}s^{-1}$ and $T_{50} > 2s$ has lognormal distribution of τ_{50} (Fig. 1. Left). *Spectral peak* (E_p) of emission spectrum represents the total spectrum of photons detected at sum of time intervals when 50% of fluence was emitted with highest power. These 218 long brightest BATSE bursts has lognormal distribution of E_p (Fig. 1. Right). *Cosmological Invariant Parameter* (CIP), as the product ($E_p \cdot \tau_{50}$), represents 218 long brightest BATSE bursts at their rest frames (Fig. 2.).

Comparison sample (CS) of 218 brightest BATSE bursts with $F_{peak} > 2 \; \gamma cm^{-1}s^{-1}$ and $T_{50} > 2s$ are compared with the *Reference sample* (RS) of 6 Z-known BATSE bursts.

The parameters τ_{50} and E_p are quite convenient for the statistical comparison of groups of bursts (see [5]). Before using the reference sample for the collective

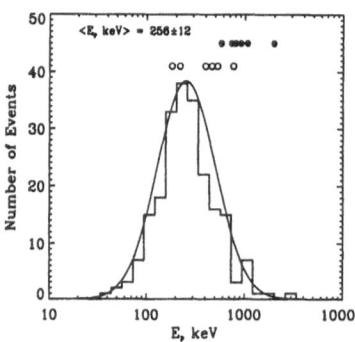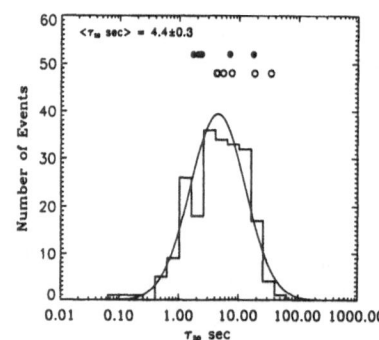

Fig. 1. Distributions of τ_{50} (Left) and E_p (Right) for 218 bright long BATSE bursts.

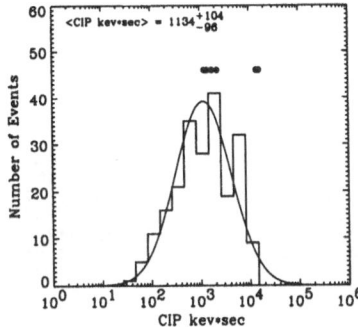

Fig. 2. Distribution of CIP for 218 bright long BATSE bursts.

estimation of the redshift of the comparison sample, we tested that the two samples of bursts are consistent in the observer's frame. We use the Pearson's χ^2, Student's t and Kolmogorov-Smirnov tests to find the probability that the six observed values E_p of the reference sample belong to the distribution observed for the comparison sample. According Pearson's χ^2, Student's t and K-S tests this probability is 0.397, 0.163 and 0.148 respectively. We also tested that properties of emitters of these two groups of bursts are consistent in the rest frames. The CIP was used for this test. According Pearson's χ^2, Student's t and K-S tests this probability is 0.146, 0.046 and 0.059 respectively. In all cases the consistency are quite good.

Therefore, we assumed that these two groups correspond to the same type of gamma-ray bursts. This assumption allowed us to use the method of "collective estimation" of the average redshift factor of the comparison group. To estimate the most probable redshift for the comparison group, were calculated the de-redshifted peak energies of spectra as $\{(1 + Z) \cdot Ep\}$ with the free redshift factor $(1 + Z)$. The same three criteria to estimate the best-fit values of Z were used (see [2]). All criteria result in similar best-fit values $Z_{(CS)} = 2.6$, the 1σ limits on $Z_{(CS)}$ are [1.8–3.8].

The comparison sample is split into three intensity subgroups using peak flux F_{max}^{64} (Table 1). We used the same method of "collective estimation" for three intensity groups. There is marginal effect of increase of z for dimmer groups (Table 1.). We used estimated redshifts to find rest frame properties for these three groups (Table 2). They are quite similar. Therefore, there is no evidence for z-dependent evolution of rest frame properties of gamma-ray bursts.

Acknowledgements

Co-authors from Space Research Institute thanks Local Organizing Committee for kind full hospitality.

Table 1. Collective Z estimation for intensity subgroups

Group	Peak Flux, F_{max}^{64} $(\frac{\gamma}{cm^2s})$	Mean Observed spectral peak energy $<\tau_{50}>$ (keV)	Mean Observed CIP (keV s)	Z	1σ limits
1	8.3–168.8	316.2^{+25}_{-23}	1074.9^{+174}_{150}	$1.9^{+0.9}_{-0.7}$	1.2–2.8
2	3.9–8.3	270.9^{+21}_{-20}	1529.8^{+227}_{-197}	$2.4^{+1.1}_{-0.8}$	1.6–3.5
3	2.2–3.9	196.7^{+16}_{-15}	893.2^{+158}_{-134}	$3.7^{+1.6}_{-1.1}$	2.6–5.3

Table 2. Intrinsic properties of bursts in rest frames

Intensity group	Luminosity $(ergs \cdot s)$	τ_{50} (sec)	CIP (keV s)
1	$2.27 \cdot 10^{52}$	1.2	1074.9
2	$1.67 \cdot 10^{52}$	1.7	1529.8
3	$1.51 \cdot 10^{52}$	0.9	893.2
Total	$1.89 \cdot 10^{52}$	1.2	1134.2

References

1. Mitrofanov I.G. et al., ApJ, 522, 1069, 1999a
2. Mitrofanov I.G. et al., "Rest Frame Properties of Gamma Ray Bursts", Gamma-Ray Bursts, 5th Huntsville Symposium
3. Mitrofanov I.G. et al., "Comparison of z-known GRBs with the main groups of bright BATSE events", subm. to ApJ

The Unique Signature of Shell Curvature in Gamma-Ray Bursts

Alicia Margarita Soderberg[1] and Edward E. Fenimore[2]

[1] DAMTP, Silver Street, Cambridge CB3 9EW, ENGLAND
[2] Los Alamos National Laboratory, Los Alamos NM 87545, USA

Abstract. As a result of spherical kinematics, temporal evolution of received gamma-ray emission should demonstrate signatures of curvature from the emitting shell. Specifically, the shape of the pulse decay must bear a strict dependence on the degree of curvature of the gamma-ray emitting surface. We compare the spectral evolution of the decay of individual GRB pulses to the evolution as expected from curvature. In particular, we examine the relationship between photon flux intensity (I) and the peak of the $\nu F\nu$ distribution (E_{peak}) as predicted by colliding shells. Kinematics necessitate that E_{peak} demonstrate a power-law relationship with I described roughly as: $I = E_{peak}^{(1-\zeta)}$ where ζ represents a weighted average of the low and high energy spectral indices. Data analyses of 24 observed gamma-ray burst pulses provide evidence that there exists a robust relationship between E_{peak} and I in the decay phase. Simulation results, however, show that a sizable fraction of observed pulses evolve faster than kinematics allow. Regardless of kinematic parameters, we found that the existence of curvature demands that the $I - E_{peak}$ function decay be defined by $\sim (1 - \zeta)$. Efforts were employed to break this curvature dependency within simulations through a number of scenarios such as anisotropic emission (jets) with angular dependencies, thickness values for the colliding shells, and various cooling mechanisms. Of these, the only method successful in dominating curvature effects was a slow cooling model. As a result, GRB models must confront the fact that observed pulses do not evolve in the manner which curvature demands.

1 Introduction to the Kinematic Model

The simulated pulses described in this study were created through code based strictly on kinematics. Simulated shells were collided with one another, thereby conserving energy and momentum and the resulting energy was distributed into standard Band function spectra. Isotropic emission from the merged shell was (initially) assumed where the entirety of the shell is modeled to be gamma-ray active. Figure 1 demonstrates the geometry of the model. Time of arrival is determined by the angle, θ, at which the emitting patch lies with respect to the line of sight. Off axis emission is received later than on axis emission by a factor of $R(1 - cos\theta)$. The emitting shell has a slight thickness defined between $R/c = t_{max}$ and $R/c = t_0$. Photons within shell volume dV contribute to the pulse shape between received times, T and $T + dT$. As a result of the relativistic motion of the shell, the volume of emitting material which contributes to the received signal at any time is constant. Emitting patches on the shell which fall between the two ellipsoids labeled T and $T+dT$ will arrive at the detector within this range of received time.

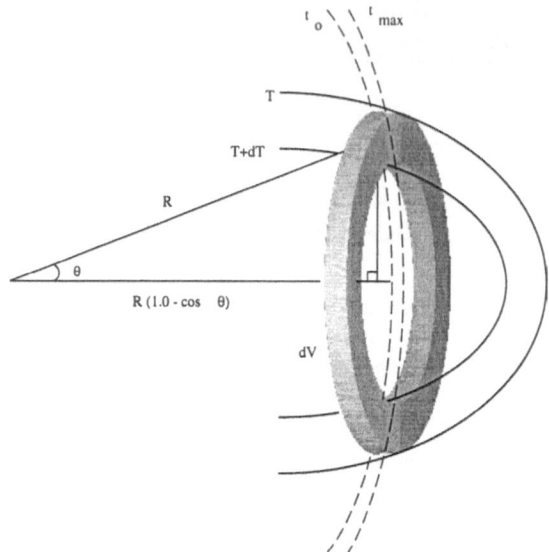

Fig. 1. Geometry of the Kinematic Model

2   Discussion: The Robust Curvature Dependency

Results of the kinematic studies demonstrate that the $I - E_{peak}$ relationship is
a robust indicator of shell curvature. The strength of this relation was analyzed
by varying both Band and kinematic parameters for the shell model. Observed
pulses, however, do not demonstrate this dependence (see Figures 2 and 3). In the
attempt to break spherical symmetry and reduce the dependence of pulse shape
on curvature effects, more complicated emission models were simulated. These
models enabled further testing to examine the possibility of additional depen-

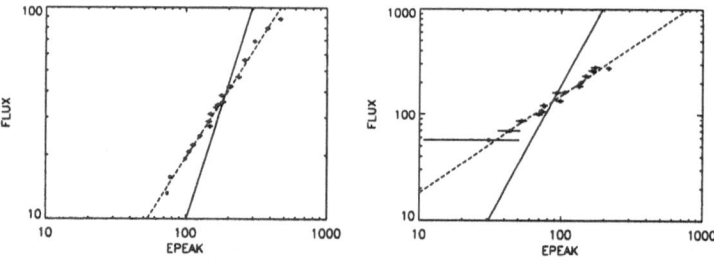

Fig. 2. Comparison with BATSE pulses. Simulations were compared with a data set of
24 pulses selected from the BATSE GRB Spectral Catalog I (Preece et al.,1999). The
top figure displays the robustness of the expected and observed $I - E_{peak}$ relationships
for GRB921207. The solid line represents the expected decay index as predicted by
colliding shells. It is evident that there is a large discrepancy between the data and
the simulations. The same is true for the bottom figure which displays the results for
GRB970201.Through these cases, the severity of this discrepancy can be clearly seen.

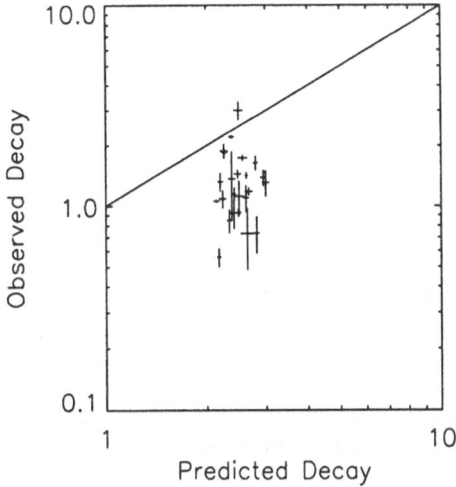

Fig. 3. Comparison of Expected $I - E_{peak}$ Decay with Observed $I - E_{peak}$ Decay from BATSE Pulses. The solid line represents the $I - E_{peak}$ decay index demanded by colliding shells. Curvature and special relativity impose such a relationship because later portions of the pulse arrive from off-axis emission. This results in an expected decay index of $I = E_{peak}^{(1-\varsigma)}$. The majority of BATSE observations lie below the solid line therefore indicating that observed $I - E_{peak}$ decay is slower than the decay predicted by kinematics.

dencies which were not included in the original kinematic code. Models explored jetting the model emission into an opening angle between 0.1-5.0 degrees and allowing for intrinsic angular dependencies of the Lorentz factor and/or E_{peak} across the skullcap. Off-axis shell collisions were simulated such that the collision time was not instantaneous in the rest frame of the central engine and the initial photon emission occurred at an angle outside the critical beaming angle. Various thicknesses were applied to the emitting shell but this only proved to distort the rise time of the pulse and did not have any effect on the shape of the pulse decay or the $I - E_{peak}$ relationship. Models also explored fast and slow cooling mechanisms. Generally, it was found that the curvature dependence is fairly difficult to break, and requires either grossly distorted geometries and/or relatively long cooling time scales. It was found that the emitting shell must cool for a time period of approximately $t_{cool} = R/c$ in the detector rest frame (where $R/c < \Gamma^2$). This corresponds to a comoving time of $t'_{cool} = R/c\Gamma$ and an arrival time of $T_{cool} = R/c\Gamma^2$ by the standard transformations. As a result, the observed cooling time was comparable in length to the duration due to curvature (e.g. 10^4 s). Such slow cooling overwhelmed the curvature dependency with cooling effects throughout the entire length of the pulse, thereby allowing for a new pulse shape evolution. It is emphasized that slow cooling was the only method included in this study which was able to break the robust curvature dependency as imposed by the kinematics of two colliding shells. Typical cooling times, however, are commonly quoted as being shorter than the duration of the pulse (e.g. $< 10^4$ s). A remedy to this situation is to minimize the timescale on which curvature effects can be detected by reducing the radius and/or increasing the bulk Lorentz factor of the emitting shell. This, in turn, allows for a relatively shorter cooling time. Long cooling times, however, face a number of problems including efficiency considerations which must be addressed.

References

1. Band, D. et al.: ApJ **413**, 281 (1993)
2. Crider, A., et al.: ApJ **519**, 206 (1999)
3. Fenimore, E. E., Madras, C. D., & Nayakshin, S.: ApJ **473**, 998 (1996)
4. Preece, R. D., et al.: ApJ Supp. **126**, 19 (1999)
5. Rybicki, G. B. & Lightman, A. P.: *Radiative Processes in Astrophysics*, 1979
6. Sari, R., & Piran, T.: ApJ **485**, 270 (1997)
7. Summer, M. C., & Fenimore, E. E.: AIP Proc. **428**, 765 (1998)

Final Results of the Off-Line Scan of the BATSE Daily Records

B.E. Stern[1,2,3] Ya. Tikhomirova[2,3] D. Kompaneets[2] and R. Svensson[3,4]

[1] Institute for Nuclear Research, Russian Academy of Sciences
[2] Astro Space Center of Lebedev Physical Institute, Moscow
[3] Stockholm Observatory
[4] Institute for Theoretical Physics, UCSB, Santa Barbara

Introduction

The sample of gamma-ray bursts (GRBs) detected by the Burst and Transient Source Experiment (BATSE) [1] on the *Compton Gamma-Ray Observatory (CGRO)* is a few times larger than the yield of all other experiments that have detected GRBs.

Nevertheless, the BATSE sample (i.e., the sample of the triggered events included in the BATSE catalogs) is smaller than what it could be at the actual sensitivity of BATSE. The reason for this is a difficult variable background. The BATSE trigger adjusted to the background conditions misses many weak but still highly significant GRBs.

Off-line searches for non-triggered bursts with less rigid off-line trigger criteria can substantially increase the sensitivity of BATSE. The first systematic search for non-triggered GRBs in 6 years of *CGRO* BATSE data was performed by Kommers et al. [2–4]. Our off-line scan of the BATSE data described in [5–8] included a procedure for calibrating the results using test bursts and covered the full period of the *CGRO* operation (1991–2000).

Here we summarize the main results of the scan.

The Data Scan and the Final Statistics

We used the continuous 1024 ms time resolution BATSE data (DISCLA, 8 detectors, 4 energy channels) from the ftp archive of the Goddard Space Flight Center. For the scan calibration, data were inserted using artificial test bursts which were prepared from a sample of 500 real triggered BATSE bursts rescaled to lower brightnesses. The average interval between inserted test bursts was 25000 seconds, their total number was ~11000. The procedure of the scan is described in [7–8].

We performed the scan of DISCLA data for the full 9.1 years of BATSE observations (TJD 8369-11690) up to the *CGRO* deorbiting. The number of events in our sample classified as GRBs is 3923, 1849 of them are non-triggered, and 2074 we identified with BATSE triggers. The total number of BATSE GRB triggers for the same period is 2704, i.e., we missed 630 triggered GRBs. We estimated that ~ 70% of these GRBs were lost due to gaps in the DISCLA data,

~ 20% were too short to be detected at 1.024 s resolution, and ~ 10% were missed due to various human mistakes. The number of detected test bursts was ~ 5400, half of the inserted test bursts.

Reference [3] scanned the time interval TJD 8600-10800 and found 873 non-triggered GRBs. We found 1132 non-triggered bursts during that time interval. 745 of them are in the catalog [3]. Kommers (1999, private communication) inspected the 387 of our non-triggered events that [3] missed. Kommers confirmed 224 of these events as probable GRBs. 90 of our events were not classified as the off-line trigger of [3] missed those events due to nearby data gaps. We checked most of the 128 events that are in the catalog [3] but that are missing in our sample. We confirm 90 of them as GRBs.

We performed a number of statistical tests for various non-GRB contaminations of the sample (e.g., a test for "negative" bursts, checking for any concentration towards known sources, the distribution of the *CGRO* geographic latitudes at the moment of GRB detection). We find no indications of any contamination and constrain it to a 1.1% fraction of the sample.

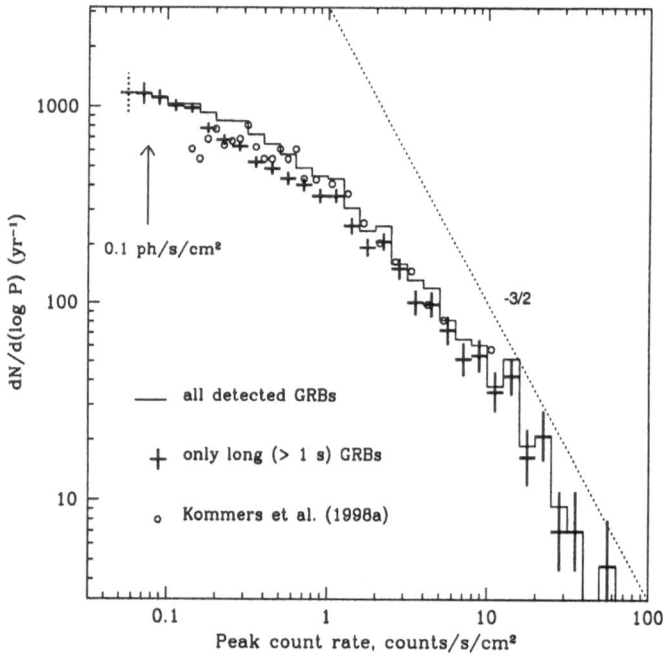

Fig. 1. The differential $\log N - - \log P$ distribution corrected for the efficiency function in absolute units for all 3923 GRBs detected in our scan (*histogram*) and for the 3300 events in our sample with a duration longer than 1 s (*crosses*). The corresponding distribution in absolute units for the 2265 GRBs (of any duration) found in [3] is also shown (*circles*)

The log N–log P Distribution and Conclusions

The log N–log P curve for our whole sample of detected events straightforwardly corrected using the efficiency function obtained as the detected/inserted test burst ratio is shown in Fig. 1. Our whole sample is not homogeneous because it contains short events where the peak flux estimate in 1 s time resolution is wrong. To eliminate the corresponding bias we excluded from the distribution all events consisting of one bin (i.e., where only one 1.024 s bin is above 0.5 of the peak flux value).

As a first result, we see the log N–log P distribution extending down to lower brightnesses almost as a straight line without any indication of a turn-over. This result differs from previous interpretations of the data, which implied that the log N–log P distribution smoothly bends down at low brightnesses (see the data points from [3] in Fig. 1).

One simple consequence is that the estimate of the number of GRBs in the visible Universe should be increased. Just the "visible" part of the log N–log P distribution implies 1200–1300 GRBs occurring per year at peak fluxes exceeding 0.1 photons s^{-1} cm^{-2} (previous versions of the log N–log P distribution implied $\sim$ 600–800 GRBs per year above 0.18 photons s^{-1} cm^{-2}). A possible extrapolation of the new log N–log P distribution to lower brightnesses would probably imply a much larger rate of up to several thousands of GRBs per year.

References

1. G.J. Fishman, et al.: 'The Burst and Transient Source Experiment (BATSE)'. In: *Proc. of the Gamma Ray Observatory Science Workshop*, ed. by W. N. Johnson (Goddard Space Flight Center, Greenbelt 1989) pp. 3-47-3-62
2. J.M. Kommers, W.H.G. Lewin, C. Kouveliotou, J. van Paradijs, G.N. Pendleton, C.A. Meegan, G.J. Fishman: Astrophys. J. **491**, 704 (1997)
3. J.M. Kommers, et al.: http://space.mit.edu/BATSE/ (1998)
4. J.M. Kommers, W.H.G. Lewin, C. Kouveliotou, J. van Paradijs, G.N. Pendleton, C.A. Meegan, G.J. Fishman: Astrophys. J. **533**, 696 (2000)
5. B.E. Stern, Ya. Tikhomirova, M. Stepanov, D. Kompaneets, A. Berezhnoy, R. Svensson: Astron. Astrophys. Suppl. **138**, 413 (1999)
6. B.E. Stern, Ya. Tikhomirova, M. Stepanov, D. Kompaneets, A. Berezhnoy, R. Svensson: 'A Search for Non-Triggered Gamma-Ray Bursts in the BATSE Continous Records: The Current Status'. In *AIP Conf. Proc. 526, Gamma Ray Bursts, 5th Huntsville Symposium*, ed. by R.M. Kippen, R.S. Mallozzi, G.J. Fishman, (AIP, New York 2000) pp. 13–17
7. B.E. Stern, Ya. Tikhomirova, M. Stepanov, D. Kompaneets, A. Berezhnoy, R. Svensson: Astroph. J. Lett. **540**, L21 (2000)
8. B.E. Stern, Ya. Tikhomirova, D. Kompaneets, R. Svensson, J. Poutanen: Astroph. J. (2000) submitted (astro-ph/0009447)

Part II

GRB Afterglows

X-Ray Afterglows and Features of Gamma-Ray Bursts

Luigi Piro

Istituto di Astrofisica Spaziale, CNR
Via Fosso del Cavaliere 100, 00133 Rome, Italy

Abstract. X-ray observations of afterglows are providing new insights on the nature of the central source of GRB. In this review I will focus on observations of temporal and spectral features, particularly on iron lines by BeppoSAX & Chandra and discuss how these observations can disclose the origin of the progenitors of GRB. I briefly review the old and new mysteries opened by recent observations and discuss observational prospects and future perspectives of GRB science.

1 Introduction

The study of afterglows of gamma-ray bursts (GRB) is an area of investigation which is yielding unique information on these mysterious objects. While we have reasonably well understood the process producing the radiation, very little is known about the nature of the progenitor, and ultimately on the process leading to the release of their huge luminosity. One reason is that the prompt and afterglow emission originate in a region which is 6 orders of magnitude larger in size than the original site of the explosion, therefore very little information remains of the original process. A way to circumvent this limitation is through the study of the circumburst environment, that should tell us – for example – whether GRB go off in dense star forming regions, and should therefore be connected with the explosion of young massive systems. In this paper I will first summarize the general properties of afterglows, and then focus on the temporal and spectral (X-ray iron features) signatures produced by the medium at the burst site.

2 Observations vs Theory: The Fireball Model

2.1 The Canonical GRB Afterglow: Power Laws in Time and Energy Domains

The typical dependence of an X-ray (and optical) afterglow on time and energy is a power law: $F \propto t^{-\delta}\nu^{-\alpha}$. In the X-ray regime, the average are $\delta = 1.4$, $\alpha = 0.9$. These properties are nicely accounted for by the fireball model (e.g. Rees & Mészáros 1992). At cosmological distances the observed fluxes correspond to a luminosity in γ-rays of $\approx 10^{53}$ erg concentrated in a region of the order of few light seconds. In this conditions a fireball develops. The initial radiative energy released by the central source is converted to kinetic energy of a relativistic shell which expands up to a size $\approx 10^{16-17}$cm, where it converts its energy back into electromagnetic radiation (i.e. the GRB and its afterglow) via shock accelerated electrons radiating by synchrotron .

2.2   Internal vs External Shock (When Does the Afterglow Start?)

The simple power law behaviour predicted by the fireball model via its interaction with an external medium (the so called external shock model) is difficult to reconcile with the chaotic variability observed in the GRB *prompt* emission (Sari & Piran 1997, Fenimore *et al.* 1996). The latter is better explained by *internal shocks* . In this variant of the model, shells with different Lorentz factors are released by the central engines. The prompt emission is then produced when the later/faster moving shell catches up with the earlier/slower shell. In this scenario the prompt and afterglow emission are then decoupled (see Fig.1).

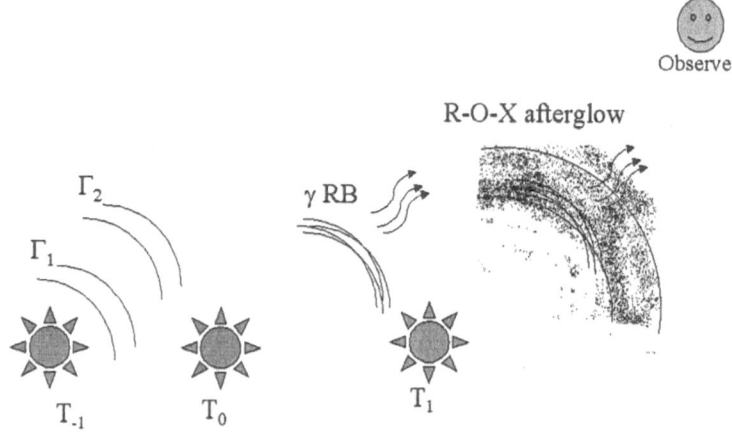

Fig. 1. The fireball scenario: internal and external shocks

This is in agreement with the behaviour exhibited e.g. by GB970228. The light curve of this event (Fig.2; Soffitta *et al.* , in preparation) shows a gap with no detected counts at about 20-30 s. The gap separates the prompt phase from beginning of the afterglow phase. The data point between 30 s and 200 s lie on the same power law connecting the measurement taken at later times and are therefore associated with the afterglow phase. Further support to this interpretation derives from a spectral analysis. The prompt emission of a GRB is characterized by large spectral variations, with a hard to soft evolution of the peak energy from gamma-ray to X-rays (Piro *et al.* 1998a, Frontera *et al.* 2000). On the contrary, during the afterglow, the spectrum does not change substantially, being consistent with a power law with a typical slope $\alpha \approx 1$. Spectral analysis of the data of GB970228 shows that this is indeed the case: a strong hard-to-soft spectral evolution is found in the data before t=20-30 sec, while the spectra accumulated in the 30-200 sec period are described by a power law with a slope consistent with that observed at much later times (Frontera *et al.* 1998).

In other bursts it is more difficult to fix the border between the prompt and the afterglow phase: in most of them there is a continuous evolution between the rapidly decaying prompt event and the slower, power-law decay of the afterglow, with the latter becoming dominant at a time corresponding to around 70% of the total prompt GRB duration (Frontera *et al.* 2000).

3   X-Ray Iron Lines and the GRB Progenitor

Information on the nature of the progenitor can be derived from the GRB environment. In the case of hypernova (Woosley *et al.* 1993, Paczyinski 1998) the massive star should die young ($\approx 10^6$ years) and therefore GRB should be preferentially hosted in regions near the center of star-forming galaxies. On the contrary, NS-NS coalescence happens on much longer time scales (billions of years) and the kick velocity given to the system by two consecutive supernova explosions should bring a substantial fraction of these systems away from their formation site (Fryer *et al.* 1999). Bloom *et al.* (1999) measured the location of optical afterglows with respect to their host galaxy. They found that GRB are predominantly localized within their host galaxies, but the present disagreement with simple calculations of the evolution of NS-NS systems needs to be confirmed by more detailed computations and a much larger sample of data.

X-ray spectral diagnostics provides an powerful tool to investigate the environment in the vicinity of the GRB. We remind that X-ray iron lines are ubiquitous features in the spectra of several classes of objects, from stars to

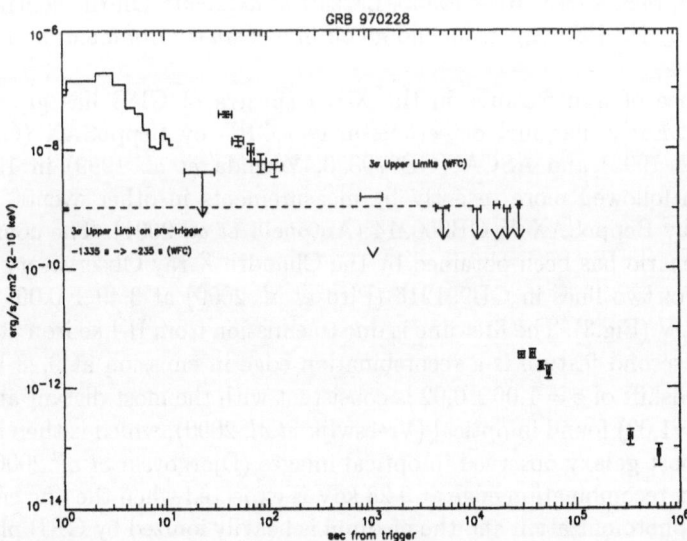

Fig. 2. X-ray (2-10 keV) light curves of GB970228 obtained with the BeppoSAX WFC and NFI (Soffitta *et al.* , in prep.)

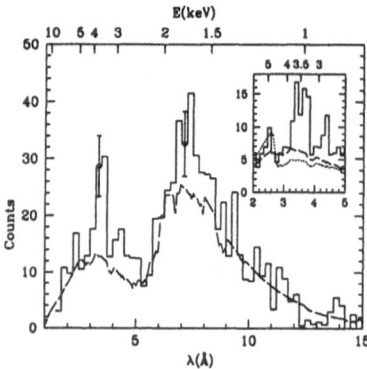
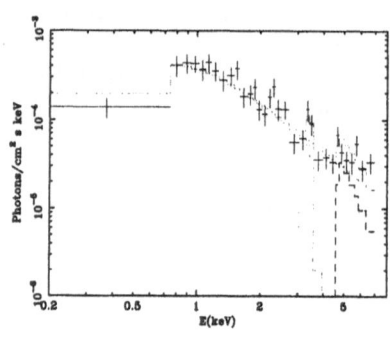

Fig. 3. The X-ray afterglow spectrum of GRB991216 obtained with the Chandra high-energy gratings (High Energy (HE) and Medium Energy (ME) summed together, (left panel) and the ACIS-S (0th order, right panel). The grating spectrum has a bin size of 0.25 Å. The dashed line represents the continuum. The peak (i.e. 2 bins) around 3.5 Å (E=3.5 keV) is the line produced by iron atoms that are completely ionized. The width of the line (see inset) tell us that the material is moving at $\approx 0.1c$. In the ACIS-S spectrum (right panel) a further feature, is present, a recombination edge at $E = 4.5 \pm 0.5$ keV (dashed line). The continuous line represents the best fit model, composed by a power law, a line at 3.4 keV and the recombination edge (Piro *et al.* 2000)

clusters of galaxies (e.g. Piro 1993). In the GRB environment, any medium lying in the vicinity of the central source would be subject to the huge ionizing flux and to the interaction with the relativistic expanding fireball. In the case of a massive progenitor, this would lead to the production of iron features (Boettcher *et al.* 1998, Mészáros & Rees 1998b, Lazzati *et al.* 1999). On the contrary, GRB produced by NS-NS mergers should go off in a *clean* environment, and no line is expected.

Evidence of iron features in the X-ray spectra of GRB has grown in the last years. Early marginal detections in two GRB by BeppoSAX (GB970508, Piro *et al.* 1999) and ASCA (GB970828, Yoshida *et al.* 1999) in 1997-1998, have been followed more recently by measurements in other events, like that observed by BeppoSAX in GB000214 (Antonelli *et al.* 2001). The confirmation of this scenario has been obtained by the Chandra X-ray Observatory with the detection of two lines in GB991216 (Piro *et al.* 2000) at 3.49 ± 0.06 keV and 4.4 ± 0.5 keV (Fig.3). The first line is due to emission from H-like iron at 6.9 keV, while the second feature is a recombination edge in emission at 9.28 keV. The implied redshift of $z = 1.00 \pm 0.02$ is consistent with the most distant absorption system (z=1.02) found in optical (Vreeswijk *et al.* 2000), which is then identified with the host galaxy observed in optical images (Djorgovski *et al.* 2000).

An iron recombination edge at 9.28 keV is expected when the line emission is driven by photoionization and the medium is heavily ionized by GRB photons. If the medium lies in the line of sight, the edge is expected to be seen in absorption at early times (e.g. Perna & Loeb 1998, Weth *et al.* 2000), when the medium is still to be fully ionized by the incident photon flux. Evidence of such a feature has been found in another GRB, GRB990507 (Amati *et al.* 2000). At later times, when the medium becomes completely ionized, recombination takes place, and the feature is seen in emission, as observed in GRB991219.

From the distance and the intensities of the detected X-ray emission lines in GB991216, the mass of the iron needed to produce the lines turns out to be approximately equivalent to at least one-hundredth that of the Sun. For solar-like abundances, the total mass of the medium would be $\approx 5M_\odot$. An estimate of the iron abundance is derived by the requirement that the medium should be optically thin for Compton scattering, which yields an iron overbundance of at least 60 times greater than the solar value. A similar result is obtained by an independent analysis of the iron absorption edge in GB990705. A further key element in our understanding of the origin of this medium is derived by the width of the emission line in GB991216, which is $\approx 10\%c$. These observations are strongly suggesting that the progenitor of the GRB was very a massive star. The large content of iron and the velocity of the ejecta also suggest that the ejection of the material was similar to a supernova explosion, as in the SupraNova model (Vietri & Stella 1999), in which the GRB arises from the delayed collapse of a neutron star formed by a supernova. One alternative to the pre-ejection scenario has been proposed by Boettcher & Fryer (2000). They suggest that a very extended torus could be produced during a common-envelope phase of a system based by a compact object spiralling in the envelope of a supergiant. The supernova-like explosion would generate – along with the GRB – a sub-relativistic material that will be shock-heated when it impacts on the torus. (A similar scenario, where the relativistic fireball material interacts with a dense medium around the GRB was introduced by Vietri *et al.* (2000) to explain the iron line observed in GB970508). The problem of this model (as well as of that of Vietri et al) is that the emission spectrum line spectrum is thermal, but in such a case *emission* from a recombination edge is negligible.

Another alternative has been proposed by Mészáros & Rees (2000). They invoke a continuing energy injection after the burst, from the "remnant" of the explosion, that continues to irradiate the line-emitting material at a very close distance ($\approx 10^{13}cm$). In this case the presence of the line 1-2 days after the GRB is explained by the duration of the continuum emission, lasting for a day or longer. Such a small radius entails higher densities, and therefore the mass of iron required to produce the line can be much less than in previous cases. From an observational point of view it is however not clear how this model can be tested, because the flux of the remnant powering the line is – by construction – much lower than that of the GRB and the afterglow.

4   Achromatic Breaks, Jets and High-Density Medium

In some of the most extreme cases, like GRB990123, the assumption of isotropic emission would imply a total energy of $1.6 \times 10^{54}erg$. This corresponds to $\sim 2M_\odot c^2$, at the limit of all models of mergers (Mészáros & Rees 1998a). This piece of evidence is lending support to the idea that, at least in some case, the emission is collimated. This would reduce the energy budget by $\sim \theta^2/4$, where θ is the angle of the jet. A typical feature of a jet expansion (vs spherical) would be the presence of an achromatic (i.e. energy-independent) break in the light curve, that appears when the relativistic beaming angle $1/\Gamma$ becomes $\approx \theta$ (e.g. Rhoads 1997, Sari et al. 1999). The presence of such a break has been claimed in some GRB's (e.g.: GRB990123 (Kulkarni et al. 1999); GRB990510 (Harrison et al. 1999), GB991216, Halpern et al. 2000). In X-rays the data are compatible with an achromatic break in the case of GB990705 (Kuulkers et al. 1999, Pian et al. 2001). We have performed a systematic analysis of the X-ray spectra and light curves on a sample of BeppoSAX afterglows observed from few hours to about 2 days after the GRB (Stratta et al. 2001). The average property of the sample are fully consistent with a spherical expansion and deviates substantially from a jet expansion *in the first two days*. Hence collimation, on average, cannot be very high, with $\theta \gtrsim 10 - 20°$.

It should be stressed that an achromatic break can be also produced when the fireball slows down to a non-relativistic expansion (NRE, Livio & Waxman (2000)). Cases where the NRE has been proposed to account for a break are GB990123 (Dai & Lu 1999), as an alternative explanation to the jet of Kulkarni et al. (1999), GRB000926, (Piro et al. 2001) – but see also Harrison et al. (2001), GRB010222, (Masetti et al. 2001, in 't Zand et al. 2001). A strong implication of these papers is that, in those GRB's, the environment is composed of a dense medium ($n \approx 10^4 - 10^6 cm^{-3}$). This high density is typical of molecular clouds in star forming regions, providing independent support to the association of some GRB with massive progenitors.

5 Recent Findings and Remaining Mysteries

5.1 GHOST: GRB Hiding Optical Source Transient/Dark GRB's

So far, about 40 GRB have been precisely localized mostly by BeppoSAX and for about 30 of them it was possible to perform fast follow-up observations. In $\approx$ 90% of the cases an X-ray afterglow was detected. On the contrary, only $\approx$ 50% counterparts were identifiedin optical (Kulkarni et al. 2000) and $\approx$ 40% in radio (Frail et al. 2000). The absence of optical counterparts is not due to an observational bias, because those searches were carried out with a similar sensitivity and promptness as those leading to a positive detection. If GRB are indeed associated with massive progenitors and therefore lie in regions of star formation, it is likely that in a large fraction of events, the optical emission is heavily absorbed by dust. On the contrary, X-ray photons are more penetrant than optical photons, and we still can see the X-ray afterglow.

Another intriguing possibility is that GHOST's are located at very high red-shift ($z > 5$), such that the absorption due to the gas in the Universe between us and the GRB will result in a cut-off below the Lyman limit. Measurement of the distance by X-ray lines will be crucial to test this hypothesis. A deep investigation in the optical band shall become also possible with the availability of arcsec position, as those provided in X-rays by Chandra (e.g. GB000210, Piro *et al.* , in preparation), or in the case where a radio counterpart (but no optical one) is present (GB970828, Djorgovski *et al.* , in preparation).

5.2 X-Ray Rich GRB's

BeppoSAX has revealed the existence of a new class of events, the so-called X-ray flashes, or X-ray rich GRB's (Heise *et al.* 2001). In those cases the bulk of the energy is not produced in γ−rays, but in X-rays. The paucity of gamma-ray emission can be explained by a fireball expanding through a medium with high density. In such a case the fireball will achieve a Lorentz factor much lower than previously considered, i.e. not high enough to boost the photons in the γ-ray range.

Another extremely intriguing possibility is that these events are located in very distant galaxies ($z > 10$). The peak of the emission, that in the rest frame is in γ-rays, should then be redshifted in the X-ray range as seen by an observer. The origin of these events remains a mystery and should be clarified with the identification of their counterparts and therefore their distance.

5.3 Short GRB's

The distribution of GRB duration appears to be bimodal, with about 30% of events lasting less than 1 sec (e.g. Kouveliotou *et al.* 1993). It is still unclear whether these events are intrinsically diverse from long bursts, i.e. if they are produced by different progenitors. So far, very little is known about these events, due to the lack of a counterpart. While the GRBM on board of BeppoSAX is detecting those events, no one has so far been observed in the X-ray range (and therefore precisely localized) by the WFC (Gandolfi *et al.* 2000). A depleted X-ray emission compared to the class of long events would explain the lack of detection in the X-ray range, but it is still unclear whether this is the case or not. If that is the case, the identification of afterglows of these events should rely only on experiments with good position accuracy in the γ-ray range, like the currently operating IPN or, in the future, SWIFT. If, instead, these events have an X-ray emission similar to long ones, we may hope to localize a few of them with BeppoSAX and HETE2.

5.4 Cosmological Implications

About 18 redshifts have been measured by optical spectroscopy. Two other redshifts have been derived by X-ray measurements. Excluding the ambiguous event of April 25, 1998, they are in the range 0.43-4.5, i.e. comparable to that of the most distant object observed in the Universe. The luminosity of GRB's is so high that they can be detectable out to much larger distances than those of the most luminous quasars or galaxies observed so far. We expect thus, in the near future, to use GRB's as beacons to probe star and galaxy formation at much earlier epochs of the Universe evolution, e.g. by studying the features imprinted over their spectrum by the gas through which they shine (e.g. Piro *et al.* 1999, Fiore *et al.* 2000, Lamb & Reichert 2000).

References

1. Amati L. *et al.* 2000, Science, 290, 953
2. Antonelli L.A., Piro L., Vietri M. *et al.* 2000, Ap.J., 545, L39
3. Bloom J.S. *et al.* . 1999a, *Mon. Not. R. Astr. Soc.*, 305, 763
4. Bottcher M., Dermer C.D., Crider A. W. & Liang E. D. 1998 A&A, 343, 111
5. Bottcher M., & Fryer 2001 A&A, 547, 338
6. Dai, Z. G. & Lu T. 1999, Ap.J.519, L155
7. Djorgovski S. G. *et al.* 2000, GCN 510
8. Fenimore E. *et al.* 1996, Ap.J., 473, 998
9. Fiore F. *et al.* 2000, Ap.J., 544, L7
10. Frail D. *et al.* , 2000, proc of the Vth Huntsville Symposium on GRB, , 1999, M. Kippen, R.S. Mallozzi & G. J. Fishman ed.s., , AIP, 526, p.298.
11. Fryier C.L., Woosley S.E. & Hartman D.H. 1999, Ap.J., 526, 152
12. Frontera F. et al., 1998 ApJ 493, L67
13. Frontera F. *et al.* 2000, *AP. J. Suppl. Ser.* , 127, 59
14. Gandolfi G. *et al.* 2000, proc of the Vth Huntsville Symposium on GRB, , 1999, M. Kippen, R.S. Mallozzi & G. J. Fishman ed.s., , AIP, 526, p.23
15. Halpern J. *et al.* 2000, Ap.J., 543, 697
16. Harrison F.A. *et al.* 1999, Apj 523, L21
17. Harrison F.A. *et al.* 2001, Ap.J., in press (astro-ph/0103377)
18. Heise J.*et al.* 2001, this conference, in press
19. in 't Zand J.J.M. *et al.* 2001, Ap.J., in press (astro-ph/0104362)
20. Kouveliotou C. *et al.* 1993, Ap.J., 413, L101
21. Kulkarni S. *et al.* 1999, Nature 398, 389
22. Kulkarni S. *et al.* 2000, proc of the Vth Huntsville Symposium on GRB, , 1999, M. Kippen, R.S. Mallozzi & G. J. Fishman ed.s., , AIP, 526, p.277.
23. Kuulkers E. *et al.* 2000, ApJ, 538, 638
24. Lamb D.Q. & Reichert R.D. 2000, Ap.J., 536, L1
25. Lazzati D., Ghisellini G. & Campana 1999, *Mon. Not. R. Astr. Soc.*, 304, L31
26. Livio M. & Waxman E. 2000, Ap.J., 538, 187 (LW)
27. Masetti N. *et al.* , 2001, *Astron. Astrophys.*, in press (astro-ph/0103296)
28. Mészáros P. & Rees M. J. 1998a, New Astronomy, (astro-ph/9808106)
29. Mészáros P. & Rees M. J. 1998b MNRAS, 299, L10.
30. Paczynski, B. 1998 ApJ, 494, L45
31. Perna R. & Loeb A., 1998, ApJ, 501, 467
32. Pian E. *et al.* 2001, *Astron. Astrophys.*, in press (astro-ph/0012107)
33. Piro L. 1993, proc.s of the UV and X-ray spectroscopy of Labroatory and Astrophysical Plasmas, E. Silver, S. Kahn, eds., (Cambridge Uni. Press), p.448
34. Piro L. et al., 1998a A&A, 329, 906
35. Piro L. *et al.* 1999, ApJ, 514, L73
36. Piro L. 1999, talk on "Probing the Early Universe with GRB and XEUS" presented at the XEUS open meeting in ESTEC, Sept. 1999, available at www.ias.rm.cnr.it/sax/xeus.html
37. Piro L. *et al.* 2000, Science, 290, 955

38. Piro L. *et al.* 2001, Ap.J., in press (astro-ph/0103306)
39. Rees M.J., Mészáros P. *Mon. Not. R. Astr. Soc.***258**, 41-43 (1992)
40. Rhoads J.E. 1997, ApJ, 478, L1
41. Sari R. & Piran T. 1997, Ap.J., 485, 270
42. Sari R., Piran T, Halpern J.P. 1999, ApJ, 497, L17
43. Stratta G., Piro L., *et al.* 2001, this conference
44. Vietri M., Perola G.C., Piro L. & Stella L. 1999, MNRAS, 308, P29
45. M. Vietri & L. Stella 1998, ApJ, 507, L45
46. Vreeswijk P.M. *et al.* 2000, GCN 496
47. Weth C. *et al.* 2000, Ap.J., 534, 581
48. Woosley S. 1993, ApJ, 405, 273
49. Yoshida A. *et al.* 1999 A&AS, 138, 433, Proc. of the Gamma-Ray Bursts in the Afterglow Era, F. Frontera &L. Piro ed.s.

Transient Spectral Features in the Prompt Emission of Gamma-Ray Bursts with BeppoSAX

Filippo Frontera[1,2], Lorenzo Amati[2], Enrico Costa[3], Cristiano Guidorzi[1], Mario Vietri[3], and Jean J.M. in 't Zand[4]

[1] Università di Ferrara, Dipartimento di Fisica, Via Paradiso, 12, 44100 Ferrara, Italy
[2] Istituto TESRE, CNR, Via Gobetti, 101, 40129 Bologna, Italy
[3] Istituto di Astrofisica Spaziale, CNR, Via Fosso del Cavaliere, 100, 00133 Roma
[4] Università di Roma Tre, Dipartimento di Fisica E. Amaldi,
    Via della Vasca Navale, 84, 00146 Roma, Italy
[5] Astronomical Institute, Utrecht University, PO Box 80 000, 3504 TA Utrecht, and
    SRON, Sorbonnelaan 2, 3584 CA Utrecht, The Netherlands

Abstract. We report here on transient features detected in the prompt emission of two Gamma Ray Bursts detected with the BeppoSAX satellite: an absorption feature from GRB 990705 and an emission feature from GRB 990712.

1 Introduction

The study of the spectral evolution of the prompt 2-700 keV radiation of Gamma-Ray Bursts (GRBs) detected with the BeppoSAX Wide Field Cameras (WFCs) and the Gamma Ray Burst Monitor (GRBM) is providing key importance astrophysical results. Among them, already reported [1], we mention the deviation from an optically thin synchrotron at early times in several GRBs, and the detection of a decreasing N_H from GRB980329. Now we report on two other outstanding results found in the spectra of two GRBs:

- a transient absorption feature in the prompt emission of GRB990705 (see also Amati et al.[2]);
- a broad transient emission feature in the prompt emission of GRB990712[3].

These results are expected to have a strong impact on theoretical models of GRBs, in particular on the radiation emission mechanisms, on the photoionization processes occurring in the circumburst medium as consequence of the GRB emission, and, more generally, on the origin of the GRBs primary events and their progenitors.

2 Observations

2.1 GRB 990705

GRB 990705 triggered the GRBM on 5 July 1999 at 16:01:25 UT. Its celestial coordinates, as provided by the WFC No. 2, are $RA_{2000} = 05^h09^m52^s$ and

$DEC_{2000} = -72°08'02"$, in a direction very close to the edge of the Large Magellanic Cloud. The time profile shows a single pulse with a duration of ~ 40 s in γ-rays (40-700 keV) and more than 60 s in X-rays (2-28 keV). The event is the strongest in gamma-rays after GRB990123. The prompt follow-on observation performed with BeppoSAX detected in the WFC error box a very weak X-ray source, 1SAX J0509.9-7207, the 2-10 keV flux of which ($1.9^{+0.6}_{-0.6} \times 10^{-13}$ erg cm^{-2} s^{-1}) faded during the following 100 ks observing time. A reddened fading counterpart in the optical and NIR bands was discovered [4]. A spiral galaxy was detected with *Hubble Space Telescope* [5] in correspondence of the GRB optical position. From its size and brightness, its redshift was estimated to be $z < 1$[6].

2.2   GRB 990712

This burst triggered the GRBM on 12 July 1999 at 16:43:02 UT. Its celestial coordinates, as provided by the WFC No. 2, are $RA_{2000} = 22^h31^m55^s$ and $DEC_{2000} = -73°24'24"$. The time profile shows two pulses of about 10 s duration, with a total time duration of about 20 s above 100 keV and about 40 s in the 2-26 keV energy band. The GRB shows the highest peak flux ever observed in X-rays (2-26 keV) with BeppoSAX [3]. Due to Sun constraints, the event was not followed-on with BeppoSAX. An optical fading counterpart was discovered with a R magnitude of 19.4 three hours after the event[7,8]. The redshift of the optical counterpart was determined (z=0.433 [9,10]). Also an optical polarization with evidence of variation was discovered[11].

3   Spectral Analysis and Results

The time profile of each GRB was divided into a given number of temporal sections and a spectral analysis in the 2-700 keV energy band was performed on the average spectrum of each section. The duration of the sections was chosen on the basis of the statistical quality of the data. A simultaneous fit to the WFC and GRBM spectra in each section was performed, using as input models a simple power-law or a smoothly broken power law[12], both photoelectrically absorbed. The Galactic column density along the GRB directions was $N_H = 4.52 \times 10^{20}$ cm^{-2} for GRB 990712 and $N_H = 7.06 \times 10^{20}$ cm^{-2} for GRB 990705.

3.1   GRB 990705

A simple power law fit the spectra of all the 7 time intervals in which we subdivided the GRB time profile, but the first two (A and B), which have a duration of 6 and 7 s, respectively. The chance probability of this deviation from a power law is particularly low for the second interval (9×10^{-6}). All possible instrumental effects that could have given origin to this deviation were considered and excluded. A spectral model that describes the B spectrum is a photoelectrically

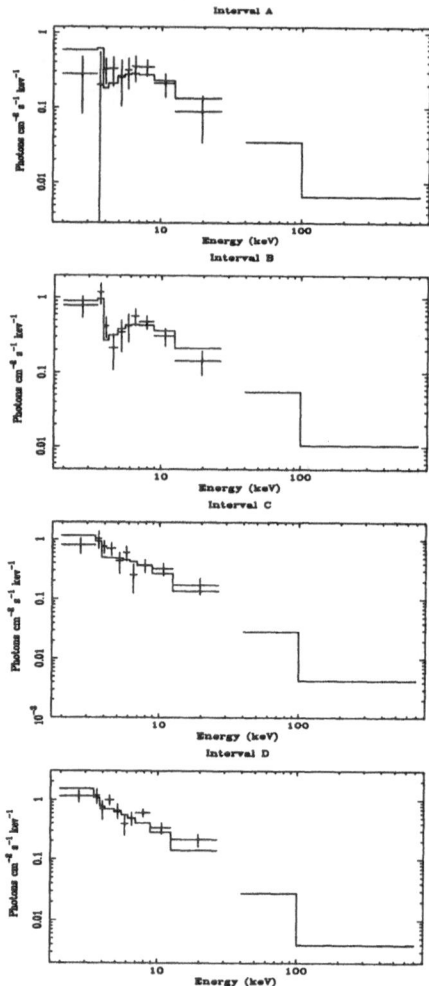

Fig. 1. Time evolution of the GRB990705 spectrum in the first 4 of the 7 time intervals in which the GRB profile was subdivided

absorbed power law plus an absorption edge. The derived column density, assuming a solar composition of the absorbing medium, is $N_{\rm H} = 3.5^{+1.0}_{-1.0} \times 10^{22}$ cm^{-2}, while the energy of the edge is $E_{edge} = 3.8 \pm 0.3$ keV and its optical depth is $\tau = 1.4 \pm 0.4$. More naturally the B spectrum is well fit by a photoelectrically absorbed power-law, where the absorbing medium, at redshift z, is assumed to have a solar composition and abundance, except Iron, the relative abundance A_{Fe} of which is left free to vary in the fit. Also the redshift is left free to vary. The parameters found are the following: $N_{\rm H} = 1.3^{+0.3}_{-0.3} \times 10^{22}$ cm^{-2}, power law photon index $\Gamma = 1.09 \pm 0.02$, $A_{Fe} = 75 \pm 19$ and $z = 0.86 \pm 0.17$. The same model fits the A spectrum. For the later GRB spectra the simple power-law provides a good fit. The time evolution of the column density and optical depth

Table 1. Time behaviour of column density and optical depth of the circumburst material, which is photoionized by the GRB990705 prompt emission[a]

Interval #	Time from GRB onset (s)	N_H ($\times 10^{22}$ cm^{-2}	τ[b]
A	0–6	1.23 ± 0.43	1.52 ± 0.53
B	6–13	1.32 ± 0.18	1.63 ± 0.22
C	13–20	0.55 ± 0.26	0.68 ± 0.32
D	20–28	0.24 ± 0.21	0.29 ± 0.26

[a] Assuming Iron relative abundance $A_{Fe} = 75$ and $z = 0.86$.
[b] From $\tau = \sigma X_{Fe} A_{Fe} N_H$ ($X_{Fe} = 4.68 \times 10^{-5}$, $\sigma = 3.53 \times 10^{-20}$ cm^2).

are shown in Table 1, assuming $A_{Fe} = 75$ and $z = 0.86$. The time evolution of the GRB spectra in the first 4 intervals is shown in Figure 1.

3.2   GRB 990712

Either a simple power law (PL) or a smoothly broken power-law (BL, Band at al. [12]) fit the spectra of the 8 time intervals in which we subdivided the GRB time profile, but the third time interval (C), that starts 6 s from the GRB onset and has 2 s duration. The probability that the deviation of the C spectrum from a power law is due to chance is 0.01. Possible instrumental causes of this deviation were considered and excluded. The deviation is apparent in the ratio between the GRB spectrum and the corresponding spectrum of the Crab Nebula, that was observed in the same pointing conditions. A good fit of the C spctrum is obtained by adding to the power law model (photon index α) either a broad Gaussian line or a blackbody. Figure 2 shows the best fit with a power law plus a Gaussian line. The best fit parameters of the Gaussian line and blackbody are shown in Table 2.

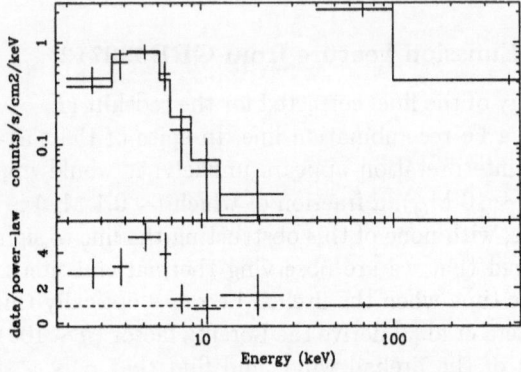

Fig. 2. *Top*: best fit of the C spectrum with a power law plus a Gaussian line. *Bottom*: residuals of the C spectrum from the best fit power law alone

Table 2. Bet fit parameters of the time interval C spectrum

Parameter	Power law + Gaussian	Power law + blackbody
α	[1.34]	[1.24]
E_{line} (keV)	4.4 ± 0.8	...
σ_{line} (keV)	1.4 ± 0.7	...
$I_{line}(\times 10^{-8}$ ergs cm^{-2} s$^{-1})$	2.7 ± 1.1	...
kT_{bb} (keV)	...	1.3 ± 0.3

4 Discussion

4.1 GRB 990705 Absorption Feature

Assuming that the absorption feature is due to a Fe K edge, its transient nature can be a consequence of the photoionization of this environment by the GRB photons. Our results point to the existence of an iron-rich environment ($A_{Fe} \sim 75$) along the line of sight to the GRB. A possible scenario[2] is that the circumburst environment (~ 0.1 pc distance from the GRB site) is uniformly enriched in Iron by the previous explosion of a supernova (I or II SN) in the same site of the GRB and the absorption feature is an Fe K edge. The transient nature of the edge is the result of the photoionization of the environment by the burst photons. A drawback of this explanation, discussed by Boettcher et al. [13], is the huge amount of Iron (tens of solar masses) required, unless the absorbing Iron is clumped and a clump is by chance along the line of sight. Another possible model is that recently discussed by Lazzati et al. [14], in which the absorption feature is a resonant absorption line due to supernova ejecta in the vicinity of the GRB, completely or quasi completely photoionized by it. The line should derive from resonant scattering of the GRB photons off H-like Iron (transition 1s-2p, $E_{rest} = 6.927$ keV). Future observations of similar features will clarify the scenario.

4.2 Transient Emission Feature from GRB990712

The centroid energy of the line, corrected for the redshift ($E_{rest} = 6.4 \pm 1.1$ keV) is consistent with a Fe recombination line. In spite of that, as discussed by us elsewhere[3], this interpretation appears unlikely: it would imply a very large mass of material (~ 10 M$_\odot$), a fraction of which (~ 0.1 M$_\odot$) is Iron, at radii of a few light seconds, with none of this obstructing the line of sight. An attractive possibility is instead that we are observing thermal emission from the fireball photosphere at the time when the fireball becomes optically thin[15]. With this assumption, Frontera et al.[3] derive the Lorentz factor ($\eta \sim 100$) and the energy ($\sim 2 \times 10^{52}$ ergs) of the fireball wind, and find that only a small fraction of this energy ($\sim$1%) is converted in electromagnetic energy during the prompt emission.

References

1. F. Frontera, L. Amati et al., Astrophys. J. Suppl. **127**, 59 (2000)
2. L. Amati, F. Frontera et al., Science **290**, 953 (2000)
3. F. Frontera, L. Amati et al., Astrophys. J. Letters **550**, L47 (2001)
4. N. Masetti, et al., Astr. and Astrophys. **354**, 473 (1999)
5. S. Holland et al., GCN Circ. 753 (2000)
6. J.G. Cohen et al., Astrophys. J. **538**, 29 (2000)
7. G. Bakos et al., IAU Circ. 7225 (1999)
8. Sahu, K.C. et al., Astrophys. J. **540**, 74 (2000)
9. Galama, T.J et al., GCN Circ. 388 (1999)
10. P.M. Vreeswijk et al., Astrophys. J. **546**, 672 (2001)
11. Rol, E. et al., Astrophys. J. **544**, 707 (2000)
12. D. Band et al., Astrophys. J. **413**, 281 (1993)
13. M. Boettcher et al., these proceedings
14. D. Lazzati et al., Astrophys. J. in press (2001) (astro-ph/0104062)
15. P. Mészáros and M. J. Rees, Astrophys. J. **539**, 292 (2000)

Discovery of a Redshifted Iron K-Line in the X-Ray Afterglow of GRB 000214

L.A. Antonelli[1], M. Vietri[2], L. Piro[3], E. Costa[3], P. Soffitta[3], M. Feroci[3], L. Amati[4], F. Frontera[4,5], E. Pian[6], J. in 't Zand[7], L. Stella[1], and G.C. Perola[2]

[1] Osservatorio Astronomico di Roma, Via Frascati, 33, 00040, Monteporzio, Italy.
[2] Universitá degli Studi "Roma Tre", Via della Vasca Navale, 84,00146, Roma, Italy
[3] IAS-CNR, Via Fosso del Cavaliere, 100, 00133, Roma, Italy
[4] ITSRE-CNR, Via Gobetti, 101, 40129, Bologna, Italy
[5] Universitá di Ferrara, Via del Paradiso, 12, 52100, Ferrara, Italy
[6] Osservatorio Astronomico di Trieste, Via G.B. Tiepolo, 11, 34131 Trieste, Italy
[7] SRON, Sorbonnelaan, 2, 3584 CA Utrecht, Netherlands

Abstract. We report on the BeppoSAX observation of the X-ray afterglow of GRB 000214 ("Valentine's Day Burst"). A strong emission line is detected (3σ significance level) in the X-ray spectrum with a centroid energy of 4.7 ± 0.2 keV. This feature, if interpreted as $K\alpha$ emission from hydrogen-like iron, corresponds to a redshift of $z = 0.47$. The intensity (EW$\sim$ 2 keV) and duration (tens of hours) of the line give information on the distance of the emitting material ($R \geq 3 \times 10^{15}$ cm) and its mass ($M \geq 1.4M_\odot$). These results are not easily reconciled with the binary merger and hypernova models for gamma ray bursts, and rather point to a SupraNova scenario.

1 Introduction

The observation of X-ray lines emitted by a Gamma Ray Burst (GRB) or its afterglow may provide a direct measurement of the redshift and a powerful diagnostic of both the nature of the central engine and the environment in which GRBs go off (*e.g.* [1] [2] and references therein). The presence of a strong iron line in the X-ray spectrum of a GRB afterglow implies an iron rich environment located very close to the GRB region and may provide an important clue in favor of collapsar models but against Neutron Star-Neutron Star mergers (*e.g.* [3] [4] and references therein). In fact, in the case of the Hypernova scenario an iron rich circumburst environment may be produced by the stellar wind before the explosion of the Hypernova [5]. A similar favorable situation is provided by the SupraNova scenario [6] where a supernova explosion precedes by a few months the GRB event ejecting a very massive iron-rich shell.

2 The "Valentine's Day Burst" X-Ray Afterglow

GRB 000214 was detected both by the Gamma Ray Burst Monitor (GRBM) and Wide Field Cameras (WFC) on-board BeppoSAX on 2000 Feb. 14. A follow-up observation with the BeppoSAX Narrow Field Instruments (NFI) began about 12 hours after the burst and lasted about 104,000 s. A previously unknown X-ray source, 1SAX J1854.4-6627, was detected in the MECS and LECS field of view within the WFC error circle[7] [8] and IPN error box [9]. The 2–10 keV source flux decreased by a factor of about two, during the BeppoSAX observation. We extracted the LECS(0.1-3.0 keV) and MECS(1.6-10.0 keV) spectra of the source from the entire BeppoSAX observation using standard procedures. We performed an accurate analysis of local background in order to check for spurious features. First we fit the data with the "canonical" model for a GRB X-ray afterglow, namely a power-law with photoelectric absorption and leaving the relative normalization of the LECS versus MECS as a free parameter. The cold absorber column density best-fit value was compatible with the Galactic value $N_H = 5.5 \times 10^{20}$ cm^{-2} [11], so we fixed it to this value. The residuals to the fit above show a clear excess around energies of $\sim 4 - 5$ keV which is highly suggestive of an emission line (see figure). The best-fit chisquare value was $\chi^2 = 27.5$ (for 15 d.o.f.). We added a narrow (with respect to MECS energy resolution), Gaussian line to the previous model inferring in this way a line centroid energy of $E = 4.7 \pm 0.2$ keV and an intensity of $I_l = (9 \pm 3) \times 10^{-6}$ photons cm^{-2} s^{-1}, translating into an E.W. of 2.1 keV. The other spectral parameters were only marginally affected by the inclusion of the Gaussian, with a power-law photon index of $\Gamma = 2.2 \pm 0.3$ and a flux of $F_x(2 - 10 \ keV) = (2.9 \pm 0.9) \times 10^{-13}$ erg cm^{-2} s^{-1}. The best fit chisquare value was $\chi^2 = 11.5$ (for 13 d.o.f.). Applying an F-test to assess the statistical significance of the Gaussian feature we obtained a chance probability of 0.27% (corresponding to 3.0σ significance level). A more detailed description of the analysis can be found in [12].

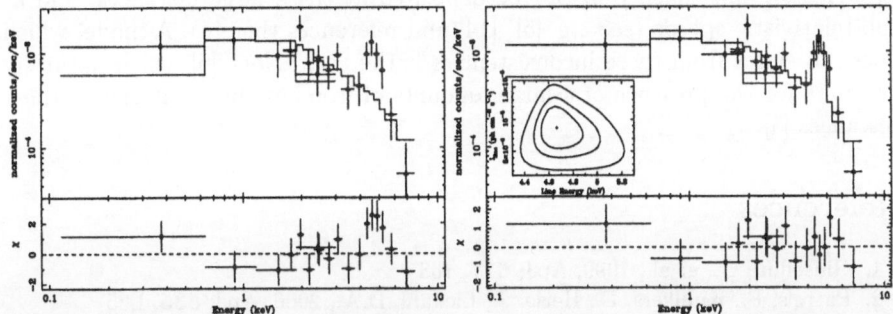

Fig. 1. BeppoSAX spectrum of GRB 000214 X-ray afterglow. LECS points (0.1–3 keV) are marked with open diamonds, MECS points (1.6–10 keV) with open triangles. Both the LECS and MECS spectra were binned so as to have at least 15 photons per bin. *Left panel:* LECS+MECS spectra fitted with an absorbed power-law; an excess around 4.7 keV is clearly seen in the residuals. *Right panel:* LECS+MECS spectra fitted with an absorbed power-law plus a narrow Gaussian line. The inset shows the contour plot of the line intensity vs energy. Contours correspond to 68%, 90% and 99% confidence levels for two interesting parameters.

3 Discussion

The only cosmologically abundant element that can produce emission lines beyond 4 keV is iron, provided one precludes Doppler shifts to shorter wavelengths. Iron emits Kα photons with an energy between 6.4 (for the lower ionization stages of iron) and 6.95 keV (for hydrogen-like iron). If we identify the emission feature in GRB 000214 as Fe Kα, the corresponding redshift is between $z = 0.37$ and 0.47. For $z = 0.47$, $H_o = 70$ km s^{-1} Mpc^{-1}, and $\Omega = 1$, we find a total energy release of $E_l = 3 \times 10^{48}$ erg in the line only, corresponding to $N_l = 3 \times 10^{56}$ photons emitted within the first 10^5 s after the burst, with the emission being nearly isotropic because the line shift implies sub-relativistic speeds. If each iron atom were to emit a single line photon, this would imply $M_{Fe} \approx 12 M_\odot$, more than even the largest stars can produce [13]. It follows that each iron atom must produce several line photons [3], and, consequently, that the time-scale t_{Fe} over which every iron atom produces a photon must satisfy $t_{Fe} \ll t_d \approx 10^5$ s, with t_d the time over which the line is observed (and does not seem to begin fading). This immediately implies that this time-scale t_d is set by geometrical time delays: the line-emitting material must be located at $R \geq 3 \times 10^{15}$ cm, and we are seeing some sizeable fraction of it. The minimum iron mass can be determined following the Lazzati et al. recipe (eq.2 in [3]) as $M_{Fe} \geq 2.4 \times 10^{-3} M_\odot$. Assuming the solar value for the iron abundance we expect, from the above equation, a minimum mass of $M \geq 1.4 M_\odot$. These values argue against a possible origin of this GRB in a neutron star–neutron star (or neutron star–black hole) binary merger or in a Hypernova. In fact, in both these models it is very difficult to explain how this much iron can be deposited at these large distances, and at sub-relativistic speeds (see e.g. [5] [14] and references therein). A model which does not suffer from these inconsistencies is the SupraNova [6] which naturally accounts for the presence of a large amounts of iron enriched material at these distances [3].

References

1. Ghisellini, G., et al., 1999, ApJ, **517**, 168
2. Paerels, F., Kuulkers, E., Heise, J., Liedahl, D.A., 2000, ApJ, **535**, L25
3. Lazzati, D., Campana, S., & Ghisellini, G., 1999, MNRAS, **304**, L31
4. Weth, C., Mészáros, P., Kallman, T., Rees, M. J., 2000, ApJ, **534**, 581
5. Mèszàros & Rees, 1998, MNRAS, **299**, L10
6. Vietri M. & Stella L., 1998, ApJ, **507**, L45
7. Paolino, A., et al., 2000, GCN Circ. #557
8. Antonelli, L.A., 2000, GCN Circ. #559
9. Hurley, K. & Feroci, M., 2000, GCN Circ. #556
10. Antonelli, L.A., et al., 2000, GCN Circ. #561
11. Dickey & Lockman, 1990, ARAA, **28**, 215
12. Antonelli, L.A., et al., 2000, ApJ, **545**, L39
13. Woosley, S.E., Weaver, T.A., 1986, ARAA, **307**, 675
14. Reisenegger, A., Goldreich, P., 1994, ApJ, **426**, 688

Observations of Iron Features with ASCA

Toshio Murakami[1], Daisuke Yonetoku[1], and Atsumasa Yoshida[2]

[1] Institute of Space and Astronautical Science,
   3-1 Yoshinodai, Sagamihara, Kanagawa 229-8510, Japan
[2] Institute of Physical and Chemical Research,
   2-1 Hirosawa Wako Saitama
   351-0198 Japan

Abstract. The origin of GRBs is an unsettled issue. Clear observational evidence can be gathered through measurements of the iron spectral feature, produced by the medium surrounding the GRB progenitor. The reported iron features with ASCA, BeppoSAX and Chandra were strong, variable and somewhat broad. The iron feature at 5.04 keV with ASCA, which was inconsistent with the OT distance, can match the OT distance, considering a recombination edge in non-equilibrium condition.

1 Introduction

Since the discovery of an X-ray afterglow [3], the distances to GRBs are known to be very remote. The observational evidences of GRBs strongly favor the fireball model [2] , which is produced by a hypernova or collapsar of a massive star as a progenitor rather than the binary merger [3]. However, clear observational evidence discriminating between these various models is lacking. The iron line is thought to be the best evidence of circumstellar matter surrounding a GRB progenitor. Although there were two early reports of iron line detections in 1998 with ASCA and BeppoSAX, these observations were statistically insignificant[3], [4] and were very much limited in the number of detections. Therefore, we require further observations to establish the existence of the iron line.

2 Iron Feature Detections

The first iron line detections were reported from BeppoSAX and ASCA and both iron lines were variable and strong [3], [4]. The iron line with ASCA in figure 1b, was detected only during the short interval, labeled B in figure 1a[4]. These lines were strong of 1 to 3 keV (EW) with a line broadening. However, neither detection was statistically significant. Moreover, the derived ASCA redshift of ~ 0.33 at the line energy of 5.04 keV, assuming an iron Kα line, did not match the OT distance of 0.9578 from the Keck observation of the host galaxy [4], [5]. Due to the low confidences and the discrepancy in the distance, the evidence for the existence of the iron line in the X-ray afterglow spectra was marginal.

Thus, ASCA has performed two more observations with better S/N than the first iron line detection of GRB 970828. However, we could not confirm the presence of an iron line, but on the contrary, could set extremely low upper limits

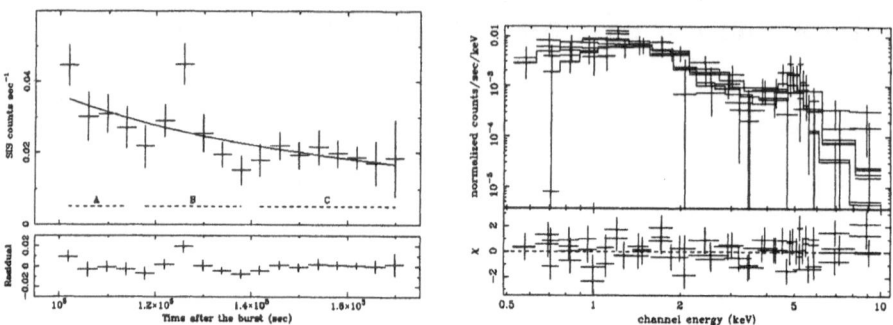

Fig. 1. (a) SIS light curve with time bins of 4000 s showing the time variation. The label B indicates the interval for which spectral study is done. (b)The X-ray spectrum for the period B, fitte   1at broadened to 0.31 keV [4].

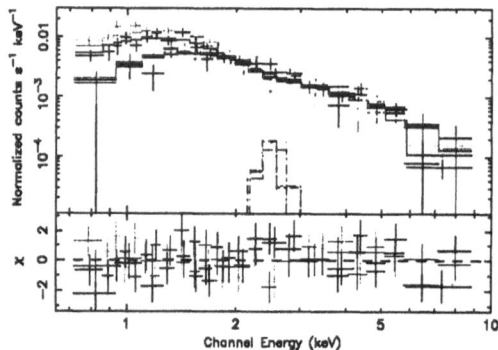

Fig. 2. X-ray spectrum of GRB 990123 combined with the four ASCA detectors into one figure. The spectrum can be explained without any iron structure at z = 1.60. The 90 % upper limit for the 6.4–6.9 keV line is 0.1 keV in EW. The line inserted in this figure is the 90% upper limit [7].

of nearly 0.1 keV (EW) for GRB 990704 and GRB 990123 in figure 2, which has a known redshift distance of z=1.60 [7].

In 1999, a new observation of GRB 991216 with Chandra resolved the question and clearly detected the iron feature at the 4.7 sigma confidence level. Moreover, Chandra detected not only the redshifted iron line, which is consistent with the OT distance, but also the recombination edge of the fully ionized Hydrogen-like iron at 9.28 keV. Soon after, BeppoSAX reported a much clearer iron line detection at the 5 sigma level from GRB 000214 [8].

3   Discussion

Although the iron lines really exist in GRBs, the iron feature is not always associated with the GRBs. The line is very rare, appearing in 10% or less of

GRBs, and also the detected lines were only present during a short interval of 10^{4-5} s in duration [3], [4].

The iron feature is confirmed, so we must solve the discrepancy in distance between ASCA and Keck. The discrepancy between the ASCA distance of $z \sim 0.33$ and $z = 0.9578$ in the OT distance can be resolved if the feature observed with ASCA was not the $K\alpha$ line but the recombination edge of Hydrogen-like iron at 9.28 keV. Although this was suggested before, it was very hard to imagine a strong recombination edge without $K\alpha$ line. In fact, the BeppoSAX observations detected the $K\alpha$ lines and the Chandra observation indicates a strong recombination edge together with a strong $K\alpha$ line [6]. However, we learned that the ASCA case can be realized in the non-equilibrium condition of a low electron temperature but a high inonization degree of irons [9]. In that condition, the recombination feature is strong but the iron $K\alpha$ line is weak. This type of model spectrum gives an acceptable fit to the ASCA iron feature (Yonetoku et al. in preparation).

Two competing ideas are proposed to explain the iron feature. One is optically thin but assumes a large emission region of about 10^{16} cm in size [10]. The other model assumes a much smaller size with a thick target of almost 10^{17} cm^{-3} in density [11]. In this case, the recombination time is so short due to the high density that we require continuous illumination where the iron should be ionized [11]. Which is better for explaining the observational facts such as the detections and many non-detections of iron feature? The short time variability favors the smaller size but requires a continuous energy source to maintain the short recombination time [11]. In summary, we can solve the discrepancy between ASCA and Keck considering the recombination edge. The iron feature's existence favors the massive collapsar type models to explain the strong iron line. However no detection of iron feature, for most of GRBs, should be explained in the same framework.

References

1. L. Piro, et al., ApJL, **514**, L73 (1999)
2. M. Rees and P. Meszaros, MNRAS, **258** S41 (1992)
3. T. Piran, Physics Report, **333**, 529 (2000)
4. A. Yoshida, et al., A&AS. **138**, 433 (1999)
5. S.G. Djorgovski, et al., ApJ. submitted, (2000)
6. L. Piro, et al., Science, **290** 953 (2000)
7. D. Yonetoku, et al., PASJ, **52**, 509 (2000)
8. L.A. Antonelli, et al., Astroph/0010221, (2000)
9. K. Masai, ApJ, **437** 770 (1994)
10. M. Vietri and L. Stella, Astroph/9910008 (1999)
11. M.J. Rees and P. Meszaros, Astroph/0010258 (2000)

Temporal and Spectral Analysis of X-Ray Afterglows of GRBs Observed by BeppoSAX

Giulia Stratta[1], Luigi Piro[1], Paolo Soffitta[1], Angelo Antonelli[2], Enrico Costa[1], Marco Feroci[1], Filippo Frontera[3], Giangiacomo Gandolfi[1], John Heise[4], Jean in 't Zand[4], Luciano Nicastro[5], and Elena Pian[6]

[1] Istituto di Astrofisica Spaziale CNR,
Via fosso del Cavaliere 100, I-00133 Rome, Italy
[2] Osservatorio Astronomico di Roma,
Via Frascati 33, 00040 Monte Porzio Catone, Italy
[3] Istituto Tecnologie e Studio Radiazioni Extraterrestri, CNR,
Via Gobetti 101, I-40129, Bologna , Italy
[4] Space research Organization in the neatherlands,
Sorbonnelaan 2, 3584 CA, Ultrecht, The Netherlands
[5] IFCAI/CNR, Via U. La Malfa 153, 90146 Palermo, Italy
[6] Osservatorio Astronomico di Trieste, Via G.B. Tiepolo 11, I-34131 Trieste, Italy

Abstract. Afterglow sources contain critical information on the origin of GRBs and on the nature of the explosion. We present a systematic study of 13 X-ray afterglows observed by BeppoSAX. The analysis aims at comparing results with standard fireball model predictions for a spherical and for a collimated fireball expansion. If the GRB energy emission is collimated we estimate a lower limit on the opening angle of the jet: $\theta > 12°$. This constrains the possibility to decrease the total GRB energy by a collimation factor of ≤ 300.

1 Introduction

The energetics of Gamma Ray Bursts (GRBs) is still one of the big quandaries concerning the nature of these phenomena. Distances derived from optical afterglow redshifts, imply enormous quantity of the total emitted energy if the emission is isotropic; otherwise, if collimated, the energy is several order of magnitude less, comparable to the most powerful Supernova. The standard fireball model [6], provides theoretical tools to distinguish jet from spherical expansion. We employ this argument in the systematic study of X-ray afterglows by fitting the observations to the model predictions of both a spherical and a collimated shock expansion.

2 How Afterglows Can Disentangle Jet vs Spherical Expansion

One of the best interpretation of GRB afterglows comes from the fireball model that interprets observed afterglow radiation as synchrotron emission. Fluxes are well described as a power law $F_\nu(\nu, t) \propto t^{-\delta}\nu^{-\alpha}$ where the temporal index δ

and the spectral index α are not independent but related by hydrodynamical considerations [4]. The standard model assumes an isotropic fireball expansion [6]. At ~ 1 day from the GRB, assuming an adiabatic evolution and a constant density environment, the model predicts for X-ray afterglows $\delta = 3\alpha/2$ for $\nu < \nu_c$ and $\delta = 3\alpha/2 - 1/2$ for $\nu > \nu_c$ where ν_c is the cooling frequency [4]. Describing the environment density n as a function of the distance from the source, $n \propto r^{-2}$, as could be in presence of a stellar wind, previous relations change only for $\nu < \nu_c$ as $\delta = 3\alpha/2 + 1/2$ [1]. An extension of the standard model analyzes afterglow behaviours in the case of a collimated expansion of the fireball. In this case $\delta = 2\alpha + 1$ for $\nu < \nu_c$ and $\delta = 2\alpha$ for $\nu > \nu_c$ [5]. By plotting δ and α obtained from X-ray afterglow sources analysis on a δ vs α plane and by evaluating which one of the expected fireball model relations better describes data distribution, we have a valid test for discriminating spherical from jet fireball expansion.

3   The Sample

We have analyzed X-ray afterglow sources observed by BeppoSAX fast follow up from 1997 January to 1999 May. In this period, 16 X-ray sources were detected and recognized as GRB afterglows by the Narrow Field Instruments (NFIs). For 10 afterglows the S/N ratio was high enough to perform an adequate temporal and spectral analysis with NFI data only, namely GRB970228, GRB970402, GRB971214, GRB980329, GRB980519, GRB980613, GRB980703, GRB981226, GRB990123 and GRB990510. In a second step of our analysis we added the Wide Field Camera (WFC) observations in temporal fitting. WFC data, integrated in the same MECS energy range, were introduced to better constrain temporal decay index measurements and to include in the analysis those sources that were too faint to be analyzed only with NFI observations as GRB970111, GRB971227, GRB980425, GRB980515 and GRB990217. The WFC plus NFI data increased the sample to 13 sources. We have excluded from the analysis GRB970508 because of the complex temporal behaviour [2].

4   Methods

A fit of spectral LECS and MECS data (0.1-10.0 keV) in each first Target Of Opportunity (TOO) has been performed using a power law function with a photoelectric absorption component. Light curves were fitted with a power law function for MECS countrates (2.0-10.0 keV) while LECS temporal data were too faint to perform a statistically significant analysis. We finally plotted the best fit spectral and decay indexes in the plane δ vs α (1σ errors).

5   Results

In Fig.1, δ and α are compared within the frame of spherical (S) and jet (J) fireball model in a constant density environment and of spherical in presence of

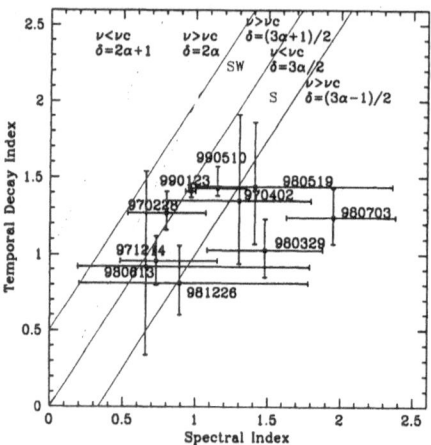

Fig. 1. X-ray afterglow distribution on the δ vs α plane

wind (SW). Afterglows are distributed in this plane along the curves predicted from the fireball model under the hypothesis of an isotropic emission with most probable case for fast cooling (0.5σ). Jet area is on the contrary leaved empty (6.7σ). Against the possibility of an instrumental selectivity effect is the fact that all the X-ray GRB/afterglows detected by the WFC, were detected by the NFIs except two. Moreover by adding WFC temporal data, decay index upper limits were measured for the fainter sources and results are consistent with the general distribution in the plane δ vs α, at least for the temporal domain.

6   Conclusions

We stress that this result does not imply an isotropic energy emission, because of the relativistic expansion of the source of X-ray afterglow. Infact, our result says that jet angle θ_{jet} can not be less than the beaming angle $1/\Gamma(t)$ where $\Gamma(t)$ is the fireball Lorentz factor value at the moment of the observed X-ray afterglow. Using the expression that describes $\Gamma(t)$ with time [4] we have $\theta_{jet} \geq 0.1[\frac{t}{6.2}(\frac{n}{E^{52}})^{1/3}]^{3/8}$ [5]. Assuming a constant density $n = 1$ cm^{-3} and a released total energy $E = 10^{52}$erg, at $\sim$48 hr from the burst the collimation angle must be $\geq 12°$. This imply a collimation factor ≤ 300, that is, GRB energy cannot be reduced more than 3 order of magnitude.

References

1. L.A. Chevalier, Z.Y. Li: ApJL **520**, 29 (1999)
2. L. Piro et al.: A&AL **331**, 41 (1998)
3. M.J. Rees, P. Meszaros: MNRAS **258**, 41 (1992)
4. R. Sari, T.Piran, R. Narayan: ApJL, **497**, 17 (1998)
5. R. Sari, T.Piran, J.P. Halpern: ApJL, **519**, 17 (1999)

Optical/Near-IR Observations
of Gamma-Ray Bursts in the Afterglow Era

Alberto J. Castro-Tirado[1,2]

[1] Instituto de Astrofísica de Andalucía (IAA-CSIC),
P.O. Box 03004, E-18080 Granada, Spain
[2] Laboratorio de Astrofísica Espacial y Física Fundamental (LAEFF-INTA),
P.O. Box 50727, E-28080 Madrid, Spain

Abstract. An overview of the optical and near-IR observations of GRBs in the After-glow Era is presented. They have allowed to a better understanding of the underlying physics as well as to constraint the progenitor models.

1 Afterglow Observations

1.1 Photometric Observations

Power-Law Declines. *BSAX* made possible to detect the first X-ray afterglow following GRB 970228 whose precise localization (1′) led to the discovery of the first optical transient (or optical afterglow, OA) associated to a GRB [1,2]. The OA was afterwards found on earlier images [3,4] and the light curve exhibited a power-law (PL) decay $F \propto t^{-\alpha}$ with $\alpha = 1.1$ [5], thus confirming the prediction of the relativistic blast-wave model [6]. PL declines have been measured for 26 OAs in 1997-2000 yielding values in the range $0.8 < \alpha < 2.3$ with $< \alpha > = 1.35$.

Breaks and Steepening in Light Curves. A break deviating from a PL decay was first observed in the GRB 990123 light curve at $T_0 + 1.5\,d$ (~ 1.5 days after the high energy event) and it was interpreted as the presence of a beamed outflow with a half opening angle $\theta_0 \sim 0.1$ [7-9], i.e. reducing the inferred energy by a factor of $\sim \theta_0^2/4$. Further breaks have been reported in another 5 GRBs: GRB 990510 [10-11], GRB 991208 [12], GRB 991216 [13,14], GRB 000301C [15,16] and GRB 000926 [17,18].

There are six possible explanations for the observed breaks: i) sideways expansion of the jet caused by the swept-up matter [19,20] although the effect might be negligible for $\theta_0 > 0.1$ [21]; ii) when the jet material propagates in a uniform density medium and the observer sees the edge of the jet, with α increasing by ~ 0.7 [22]; iii) when the jet material propagates in a medium with a PL density profile and the observer sees the edge of the jet, α increases by a factor of Γ^2 [22] with Γ the bulk Lorentz factor; iv) when the jet material propagates within a pre-ejected stellar wind ($\rho \propto r^{-2}$) and the observer sees the edge of the jet, α increases by ~ 0.4 [22]; v) when in both the relativistic and non-relativistic cases (if ρ is high) the inverse Compton scattering is important, the light curves can be flattened or steepened [21]; and vi) when the transition from the relativistic phase to the non-relativistic phase of an isotropic blastwave takes place in a dense medium ($\rho \sim 10^6$ cm^{-3}) [23].

Rapid fading ($\alpha > 2.0$) has been observed in 3 GRBs: GRB 980326 [24], GRB 980519 [25,26] and GRB 991208 [12]. Two possible causes can explain such behaviour: the synchrotron emission during the mildly relativistic and non-relativistic phases [23] and the interaction of a spherical burst with a pre-burst Wolf-Rayet star wind [27,28].

"Plateau" States. For GRB 970508, a "plateau" ($\alpha = 0$) was observed between $T_0 + 3$ hr and $T_0 + 1$ d [29,30]. The optical light curve reached a peak in two days [31,32] and was followed by a PL decay $F \propto t^{-1.2}$. The "plateau" has been explained by several plasmoids with different fluxes occurring at different times [33]. Another "plateau" was detected in the near-IR light curve of GRB 971214 between $T_0 + 3$ hr and $T_0 + 7$ hr [34].

Short-Term Variability. Flux fluctuations are expected due to inhomogeneities in the surrounding medium as a consequence of interstellar turbulence or by variability and anisotropy in a precursor wind from the GRB progenitor [35]. However, short-term variability was found neither in GRB 970508 [30] nor in GRB 990510 [11]. In GRB 000301C, the high variability observed at optical wavelengths [14,15] can be due to several reasons: i) refreshed shock effects; ii) energy injection by a strongly magnetic millisecond pulsar born during the GRB [36]; iii) an ultra-relativistic shock in a dense medium rapidly evolving to a non-relativistic phase [37]; and iv) a gravitational microlens [38].

The SN-GRB Association. A peculiar type Ib/c supernova (SN 1998bw) was found in the error box for the soft GRB 980425 [39] coincident with a galaxy at $z = 0.0085$, but this SN/GRB relationship is still under debate.

In any case, "SN-like" bumps have been detected in other GRBs: GRB 970228 [40,41], GRB 970508 [42], GRB 980326 [43,44], GRB 980703 [45], GRB 991208 [12] and GRB 000418 [46]. There are some alternative explanations for the existence of such a bump in the OA light curves: i) scattering of a prompt optical burst by 0.1 M$_\odot$ dust beyond its sublimation radius 0.1-1 pc from the burst, producing an echo after 20-30 d [47]; ii) delayed energy injection by shell collision [48]; and iii) an axially symmetric jet surrounded by a less energetic outflow [49]. But this is certainly not the case for all GRBs: GRB 990712 provided the first firm evidence that an underlying SN was not present [50].

1.2   Spectroscopy

GRB 970508 was the clue to the distance: optical spectroscopy obtained during the OA maximum brightness allowed a direct determination of a lower limit for the redshift ($z \geq 0.835$), implying $D \geq 4$ Gpc (for $H_0 = 65$ km s^{-1} Mpc^{-1}) and $E \geq 7 \times 10^{51}$ erg. It was the first proof that GRB sources lie at cosmological

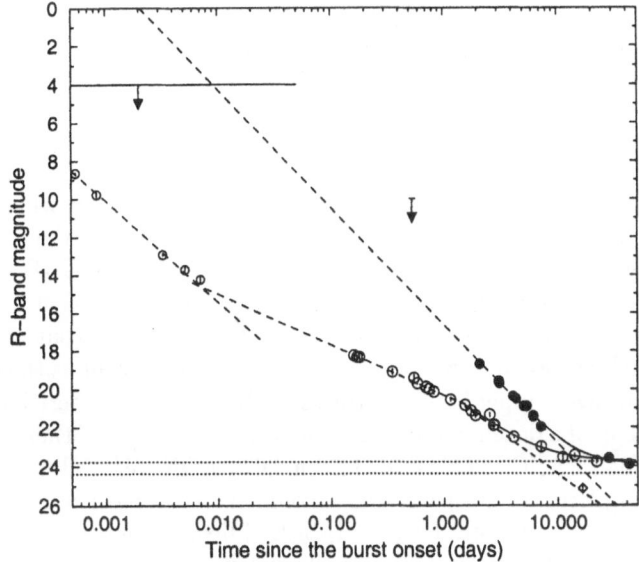

Fig. 1. The two brightest optical GRB afterglows detected so far: GRB 990123 (empty circles) and GRB 991208 (filled circles). The dotted lines are the constant contribution of the two host galaxies, R ~ 23.9 and 24.3 respectively. The dashed-lines are the pure OAs contributions to the total fluxes. The solid lines (only shown here after T > 5 d) are the combined fluxes (OA plus underlying galaxy on each case). Upper limits are for GRB 991208. Adapted from [12].

distances [51]. The flattening of the decay at T_0 + 100 d [30,52] revealed the contribution of a constant brightness source – the host galaxy – seen in late-time imaging at T_0 + 1 yr. The 15 GRB redshifts measured so far are in the range $0.430 \leq z \leq 4.50$ [53] with $< z > = 1.5$ and they were derived either from absorption lines in the OA spectrum, from the Ly-α break, or from emission lines arising in the host galaxy.

1.3 Polarimetry

As synchrotron radiation under favourable conditions can be up to 70 % polarized, the first polarimetric observations were attempted for GRB 990123, but only an upper limit was established ($\Pi < 2.3\%$) [54]. Polarized optical emission was first detected in GRB 990510 ($\Pi = 1.7 \pm 0.2\%$) by means of observations performed at T_0 + 18.5 hr [55], T_0 + 21 hr and T_0 + 43 hr [56]. This confirmed the synchrotron origin of the blast wave itself and represented another case for a jet-like outflow [11]. Further polarization measurements were carried out in GRB 990712 during a 1 d time interval (at T_0 + 10 hr, T_0 + 17 hr and T_0 + 35 hr). In that case, the polarization angle did not vary significantly but the degree of polarization was not constant [18] which can be explained by a laterally expanding jet [57]. See also [58,59].

2   Near-Simultaneous GRB Observations

Significant early optical emission may arise from the reverse shock [27,60], i.e. strong optical/near-IR flashes accompanying gamma-ray emission should be a generic characteristic (at least for typical GRBs, with $E \sim 10^{53}$ erg and $\rho \sim 1$ cm^{-3} [61]). Such observations will allow: i) to derive Γ by the relative timing of optical and γ-ray emission, ii) to pinpoint the process by which the shells responsible for the external shock arise, and iii) to constraint the environment.

ROTSE [62] achieved the detection simultaneously to the GRB of the bright optical emission from GRB 990123 [63]: the most luminous object ever recorded (Fig. 1), with $M_V = -36$ (peaking at $m_V = 8.9$), implying that at least some subsets of GRBs do exhibit variable optical emission as violent as the gamma-ray variations. However, upper limits (in the range $R = 4\text{-}15$) were derived for prompt optical emission of a dozen of bursts by means of ROTSE and other experiments, like LOTIS [64], TAROT [65], BOOTES [66,67], EON [68] and CONCAM [69]. Thus, bright optical counterparts are uncommon. Why most optical flashes are not detected? Due to several reasons: i) lack of deeper coverage, given the wide GRB luminosity function [70]; ii) fireballs in low-density environments ($\rho \ll 1$ cm^{-3}) would not be expected to produce strong prompt emission; iii) the reverse shock energy is radiated at a non-optical frequency, with the synchrotron peak frequency $\nu_m \gg \nu_{opt}$ or $\nu_m \ll \nu_{opt}$ [71]; and iv) highly absorbed GRB by dust in their host galaxies.

3   "Dark" GRBs

The first "dark" event (GRB 970828) was detected as a fading X-ray source [72], although no optical counterpart down to $m_R = 24$ at $T_0 + 4\ hr$ was detected [73]. At least in another three cases (GRB 981226, GRB 990506 and GRB 001109), radiotransients were detected without accompanying optical/IR transients. For GRB 000210, the X-ray position ($1''$ accuracy) coincides with a faint galaxy [74].

About 40% of the GRBs with X-ray counterparts do not show OAs, and this could be due to: i) intrinsic faintness because of a low ambient medium; ii) Lyman limit absorption in high redshift galaxies ($z > 7$); and iii) high absorption in a dusty enviroment: if GRBs are tightly related to star-formation, a substantial fraction of them should occur in highly obscured regions.

4   Summary

The first optical/near-IR counterparts have been found for ~ 30 precisely localized GRBs in 1997-2000 although they should have been discovered prior to the *BSAX* launch (Fig. 2). In any case, only the population of GRBs with durations of few seconds has been explored (see [76] for a more extensive review). Short bursts lasting less than 1 s, that follow the $-3/2$ slope in the log N-log S diagram (in contrast to the longer bursts) remain to be detected at longer wavelengths. Future missions should be able to address some of the issues still to be solved, i.e. prompt optical and near-IR observations should be persued !

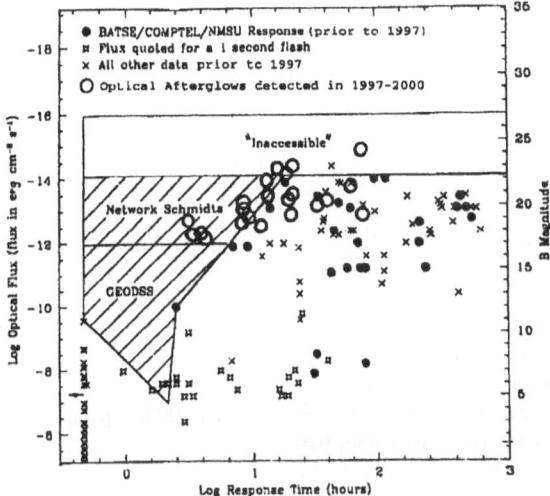

Fig. 2. The GRB follow-up response prior to 1997 (based on [75]) compared to the detection of optical counterparts in the Afterglow Era (since 1997). It is clearly seen that the fact the no optical afterglows were discovered prior to the BSAX launch was just a matter of bad luck (i.e. the ∼ 10 events for which prompt deep follow-up optical searches were performed seemed to be either "rapidly-fading" or "dark" GRBs).

References

1. Groot P. J. et al. 1997, IAU Circ. 6584
2. van Paradijs, J. et al. 1997, Nat 386, 686
3. Pedichini, F. et al. 1998, A&A 327, L36
4. Guarnieri, A. et al. 1997, A&A 328, L13
5. Galama, T. et al. 1997, Nat 387, 479
6. Sari, R., Piran, T. and Narayan, R. 1998, ApJ 497, L17
7. Castro-Tirado, A. J. et al. 1999, Sci 283, 2069
8. Fruchter, A. et al. 1999, ApJ 519, L13
9. Kulkarni, S. et al. 1999, Nat 398, 389
10. Harrison, F. et al. 1999, ApJ 523, L121
11. Stanek, K. Z. et al. 1999, ApJ 522, L39
12. Castro-Tirado, A. J. et al. 2001, A&A, in press (astro-ph/0102177)
13. Halpern, J. P. et al. 2000, ApJ 543, 697
14. Sagar, R. et al. 2000, BASI 28, 499
15. Masetti et al. 2000a, A&A 359, L23
16. Jensen, B. L. et al. 2001, A&A, in press (astro-ph/0005609)
17. Fynbo, J. P. et al. 2001, A&A, in press (astro-ph/0102158)
18. Rol, E. et al. 2000, ApJ, 544, 707
19. Rhoads, J. E. 1999, ApJ 525, 737
20. Sari, R. et al. 1999, ApJ 519, L17
21. Wei, D. M. and Lu, T. 2001, MNRAS 320, 37
22. Kumar, K. and Panaitescu, A. 2000, ApJ 541, L51
23. Huang, Y. F., Dai, Z. G. and Lu, T. L. 2000, MNRAS, 316, 943
24. Groot P. J. et al. 1998, ApJ 502, L123

25. Halpern, J. P. et al. 1999, ApJ 517, L105
26. Vrba, F. J. et al. 2000, ApJ 528, 254
27. Mészáros, P. and Rees, M. J. 1997, ApJ 476, L232
28. Chevalier, R. A. and Li, Z.-Y. 1999, ApJ 520, L29
29. Pedersen, H. et al. 1998, ApJ 496, 311
30. Castro-Tirado, A. J. et al. 1998, Sci 279, 1011
31. Djorgovski, S. G. et al. 1997, Nat 387, 876
32. Galama, T. et al. 1998a, ApJ 497, L13
33. Dar, A. and De Rújula, A. 2001, A&A, in press (astro-ph/0008474)
34. Gorosabel, J. et al. 1998, A&A 335, L5
35. Wang, X. and Loeb, A. 2000, ApJ 535, 788
36. Dai, Z. G. 2000, in *Explosive Phenomena in Astrophysical Compact Objects*, Seoul, Korea (May 2000), in press (astro-ph/0008304)
37. Dai, Z. G. and Lu, T. 2001, A&A 367, 501
38. Garnavich, P. M., Loeb, A. and Stanek, K. Z. 2000, ApJ 544, L11
39. Galama, T. et al. 1998, Nat 395, 670
40. Reichart, D. E. 1999, ApJ 521, L111
41. Galama, T. et al. 2000, ApJ 536, 185
42. Sokolov, V. V. et al. 2001, These Proceedings
43. Castro-Tirado, A. J. and Gorosabel, J. 1999 A&AS 138, 449
44. Bloom, J. et al. 1999, Nat 401, 453
45. Holland, S. et al. 2001, A&A, in press (astro-ph/0103058)
46. Klose, S. et al. 2000, ApJ 545, 271
47. Esin, A. A. and Blandford, R. 2000, ApJ 534, L151
48. Kumar, P. and Piran, T. 2000, ApJ 535, 152
49. Panaitescu, A., Mészáros, P. & Rees, M. J. 1998, ApJ 503, L314
50. Hjorth, J. et al. 2000, ApJ 539, L147
51. Metzger, M. R. et al. 1997, Nat 387, 878
52. Sokolov, V. V. et al. 1998, A&A 334, 117
53. Andersen, M. I. et al. 2000, A&A 364, L54
54. Hjorth, J. et al. 1999, Sci 283, 2073
55. Covino, S. et al. 1999, A&A 348, L1
56. Wijers, R. A. M. J. et al. 1999, ApJ 523, L33
57. Björnsson, G. and Lindfors, E. J. 2000, ApJ 541, L55
58. Ghisellini, G. and Lazzati, D. 1999, MNRAS 309, L7
59. Sari, R. 1999, ApJ 524, L43
60. Sari, R. and Piran, T. 1999, ApJ 520, 641
61. Waxman, E. and Draine, B. T. 2000, ApJ 537, 796
62. Akerlof, C. et al. 2000, ApJ 532, L25
63. Akerlof, C. et al. 1999, Nat 398, 400
64. Park, H.-S. et al. 1999, A&A 138, 577
65. Boer, M. et al. 1999, A&AS 138, 579
66. Castro-Tirado, A. J. et al. 1999, A&AS 138, 583
67. Castro Cerón, J. M. et al. 2001, These Proceedings
68. Hudec, R. et al. 1998, in *The BL Lac Phenomenon*, Turku (Jun 1998), p. 101.
69. Nemiroff, R. J. and Bruce, R. J. 1999, PASP 111, 886
70. Hogg,D. W. and Fruchter, A. S. 1999, ApJ 520, 54
71. Kobayashi, S. 2000, ApJ 545, 807
72. Yoshida, A. et al. 1999, A&AS 138, 433
73. Groot P. J. et al. 1998, ApJ 493, L27
74. Gorosabel, J. et al. 2000, GCN Circ. 783
75. McNamara, B. J. et al. 1995, Ap&SS 231, 251
76. Castro-Tirado, A. J. 2001, ESA-SP Conf. Proc., in press (astro-ph/0102122)

The GRB Followup Euro–US Consortium: Results from the ESO Telescopes

Nicola Masetti, on behalf of a large collaboration

Istituto TeSRE, CNR, via Gobetti 101, I-40129 Bologna, Italy

Abstract. In October 1997, the Italian and Dutch GRB teams started a collaboration on ESO optical follow-up of rapidly and accurately localized GRBs. Subsequently, starting April 1, 2000, this collaboration was extended to astronomers from other countries, who contributed their expertise for the creation of a Consortium committed to the study of GRB counterparts and host galaxies at optical and near-infrared wavelengths. The collaboration aims at the joint exploitation of the observations taken within an ESO Large Programme approved for the two-year period April 1, 2000–March 31, 2002. Here we describe history and organization of this Consortium, the goals of the ESO Large Programme, and the main results obtained up to now with ESO telescopes.

1 Introduction and History

The availability of fast (few hours after the GRB trigger) and precise (arcminsized) GRB localizations, first afforded by the Italian-Dutch X-ray satellite *BeppoSAX* in 1996 and subsequently by other spacecraft, allowed astronomers to pinpoint Optical Transients (OTs) associated with the high energy events and to effectively explore the physics behind these phenomena. In this contribution we briefly outline history, status and results of the search and followup of GRB optical and near-infrared (NIR) counterparts at ESO since the beginning of the *BeppoSAX* afterglow era.

The first ESO ToO observation of a fast and precise GRB position was activated on January 1997 at NTT on the GRB970111 error box, but no optical counterpart was detected. The first detection of an OT associated with a GRB was achieved at ESO about two months later, on 1997 March 13, with NTT observations of the GRB970228 error box [6], the first GRB for which X-ray and optical afterglows were discovered.

Subsequently, thrusted by this outstanding scientific achievement, the Italian and Dutch GRB search and followup teams, led by Filippo Frontera and Jan van Paradijs, respectively, independently submitted regular ESO proposals for Period 60 (October 1997–March 1998) for the activation of ToO observations at several ESO telescopes on GRB error boxes observable from Chile. Both proposals were approved and, following the suggestion from ESO, the Italian and Dutch groups decided to start collaborating in this search and to form a single team.

The Italian-Dutch collaboration in the GRB follow-up at the ESO telescopes was organized in a way that the two groups alternated in the program lead at every trigger of an accurately and rapidly localized GRB, independent of spacecraft. The leadership of the group 'on duty' encompassed every step from alerting the telescopes to taking responsibility of results publication.

Besides the Italian-Dutch collaboration, a parallel proposal set up by Danish, Spanish and German teams was active at several ESO telescopes.

During Periods 60 to 64 (October 1997–March 2000) ESO observations placed several milestones in the study of GRB optical afterglows, in particular: the detection of SN 1998bw in the error box of GRB980425, the only case in which a SN was clearly detected in the field of a GRB within stringent temporal and spatial limits [2]; the discovery and monitoring of optical polarization in the OTs of GRB990510 [9] and GRB990712 [5]; the determination of the redshift of GRBs 990510, 990712 [7] and 991216 [8] using VLT-Antu; the NTT detection of two very red afterglows associated with GRB980329 [4] and GRB990705 [3], their color being most likely due to high local absorption in the host galaxy or to high redshift; and the discovery of the farthest GRB observed so far, located at redshift $z = 4.5$ [1]. Also, in several cases the detection and the spectrophotometric observations of host galaxies associated with GRBs were accomplished with VLT-Antu and NTT ESO telescopes.

2   The ESO Large Programme

In the fall of 1999, the Italian-Dutch collaboration was extended to three more European countries, i.e. Denmark, Germany and Spain, where groups committed to the search and followup of GRB OTs were operating and on several occasions had already worked together with the Italian and Dutch teams.

This led to the formation of a Consortium and to the submission of a joint ESO proposal for a Large Programme (LP) of GRB follow-up, spanning 2 years. This programme, formerly conceived and fostered by Jan van Paradijs, has been approved and became effective on April 1, 2000 (start of Period 65), with Ed van den Heuvel as Principal Investigator (PI).

In order to exploit the high detection rate and the unprecedented localization capabilities of HETE-II, which will allow increasing the number of accurate GRB locations observable from ESO, the Large Programme has been structured in order to achieve the following aims: (i) determining the redshift distributions of GRBs; this programme may increase the redshift sample by a factor of two; (ii) studying the nature of circumsource environment, and probing the 'dark' afterglow population; (iii) establishing the reality and nature of GRB-supernova connection; (iv) constraining the physics of the 'fireball' model; (v) performing the first-ever follow-up observations of the subclass of short-duration GRBs.

In the European Collaboration each of the 5 nodes has a contact person, or 'captain', who acts as a link among his/her node, the PI and the other captains. These persons are: Alberto Castro-Tirado (Spanish node), Jochen Greiner (German node), Jens Hjorth (Danish node), Elena Pian (Italian node) and Paul Vreeswijk (Dutch node).

Besides the ESO LP, an informal agreement among the nodes of the European Consortium also exists for collaborating at other (non-ESO) telescopes, mainly located in the Northern Hemisphere and at which each single node has an approved GRB ToO proposal. This is to guarantee a coverage as complete as possible in terms of wavelength and of temporal baseline for each single GRB.

Concerning the guidelines for the activation of ESO telescopes, it was agreed that the 5 nodes rotate so that the node on duty activates the ToO request soon after the notification of a fast and precise GRB localization and after having consulted the PI and the captains of the other nodes. Each node on duty is supported by a 'backup' node in case of need (e.g. when extra manpower is required due to a very large and concentrated data flow).

As a rotation criterion, it was first decided that the turn of the ESO ToO activation was shifted from one node to the following after having detected an OT with VLT; then the policy was changed so that, effective November 1, 2000, each node is on duty for 15 days. Then, the turn passes to the next node in line. Again, the node on duty has the direct responsibility of the data reduction, analysis and publication. In December 2000 the European Consortium was extended to a sixth, 'North-Atlantic', group (with Andy Fruchter as captain) formed by astronomers from US and British institutes who were already included in the LP as co-investigators.

3 Results from ESO LP

Up to the end of December 2000, ESO LP observations were activated 9 times. In 5 cases (GRBs 000528, 000529, 000607, 000801 and 001204) no afterglow was detected in optical or NIR; in one case (GRB000830) a faint and very red fading object was discovered, but its nature still needs confirmation; in three cases (GRBs 000911, 001007 and 001011) an OT was detected and monitored. Work on these objects is still in progress.

Besides the search and monitoring of OTs, the Euro-US GRB consortium is also involved in the study of the host galaxies of GRBs detected so far and observable from La Silla and Paranal. Indeed, up to one third of the total time allocated to the LP is devoted to spectrophotometry of hosts, in order to determine their main parameters (redshift and star-formation rate) and to construct their optical/NIR broadband spectral energy distributions.

References

1. M.I. Andersen, J. Hjorth, H. Pedersen et al.: Astron. Astrophys. 364, L54 (2000)
2. T.J. Galama, P.M. Vreeswijk, J. van Paradijs et al.: Nature 395, 670 (1998)
3. N. Masetti, E. Palazzi, E. Pian et al.: Astron. Astrophys. 354, 473 (2000)
4. E. Palazzi, E. Pian, N. Masetti et al.: Astron. Astrophys. 336, L95 (1998)
5. E. Rol, R.A.M.J. Wijers, P.M. Vreeswijk et al.: Astrophys. J. 544, 707 (2000)
6. J. van Paradijs, P.J. Groot, T.J. Galama et al.: Nature 386, 686 (1997)
7. P.M. Vreeswijk, A.S. Fruchter, L. Kaper et al.: Astrophys. J. 546, 672 (2001)
8. P.M. Vreeswijk, A.S. Fruchter, E. Palazzi et al.: Astrophys. J., submitted (2001)
9. R.A.M.J. Wijers, P.M. Vreeswijk, T.J. Galama et al.: Astrophys. J. 523, L33 (1999)

GRB 000301C: A Possible Short/Intermediate Duration Burst Connected to a DLA System

J. Gorosabel[1,2], J.U. Fynbo[2], B.L. Jensen[3], P. Møller[2], H. Pedersen[3], J. Hjorth[3], M.I. Andersen[4], and K. Hurley[5]

[1] Danish Space Research Institute,
    Juliane Maries Vej 30, DK-2100 Copenhagen Ø, Denmark
[2] European Southern Observatory,
    Karl-Schwarzschild-Straße 2, D-85748 Garching, Germany
[3] Astronomical Observatory, University of Copenhagen,
    Juliane Maries Vej 30, DK-2100 Copenhagen Ø, Denmark
[4] Division of Astronomy, P.O. Box 3000, FIN-90014 University of Oulu, Finland.
[5] Space Science Laboratory, University of California at Berkerley, USA.

Abstract. We discuss two main aspects of the afterglow GRB 000301C [1,2]; its short duration and its possible connection with a Damped Lyα Absorber (DLA). GRB 000301C falls in the short class of bursts, though it is consistent with belonging to the proposed intermediate class or the extreme short end of the distribution of long-duration GRBs. Based on two VLT spectra we estimate the H I column density to be log(N(H I))= 21.2 ± 0.5. This is the first direct indication of a connection between GRB host galaxies and Damped Lyα Absorbers.

1 Introduction

GRB 000301C was localised by the Inter Planetary Network (IPN) and RXTE to an area of $\sim$50 arcmin^2. A fading optical counterpart was subsequently discovered with the Nordic Optical Telescope (NOT) about 42 h after the burst. The GRB was recorded by the Ulysses GRB experiment and by the NEAR X-Ray/Gamma-Ray Spectrometer. From the NEAR data we estimate the 150–1000 keV fluence to be approximately 2×10^{-6} erg cm^{-2}. The IPN/RXTE errorbox of GRB 000301C [3] was observed with the 2.56-m Nordic Optical Telescope (NOT) on 2000 March 3.14–3.28 UT ($\sim$1.8 days after the burst) using ALFOSC. Comparing with red and blue Palomar Optical Sky Survey II exposures, a candidate Optical Transient (OT) was found at the position $(\alpha, \delta)_{2000}$=(16^h 20^m 18.56^s, +29° 26′ 36.1″).

2 The First Short GRB Optical Counterpart Detection?

As measured by both Ulysses and NEAR, in the >25 keV energy range, the duration of this burst was approximately 2 s. GRB 000301C falls in the short class of bursts, though it is consistent with belonging to the proposed intermediate class or the extreme short end of the distribution of long-duration GRBs. We obtain a hardness ratio of 2.7±0.6(cutoff)±30%(statistical error, see [2] for

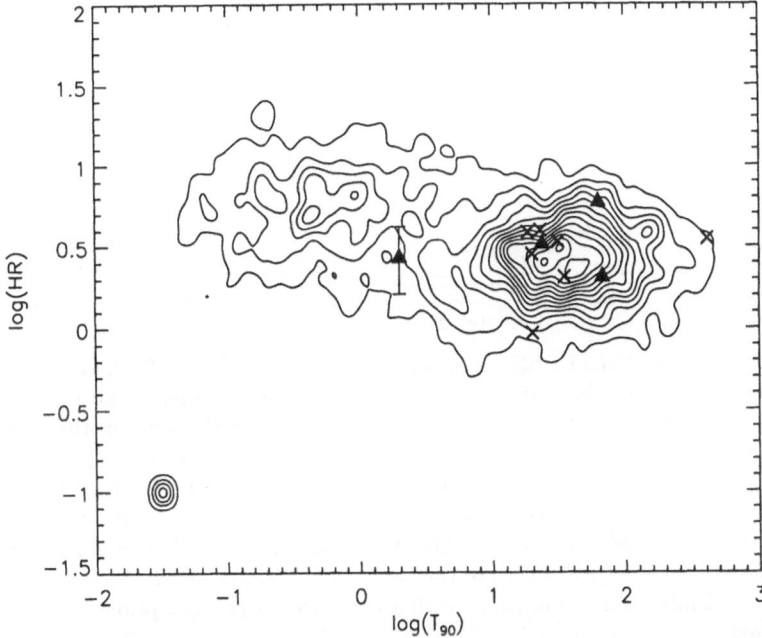

Fig. 1. A contour plot showing the duration-hardness ($\log(T_{90})$–$\log(HR_{32})$) distribution of 1959 BATSE bursts. The triangle with an error-bar near the center of the plot represents GRB 000301C. As seen, the burst was located in the short/intermediate duration part of the distribution. Other symbols represent 10 other BATSE bursts with identified optical counterparts for which data on fluence and duration are available. Triangles are bursts which have a break in their optical light curves. Errors in the BATSE data are smaller than the symbol size. Contour levels scale linearly. The point in the lower left corner illustrates the resolution of the contours.

details on the calculation of the hardness ratio). Fig. 1 shows the location of GRB 000301C in a hardness vs. duration plot.

3 The First GRB-DLA Connection?

Spectroscopic observations were carried out on 2000 March 5 and 6 UT with VLT-Antu equipped with FORS1. Fig. 2 shows the normalized spectrum of the OT. Following the procedure explained in [2] we obtained a redshift of $z_{abs} = 2.0404 \pm 0.0008$. The oscillator strength weighted mean observed equivalent width of the Fe II lines is 2.56 Å, which is strong enough that by comparison to known quasar absorbers one would expect this to likely have a column density of neutral Hydrogen in excess of $2 \times 10^{20} \mathrm{cm}^{-2}$. Such absorbers are known as Damped Lyα Absorbers (DLAs), and hold a special interest because of the large amounts of cold gas locked up in those objects [4]. On the other hand we found that the spectrum drops steeply before the expected central position of the Lyα line, and

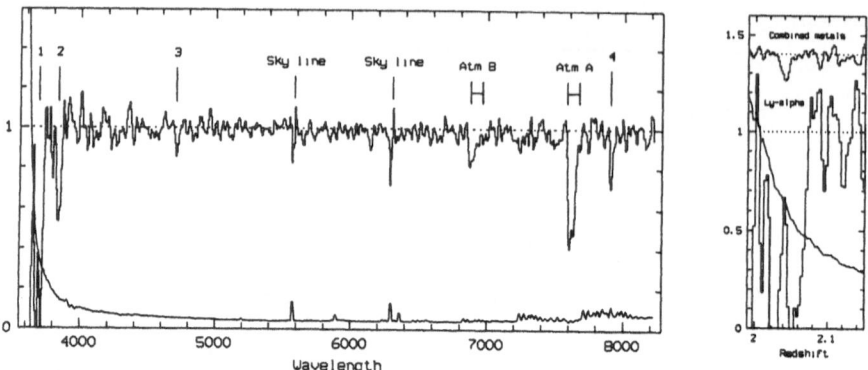

Fig. 2. Combined VLT+FORS1 spectrum of GRB 000301C from 2000 March 5+6 UT. The spectrum is normalized to 1 in the continuum. The atmospheric absorption bands and residuals from strong sky-lines are marked, as well as the 4 absorption lines. The spectrum is binned to 7 Å pixels, and the lower curve shows the noise (per pixel). The detected lines are: Lyα (1), SI II (2), C IV (3) and Fe II (4). On the right side we can see a blow up of the Lyα region in the redshift space. We have overplotted the oscillator strength weighted mean of metal lines lines (Fe II, Si II and C II). The lower curve shows the noise per bin. Note the very sharp onset of absorption well above the expected redshift. This is consistent with a very broad Lyα absorption line as detailed in the text.

well before the S/N drops below detection (see right side plot of Fig. 2). One likely explanation for this is the presence of a very broad Lyα absorption line. To quantify this we have modelled several Lyα absorption lines, all at redshift 2.0404. The formal χ^2 minimum is found at N(H I)= 1.5×10^{21}cm^{-2} (χ^2 per DOF = 0.86), but any value within a factor 3 of this is acceptable.

4   Conclusions

GRB 000301C is so far the GRB of shortest duration, for which a counterpart has been detected. The high-energy properties of the burst are consistent with membership of the short-duration class of GRBs, though GRB 000301C could belong to the proposed intermediate class of GRBs or the extreme short end of the distribution of long-duration GRBs. We argue that there may be a connection between the host galaxy of GRB 000301C and DLAs.

References

1. Fynbo, J., Jensen, B.L., Hjorth, J., et al., 2000, GCN 570.
2. Jensen, B.L., Fynbo, J., Gorosabel, J., et al., 2001, A&A 370, 909.
3. Smith D.A., Hurley K., Cline T., 2000, GCN 568.
4. Storrie-Lombardi L.J., Irwin M.J., and McMahon R.G., 1997, MNRAS 283, L79

Hunting Gamma-Ray Bursts
in the Lyman-Forest; GRB 000131 at $z = 4.50$*

Michael I. Andersen[1], Jens Hjorth[2], Holger Pedersen[2], Brian L. Jensen[2],
Leslie K. Hunt[3], Javier Gorosabel[4], Palle Møller[5] Johan Fynbo[5],
and Bjarne Thomsen[6]

[1] Division of Astronomy, P.O. Box 3000, FIN-90014 University of Oulu, Finland
[2] Astronomical Observatory, University of Copenhagen,
   Juliane Maries Vej 30, DK-2100 Copenhagen Ø, Denmark
[3] Centro per l'Astronomia Infrarossa e lo Studio del Mezzo Interstellare, CNR,
   Largo E. Fermi 5, 50125 Firenze, Italy
[4] Danish Space Research Institute,
   Juliane Maries Vej 30, DK-2100 Copenhagen Ø, Denmark
[5] European Southern Observatory,
   Karl-Schwarzschild-Straße 2, D-85748 Garching, Germany
[6] Institute of Physics and Astronomy, University of Aarhus,
   DK-8000 Århus C, Denmark

Abstract. We report the discovery and follow-up observations of the afterglow of
GRB 000131. The optical identification was made with the VLT, 84 hours after the
burst following a BATSE detection and an Inter Planetary Network localization. Broad-
band and spectroscopic observations of the spectral energy distribution reveals a sharp
break at optical wavelengths which is interpreted as a Lyα absorption edge at 6700 Å.
This places GRB 000131 at a redshift of 4.500 ± 0.015.

1 Introduction

Since the first detections of soft X-ray [1], optical [2], and radio [3] afterglows
of gamma-ray bursts (GRBs), about 20 optical afterglows have been detected.
Of these, about 12 have well-determined spectroscopic redshifts, with a median
of 1.1 and the highest value being $z = 3.418$ for GRB 971214 [4]. The highest
redshift determined directly from a spectrum of the afterglow emission is $z =
2.0404 \pm 0.0008$ for GRB 000301C [5,6]. Several authors have speculated about
the possibility of detecting GRBs at even higher redshifts than achieved so far.
Wijers et al. [7] and Lamb & Reichart [8] claim that GRBs should be detectable
at very high redshifts ($z \gtrsim 5$) based on the assumption that GRBs are associated
with star formation and by using models for the cosmic star formation rate as a
function of redshift.

We present the first attempt to identify the afterglow of a GRB with an
8-m class telescope. The attempt was successful as the observations led to the
discovery of the afterglow and the subsequent determination of the decay slope,
spectral energy distribution and record high redshift.

* Based on observations collected at the European Southern Observatory, La Silla and
  Paranal, Chile (ESO Programmes 64.H-0573, 64.H-0580, 64.O-0187, and 64.H-0313)

2   Photometric Observations of GRB 000131

GRB 000131 was observed by Ulysses, Konus/Wind, NEAR, and CGRO-BATSE on 2000 Jan. 31.624 UT and localized via the InterPlanetary Network (IPN) to a 55 sq. arcmin error box 56 hours later [9], [10]. Images of the error box were obtained with the FORS1 instrument on Antu (ESO VLT UT1) on Feb. 4. and Feb. 6. Comparing these two epochs, one source was found to have declined by about 1.1 mag in the R-band. In the first epoch images this source was also detected in the V-band, but not in the B-band. Its proximity (1".1) to the center of the error box, made this object the likely afterglow of GRB 000131 [11]. In addition to the BVR observations at the VLT, images were also obtained in the I,J,H and K bands using telescopes at the ESO La Silla observatory. This allowed a reconstruction of the broad band spectral energy distribution, as shown in Fig. 1. For B and J, upper limits corresponding to our non-detections in these bands are given.

The spectral energy distribution does not resemble a power-law with a single index β. It is not possible to explain the strongly curved shape by reddening in the host galaxy by plausible reddening laws. The most likely interpretation of the spectral break is therefore the onset of Lyman forest blanketing, hence implying that GRB 000131 was at a high redshift ($z \gtrsim 4$). To examine this interpretation further, the spectral energy distribution was fitted with power-laws modified by the effects of Lyman forest blanketing and internal reddening in the host galaxy. Using the SMC extinction law, we are able to fit the spectral energy distribution for values of β from 0.5 to 1.1, with redshifts in the range $4.38 < z < 4.70$.

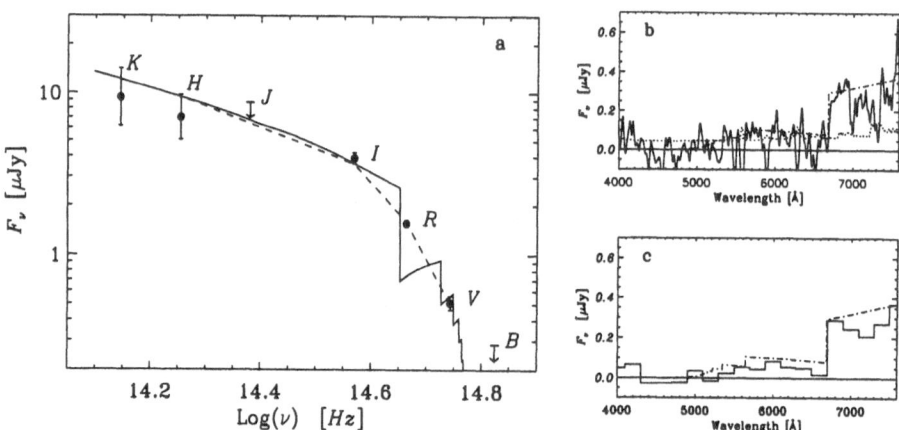

Fig. 1. a) The broad band spectral energy distribution, as derived from photometry. The dashed line is a fit by a power law spectrum with Lyman forest absorption and SMC-type reddening. This yields $A_V = 0.18$, when an intrinsic spectral slope, $\beta = 0.70$, and a redshift of 4.5 is assumed. The solid line shows the corresponding spectrum with its Lyman absorption edges. **b)** The spectrum of the GRB 000131 afterglow smoothed to a resolution of 30 Å and **c)** rebinned to a resolution of 200 Å. The dotted curve shows the noise per bin, while the dot-dashed line is the model spectrum from **a)**.

3   The Spectroscopic Redshift of GRB 000131

On Feb. 8, 3 hours of spectroscopic observations were acquired with FORS1 at Antu. The resulting spectrum is shown in Fig. 1. Our analysis of the photometric observations implies the presence of a Lyα absorption edge in the range 6500 Å to 6900 Å. This implication is confirmed by the spectroscopic observations. The recorded spectrum was very faint, at the level of less than 2% of the continuum of the night sky, which is dominated by emission lines in the red part of the spectrum. The noise in the spectrum is therefore dominated by uncertainty in the sky subtraction. An improved representation of the continuum was obtained by rebinning to a resolution of 200 Å, omitting spectral bins coincident with sky lines. The model spectrum was obtained assuming a redshift of 4.5 and using the parameters for the best fit to the broad band spectral energy distribution. It is normalized to the R-magnitude of the afterglow, obtained immediately before the spectroscopic observations began, and corrected for slit losses. The model spectrum is in very good agreement with the binned spectrum for wavelengths below 7200 Å, beyond which it is dominated by an atmospheric band and strong sky lines. Although the signal-to-noise ratio of the spectrum is very low, the absorption edge is very well defined. Taking into account that the afterglow may be associated with a strong damped Lyα absorption line at the redshift of the GRB, we find that the redshift of GRB 000131 is $z = 4.500 \pm 0.015$.

4   Conclusions

GRB 000131 has demonstrated that GRBs will be observable at very high redshifts. GRB 000131 therefore indicates that it is possible to obtain high resolution spectra of GRBs at very high redshift, if the afterglow is identified at an early stage. High-redshift GRBs will no doubt prove to be a unique tool for understanding the early Universe. For a detailed discussion of GRB 000131 the reader is referred to [12].

References

1. Costa, E. et al.: Nature **387**, 783 (1997)
2. van Paradijs, J., et al.: Nature **386**, 686 (1997)
3. Frail, D. A. and Kulkarni, S. R.: IAU Circular **6662** (1997)
4. Kulkarni, S. R. et al.: Nature **393**, 35 (1998)
5. Smette, A. et al.: Astrophys. Journ., submitted (2001)
6. Jensen, B. L. et al.: Astron. Astrophys., in press (2001)
7. Wijers, R. A. M. J. et al.: MNRAS **294**, L13 (1998)
8. Lamb, D. Q. and Reichart, D. E.: Astrophys. Journ. **536**, 1 (2000)
9. Hurley, K. et al.: GCN Circular No. 529 (2000)
10. Kippen, R. M.: GCN Circular No. 530 (2000)
11. Pedersen, H. et al.: GCN Circular No. 534 (2000)
12. Andersen, M. I. et al.: A&A **364**, L54 (2000)

Evidence for a Supernova in the Ic Band Light Curve of the Optical Transient of GRB 970508

V.V. Sokolov

Special Astrophysical Observatory of RAS,
Karachai-Cherkessia, Nizhnij Arkhyz, 369167 Russia

Abstract. Unique data on $BVRI$ brightness curves of the OT of GRB 970508 obtained with the 6-m telescope have been interpreted in the framework of the GRB-SN (supernovae) connection. The effect must be maximal in the I_c band as OT GRB 970228 [5]. The peak absolute (M_B) magnitude of the suggested SN must be around −19.5 for the OT of GRB 970508. If all or the main part of long GRBs are associated with SNe, the GRB host galaxies (for ground-based observations, at least) must be dimmer than the peak magnitude of the SN [9]).

The I_c band light curve of GRB 970508 afterglow exhibits a rebrightening phase, or a "shoulder" [7]. This spectral property lasts for not less than 51 days [8]. Over this period, the color index V-I increases by 1.6 mag, while $I_c = 23.04+/-0.14$ for the OT. The deviation from the "average" power-law exceeds 1 magnitude. Recent papers show that GRB optical afterglows have unusual temporal and VRI-spectral properties: GRB 970228 [5], GRB 980326 [2] , and possibly GRB 990712, and GRB 991208 (Castro-Tirado A., in these proceedings). All these OTs have temporal and spectral characteristics similar to those of GRB 970508, for which the I_c band light curve is consistent with the assumption that the SN emission overtakes the OT flux at late times, i.e. nearly 31 days after the GRB. It should be noted that the emission of all type-I SNe shows a strong UV deficit owing to absorption lines below 3900Å. Thus, a weak flux is expected from the suggested SN in the R_c band for a redshift z of 0.835 corresponding to the U_{rest} (3652Å) in the rest frame. In contrast, the SN brightness must be maximal at larger wavelengths (Fig1), i.e. in the I_c band as observed for GRB 970508 (and for GRB 970228 with $z = 0.695$ [5]), plus a time profile stretching for the light curve of the SN by a factor of $1 + z$. As $z = 0.835$, the I_c band in the observer frame corresponds to the B_{rest} (4448Å) band in the rest frame, the observed I_c flux is thus used to derive the peak absolute magnitude (M_B) of the assumed SN. At the "shoulder" the brightness of GRB 970508 OT corresponds to the maximum luminosity of a peculiar SN with M_B around −19.5 if for GRB990708 host we have $M_{B_{rest}}$ around −18.5 [8] (Fig.2).

We can consider the type Ib/c SNe as a preliminary model of GRB/SN gamma-ray bursters, but the peak magnitudes of Type Ib/c (or core collapse) SNe are not constant: M_B is from −16 to −19.5. Thus, the GRB host galaxies (for ground-based observations, at least) must be dimmer than the peak magnitude of the SN (see Table 1 in [9]). SN1997ef has been recognized as a peculiar SN Ic from its light curve (Mazzali P., et al., in these proceedings). It shows a very

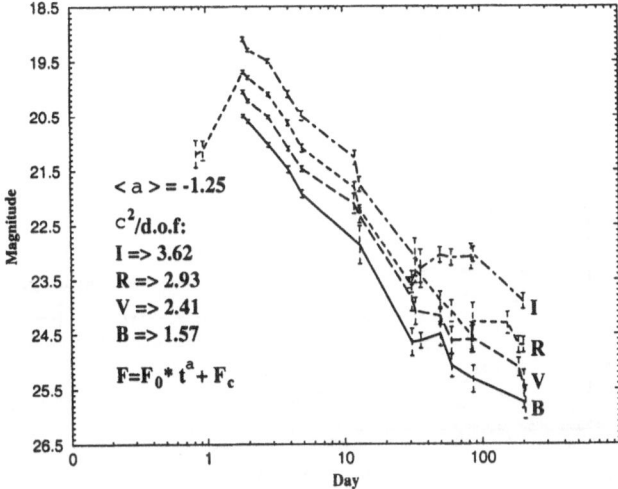

Fig. 1. The maximum of the effect is in I-band and decreases toward the blue part of spectrum as it should be for SN Ib/c. The data from the SAO-6m telescope (May 1997–Aug 1998) were used. [7], [8]. The deviation from an "average" power-law is shown.

broad/flat peak and a slow tail. But the luminosity is lower than that of SN 1998bw. However, for the purposes of this report, the shape of SN Ic light curve is sufficient as an example of Type Ib–Ic SN. We do not know exactly the light curve (and the spectrum) of the true SN accompanying GRBs; may be in the OT GRB 970508 case it was a peculiar SN by its luminosity and by the shape of the light curve (Fig.2).

Let us suppose that all or at least the main part of long **GRBs are associated with SNe** (but not all SNe are associated with GRBs). From the observations of 1997-2000 for known GRB host galaxies we have: $m_{HostGal}$ is from 22 to 28.5 mag [3]. It leads to the fact that the search for direct GRB-SN associations in galaxies close to us is a challenge, because the majority of the SNe related to GRBs will be faint (22-26 mag) and in very distant galaxies with $z \geq 0.4-4.5$. For a number of galaxies brighter than 26 mag in one square degree of the sky we have $N_{gal<26mag} \approx 2 \cdot 10^5 \cdot gal \cdot deg^{-2}$ [4]. Taking into account the observations with BATSE $n_{GRB} \approx 0.01 \cdot deg^{-2} yr^{-1}$, we conclude that the rate of GRB events is $N_{GRB} \approx 0.5 \cdot 10^{-7} \cdot yr^{-1} galaxy^{-1}$. The SNe rate for the local starburst galaxies is $N_{SN} \approx 0.02 \cdot yr^{-1} galaxy^{-1}$ for all types of SNe combined. It should be kept in mind that GRB are not observed in all galaxies and that not all types of SNe can be related to GRBs, but only Type Ib/c [6]. So we have $N_{SN} \approx 0.001 \cdot yr^{-1} galaxy^{-1}$. But if one looks at the matter the other way round, we must remember: the massive star-forming in very distant (and dusty) GRB galaxies with $z \sim> 1$ is more vigorous than in local starburst galaxies or than in galaxies like the Milky Way [1], giving rise of SN bursts to 10 times. Thus, with due regard for these reservations, an estimate for the number of GRBs which

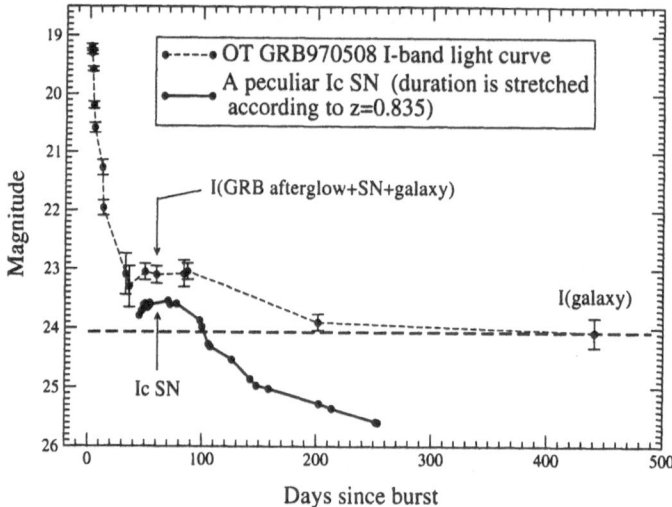

Fig. 2. OT GRB 970508 *I*-band light curve plus a kind of a pecular Ic SN (by the shape of the light curve) similar to SN 1997ef, but with the peak luminosity like SN 1998bw.

could be related to SNe is of order of $N_{GRB}/N_{SN} \approx 5 \cdot (10^{-5} - 10^{-6})$. So, if some GRB–SN relation really exists, then we have two possibilities: **1)** Either we deal with a very very rare type of SNe (hypernova) – GRBs related only to the $10^{-5} - 10^{-6}$-th part of all observed SNe in distant galaxies (up to 28th mag). **2)** Or we have a very strong **γ-ray beaming** with a solid angle up to $\Omega_{beam} \approx 5 \cdot (10^{-5} - 10^{-6}) \cdot 4\pi$ if GRBs are associated with an **asymmetric** SN explosion and the γ-ray beam is directed towards an observer on Earth.

Thus, if all (or the main part) of long GRBs are associated with SNe, we need more examples to test the GRB-SN link. But from Earth we can see only peculiar SNe (by their luminosities & by light curves) connected with GRBs. The SNe must be brighter than their hosts. The luminosity of these dusty galaxies M_B are between about −18.6 and −21.3 [9].

Acknowledgements: This work was supported by INTAS N96-0315, "Astronomy" Foundation (grant 97/1.2.6.4) and RFBR N98-02-16542.

References

1. Blain A.W. and Priyamvada Natarajan 2000, MNRAS, 312, L35-L38
2. Bloom J. S. et al. 1999a, Nature, 401, 453
3. Bloom J. S., Kulkarni S. R., Djorgovski S. G. 2000, astro-ph/0010176
4. Casertano, S. et al. 2000, astro-ph/0010245
5. Galama , T. J. et al. 2000, ApJ, 536, 185
6. Capellaro, Evans, and Turatto, 1999, A&A 351, 459
7. Sokolov V.V., et al., 1998, A&A , 334, 117
8. Sokolov V.V., et al., 1999, A&A , 344, 43
9. Sokolov V.V., et al., 2001, in these proceedings

Hypernovae and Gamma-Ray Bursts

P.A. Mazzali[1], K. Nomoto[2,3], K. Maeda[2], and T. Nakamura[2]

[1] Osservatorio Astronomico, via G.B. Tiepolo, 11, Trieste, Italy
[2] Department of Astronomy, University of Tokyo, Tokyo, Japan
[3] Research Center for the Early Universe, University of Tokyo, Tokyo, Japan

Abstract. Recently, there have been a number of candidates for the gamma-ray burst (GRB)/supernova (SN) connection (see Nomoto et al. 2001 for references). Among the SNe with a possible GRB counterpart, the Type Ic SNe 1998bw and 1997ef are characterised by a very large kinetic explosion energy, $E_K \gtrsim 10^{52}$ erg. This is more than one order of magnitude larger than in typical SNe, so that these objects may be called "hypernovae".

1 SN 1997ef

SN 1997ef was noticeable for its unique light curve and spectra. At early times, the spectra were dominated by broad oxygen and iron absorptions, but did not show any clear feature of hydrogen or helium (Garnavich et al. 1997; Filippenko et al. 1997), and the SN was classified as Type Ic (SN Ic). The most striking feature of the spectra of SN 1997ef is the broadness of the line features, which suggests that the SN may have had a large explosion energy.

Iwamoto et al. (2000) constructed hydrodynamical models for an ordinary SN Ic and for a hypernova. For the ordinary SN Ic, a C+O star with a mass $M_{CO} = 6.0 M_\odot$ (the core of a 25 $M_\odot$ star) was exploded with a kinetic energy $E_K = 1.0 \times 10^{51}$ ergs and an ejecta mass $M_{ej} = 4.6 M_\odot$ (model CO60). For the hypernova model (CO100), a C+O star of $M_{CO} = 10.0 M_\odot$ (the core of a 30–35 $M_\odot$ star) was exploded with $E_K = 8.0 \times 10^{51}$ ergs and $M_{ej} = 8.0 M_\odot$. The mass of ^{56}Ni was set to be 0.15 $M_\odot$ for both models to explain the observed peak brightness.

It is difficult to distinguish clearly between the ordinary SN Ic and the hypernova model from the light curve alone, since models with different values of M_{ej} and E_K can display a similar light curve if $M_{ej}^{3/4} E_K^{-1/4}$ = const. In fact, the synthetic V light curves for both models reproduce the light curve of SN 1997ef. However, these models can be expected to produce different spectra, because of the different E_K. Therefore, spectrum synthesis can distinguish between them.

Spectra computed with model CO60 for the early epochs show narrow lines, much narrower than the observations. This clearly indicates a lack of material at high velocity in model CO60, and suggests that the kinetic energy of this model is too small. Synthetic spectra computed with model CO100 have much broader lines and are in good agreement with the observations (Mazzali et al. 2000). We therefore conclude that SN 1997ef was a hypernova.

2 SN 1998bw

SN 1998bw was discovered as the optical counterpart of GRB980425 (Galama et al. 1998). The absorption lines are so broad in SN 1998bw that they blend together, giving rise to broad absorption features. Velocities in the Si II 6355Å line are as high as $30,000$ km s^{-1}. Also, the SN was very bright for a SN Ic: the observed peak luminosity, $L \sim 1.4 \times 10^{43}$ erg s^{-1}, is almost ten times higher than that of previously known SNe Ib/Ic (Woosley et al. 1999).

Iwamoto et al. (1998) computed light curves and spectra for various C+O star models with different values of E_K and M_{ej}. The best model is the explosion of a 13.8 $M_\odot$ C+O star with a large kinetic energy ($E_K = 6 \times 10^{52}$ erg). Such a massive C+O star progenitor is the product of a main-sequence star of ~ 40 $M_\odot$ which lost its H/He envelope via a stellar wind or binary interaction. A ^{56}Ni mass of $\sim 0.63 M_\odot$ is necessary to reproduce the light curve maximum. The broad lines are well reproduced with model CO138 because there is enough material at high velocities in this model.

Despite the success of our 'hypernova' model in reproducing the early light curve of SN 1998bw (t $\lesssim$ 60 days), the model light curve tail declines too rapidly (Nakamura et al. 2001). This suggests that there might be a high density region in the ejecta of SN 1998bw, where the γ-rays are efficiently trapped. Another peculiarity appeared in the late phase spectra. Measuring the velocity of the various elements from the width of their emission lines, Patat et al. (2001) noted that iron expands faster than oxygen, which is contrary to expectations.

3 2D Models of an Asymmetric Explosion

The need for a high density region and the velocity inversion might indicate that the explosion was aspherical (Höflich et al. 1999; MacFadyen & Woosley 1999). The outburst in SN 1998bw may have taken the form of a prolate spheroid, as also indicated by polarization measurements (Patat et al. 2001).

We computed 2D models to follow the explosive nucleosynthesis in an axisymmetric explosion (Maeda et al. 2001). The progenitor model was CO138. The hydrodynamical simulation was started by depositing the energy as 50% thermal and 50% kinetic below the mass cut which divides the ejecta from the collapsing core. More energy was deposited in the direction of the jet than in other directions.

The shock is stronger and the post-shock temperatures are higher along the jet direction (z), so that explosive nucleosynthesis takes place in a more extended, lower density region compared with the perpendicular direction (r). A large amount of ^{56}Ni is produced in the jet direction and ejected at high velocity. In addition, elements produced by α-rich freezeout are enhanced because nucleosynthesis proceeds at higher entropies than in the region away from the jet. In contrast, little ^{56}Ni is produced in the r direction and the expansion velocities are lower than in the z-direction. Therefore, the Fe velocities (mostly in the z-direction) can exceed the O velocities (in the r-direction), as observed in

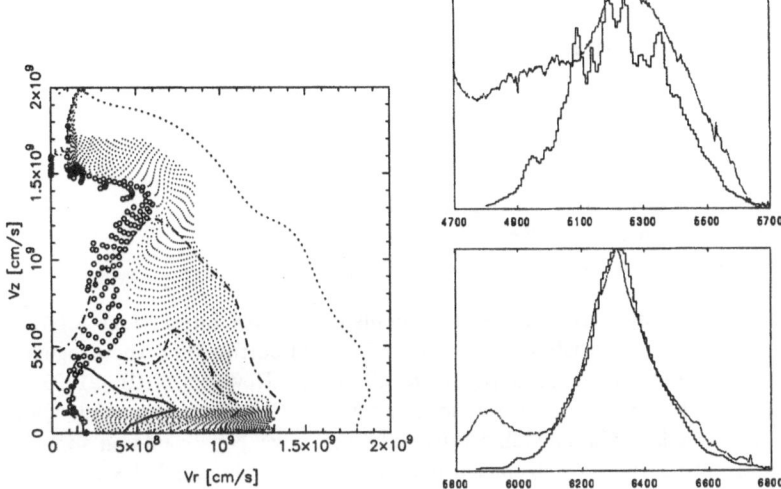

Fig. 1. Left: Distribution of ^{56}Ni (circles) and ^{16}O (dots) in (r, z) space. The lines mark iso-density contours. Right: Observed spectra of SN1998bw (full lines) and synthetic Fe and O emission lines computed using the aspherical model (dashed lines).

SN 1998bw. The density in the r direction may be high enough that γ-rays are trapped even at advanced phases, thus giving rise to the slowly declining tail. The distribution of ^{56}Ni and ^{16}O in (r, z) space is shown in Fig. 1 (left). The estimated kinetic energy is smaller in the aspherical model, but the value we derived ($E_K = 1 \times 10^{52}$erg) is still very large for a normal SN.

Maeda et al. (2001) computed nebular emission lines of Fe and O based on the distribution shown in Fig.1 (left). If the explosion is viewed from about 15° from the jet axis, the profiles of the Fe blend near 5200Å and of O I] 6300Å are correctly reproduced, as shown in Fig.1 (right).

References

1. Filippenko, A.V. et al. 1997, IAU Circ. No.6783, 6809
2. Galama, T.J. et al. 1998, Nature, 395, 670
3. Garnavich, P. et al. 1997, IAU Circ. No. 6778, 6786, 6798
4. Höflich, P., Wheeler, J.C., Wang, L., 1999, ApJ, 521, 179
5. Iwamoto, K. et al. 1998, Nature, 395, 672
6. Iwamoto, K. et al. 2000, ApJ, 534, 660
7. MacFadyen, A.I. & Woosley, S.E. 1999, ApJ, 524, 262
8. Maeda, K., et al., 2001, ApJ, submitted (astro-ph/0011003)
9. Mazzali, P.A. et al., 2000, ApJ, 545, 407
10. Nakamura, T., et al., 2001, ApJ, 550 (astro-ph/0007010)
11. Nomoto, K., et al., 2001, in "Supernovae and Gamma-Ray Bursts" (CUP) (astro-ph/0003077)
12. Patat, F. et al., 2001, ApJ, submitted
13. Woosley, S.E., Eastman, R.G., Schmidt, B.P. 1999, ApJ, 516, 788

Theoretical Implications of the Gamma-Ray Burst–Supernova Connection

Roger A. Chevalier

Department of Astronomy, University of Virginia,
Charlottesville, VA 22903, USA

Abstract. Although observations of circumstellar shock interactions around supernovae are generally consistent with a $\rho \propto r^{-2}$ wind surrounding the progenitor star, this is not true for GRB (gamma-ray burst) afterglows. However, GRB 991208 and GRB 000301C may be consistent with wind interaction if the injection particle spectrum is a broken power law. Circumstellar dust echos can place constraints on supernova and GRB progenitors, but have been clearly observed only around SN 1987A. Excess emission observed in two GRB afterglows is more likely to have a supernova origin. An interstellar dust echo, causing the light curve to flatten out, is a possibility for GRB afterglows, but is not likely to be observable.

1 Introduction

A year ago, there were a number of pieces of evidence that pointed to a connection between GRBs (gamma-ray bursts) and SNae (supernovae) (see [1] for a review). The positions of GRBs in galaxies, near star forming regions, were consistent with a massive star origin. In addition to the association of GRB 980425 with SN 1998bw, two distant GRB afterglows showed evidence for supernovae in their light curves and spectra. In some cases, the afterglow evolution suggested interaction with a freely expanding stellar wind, as expected in the immediate vicinity of a massive star.

Although the evidence for an association with star forming regions has held up [2], the other topics have not provided further support. There have been no further convincing associations of supernovae with GRB afterglows. Dust echos have been proposed as the mechanism for the two apparent distant supernova cases that have been observed [3,4]. The two best observed recent afterglows, GRB 991208 and GRB 000301C, have been interpreted in terms of expansion in a constant density medium, with jet effects [5,6]. All these points weaken the case for a GRB–SN connection. Here, I discuss these recent developments in the context of studies of supernovae and GRB afterglows.

2 Supernova and GRB Shock Interactions

The shock interactions of supernovae with circumstellar matter can be observed through radio synchrotron emission (analogous to the synchrotron emission in GRB afterglows) and through X-ray thermal radiation. The radio light curves are characterized by an initial rise followed by a power law decline. A plot of the time of the light curve peak vs. the peak luminosity at that time sorts the observed supernovae into various types [7,8]. These properties are related to the density of the gas into which the shock wave is propagating. Except for SN 1987A, Type II supernovae are thought to have red supergiant progenitors, which have wind velocities of ~ 10 km s^{-1}. The inferred mass loss rate from the progenitor of SN 1999em, a recently detected radio supernova [9], is $\dot{M} \sim 10^{-6}$ $M_\odot$ yr^{-1}. This may be typical of SN II progenitors; the well-observed Type II radio supernovae are probably at the upper end of a distribution of circumstellar densities. The most luminous radio supernovae are inferred to have $\dot{M} \sim 10^{-4} - 10^{-3}$ $M_\odot$ yr^{-1}.

Although some deviation from standard evolution has been found in radio supernovae with red supergiant progenitors [10,11] the variations are small compared to those in SN 1987A. The well-known equatorial ring is at a radius of 6.3×10^{17} cm, but moderately dense gas probably extends in from this radius because of photoionization by the progenitor star [12]. The radio luminosity was initially low compared to the red supergiant case because of interaction with the low density progenitor wind, but started to increase in 1990 and rapidly rose to > 100 times an extrapolation of the early decline [13]. During the last few years, the supernova shock wave has started to interact with parts of the equatorial ring that are closest to the supernova [14]. The changes in radio flux, which are more dramatic than any observed for other Type II supernovae, show what happens when a supernova shock wave runs into dense gas from an earlier evolutionary stage. At the present time, SN 1987A remains unique among the SN IIae and the precise causes for the progenitor evolution to the blue are not clear; they may have to do with binary evolution.

Except for SN 1998bw, the SN Ib/c that have been observed in the radio have similar properties [8,15]. If they are able to efficiently radiate synchrotron emission, the radio flux is approximately in accord with $\dot{M} \sim 10^{-5} - 10^{-4}$ $M_\odot$ yr^{-1} and $v_w \approx 1,000$ km s^{-1}, as expected for a Wolf-Rayet star [7]. Except for SN 1998bw, the radio light curves show a smooth evolution, although the data are sparse. In the case of SN 1998bw, the observed rise in flux between days 20 and 30 [15] is more likely to be due to an increase in the energy of the explosion than to an encounter with denser gas [8]. The evolution is consistent with interaction with a $\rho \propto r^{-2}$ wind extending out to at least 4×10^{17} cm.

If GRBs have massive star progenitors, they are likely to be Wolf-Rayet stars. Arguments in favor of this assumption include: (1) SN 1998bw, the best case of a SN–GRB association (GRB 980425), was of Type Ic, with a probable Wolf-Rayet progenitor. (2) The high energy of GRBs suggest that a moderately massive black hole is involved, which, in turn, requires a massive, $\gtrsim 20 - 25$ $M_\odot$, progenitor [16]. These stars are likely to be Wolf-Rayet stars at the end of their lives [17]. (3) The relativistic flow from a central object may be able to penetrate a relatively compact Wolf-Rayet star, but probably cannot penetrate an extended red supergiant star [18]. None of these arguments is definitive, but they are suggestive. The GRB rate is a small fraction of the SN rate so that a peculiar kind of star with an unusual surroundings could be involved. For example, a binary merger may be needed so that the central black hole has a large rotational energy. The merger process may be associated with a strong equatorial outflow of the stellar envelope. However, the relativistic flow is likely to be along the polar axis, where the mass loss properties may be normal.

The best tests for afterglow models are sources with extensive data; radio data are especially useful because they give information on the absorption frequency and the peak flux and its frequency. The extensive radio data [19] on GRB 970508 can be fitted by a model afterglow interacting with an r^{-2} wind, extending out to 3×10^{18} cm [3]. Frail et al. [19] fit the same data by jet expansion in a uniform medium. In their model, jet effects become important at day 25 and there is a transition to spherical nonrelativistic flow at day 100. The radio data on GRBs 991208 and 000301C can also be approximately fitted by a wind model [21], but in these cases a broken power law spectrum is needed for the particles in order to fit the higher energy light curves. The jet in a uniform medium models for these objects can assume a single power law spectrum; however, the evolution of GRB 991208 is taken to be in the jet transition phase during the period of observation ($\sim 3 - 20$ days) [5], whereas GRB 000301C is taken to have a sharp transition to asymptotic jet evolution on day 7.3 [6]. A problem with the jet models is that the evolution is not well defined. The broken power law, wind models can be tested by late observations because different time dependences are expected at radio and optical wavelengths. The wind interaction models [3,21] have not included jet effects, although a collimated flow is expected if it must escape from the center of a star. To some extent, this can be justified by the slow apparent evolution of a jet in a wind [14,23].

The facts that many afterglows can be fitted by model evolution in a constant density medium and that massive stars are attractive progenitors have led to the suggestion [24] that the expansion is into the constant density medium that is expected downstream from the termination shock of the massive star wind [2]. The outer radius of such a region is $\sim 2 - 2.5$ times the inner radius when the fast wind from a Wolf-Rayet star runs into a slow wind from a previous evolutionary stage [2,26]. Models and observations of Galactic Wolf-Rayet stars show that the swept-up shell of red supergiant material at the outer radius is at a distance $\gtrsim 3$ pc from the star [17,2]. This radius is sufficiently large that interaction with the free $\rho \propto r^{-2}$ wind is expected over the typical period of observation of afterglows. The time for the wind to reach the termination shock is relatively short (~ 300 years to reach 10^{18} cm), so that the assumption of constant wind properties is plausible. In a massive star progenitor model, a high density in the immediate vicinity of the explosion is expected even if the wind passes through a termination shock at a relatively small radius. The prompt optical emission from a GRB gives the opportunity to investigate the immediate surroundings; in the case of GRB 990123, low density (interstellar medium) interaction gives a better explanation for the observations [3,27].

Despite the plausibility of free wind interaction, the uncertainty in the evolution of massive stars leaves open the possibility of interaction with denser material at early times; Ramirez-Ruiz et al. [28] have investigated such a case.

The interaction of a GRB flow with a dense shell could have different properties compared to a supernova, as exemplified by SN 1987A. In the supernova case, the shock front is continually driven by the lower velocity supernova ejecta with most of the kinetic energy. Most GRB afterglow models have constant energy, so that although an initial overpressure can be expected upon a collision with a shell, it is not as marked as in the supernova case. In addition, the GRB flow is likely to be in a jet, unlike the supernova case. When the beamed flow interacts with a large density jump, it is rapidly decelerated and lateral spreading can occur. The afterglow could make a rapid transition to the asymptotic behavior for a jet flow [15]. The afterglow of GRB 000301C did show a bump followed by a transition to a steeper decline that could be lateral jet expansion, although the bump can be interpreted as a microlensing event [4].

3   Dust Echos

The possibility of circumstellar dust echos arises naturally for Type II supernovae with red supergiant progenitors because their cool winds are known to contain dust. For the higher estimated mass loss rates, e.g., $\sim 10^{-4}$ $M_\odot$ yr^{-1} for SN 1979C, the progenitor star is expected to be entirely enshrouded in dust. Yet when we observe Type II supernovae, there is little evidence for dust absorption in the spectra of the supernovae, including SN 1979C. The probable solution is that the dust near the star is evaporated by the radiation from the supernova and the remaining dust does not have a large optical depth. The radius at which dust becomes so hot that it is evaporated, r_v, occurs at $\sim 3 \times 10^{17}$ cm for typical supernova and dust parameters [31].

The circumstellar dust beyond r_v can give rise to infrared and scattered light echos. Both SN 1979C and SN 1980K showed evidence for infrared excesses that might be attributable to dust echos [31]. The late optical light from these supernovae appeared to be dominated by emission from the circumstellar shock wave interactions and did not show dust scattering effects [32]. Roscherr and Schaefer [33] have recently examined the emission from SN IIn to search for scattered light echos, but again found that the late emission is dominated by circumstellar shock interactions. Circumstellar scattered light echos have been observed around SN 1987A [34] because it was possible to spatially resolve the echos away from the supernova; it would not have been possible to observe the echos for a supernova at a typical distance.

No evidence for circumstellar dust echos has been found in Type Ib/c supernovae, which is not surprising considering their presumed Wolf-Rayet star progenitors. Wolf-Rayet stars atmospheres are too hot for dust formation. Despite this, circumstellar dust has been found in a small fraction of these stars. The source appears to be the compressed interaction region where the Wolf-Rayet star wind collides with the wind from a close companion, as indicated by the pinwheel dust patterns observed around two Wolf-Rayet stars [35]. Dust may be more commonly present in circumstellar shells at radii $\gtrsim 3$ pc, but such shells would have only a small dust optical depth.

Circumstellar dust echos have been of recent interest for GRB afterglows because of the possibility that they could explain the excess emission observed in two sources at ages $10 - 60$ days without the need for a supernova [3,4]. An echo from thermal re-radiation by dust is too red to explain the observations, so the scattered light explanation is considered here [4]. The timescale for the emission sets the dust at a radial distance of 4.5×10^{17} cm and its color requires an optical depth at 3000 Å of 7 [4]. The required dust mass is $\sim 0.1\ M_\odot$ or $\sim 10\ M_\odot$ total mass in the shell. If the gas is moving out as part of the stellar wind, special timing is needed to have much of the envelope mass end up at the appropriate radial distance. Thus, on the basis of the fact that the most plausible massive star progenitors for GRBs (Wolf-Rayet stars) do not typically have dusty winds plus the special conditions that are needed for the dust if present, the supernova hypothesis appears to be a more natural explanation for the excess light than a dust echo. Reichart [36] has recently examined detailed models for circumstellar echos and found that the colors of the excess light of GRB 970228 are inconsistent with an echo. Even if the excess light could be explained as a scattered light echo, the implication would be that the progenitor object was a massive star which was a red supergiant at the time of the explosion or soon before the explosion.

In addition to circumstellar dust echos, dust in interstellar clouds can give rise to echo phenomena. In the supernova case, the well-known rings around SN 1987A are due to clouds that are 100 pc or more in front of the supernova [37]. Similar scattered light echos have been observed around the Type Ia supernovae SN 1991T and SN 1998bu [38,39]. The echo phenomenon is characterized by a flattening of the supernova light curve, a late spectrum related to the spectrum near maximum light, and a significant spatial size related to the distance of the interstellar cloud in front of the supernova. If the peak luminosity, L_p, of the supernova or GRB lasts for a time Δt, the ratio of the echo luminosity to L_p is $\tau c F \Delta t / 2d$, where τ is the dust optical depth, c is the speed of light, F is the phase function (forward scattering is expected), d is the distance of the cloud in front of the explosion, and single scattering is assumed [37]. This can explain the $9 - 10$ magnitude decline from peak to flattening of the light curve in the 3 observed echos. GRB 990123 has the best defined afterglow optical light curve; it peaked at ~ 9 mag. and subsequently declined more rapidly than t^{-1} so that most of the radiated energy appeared near the peak. If a similar cloud were placed in front of GRB 990123 to those in front of SNae 1991T and 1998bu, the fact that $\Delta t \sim 30$ sec for GRB 990123, but is ~ 20 days for the supernovae would make the luminosity $\sim 6 \times 10^4$ fainter compared to the peak luminosity. This would give an echo magnitude ~ 30, which is undetectable. The angular radius of the echo from the explosion is $0.08(t/\mathrm{yr})^{1/2}(d/100\ \mathrm{pc})^{-1/2}$ rad, which could be outside the beamed emission from a GRB, further reducing the echo brightness. It is only for a relatively nearby GRB with a favorable distribution of dust that there is any chance of detecting an echo. The best candidates for an echo would be sources that show evidence for dust in the host galaxy, e.g.,

GRB 000418 [40]. Such a detection would be of interest in that it would give information on the very early optical luminosity of the burst.

I am grateful to C. Fransson and Z.-Y. Li for discussions and collaboration, and to NASA grant NAG5-8130 for support.

References

1. S. R. Kulkarni, E. Berger, et al.: 'The Afterglows of Gamma-Ray Bursts'. In: *Gamma-Ray Bursts: 5th Huntsville Symposium.* ed. by R.M. Kippen, R.S. Mallozzi, G.J. Fishman (AIP, New York 2000) pp. 277–297
2. J.S. Bloom, S.R. Kulkarni, S.G. Djorgovski: Astron. J. submitted (astro-ph/0010176) (2000)
3. E. Waxman, B.T. Draine: Astroph. J. **537**, 796 (2000)
4. A.A. Esin, R. Blandford: Astroph. J. **534**, L151 (2000)
5. T.J. Galama, M. Bremer, et al.: Astroph. J. **541**, L45 (2000)
6. E. Berger, R. Sari, et al.: Astroph. J. **545**, 56 (2000)
7. R.A. Chevalier: Astroph. J. **499**, 810 (1998)
8. Z.-Y. Li, R.A. Chevalier: Astroph. J. **526**, 716 (1999)
9. C.K. Lacey, S.D. Van Dyk, et al.: *IAUC* No. 7336 (1999)
10. M.J. Montes, S.D. van Dyk, et al.: Astroph. J. **506**, 874 (1998)
11. M.J. Montes, K.W. Weiler, et al.: Astroph. J. **532**, 532 (2000)
12. R.A. Chevalier, V.V. Dwarkadas: Astroph. J. **452**, L45 (1995)
13. L. Ball, D. Campbell-Wilson et al.: Astroph. J. **453**, 864 (1995)
14. S.S. Lawrence, B.E. Sugerman, et al.: Astroph. J. **537**, L123 (2000)
15. S.R. Kulkarni, D.A. Frail, et al.: Nature **395**, 663 (1998)
16. E. Ergma, E.P.J. van den Heuvel: Astron. & Ap. **331**, L29 (1998)
17. G. García-Segura, N. Langer, M.-M. MacLow: Astron. & Ap. **316**, 133 (1996)
18. A.I. MacFadyen, S.E. Woosley, A. Heger: Astroph. J. submitted (astro-ph/9910034) (1999)
19. D.A. Frail, E. Waxman, S.R. Kulkarni: Astroph. J. **537**, 191 (2000)
20. R.A. Chevalier, Z.-Y. Li: Astroph. J. **536**, 195 (2000)
21. Z.-Y. Li, R.A. Chevalier: Astroph. J. in press (astro-ph/0010288) (2001)
22. P. Kumar, A. Panaitescu: Astroph. J. **541**, L9 (2000)
23. L.J. Gou, Z.G. Dai, et al.: Astron. & Ap. in press (astro-ph/0010244) (2000)
24. R.A.M.J. Wijers: these proceedings
25. R.A. Chevalier, Z.-Y. Li: Astroph. J. **520**, L29 (1999)
26. R.A. Chevalier, J.N. Imamura: Astroph. J. **270**, 554 (1983)
27. R. Sari, T. Piran: Astroph. J. **517**, L109 (1999)
28. E. Ramirez-Ruiz, L.M. Dray, et al.: MNRAS, submitted (astro-ph/0012396) (2000)
29. R. Sari, T. Piran, J.P. Halpern: Astroph. J. **519**, L17 (1999)
30. P.M. Garnavich, A. Loeb, K.Z. Stanek: Astroph. J. **544**, L11 (2000)
31. E. Dwek: Astroph. J. **274**, 175 (1983)
32. R.A. Chevalier: Astroph. J. **308**, 225 (1986)
33. B. Roscherr, B.E. Schaefer: Astroph. J. **532**, 415 (2000)
34. A.P.S. Crotts, W.E. Kunkel, S.R. Heathcote: Astroph. J. **438**, 724 (1995)
35. J.D. Monnier, P.G. Tuthill, W.C. Danchi: Astroph. J. **525**, L97 (1999)
36. D.E. Reichart: Astroph. J. in press (astro-ph/0012091) (2001)
37. R.A. Chevalier, R.T. Emmering: Astroph. J. **331**, L105 (1988)
38. W.B. Sparks, F. Machetto, et al.: Astroph. J. **523**, 585 (1999)
39. E. Cappellaro, F. Patat, et al., Astroph. J. in press (astro-ph/0101342) (2001)
40. S. Klose, B. Stecklum, et al.: Astroph. J. **545**, 271 (2000)

Properties of GRB Optical Afterglows: Colors and Luminosities

Vojtěch Šimon[1], Nicola Masetti[2], René Hudec[1], and Graziella Pizzichini[2]

[1] Astronomical Institute, Academy of Sciences, 251 65 Ondřejov, Czech Republic
[2] Istituto TeSRE-CNR, via Gobetti 101, I-40129 Bologna, Italy

1   Introduction

The color indices of the optical afterglows (OAs) of the GRBs are a powerful tool for the study of such events. They might help us to: (a) search for common properties of these events; (b) understand the related physical processes using the colors as discriminant parameters; (c) find out whether an optical event is related to a GRB even without available γ-ray detection by using the color indices of the OAs; (d) search for relations among colors, luminosities and the decay rates of the OAs (if the redshift z is known); (e) constrain the properties of the local interstellar medium of the GRBs. This work is however preliminary: a more complete description of it can be found elsewhere [5].

2   Collection and Analysis of the Data

The analysis presented here makes use of the data published in the GCN circulars archive [1], in Jochen Greiner's Web page [2] and in the journals. Because of space limitations, the reader is referred to these web pages for full bibliographic references on each OA. Suitable *multicolor* photometry is available for 16 OAs but it often comprises unorganized observations. The light curves of the OAs were therefore plotted and critically examined. Mostly they were found to be free of complicated rapid changes (hours to a few days), so a meaningful interpolation between the neighbouring measurements was possible, at least in the early parts of the light curves ($t - T_0 < 10$ days, where T_0 is the time of the GRB). It enabled us to obtain at least one color index for each OA. The typical standard deviations of the indices, derived from the errors quoted in the original literary sources, lie within 0.04–0.3 mag. The indices were corrected for the Galactic reddening (computed from the maps by [4]). The light contribution of the host galaxies was quite small for $t - T_0 < 10$ days and was therefore neglected.

The color indices $(R-I)_0$, $(V-R)_0$ and $(B-V)_0$, when plotted versus $t-T_0$, were found to occupy narrow belts with negligible evolution, so the common color-color diagrams could be built even if the observations of the various OAs came from different epochs after the GRB. The $V - R$ vs. $R - I$ diagram is shown in Fig. 1a; the $B - V$ vs. $V - R$ diagram looks very similar.

The observed passbands of the OAs differ from those in the rest frame because of the effects introduced by the redshift z. The OAs considered here have $z =$

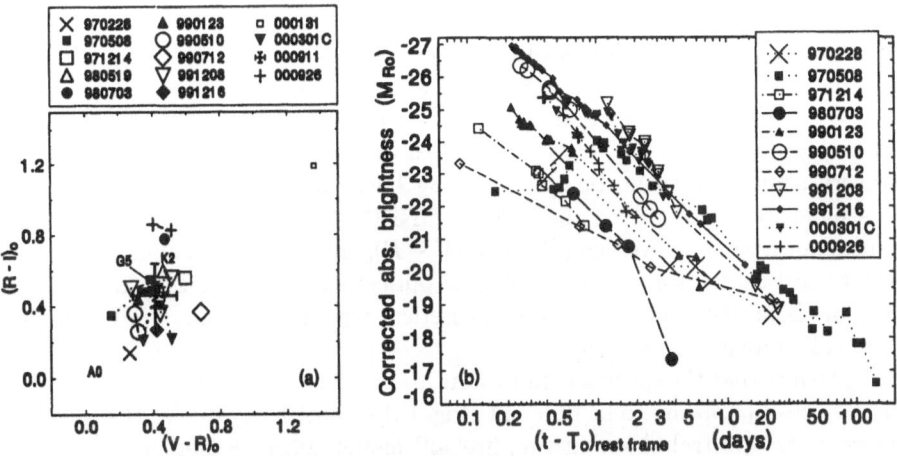

Fig. 1. (a) $V - R$ vs. $R - I$ diagram of OAs ($t - T_0 < 10$ days). Multiple indices of the same OA are connected with a line for convenience. The mean colors (centroid) of the whole ensemble of OAs (except the largely outlying GRB000131), including the standard deviations, are marked by the large cross. Positions of main-sequence stars are included for comparison. (b) Absolute brightness of OAs in the R-band corrected for the Galactic reddening, the light contribution of the host galaxy, and a 'zeroth order' k-correction (see text). The time intervals are corrected for relativistic time dilation and expressed in rest-frame time $(t - T_0)_{rest}$.

0.43–4.5 (with typical $z \sim 1$). The color indices of all OAs with known z were plotted versus z. This plot confirms that any dependence of the color on z is quite weak and within $1\,\sigma$ for $(R - I)_0$, $(V - R)_0$ and $(B - V)_0$, at least for the OAs with $z < 3.5$ (i.e. except GRB000131, located at $z = 4.5$). The only exception may be $(U - B)_0$ but this index could be determined just for a few OAs.

Absolute brightness in the R-band (M_{R_0}; Fig. 1b) of each OA was calculated from the redshift z (Eq.15.3.25 of [6]) and using $H_0 = 60$ km s^{-1} Mpc^{-1}. M_{R_0} has then been corrected assuming OA power law spectra with $\beta \sim 1$ (see below) and keeping into account which rest-frame band is corresponding to the R band as seen in the observer frame, We can consider this as a sort of 'zeroth order' k-correction.

3 Conclusions

The typical color indices of the OAs between $0.14 < t - T_0 < 10$ days are $(R - I)_0 = 0.45 \pm 0.19$, $(V - R)_0 = 0.40 \pm 0.13$, $(B - V)_0 = 0.47 \pm 0.18$ and their changes are negligible although the brightness of all OAs declines by several magnitudes during this time interval. This implies that the shape of the spectrum does not change significantly while the luminosity of the OAs decreases by a large amount. On the other hand, the $(U - B)_0$ index (not shown here) presents a large scatter and the position of the OAs in the $U - B$ vs. $B - V$ diagram is often quite different from other sources.

The strong concentration of most OAs in the $V - R$ vs. $R - I$ diagram and the $B - V$ vs. $V - R$ diagram, despite the smearing introduced by the shifts of the passbands due to various z, implies that the spectral shape of the OAs is very smooth, with no bumps or strong lines, between the observed B to I passbands (that is between about 2000 and 5600 Å in the rest frame for the OAs with known z). The average $(R - I)_0$, $(V - R)_0$ and $(B - V)_0$ colors of the OAs are consistent with a power-law shaped optical spectral distribution with $\beta \sim$ 1. This is in accordance with the theoretical treatment of the GRB afterglow 'fireball' emission model [16].

The fact that the spectra of most OAs are similar although their luminosity in a given $t - T_0$ appears to be different (Fig. 1b) is most likely due to the following reason: the spectral shape in the 'fireball' model [16] does not depend on the input energy, while the luminosity of the afterglow at a particular epoch does depend on it. So, the higher the GRB input energy is, the brighter the OA is; this would also suggest that GRBs and their afterglows *are not* standard candles.

The strong concentration of most color indices also points to the fact that in-trinsic reddening (inside their host galaxies) must be quite similar and relatively small for these events (except for GRB000131). In the case of a large reddening it would be quite unlikely to obtain such similar values of absorption in all cases. These GRBs are therefore unlikely to come directly from the inner (densest) parts of the star-forming regions (but they may lie on the earth-watching side of a star-forming region). Alternatively, the density and the dust abundance of the local ISM might be reduced by the high-energy radiation of the GRB.

The absolute brightness of the OAs, corrected for the host galaxy, lies within $M_{R_0} = -26.5$ to -22.2 for $(t - T_0)_{rest} = 0.25$ days. This spread of M_{R_0} is not significantly influenced by the shifts of λ, caused by the different redshifts of OAs. The general decline rate of most OA considered here seems to be independent on the absolute optical brightness of the OA as measured at some fixed $t - T_0$ for all OAs, and the light curves of all events are almost parallel, when corrected for the redshift-induced time dilation.

Acknowledgements: This study was supported by the project KONTAKT ME 137 and ES002 by the Ministry of Education and Youth of the Czech Republic and the grant 205/99/0145 of the Grant Agency of the Czech Republic. We also acknowledge the CNR-AVČR Joint Research Program No. 3 (1998/2000).

References

1. S. Barthelmy: http://gcn.gsfc.nasa.gov/gcn/gcn3_archive.html
2. J. Greiner: http://www.aip.de/ĵcg/grbgen.html
3. R. Sari, T. Piran, R. Narayan: ApJ **497**, L17 (1998)
4. D.J. Schlegel, D.P., Finkbeiner, M. Davis: ApJ **500**, 525 (1998)
5. V. Šimon, N. Masetti, R. Hudec, G. Pizzichini: A&A submitted (2001)
6. S. Weinberg: *Gravitation and cosmology: principles and applications of the general theory of relativity* (Wilby, New York 1972) p. 485

GRB Optical Afterglows:
Correlation between Pair of Parameters

Corrado Bartolini[1], Gregory Beskin[2,3], Giuseppe Cosentino[1],
Adriano Guarnieri[1], Adalberto Piccioni[1], and Aleksej Pozanenko[4]

[1] Dipartimento di Astronomia, Università di Bologna, Italy
[2] Special Astrophysical Observatory, Russia
[3] Isaac Newton Institute of Chile, SAO Branch
[4] Space Research Institute, Russia

Abstract. Parameters of 22 GRBs are analysed. Correlations between energies and luminosities in optical and γ-ray range for 14 objects with measured red shift are found.

The first observed optical afterglow of a GRB was observed on February 28, 1997 (Guarnieri et al., 1997; van Paradijs et al., 1997). From this date and up to September 2000, 26 afterglows have been detected. The cosmological hypothesis is strongly supported by these observations, mainly from the measure of the large redshifts of spectral lines of underlying galaxies at the optical transient (OT) locations. The redshift determinations are crucial because without this information the evaluation of the energy release is impossible. Assuming a pure phenomenological approach, we believe that many possible informations are coded in the data obtained from the afterglow optical observations, so the problem is to find a key to decodify these data. A possible manner to extract information is to search for possible correlations between the measured quantities obtained from the GRBs and the up to now observed afterglows.

We used for statistical analysis the data collected in Piccioni et al. (2001): $L\gamma$ – peak luminosity in the γ-ray range, L_{opt} – peak luminosity in R band, E_γ – energy released in the γ-ray range, E_{opt} – energy released in R band, $\Delta T_{\gamma o}$ – delay between gamma event and afterglow detection in the host galaxy system, T_o – afterglow duration in the host galaxy rest system, z – redshift measured from underlying galaxy, α – slope of the afterglow fading power law (R band).

Our sample of date shows that:

1) concerning peak γ luminosity, there are two groups of GRBs with:
a) high γ luminosity: $10^{53} - 5 \times 10^{53}$ erg/s, 7 events
b) low γ luminosity: $5 \times 10^{50} - 10^{52}$ erg/s, 6 events

2) concerning peak optical luminosity, there are two groups of GRBs with:
a) high optical luminosity: $3 \times 10^{44} - 3 \times 10^{45}$ erg
b) low optical luminosity: $10^{43} - 3 \times 10^{44}$ erg

3) concerning the power law index of the light curve decay (R band), there are two groups of GRBs:

a) one with α-indices in the range 0.7–1.7 (17 objects)

b) one with α-indices in the range 1.9–2.5 9 (5 objects)

   4) there is a correlation between:

a) E_γ and L_γ. The correlation coefficient $R \approx 0.95$, the significance level $SL \approx 10^{-7}$; we included in the calculation GRB 980425, which is suspected to be connected with a supernova (Fig. 1). If we remove it, we obtain $R \approx 0.8$, $SL \approx 10^{-3}$ (Fig. 2)

b) E_{opt} and L_{opt}. We find $R \approx 0.75$, $SL \approx 10^{-3}$ in both cases

   5) concerning the connections between γ and optical parameters:

a) there is no dependence of E_{opt} from E_γ and L_γ, indicating that the energy of the whole optical event "forget" the γ event

b) there is a dependence of L_{opt} from E_γ (Fig. 3, without SN; Fig. 4, with SN) and from L_γ (Fig. 5, without SN; Fig. 6, with SN). This suggests that the optical emission at the very first time reflects the energetic of the γ event.

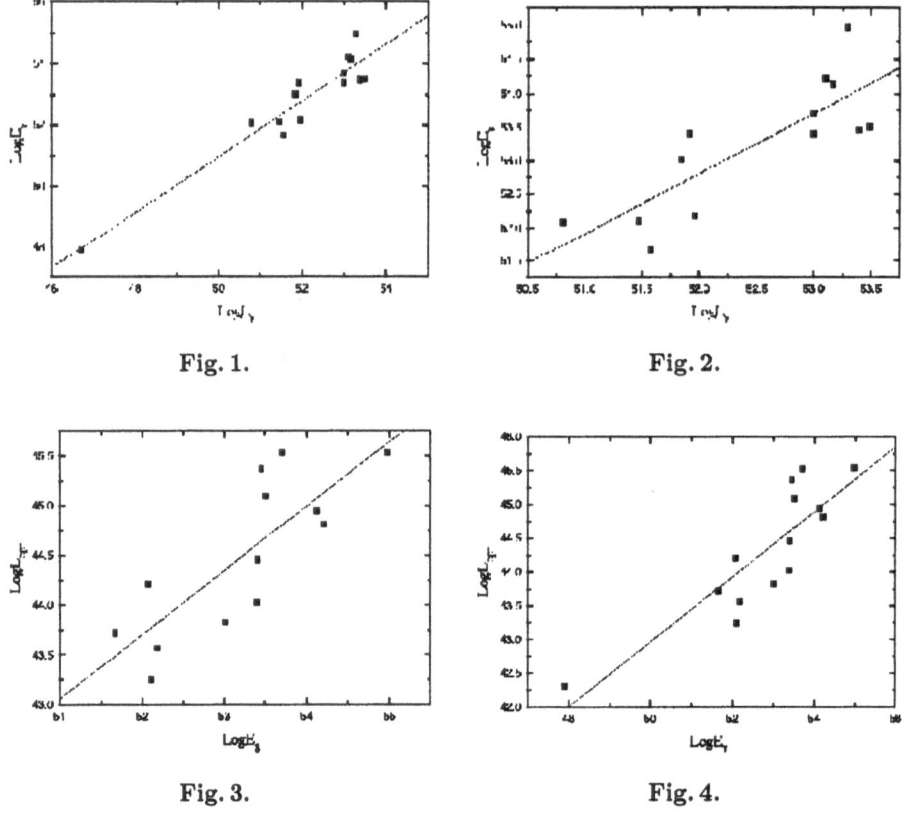

Fig. 1.   Fig. 2.

Fig. 3.   Fig. 4.

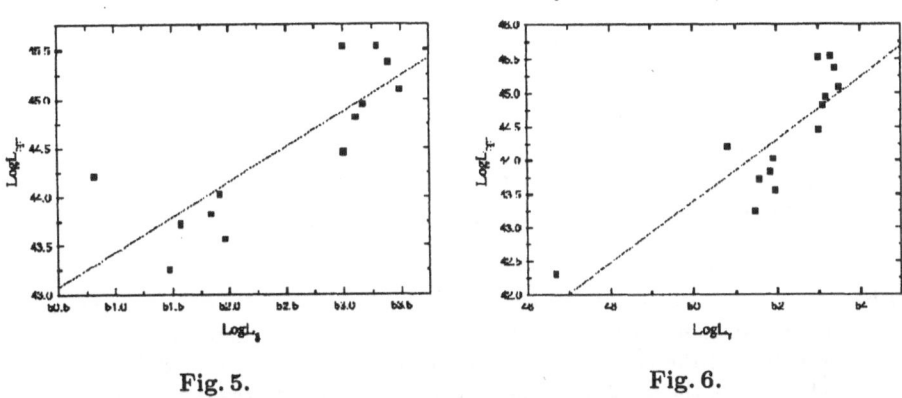

Fig. 5.   Fig. 6.

Other combinations of pairs of parameters do not show significant correlations.

Correlations between E_γ vs. L_γ, E_{opt} vs. L_{opt}, L_{opt} vs. L_γ, were already found by Beskin et al. (2000), in a sample of 7 GRBs.

It is worthwhile to note that the results shown in Figures 1-6 strengthen the identification of GRB 980425 with Supernova 1998 bw, because the slopes of the lines are very similar with and without the SN.

We believe these results can be a starting point for further analysis of GRBs physical properties.

This investigation was supported by the Italian Ministry of Foreign Affairs, by the University of Bologna (Funds for Selected Research Topics) and by the Russian Foundation of Fundamental Researches (grant 98-02-17570).

References

1. G. Beskin et al.: 'Connection between Parameters of GRB Afterglows'. In:Gamma-ray Bursts: 5th Huntsville Symposium, 2000, ed. by R.M. Kippen (American Institute of Physics, New York 2000) pp. 355-358
2. A. Guarnieri et al.: Astron. Astrophys. **328**, L13 (1997)
3. J. van Paradijs et al.: Nat. **386**, 686 (1997)
4. A. Piccioni et al.: in Proceedings of Vulcano Workshop, in press (2001)

Broad-Band Modelling of GRB Afterglows

Edo Berger[1], Re'em Sari[1], Dale Frail[2], and Shri Kulkarni[1]

[1] Division of Physics, Mathematics & Astronomy 105-24,
   California Institute of Technology, Pasadena, CA 91125, USA
[2] National Radio Astronomy Observatory, P. O. Box O, Socorro, NM 87801, USA

Abstract. Observations of GRB afterglows ranging from radio to X-ray frequencies generate large data sets. Careful analysis of these broad-band data can give us insight into the nature of the GRB progenitor population by yielding such information like the total energy of the burst, the geometry of the fireball and the type of environment into which the GRB explodes. We illustrate, by example, how global, self-consistent fits are a robust approach for characterizing the afterglow emission. This approach allows a relatively simple comparison of different models and a way to determine the strengths and weaknesses of these models, since all are treated self-consistently. Here we quantify the main differences between the broad-band, self-consistent approach and the traditional approach, using GRB 000301C and GRB 970508 as test cases.

1 Introduction

The quest for an understanding of GRB and afterglow physics, as well as the parameters that characterize the burst has recently led us to a new approach to the modelling of afterglow data. In principle, by modelling the afterglow data it is possible to extract the five parameters characterizing the synchrotron spectrum (ν_a, ν_m, ν_c, p, and F_{ν_m}) from which we can calculate the burst energy, the ambient medium density, and the fractions of energy in the magnetic fields and electrons [5]. At the same time, with accurate modelling it is possible to distinguish between the different models of afterglow emission, i.e. ISM vs. wind, and spherical vs. collimated outflow [1] [15] [7] [2] [3].

2 The Shortcomings of the Traditional Approach

Since the discovery of afterglow emission from GRBs in the late 1990s, the general approach to afterglow modelling has consisted of the following steps [10]. The data set collected for a particular burst was broken up into lightcurves and spectra, which were fitted separately. The spectra were modelled using the broken synchrotron spectrum in order to extract the value of p, and possibly the break frequencies. The lightcurves were each fitted separately to solve for the temporal decay slopes, α_i, which were then compared for consistency, and in the optical band to extract any host galaxy extinction. The temporal decay slopes were also used to distinguish between the different models of afterglow emission, and breaks were used to infer the existence of a jet geometry. This approach has several serious drawbacks:

- Only a few data points are modelled at a time, and the uncertainty in the derived parameters is large.
- The deduced model parameters and power-law indices are not always physically meaningful (e.g. can give $\epsilon_B > 1$).
- Since this approach employs the broken synchrotron power-law, the modelling of lightcurves and spectra near the break frequencies is inaccurate.
- It is extremely difficult to account for the changes in the spectrum and time dependences when the order of the break frequencies changes.

3   The Advantages of the Broad-Band Approach

Our approach attempts to remedy the aforementioned problems, and in addition to clearly identify the present shortcomings of afterglow studies. The procedure we use in modelling the data is significantly different. We use a broad-band data set ranging from radio to X-rays and fit it simultaneously [1]. We therefore give equal weight to all data points, and do not disregard scattered data points, which in the traditional approach are useless. Our approach also tests a complete model with all its different early and late time variations, including the transition to the sub-relativistic phase. It is therefore self-consistent since it does not include or exclude any assumptions and constraints that are part of the complete model. With this approach we gain the following advantages:

- All data points are used simultaneously since the model includes both the temporal and frequency dependence of each parameter.
- We use the Granot, Piran and Sari smoothed synchrotron spectrum [5] [6], which is a much more accurate and realistic representation of the actual data.
- We can easily include all special cases of the spectral and temporal evolution; therefore, any significant deviation of the data from the predicted models can be interpreted as a possibly new phenomenon (e.g. GRB 000301C [1] [4]).
- We can easily extract the values of the burst energy, ambient density and fractions of energy in the magnetic fields and electrons.
- We can directly determine which model (e.g. ISM vs. Wind) gives the most accurate description of the data using a simple χ^2 statistic.

4   Conclusion

The study of GRB afterglows and the extraction of the burst characteristics from the observations can be severly limited if a narrow-band approach is used. The problems of this traditional approach to modelling are numerous, but they can be easily solved if a broad-band, self-consistent approach is used instead. We have shown that the overall behavior of the afterglow emission can be easily studied within this approach, that the correct emission model can be unambiguously identified if the data set is large enough, and that the parameters characterizing the burst (e.g. energy, ambient density) can be easily solved for.

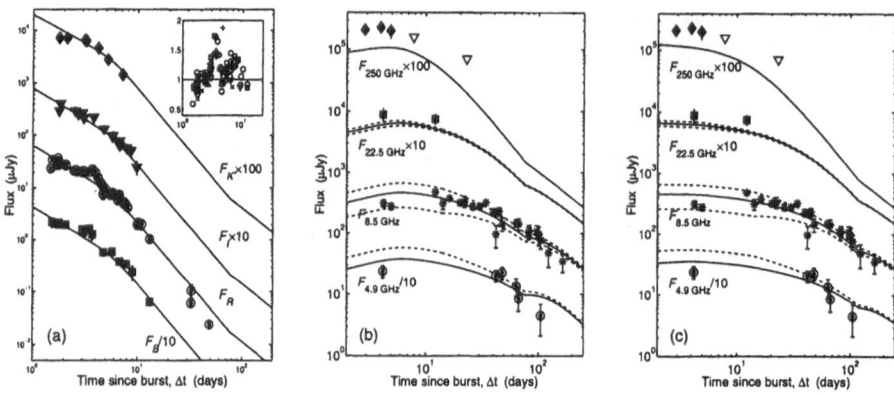

Fig. 1. (a) Optical and (b) radio lightcurves of GRB 000301C for the ISM+jet model and (c) the wind+jet model. The dashed lines indicate flux variation due to scintillation. The models include a jet break and a non-relativistic phase. The insert shows the achromatic bump which was only evident as a result of the global fitting [1].

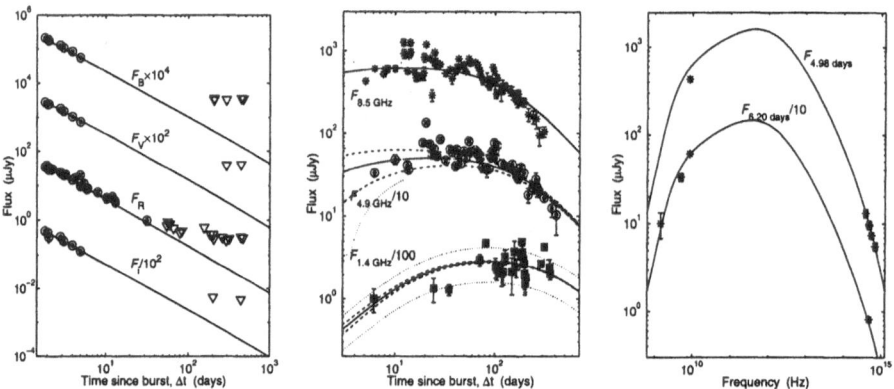

Fig. 2. Optical and radio lightcurves and spectra of GRB 970508 for the wind model. The Modelling includes the effect of host galaxy extinction and host flux density. Upper limits in the optical indicate measurements in which the host galaxy flux dominates.

References

1. Berger et al., ApJ, **545**, 56 (2000)
2. Chevalier, R. A., Li, Z.-Y., ApJ, **520**, L29 (1999)
3. Chevalier, R. A., Li, Z.-Y., ApJ, **536**, 195 (2000)
4. Garnavich, P. M., Loeb, A., Stanek, K. Z., ApJ, **544**, L11 (2000)
5. Granot, J., Piran, T., Sari, R., ApJ, **513**, 679 (1999)
6. Granot, J., Piran, T., Sari, R., ApJ, **527**, 236 (1999)
7. Rhoads, J. E., ApJ, **525**, 737 (1999)
8. Sari, R., Piran, T., Narayan, R., ApJ, **497**, L17 (1998)
9. Sari, R., Piran, T., Halpern, J. P., ApJ, **519**, L17 (1999)
10. Wijers, R. A. M. J., Galama, T. J., ApJ, **523**, 177 (1999)

The Jet and the Supernova in GRB 990712

G. Björnsson[1], J. Hjorth[2], P. Jakobsson[2], L. Christensen[2],
E.J. Lindfors[1], and S. Holland[3]

[1] Science Institute, University of Iceland
[2] Astronomical Observatory, University of Copenhagen
[3] Department of Physics, University of Notre Dame

Abstract. We show that the polarization measurements of GRB 990712 can be interpreted as originating in a modestly spreading jet. For that interpretation to be self-consistent, a break should be present in the optical light curve. We re-analyze the optical light curve and show that not only is it consistent with such a break, it also shows a strong supernova-like component.

1 The Polarization Data and the Light Curves

The polarization measurements of the optical afterglow of GRB 990712 showed it to vary from 2.9%±0.4% (at burst age of 10.6 hr) to 1.2%±0.4% (16.7 hr) and 2.2%±0.7% (34.7 hr) [6]. The position angle remained constant. The varying degree of polarization at a constant position angle may seem to be in conflict with existing models of polarization in gamma-ray burst afterglows.

The optical light curve of GRB 990712 was shown by to be well fitted by a decaying power law, t^{-1}, plus a constant host contribution [7]. Adding a SN1998bw component resulted in a worse fit. Reference [5] also concluded that a supernova component was not needed to interpret the light curve of GRB 990712.

The polarization data may be explained by the model of [4]. If the observer's line of sight makes a fixed angle, θ_0, with the symmetry axis, a spreading jet causes $f = \theta_0/\theta_c$, to decrease with time (θ_c is the jet opening angle). The degree of polarization is very sensitive to f. Figure 1 a) shows polarization light curves of a jet of fixed opening angle for 2 different values of f. Note that the polarization light curve has 2 maxima, with the otherwise constant position angle changing by 90° at the minimum. Note also that the minimum occurs typically less than 10 hours after the burst and so does the 90° change in position angle.

Holding θ_0 fixed, allowing the opening angle to increase, results in a decreasing f and thus a decrease in the amplitude of the corresponding polarization light curve. As the first polarization point is obtained 11 hr after the burst [6], we assume the position angle has already changed by 90°. The light curve evolution then takes place entirely under the second maximum shown in Fig. 1 b). A modest change in opening angle of 1° during the 6 hours between the first two data points is sufficient to explain the observed polarization behaviour [1].

We re-reduced all publicly available ESO data in the V-band for consistency and independence. We also included additional V-band data [7] and the *HST* point [3]. We also consider the entire R-band light curve [7]. In addition, we used

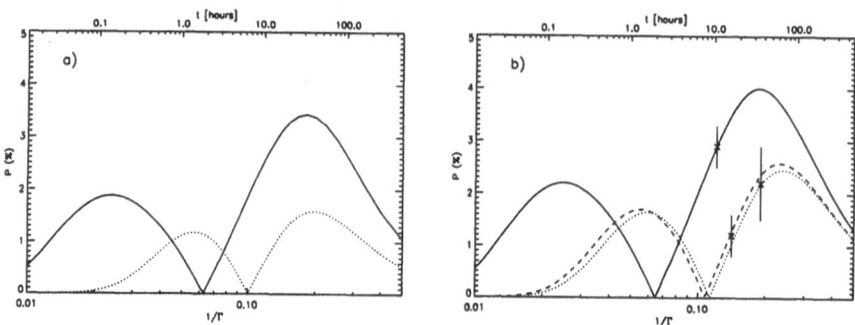

Fig. 1. a) Typical evolution of polarization light curves for constant $\theta_c = 5°$, $f = 0.9$ (solid) and $f = 0.67$ (dotted). Note that the 1st max and the min shifts to the right as f decreases, the 2nd max only decreases in amplitude. b) Polarized light curves of an outflow with a fixed line of sight angle. Increasing θ_c, decreases f. The solid curve has $\theta_c = 5.1°$ and $f = 0.9$, the dashed has $\theta_c = 6.0°$ and $f = 0.77$, the dotted has $\theta_c = 6.2°$ and $f = 0.74$. As in a), the 1st max and the min move to the right while the 2nd max shifts to the right by a factor of 2 in time but does not cross the initial curve (solid). The polarization values for GRB 990712 are superimposed. (From [1]).

the ground based measurements of the host magnitudes, $V = 22.40 \pm 0.04$ and $R = 21.91 \pm 0.04$ [5], reducing the number of free parameters in our fits. Our host subtracted light curves, in V and R are shown in Fig. 2.

A broken power law fit to the first 2.5 days of V-band data gives $\alpha_1 = -0.83 \pm 0.03$, steepening to $\alpha_2 = -3.06 \pm 1.28$ at $t_b = 1.61 \pm 0.91$ days ($\chi_4^2 = 0.434$). The evidence for a break is not strong, however, as it hinges mostly on one data point. There is no evidence for a break in the R-band, that decays at $\alpha = -0.94 \pm 0.02$ ($\chi_{11}^2 = 1.39$), that is different from the early V-band decay.

2   The Interpretation

A break at $t_b \approx 1.6$ days would be in agreement with the interpretation of [1], that the observed early light curve results from a sideways expanding jet. The light curve would then steepen by $\Delta\alpha = 1 - \alpha_1/3 = 1.28 \pm 0.01$, quite consistent with the observed steepening of 2.23 ± 1.28. If the break at 1.6 days in the V-band is real, the late time light curve is seen to be dominated by a component that is first clearly detected at a burst age of about 7 days. Although no break is apparent in the R-band, this late time component is also clearly seen there.

It is simplest to interpret this late time component as being due to a supernova. We also show in Fig. 2, the light curve evolution based on the SN1998bw light curve at a redshift of $z = 0.434$ [5,9], assuming $H_0 = 65$ km/s/Mpc, $\Omega_0 = 0.2$ and $\Omega_\Lambda = 0.0$. We have not attempted a fit to the SN1998bw light curve to the data, as it is unknown if it is typical of GRB associated supernovae.

The lack of V-band data from days 1–7 makes it difficult to quantify the significance of a break in the light curve. The strongest evidence for it may come

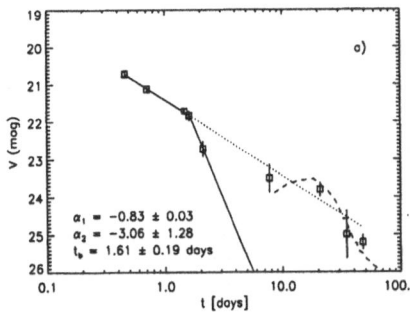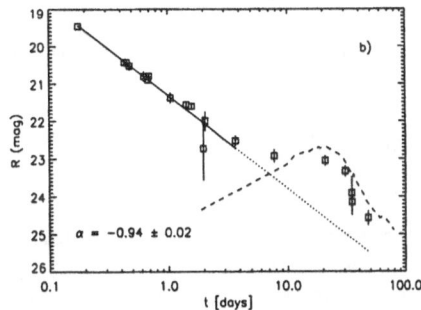

Fig. 2. Left panel shows the V-band light curve of GRB 990712. Note the break after about 1.6 d and the prominent supernova component. The light curve of SN1998bw at $z = 0.434$ is shown dashed. The dotted line is an extrapolation of the early light curve. Right panel shows the R-band light curve, again with the SN1998bw light curve superimposed. Solid line is a fit to the first 3 days. No break is seen in this case, but the supernova component rises well above the extrapolated power law fit. (From [2]).

from the light curve and the polarization data *together*, as the same physical model then accounts for both measurements, implying that the early afterglow is produced in a collimated outflow.

The R-band light curve does not show a break, but shows a late time light curve behavior similar to the V-band. A supernova-like component is then a necessary ingredient to explain the late time light curves. The argument may be reversed, because accepting the late time behavior in the R-band as being due to a supernova, we should expect similar behavior in the V-band. A break in the V-band light curve is then demanded by the data.

The case of GRB 990712 provides independent evidence from two different sets of measurements for a jet in a GRB, i.e. the polarization measurements and the light curve properties. As complete time coverage as possible should be attempted for future optical afterglows for at least a full month. The first few days are most demanding as a break in the light curve is expected to fall within this burst age. It is also important to follow the late time behavior closely to look for a possible supernova accompanying the burst.

References

1. G. Björnsson, E. J. Lindfors: ApJ, **541**, L55 (2000)
2. G. Björnsson et al.: ApJL, submitted (2001)
3. A. Fruchter et al.: GCN 752 (2000)
4. G. Ghisellini, D. Lazzati: MNRAS, **309**, L7 (1999)
5. J. Hjorth et al.: ApJ, **534**, L147; **539**, L75 (2000)
6. E. Rol et al.: ApJ, **544**, 707 (2000)
7. K. C. Sahu et al: ApJ, **540**, 74 (2000)
8. R. Sari: ApJ, **524**, L43 (1999)
9. P. M. Vreeswijk et al.: ApJ, **546**, 672 (2000)

On the Transient Fe K-Edge
in the Prompt Emission of GRB 990705

Markus Böttcher[1,2], Charles D. Dermer[3],
Lorenzo Amati[4], and Filippo Frontera[4,5]

[1] Rice University, Houston, TX, USA
[2] Chandra Fellow
[3] Naval Research Lab, Washington, D.C., USA
[4] ITESRE, CNR, Bologna, Italy
[5] Dip. di Fisica, Università di Ferrara, Ferrara, Italy

Abstract. A transient Fe absorption edge during the prompt phase of a GRB has
recently been observed for the first time. This is evidence for photoionization of the
GRB environment by the burst radiation and allows diagnostics of the density structure
and element abundances in the vicinity of the burst. We model the time-dependent
photoionization and X-ray radiation transport, and deduce constraints on the size and
composition of the photoionized region responsible for the absorption feature. We find
that the intervening material must contain $\sim 44\,\Omega\,M_\odot$ of iron within ~ 1.3 pc of the
burst source, assuming the measured best-fit ~ 75-fold overabundance of iron. Possible
solutions to reduce the required amount of iron are briefly discussed.

1 Introduction

Recently, the detection of a transient Fe K edge in the prompt X-ray emission of
GRB 990705 has been reported [1]. The transient nature of the absorption feature
is evidence for photoionization of the absorber by the prompt burst emission. The
corresponding time-dependent photoionization and radiation transfer problem
had been simulated before [3], predicting that excess absorption features are
either transient on time scales of $\lesssim 1$ minute, or do not change appreciably as
the burst evolves, depending primarily on the average distance of the absorbing
material from the GRB. In this paper, we apply the formalism developed in [3]
to model the observed time-dependence of the Fe K edge in GRB 990705 and
deduce constraints on the radial extent and the composition of the absorbing
material.

2 Model Setup

We model the absorbing material as a uniform shell with density n_{sh} and inner
and outer radii $r_{in,out}$. We note, however, that from the absorption feature alone
we can not make any statement concerning the isotropy or anisotropy of the
circumburster material (CBM). From their X-ray spectroscopy of GRB 990705
and analytical estimates of the photoionization of the CBM due to the prompt
burst radiation, Amati et al. [1] found $n_{sh} \sim 10^5$ cm^{-3}, $r_{out} \sim 0.1$ pc, and 75-fold

iron overabundance w.r.t. standard solar-system abundances. Since the available spectroscopic data do not allow to put stringent constraints on elements other than iron, we fix the element abundances within the shell to the solar-system values, except for iron, for which we assume a 75-fold overabundance.

The prompt burst emission from GRB 990705 can be well represented by a FRED-type single pulse. We can thus describe the GRB spectral evolution by the parametrization of [5], based on the external shock model for GRBs. The observables of GRB 990705 [1] can be reproduced in the framework of the parametrization of [5] using $E_0 = 10^{54}$ erg, $\Gamma_0 = 247$, $n_0 = 100$ cm^{-3}, $v = 0.9$, $\delta = 0.5$, $g = 1.8$, $q = 2.5 \cdot 10^{-3}$.

We are using the code developed in [3] and [2] to simulate the time-dependent photoionization and radiation transfer problem in the CBM. The modeling is done by varying the shell density n_{sh}, the hydrogen column density N_H (sh) through the shell, and the radius r_{in} of the shell.

3 Modeling Results

Fig. 1 shows sample results of some of our simulations for various values of the column depth N_H and the radius r_{in} of the CBM. The shaded regions indicate the measured depth of the iron line (with 1 σ error) during four time intervals after the onset of GRB 990705. We find that the measurements place rather tight constraints on the parameters N_H and r_{in}. Good agreement is achieved for $N_H = 1.8 \times 10^{22}$ cm^{-2}, $r_{in} = 4 \times 10^{18}$ cm.

Our value for the shell radius r_{in} differs by ~ 1 order of magnitude from the one obtained analytically by Amati et al. [1]. Our column density refers to

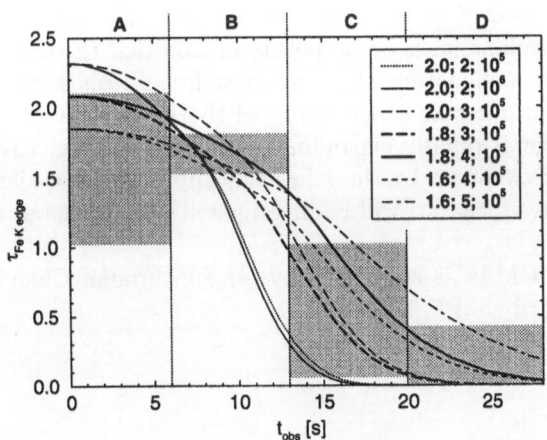

Fig. 1. Simulated time evolution of the depth of the Fe K edge in GRB 990705 for varying column density, N_H, radius r_{in}, and density n_{sh} of the absorbing material. The shaded regions show the values allowed by the measurements [1]. The numbers in the legend are N_H [10^{22} cm^{-2}], r_{in} [10^{18} cm], and n_{sh} [cm^{-3}], respectively.

the initial column density, before any photoionization occurs. Since significant photoionization already affects the environment during time segment A, the N_H value for the fit to this time segment in [1] has to be lower than our initial value. We deduce a total mass of $M_{\mathrm{CBM}} = 365\,\Omega\,M_\odot$ in the shell, where Ω is the solid angle subtended by the absorber. This corresponds to a mass of iron of $M_{\mathrm{Fe}} = 44\,\Omega\,M_\odot$. This seems to be strong evidence for an anisotropic distribution of the absorbing material around the burster ($\Omega \ll 1$).

4   Alternative Scenarios

Within the scenario investigated here, in which the recombination time scale in the absorber is long compared to the photoionization and burst evolution time scales, the transient absorption feature in the prompt X-ray emission from GRB 990705 implies that a large amount of strongly metal-enriched material must be concentrated in a volume of $r \sim 1.3$ pc around the source of the burst. If the CBM were isotropic around the burst source, then the CBM would have to contain $\sim 560\,M_\odot$ of iron. This appears to be strong evidence for an anisotropic distribution of the absorbing material around the burst source.

A possible alternative to our scenario could assume that the Fe is concentrated in small dense pellets of radius $R \sim$ a few $\times 10^8$ cm, which would need to have a covering factor of ~ 1 over the effectively emitting surface of the burst source and be located at $\sim 10^{17}$ cm from the burster. In such a scenario, the required amount of Fe is reduced to $\lesssim 1\,M_\odot$, consistent with current supernova-related GRB progenitor models. A critical prediction of this alternative would be a rapid decline from a large initial value of the depth of the absorption edge ($\tau_{\mathrm{Fe\,K}} \gtrsim 5$) within the first ~ 3 seconds of the burst, which would have remained undetectable for the BeppoSAX WFC due to the low flux level during this phase. This prediction should be testable with the SXC on board HETE II. However, the scenario invoking such dense pellets of iron-rich material would also lead to very rapid recombination after the most intense phase of the burst, which would lead to an increase of the depth of the Fe K edge after ~ 15–20 s, unless the absorber is rapidly expanding due to efficient radiative heating of the pellets, or it is swept up by the relativistically expanding ejecta of the burst. These alternative scenarios will be investigated in more detail in a forthcoming publication.

The work of M.B. is supported by NASA through Chandra Postdoctoral Fellowship Award no. PF 9-10007.

References

1. Amati, L., et al., Science, **290**, 953 (2000)
2. Böttcher, M., ApJ, **539**, 102 (2000)
3. Böttcher, M., et al., A&A, **343**, 111 (1999)
4. Böttcher, M., & Fryer, C. L., ApJ, **547**, in press (2001)
5. Dermer, C. D., et al., ApJ, **513**, 656 (1998)

On the Detectability of Delayed X-Ray Flashes from GRBs

Markus Böttcher[1,2], Chris L. Fryer[3,4], Edison P. Liang[1], and Ian A. Smith[1]

[1] Rice University, Houston, TX, USA
[2] Chandra Fellow
[3] Los Alamos National Lab, Los Alamos, NM, USA
[4] Feynman Fellow

Abstract. We investigate the observability of delayed X-ray flashes predicted in currently popular GRB models as a result of the interaction of a non-relativistic blast wave with a disk of pre-ejected material from the GRB progenitor. We have recently calculated the quasi-thermal X-ray line features produced in such a scenario. In the Hypernova/Collapsar model, delayed (a few days–several months after the GRB) bursts of line-dominated, thermal X-ray emission may be expected. The He-merger scenario predicts similar X-ray emission line bursts $\lesssim$ a few days after the GRB. Deep observations and monitoring of X-ray afterglows with *Chandra* and *XMM-Newton* will be able to detect such delayed X-ray flashes from GRB sources out to at least $z \sim 1$.

1 Introduction

The recent discoveries of time-dependent X-ray spectral features in GRB 990705 [1], GRB 991216 [10], and GRB 000214 [2] by the *BeppoSAX* and *Chandra* satellites demonstrate the power of the new generation of pointed X-ray telescopes for GRB research. While early modeling work on emission line features in GRB afterglows focused on photoionization scenarios [6,7,3,14], a re-analysis of the BeppoSAX spectrum of GRB 970508 favored a collisionally ionized over a photoionized plasma [9]. In addition, photoionization scenarios require extreme conditions, which current GRB progenitor models are very unlikely to produce.

We have recently calculated the expected X-ray signatures in two of the currently popular GRB progenitor models [4]. Prominent X-ray spectral features result from the hydrodynamical interaction of ejecta of a SN associated with the GRB explosion with a disk-like structure of material which has been pre-ejected during a common-envelope phase. Here we discuss the prospects of detecting the predicted X-ray spectral features with *Chandra* and *XMM-Newton*.

2 Disk Formation in the Collapsar and He-Merger Scenarios

Nearly all of the formation scenarios of BHAD GRBs invoke a "common-envelope" phase. Friction and/or tidal forces cause the companion to spiral in towards the giant's helium core, ejecting the hydrogen envelope. The hydrogen envelope carries away much of the orbital angular momentum and is preferentially ejected

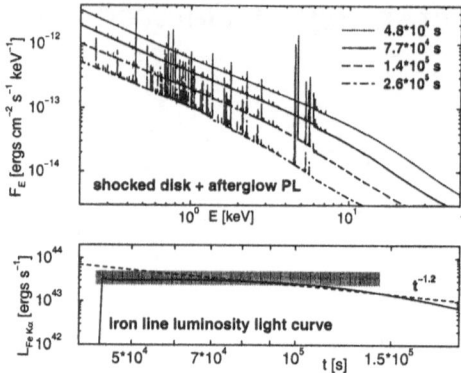

Fig. 1. Model simulation to reproduce the iron line in GRB 000214 [2]. Top: Energy spectra at different times. Bottom: Time-evolution of the luminosity in the iron line (shaded area indicates the luminosity and time interval of the line detection).

in the orbital plane, forming a disk around the GRB engine. Double NS, BH-NS, and BH-WD mergers go through a common envelope phase long before they actually merge. Collapsar and He-merger GRBs, on the other hand, occur shortly after their common-envelope phase. In the He-merger scenario [5], one expects $\gtrsim 1\,M_\odot$ of pre-ejected material in a disk-like structure with inner radii $r_{in} \lesssim 10^{13}$ cm. In the collapsar scenario [5], the disk inner radius is expected to be located at $r_{in} \gtrsim 10^{14}$ cm. In both scenarios, it is likely that the GRB event is associated with a SN, which, in directions other than the symmetry axis, appears like a "regular" supernova, ejecting $\sim 1\,M_\odot$ of material at $v_s \sim 10^9$ cm s^{-1}.

3 Predicted X-Ray Spectral Features

When the SN ejecta hit the pre-ejected disk, a shock is induced in the disk. The shocked material is heated to temperatures of $T \sim 10^7$–10^8 K, resulting in line-dominated, thermal X-ray emission [4]. Fig. 1 shows a simulation appropriate to reproduce the iron line feature recently observed in GRB 000214 [2]. Here, we have assumed a disk density profile appropriate for a $25\,M_\odot$ progenitor and $r_{in} = 6 \times 10^{13}$ cm. We find that the shock wave / disk interaction expected for the He-merger scenario can plausibly produce an Fe Kα line of isotropic luminosity $L_{\mathrm{Fe\,K\alpha}} \sim 10^{44}$ ergs s^{-1} maintained over $\lesssim (1 + z) \times 10^4$ s after the ejecta begin to interact with the disk. In the collapsar case, maximum iron line luminosities in excess of $\sim 10^{42}$ ergs s^{-1} are possible, while the typical time delay between the GRB and the onset of the X-ray flash is now $\sim (1 + z) \times 10^6$ s.

4 Observational Prospects

In both the He-merger and the collapsar/hypernova scenarios, a quasi-thermal X-ray flash from the shock-heated disk is expected. The maximum X-ray luminosity and the onset and duration of this X-ray flash depend on the inner disk

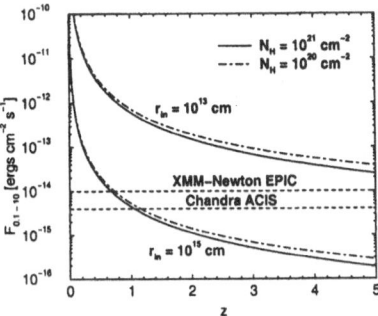

Fig. 2. Peak 0.1–10 keV X-ray fluxes as a function of redshift, for representative cases of He-merger and collapsar GRBs. Horizontal lines show the nominal point-source sensitivitiess of *Chandra* ACIS and *XMM-Newton* EPIC for a 10 ksec observation.

radius, primarily determined by the duration of the common-envelope phase. For the He-merger, we expect $r_{in} \sim 10^{13}$ cm, while for the collapsar/hypernova, 10^{14} cm $\lesssim r_{in} \lesssim 10^{17}$ cm. The onset and decay time scale of the resulting secondary X-ray flash scale as $\Delta t_X \propto r_{in}$, while $L_{Fe\,K\alpha} \propto r_{in}^{-1}$.

Fig. 2 shows the absorbed 0.1–10 keV peak fluxes resulting from representative simulations for the He-merger and the collapsar case, as a function of redshift. The predicted X-ray flashes may be detectable out to redshifts of at least $z \sim 1$ for an inner disk radius of $r_{in} \sim 10^{15}$ cm. For such a source at $z = 1$, we expect the onset of the secondary X-ray burst ~ 3 weeks after the GRB.

If long-duration GRBs are caused by BHAD models, we predict that there may be many GRBs in which a secondary X-ray flash should emerge even long after the continuum X-ray afterglow has faded away. Late-time X-ray monitoring of GRB afterglows over weeks or even months after the GRB with *Chandra* and/or *XMM-Newton* might be able to detect these delayed X-ray flashes, providing a crucial test for the currently most popular classes of GRB progenitor models.

References

1. Amati, L., et al., Science, **290**, 953 (2000)
2. Antonelli, L. A., et al., ApJ, **545**, L39 (2000)
3. Böttcher, M., et al., A&A, **343**, 111 (1999)
4. Böttcher, M., & Fryer, C. L., ApJ, **547**, in press (2001)
5. Fryer, C. L., et al., ApJ, **526**, 152 (1999)
6. Ghisellini, G., et al., ApJ, **517**, 168 (1999)
7. Lazzati, D., et al., MNRAS, **304**, L31 (1999)
8. MacFadyen, A. I., & Woosley, S. E., ApJ, **524**, 262 (1999)
9. Paerels, F., et al., ApJ, **535**, L25 (2000)
10. Piro, L., et al., Science, **290**, 955 (2000)
11. Weth, C., et al., ApJ, **534**, 581 (2000)

Wolf-Rayet Stars and GRB Connection

Anatol Cherepashchuk and Konstantin Postnov

Sternberg Astronomical Institute, 119899 Moscow, Russia

Abstract. Arguments are given favoring possible connection of GRB with core collapse of massive Wolf-Rayet stars. The possibility of GRB to be a transient phenomenon in the early history of galactic star formation related to evolution of very massive metal-free stars is discussed.

Cosmic gamma-ray bursts (GRB) can be produced by several classes of astrophysical sources: coalescencs of binary neutron stars (NS) and/or black holes (BH) [1], core collapses of massive stars [2], or by electromagnetic process near the young magnetized NS [3]. There is a growing evidence that cosmic GRBs may be associated with star-forming regions [4]. If confirmed, this favors GRB relation to massive star evolution and makes other GRB progenitors (such as double neutron star coalescences) less likely.

At late evolutionary stages, very massive stars lose their hydrogen-rich envelopes (via stellar wind or in a binary system) and are observed as helium-rich Wolf-Rayet (WR) stars. They are considered as progenitors of Ibc type supernovae. If GRB are directly related to collapses of massive stars, there should be some links between properties of WR stars before collapse and GRB.

The energetics of GRB with known redshift spans from $\approx 7 \times 10^{51}$ ergs to $\approx 2 \times 10^{54}$ ergs (Fig. 1). Adding GRB980425, which is possibly associated with peculiar type Ic supernova SN 1998bw in a nearby galaxy ESO 184-G82, either evidences for a *bimodality* of GRB energy distribution or for an extremely broad luminosity function (more than 5 orders of magnitude!).

We suggest that observed broad (and, possibly, bimodal) distribution of GRB energetics is related (1) to the broad distribution of final masses of CO cores of *observed* WR stars before collapse and (2) to the *observed* bimodal distribution of masses of relativistic remnants of stellar evolution (NS and BH). The arguments favoring the link with WR stars are:

(1) *Energetics* The observed GRB energy release $\Delta E \simeq 10^{51} - 10^{54}$ erg is roughly comparable with the wide range of CO cores of WR stars before collapse, $M_{CO}^f \simeq 2 - 40 M_\odot$, Fig 2. During BH formation without mass ejection the maximum available energy $\propto M_{CO}$ can reach $10^{53} - 10^{56}$ ergs for the observed mass range of compact remnants, i.e. conversion of 1% of the available energy into kinetic energy of shocks with subsequent radiation would be sufficient to explain the broad luminosity function.

(2) *Bimodality of GRB energy distribution and stellar remnant mass distribution.* The known masses of NS and BH in binaries are shown in Fig.2. NS

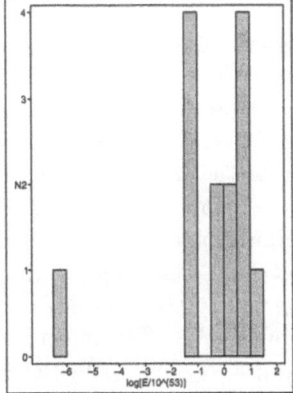

Fig. 1. Total energy release of GRB with known redshifts

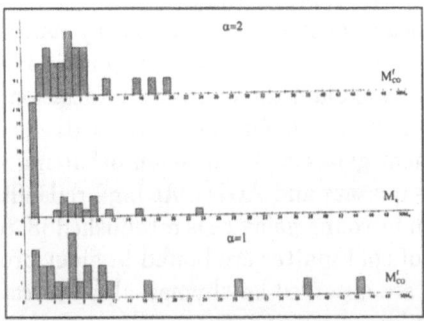

Fig. 2. Masses of final CO cores M_{CO} of WR stars as calclated using observed stellar wind mass loss $\dot{M}_{WR} \propto M^\alpha$. Masses of known NS and BH candidates in binary systems are also shown. Figure from [5]

masses are grouped in a narrow interval is $M_{NS} = (1.35 \pm 0.15)M_\odot$, while BH masses fall in a broader range $M_{BH} = (5 \div 15)M_\odot$. A real gap between NS and BH masses is observed. GRB with low energetics associated with peculiar supernovae type Ic (such as GRB980425) can be explained by collapses of bare CO cores of massive stars with significant rotation which causes most envelope to be ejected and neutron star to be formed, while collapses of slower rotating cores do not accompanied by a significant envelope ejection and lead to black hole formation. In the latter case an energetic GRB can be generated with energy proportional to the pre-collapse core mass.

(3) *Association of GRB with star-forming regions*

(4) *A diversity of the observed afterglows* A GRB in a binary system can induce different optical phenomena due to illumination of the companion's atmosphere by hard X-ray and gamma-radiation. These effects can occur in a time interval $\Delta t_{opt} > D/c$ after the burst (D is the distance to the optical star from GRB, c is the speed of light). About 50% of all WR stars in our Galaxy can be in binaries with O-star or A-M-star. E.g., for WR+O system V444 Cyg with an

orbital period of $P = 4^d.2$ $D \approx 40R_\odot$ and the time delay $\Delta t_{opt} \approx 100$ s, and for parameters of WR+O binary system CV Ser $\Delta t_{opt} \approx 300$ s. An extremely bright optical emission ($V \simeq 9^m$) was observed in GRB990123 only 50 s after the burst beginning. Another example is a peculiar shape of achromatic optical afterglow observed in GRB000301c. The observed several peaks separated by 2-3 days can be a manifestation of the underlied binary system, e.g., through the binary-period shaped mass loss before collapse. Orbital periods of $\sim$ several days perfectly fit the observed period range $1^d.6 \div 2900^d$ in WR+O binary systems.

The GRB event rate per unit comoving volume using BATSE data with $F_{tr} = 0.1$ ph/cm^2 [6] is $\mathcal{R}_{GRB} \sim 10^{-9}$GRB/yr/Mpc^3, about 10^{-7} per year in the average galaxy with a mass of $10^{11} M_\odot$. The mean formation rate of all types of WR stars in the Galaxy is $\mathcal{R}_{WR} = R_\odot(N_{WR}/N_\odot) \times (\Delta t_\odot / \Delta t_{WR}) \sim 1/1000\text{yr}^{-1}$, i.e. by a factor of 1000 exceeds that of GRB. However, for the most compact WR stars of type WO (3 out of 200 known), the difference is smaller. This issue can be solved either by postulating moderately thin gamma-ray jets or, admitting quasi-spherically symmetric emission, by assuming the existence of some "hidden" collapse parameters (rotation, magnetic field, etc.) [7].

Another possibility to explain the observed association of cosmic GRB with star-forming regions at high redshifts and their extreme rarity is that GRB may represent a transient galactic phenomenon occurring at the early stages of galactic evolution, like quasars and AGNs. At high redshifts $z \sim 1-2$ a violent epoch of star formation in young galaxies is established [8] to have occurred. It is also known that a lot of cold matter are bound in giant proto-galactic clouds at redshifts $z > 2$, which are observed as "Lyman-alpha forest" of absorption lines in quasar spectra. The formation of very massive stars 100-500 $M_\odot$ which final collapse into massive black holes took place at that epoch. Such massive star can not form from matter enriched with metals because of pulsational instabilities (e.g. [9]). At low metallicity at the epoch of violent star formation, however, they could have formed. The weakness of GRB980425 in a nearby galaxy can be a natural consequence of smaller upper masses of stars at the present epoch.

References

1. S.I. Blinnikov, I.D. Novikov et al.: Sov. Astron. Let. **10**, 177 (1984)
2. S.E. Woosley: ApJ **405**, 273 (1993)
3. V.V. Usov: Nature **357**, 472 (1992)
4. P. Vreeswijk et al.: (2000) T. Galama, R.A.M.J. Wijers: astro-ph/0009367 (2000)
   J. Bloom, S.R. Kulkarni, S.G. Djorgovski: ApJ in press (astro-ph/0010176) (2000)
   L.A. Antonelli, L.Piro et al.: ApJ in press (2000)
5. A.M. Cherepashchuk: Astron. Rep. in press (2001)
6. B.E. Stern, Ya. Tikhomirova et al.: ApJ **540**, L21 (2000)
7. E. Ergma, E.P.J. van den Heuvel: Astron. Astrophys. **331**, L29 (1998)
8. P. Madau et al.: MNRAS **283**, 1388 (1996)
9. I. Baraffe, A. Heger, S.E. Woosley: ApJ in press (astro-ph/009410) (2000)

Follow-Up Observations from Observatories Based in Spain

Javier Gorosabel[1], Alberto J. Castro-Tirado[2,3], Jochen Greiner[4], José María Castro Cerón[5], Sylvio Klose[6], and Niels Lund[1]

[1] Danish Space Research Institute,
   Juliane Maries Vej 30, DK-2100 Copenhagen Ø, Denmark
[2] Instituto de Astrofísica de Andalucía (IAA-CSIC),
   P.O. Box 03004, Granada, Spain.
[3] Laboratorio de Astrofísica Espacial y Física Fundamental (LAEFF-INTA),
   P.O. Box 50727, E-28080 Madrid, Spain.
[4] Real Instituto y Observatorio de la Armada, Sección de Astronomía,
   11100 San Fernando-Naval, Cádiz, Spain
[5] Astrophysikalisches Institut Potsdam,
   An der Sternwarte 16, 14482 Potsdam, Germany
[6] Thüringer Landessternwarte, Sternwarte 5, D-07778 Tautenburg, Germany

Abstract. We present a review of the follow-up observations carried out from observatories located in Spain; Calar-Alto, Izaña and Roque de Los Muchachos. It summarizes the observations carried out by our group for 27 GRBs occurred in the period 1999-2000, spanning from GRB990123 to GRB001007.

1 Introduction

Since the discovery of the GRB optical counterparts in 1997 [26] a great effort in the field has been carried out from many ground-based observatories. Here we present a summary of the optical/IR follow-up observations performed in 1999-2000 from several observatories based in Spain; La Palma, Izaña and Calar Alto (CAHA).

2 Observations and Results

We have performed optical and IR observations in 1999-2000 for many SAX, XTE and IPN GRB error boxes.

For the optical observations we have used the following telescopes: the 1.5OSN, INT, IAC80, JKT, NOT, TCS, 1.23-m CAHA and specially the 2.2-m CAHA Telescope. The observations at JKT, NOT, IAC80 and INT were performed at Spanish time. The CAHA observations were done by means of override programs either at German Time (P.I.: J. Greiner and S. Klose) or Spanish time (P.I.: A.J. Castro-Tirado).

Concerning the IR follow-ups, the observations have been mostly performed from CAHA, which is equipped with IR instrumentation very suitable for GRB follow-ups. Among the IR instruments mounted on the CAHA telescopes we

Table 1. Summary of the observations performed from Calar Alto, Izaña and La Palma.

Name	OT	Telescope	Filter	Reference
GRB990123	YES	NOT, 2.2m(CAFOS), 1.23m, 3.5m, TCS	BVRIJHK'	[1]
GRB990520	NO	2.2m(CAFOS), 3.5m(OMEGA)	RH	[2]
GRB990704	NO	NOT, IAC80(CCD), 2.2m(CAFOS)	BIR	[3]
GRB991106	NO	INT(WFC), 1.23m(CCD), 1.5OSN(CCD)	RI	[4–6]
GRB991208	YES	INT(WFC), 2.2m(CAFOS), 1.23m(CCD)	BVRI	[7]
GRB991216	YES	2.2m(CAFOS)	VR	[25]
GRB000115	NO	IAC80(CCD), 1.23m(CCD)	BVR	[19]
GRB000301A	NO	1.23m(CCD)	R	–
GRB000301C	YES	1.23m(CCD), 2.2m(CAFOS)	BVRI	[20,23]
GRB000313	NO	1.23m(MAGIC), JKT(CCD)	BVRIK'	[8]
GRB000315	NO	NOT(ALFOSC)	R	–
GRB000408	NO	IAC80(CCD)	R	[18]
GRB000418	YES	1.23m(MAGIC), 3.5m(OMEGA)	JK'	[21]
GRB000424	NO	1.23m(MAGIC)	K'	–
GRB000508B	NO	NOT(ALFOSC)	R	–
GRB000519	NO	1.23m(MAGIC)	JK'	–
GRB000604	NO	2.2m(CAFOS)	R	–
GRB000607	NO	2.2m(CAFOS)	R	–
GRB000615	NO	IAC80(CCD)	R	[22]
GRB000620	NO	2.2m(CAFOS)	R	[14]
GRB000623	NO	2.2m(CAFOS)	R	[15]
GRB000630	YES	2.2m(CAFOS)	R	[17,12]
GRB000830	NO	2.2m(CAFOS)	V	–
GRB000911	NO	2.2m(CAFOS)	R	[24]
GRB000925	NO	2.2m(CAFOS)	R	–
GRB000926	YES	2.2m(CAFOS)	BVRI	[16,13]
GRB000107	YES	IAC80(CCD)	BVR	[9]

would like to remark Omega-Cass and Omega-Prime mounted on the 3.5-m Telescope. The field of view (FOV) of Omega-Prime (6.8' × 6.8') and Omega-Cass (up to 5' × 5') allow to cover SAX (and even IPN) error boxes with single pointings, avoiding inconvenient mosaics. Also the MAGIC IR camera mounted on the 1.23-m Telescope is a very important support of the observations performed at the 3.5-m Telescope, as occurred for the discovery of the GRB000418 counterpart [21].

As it can be seen in Table 1 the most used telescope/instrumentation is the 2.2m(+CAFOS) configuration. The large FOV of CAFOS (diameter of 16') is specially useful to cover IPN and XTE error boxes. We have not included in Table 1 the observations carried out by the BOOTES alerting system (see [10]).

3 Conclusion

Among the fifteen optical counterparts discovered in 1999–2000 nine were visible from Spain, begin seven of them detected (only GRB990308 and GRB000911 were not detected). Two of these seven afterglows were discovered from CAHA (GRB991208 and GRB000926, this last one co-discovered jointly with the NOT [11]). These numbers show the relevant role that observations from Spain (and specially from CAHA) have played in the GRB field.

Acknowledgments

We are very grateful to the TAC of CAHA and IAC for granting time for our GRB ToO programs. Javier Gorosabel acknowledges the receipt of a Marie Curie Research Grant from the European Commission.

References

1. Castro-Tirado, A.J., et al., 1999, Science 283, 2069.
2. Castro-Tirado, A.J., et al., 2000a, GCN # 336.
3. Castro-Tirado, A.J., et al., 2000b, GCN # 362.
4. Castro-Tirado, A.J., et al., 2000c, GCN # 436.
5. Castro-Tirado, A.J., et al., 2000d, GCN # 439.
6. Castro-Tirado, A.J., et al., 2000e, GCN # 447.
7. Castro-Tirado, A.J., et al., 2000f, A&A in press (astro-ph/0102177).
8. Castro-Tirado, A.J., et al., 2000g, GCN # 612.
9. Castro-Tirado, A.J., et al., 2000h, GCN # 845.
10. Castro Cerón, J.M., et al., 2000, these proceedings.
11. Dall, T., et al., 2000, GCN# 804.
12. Fynbo, J.P.U., et al. 2000a, A&A in press (astro-ph/0101425).
13. Fynbo, J.P.U., et al. 2000b, A&A in press (astro-ph/0102158).
14. Gorosabel, J., et al., 2000a, GCN# 734.
15. Gorosabel, J., et al., 2000b, GCN# 735.
16. Gorosabel, J., et al., 2000c, GCN# 803.
17. Greiner, J. et al., 2000, GCN# 743.
18. Henden, A., et al., 2000, GCN# 633.
19. Jensen, B.L., et al., 2000a, GCN # 524.
20. Jensen, B.L., et al., 2000b, A&A in press, (astro-ph/0005609).
21. Klose, S., et al., 2000a, ApJ 545, 271.
22. Klose, S., et al., 2000b, GCN# 713.
23. Masetti, N., et al., 2000a, A&A 359, 941.
24. Masetti, N., et al., 2000b, A&A in preparation.
25. Rol, E. et al., et al., 2000, ApJ in preparation.
26. Van Paradijs, J., et al., 1997, Nature 386, 686.

Optical Observations
of the Dark Gamma-Ray Burst GRB 000210

Javier Gorosabel[1], Jens Hjorth[2], Holger Pedersen[2], Brian L. Jensen[2],
Lisbeth F. Olsen[2], Lise Christensen[2], Evencio Mediavilla[3], Rafael Barrena[3],
Johan U. Fynbo[4], Michael I. Andersen[5], Andreas O. Jaunsen[4],
Stephen Holland[6], and Niels Lund[1]

[1] Danish Space Research Institute,
    Juliane Maries Vej 30, DK-2100 Copenhagen Ø, Denmark
[2] Astronomical Observatory, University of Copenhagen,
    Juliane Maries Vej 30, DK-2100 Copenhagen Ø, Denmark
[3] Instituto de Astrofísica de Canarias,
    E-38200 La Laguna, Tenerife, Canary Islands, Spain
[4] European Southern Observatory,
    Karl-Schwarzschild-Straße, D-85748 Garching, Germany
[5] Division of Astronomy, P.O. Box 3000, FIN-90014 University of Oulu, Finland
[6] Department of Physics, University of Notre Dame,
    Notre Dame, IN 46556-5670, USA

Abstract. We report on optical observations on GRB 000210 obtained with the 2.56-m Nordic Optical Telescope and the 1.54-m Danish Telescope starting 12.4 hours after the gamma-ray event. The content of the X-ray error box determined by the Chandra satellite is discussed.

1 Introduction

The BeppoSAX Gamma Ray Burst Monitor and the Wide Field Camera (unit 1) observed a strong gamma-ray burst on 2000 February 10.36396 UT. It exhibited an X-ray flux of 7.4 Crab (2-26 keV) and a duration of $\sim$ 20s [11]. The field was observed by the Chandra X-ray satellite approximately 21 hours after the GRB [3,4] which localised an uncataloged X-ray source to within 2″. The position was consistent with the one derived independently by BATSE, IPN and the Narrow-Field Instruments (NFI) on-board BeppoSAX [9,8,2]. The Chandra position was improved and the X-ray error box reduced to a circle of 1.6″ radius [5]. Optical observations obtained with the 1.54-m Danish Telescope (1.54D) on 2000 Feb. 11.03-11.08 UT ($\sim$ 16 hours after the burst) revealed an object coincident with the X-ray error box determined by Chandra [6]. These optical observations are part of the results reported in this paper. Deep radio follow-up observations starting 14.8 hours after the gamma-ray event, did not find any radio emission associated to the afterglow above 55 microJy [1,10].

Table 1. List of the observations obtained with NOT and 1.54D Telescopes.

Date 2000(UT)	Filter	Exposure Time	Seeing	Telescope
Feb. 10.88–10.90	R	3x300	1.2	NOT
Feb. 11.03–11.08	R	10x300	1.6	1.54D
Feb. 14.02–14.03	R	600	1.9	1.54D
May 5.42– 5.44	R	2x600	2.3	1.54D
Aug. 22.29–22.41	R	7x900	2.2	1.54D
Aug. 23.23–23.29	R	5x900	2.3	1.54D
Aug. 24.23–24.30	R	4x900	3.0	1.54D
Aug. 26.29–26.43	V	9x900	1.5	1.54D
Aug. 27.21–27.24	I	2x900	1.4	1.54D
Aug. 28.21–28.24	I	2x1200	1.4	1.54D
Aug. 29.21–29.30	I	7x1200	1.1	1.54D
Aug. 30.22–30.24	I	1200	1.1	1.54D
Aug. 31.21–31.24	B	2x1200	1.5	1.54D

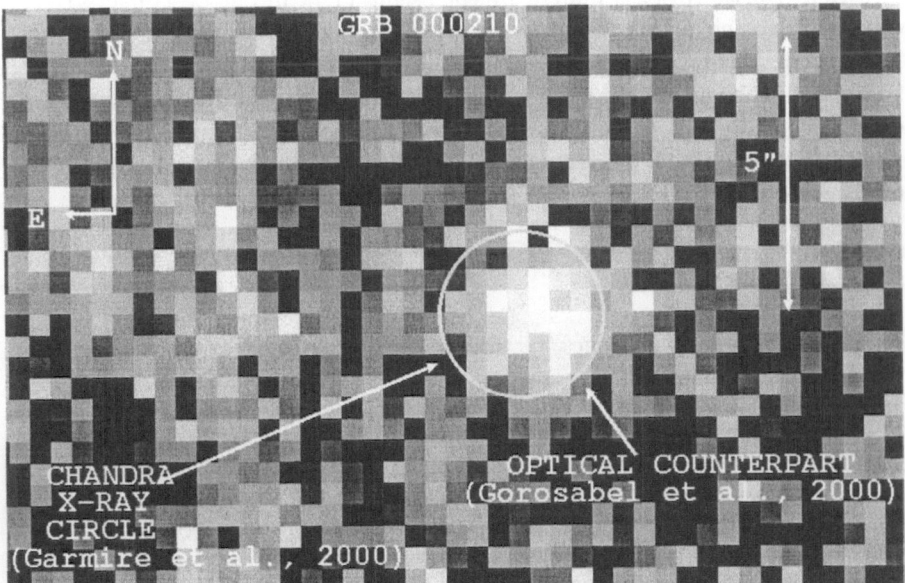

Fig. 1. The figure shows a blow up of the co-added R-band image taken on Aug 22.29–24.30 UT. The plot shows the position of the improved Chandra X-ray circle [5] and the optical candidate [7].

2   Observations

We obtained optical observations with the 2.56-m Nordic Optical Telescope (NOT) and the 1.54D starting 12.4 hours and 16 hours after the burst, respectively. The NOT observations were carried out with HIRAC and at the 1.54D with the DFOSC instrument. Table 1 displays the observing log.

The observations carried out in August improved the previous position reported in February [6], yielding $\alpha_{J2000} = 01^h59^m15.60^s$, $\delta_{J2000} = -40°39'32.8''$ with an uncertainty of $\pm 1''$ [7]. This position is fully consistent with the improved Chandra X-ray circle position [5] (see Fig 1). The first images taken at NOT were not deep enough to detect the object. A comparison of the co-added R-band image taken on Aug 22.29–24.30 UT with the R-band image taken 16 hours after the burst gives a magnitude difference of 0.03 ± 0.30 mag. Therefore, the object remains constant in brightness within the photometric errors. We derive a magnitude of R = 23.5 ± 0.2 for the object.

3   Conclusion

Although the object is very faint in our images its appearance is not stellar. The photo-profiles shows an object slightly elongated in the North-East direction with an angular extension of ~1.5''. The angular size in the orthogonal direction (North-West) is limited by the seeing (1.1'' in our best images). The object did not change in brightness since the first detection carried out just 16 hours after the burst. GRB 000210 appears to be one of the best candidates to study the problem of the "dark burst", resembling GRB 970828. It would be extremely important to perform deep optical/IR observations aimed to determine the redshift, the spectral energy distribution and the morphology of this enigmatic object.

Acknowledgments

Javier Gorosabel acknowledges the receipt of a Marie Curie Research Grant from the European Commission.

References

 1. Berger, E., et al. 2000, GCN 546.
 2. Costa, E., et al. 2000, GCN 553.
 3. Garcia, M., et al. 2000, GCN 544.
 4. Garcia, M., et al. 2000, GCN 548.
 5. Garmire, G., et al. 2000, GCN 782.
 6. Gorosabel, J., et al. 2000, GCN 545.
 7. Gorosabel, J., et al. 2000, GCN 783.
 8. Hurley, K., et al. 2000, GCN 543.
 9. Kippen, R.M., et al. 2000, GCN 549.
10. McConnell, D., et al. 2000, GCN 560.
11. Stornelly, M., et al. 2000, CCN 540.

The Light Curves of GRB 990123 and GRB 990510

G. Björnsson[1], S. Holland[2,4], J. Hjorth[3], and B. Thomsen[2]

[1] Science Institute, University of Iceland
[2] Institute for Physics and Astronomy, University of Aarhus
[3] Astronomical Observatory, University of Copenhagen
[4] Department of Physics, University of Notre Dame

Abstract. We have collected all published photometry for GRBs 990123 and 990510, the first two gamma-ray bursts where breaks were observed in their afterglow light curves. We show that the light curve of GRB 990123 is consistent with a collimated outflow with a fixed opening angle, whereas GRB 990510 may have originated in an initially radiative jet that started to spread about 1 day after the burst.

1 Introduction

The optical afterglows following GRB 990123 and GRB 990510 were bright and enabled extensive follow-up observations for several weeks. They are the first bursts to show strong evidence for a break in their light curves. We examine three fitting functions for broken light curves and investigate the effects of the different functions on the time of the break and the pre- and post-break decay slopes. Our goal is to determine reliable light curve parameters for each GRB and to use these parameters to distinguish between different physical models for the nature the afterglows. An extensive account of this work is presented in [6].

We use three different fitting functions to parameterize the light curves, a broken power law, and two smooth fitting functions [4,12]. All three fitting functions result in a pre-break decay slope α_1, a post-break slope α_2, the time of the break t_b and the flux, $f_\nu(t_b)$, in a given passband, at that time. None of these functional forms has been derived from any physical principles and they are not based on any model for the afterglow of a GRB.

We use three different models to interpret the optical light curves:

i) A spherically symmetric fireball in a dense medium. A break in the light curve is expected when the early relativistic expansion becomes non-relativistic [13,2].

ii) A collimated outflow of a constant opening angle, θ_0. In this case the light curve steepens when the bulk Lorentz factor, Γ, decreases to the $\Gamma < 1/\theta_0$ regime [8,7]. This is a purely geometrical effect, the amount of steepening being $\Delta\alpha = (\alpha_1 - \alpha_2) = 3/4$.

iii) A collimated outflow expanding sideways [11,10]. The steepening in this case can be shown to be $\Delta\alpha = 1 - \alpha_1/3$ [11]. Detailed numerical simulations show the steepening to be gradual both in time and in slope [9].

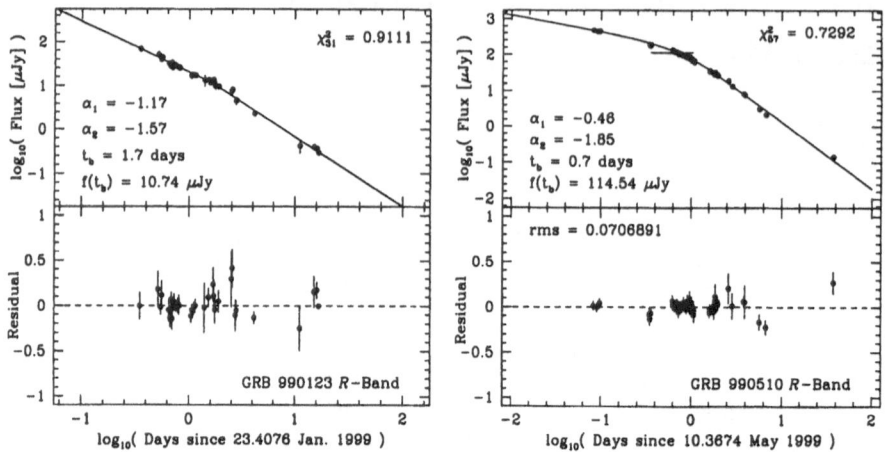

Fig. 1. a) The upper panel shows the best-fitting broken power law for the GRB 990123 R-band photometry. b) The upper panel shows the best-fitting smooth function for the GRB 990510 R-band photometry. In both cases the horizontal bar shows the 1σ uncertainty in the time of the break. The magnitudes have been corrected for Galactic extinction in the direction of each burst. The lower panels show the residuals in the fit in the sense ($R_{fit} - R_{obs}$).

2   The Light Curves

We collected the published optical observations associated with GRB 990123 and subtracted a uniform estimate of the light from the galaxy [5]. The data consists of 88 observations in the Johnson B-, and V-bands, Kron-Cousins R-, and I-bands, and the Gunn r-band. In this case a broken power law provides the best fit with the weighted mean values in the R-band for each parameter shown in Fig. 1a.

Similarly, we collected the published optical observations associated with GRB 990510. These data consists of 182 observations in the Johnson V-band, and in the Kron-Cousins R-, and I-bands. No corrections were made for a host galaxy since a very faint candidate host galaxy has only recently been discovered [1,3]. In this case a smooth function [4] provides the best fit to the data, an example is shown in Fig. 1b.

3   Interpretation

In both cases, GRB 990123 and GRB 990510, the parameters of the light curve (α_1, α_2, and t_b), are not strongly dependent on the fitting function. This means that our interpretations are not strongly dependent on the type of fitting function used to derive the slopes and break time. However, the choice of fitting function will influence the details of the interpretation and may influence the interpretation of the light curve for some GRBs.

GRB 990123: Although a spherical fireball entering a non-relativistic regime at the time of the light curve break appears in best overall agreement with the data, our preferred model to interpret the light curve is that of a collimated outflow of a fixed opening angle [6]. The reason is that the spherical model cannot account simultaneously for the optical and radio observations [7]. Our interpretation for the optical light curve of GRB 990123 then implies an opening angle of $\theta_0 \approx 5°n^{1/8}$, that is very insensitive to the number density, n, of the ambient medium. The observed break in the light curve is a geometric effect that occurs when the relativistic expansion of the fireball slows down to $\Gamma \approx 1/\theta_0$. This model reduces the energy released by the GRB by a factor of ~ 300 to $E_{52} = 1.1$. The magnetic field strength is $\epsilon_B \approx 2.6 \times 10^{-7}$ if the local number density is ~ 1 cm^{-3} and $\epsilon_B \approx 5.6 \times 10^{-5}$ if the local number density is ~ 1000 cm^{-3}. This is well below the equipartition energy. Assuming a power law spectrum, $f_\nu(\nu) = \nu^\beta$, allows us to estimate the average spectral index, $\beta = -0.750 \pm 0.068$, directly from the observed light curves. We find no evidence for a variable β.

GBR 990510: In this case the initial slow decay is incompatible with all adiabatic fireball models. It is, however, consistent with a radiative fireball. Only the sideways expanding jet model is consistent with the light curve behavior at late times and is able to account for the size of the observed break [6].

The optical light curve of GRB 990510 is consistent with a collimated outflow that begins to undergo a sideways expansion at the time of the break. The outflow had an opening angle of $\theta_0 \approx 5°$ at that time. We are unable to estimate reliably the magnetic field strength because no measurements are available of the flux at frequencies above ν_c, the electron cooling frequency. We obtain the mean spectral index in the same manner as for GRB 990123, $\beta = -0.531 \pm 0.019$. There is weak evidence for a variable β before the break in the light curve.

References

1. J. S. Bloom et al.: GCN Circ. 756, (2000)
2. Z. G. Dai, T. Lu: ApJ, **519**, L155 (1999)
3. A. S. Fruchter, R. Hook, E. Pian et al.: GCN Circ. 757, (2000)
4. F. A. Harrison, J. S. Bloom, D. A. Frail et al.: ApJ, **523**, L121 (1999)
5. S. Holland, J. Hjorth: A&A, **344**, L67 (1999)
6. S. Holland, G. Björnsson, J. Hjorth, B. Thomsen: A&A, in press (2001)
7. S. R. Kulkarni et al.: Nature, **398**, 389 (1999)
8. P. Mészáros, M. J. Rees: ApJ, **530**, 292 (1999)
9. R. Moderski, M. Sikora, T. Bulik: ApJ, **529**, 151 (1999)
10. A. Panaitescu, P. Mészáros: In: *19th Texas Symposium on Relativistic Astrophysics and Cosmology* ed. by J. Paul, T. Montmerle, E. A. Aubourg, p44 (1998)
11. J. E. Rhoads: ApJ, **525**, 737 (1999)
12. K. Z. Stanek, P. M. Garnavich, J. Kaluzny, W. Pych, I. Thompson: ApJ, **522**, L39 (1999)
13. R. A. M. J. Wijers, M. J. Rees, P. Mészáros: MNRAS, **288**, L51 (1997)

GRB/SNe Correlations: Search for Unrecognized OAs of GRBs among Detected SNe

Věra Hudcová[1], René Hudec[1], Nicola Masetti[2],
Graziella Pizzichini[2], and Eliana Palazzi[2]

[1] Astronomical Institute, Academy of Sciences, 251 65 Ondřejov, Czech Republic
[2] Istituto TeSRE/CNR, via Gobetti 101, I-40129 Bologna, Italy

1 Introduction

The possible Supernovae-GRBs correlations have been intensively debated from various aspects recently. The recent detections of optical afterglows (OAs) of Gamma Ray Bursts (GRBs) confirm the low energy emission of these events. There have been optical afterglows observed for nearly half of GRBs so far (from subset of GRBs with good optical data) hence limited information is available now for their light curves, magnitudes as well as other parameters. These general properties and behavior of OAs resemble in some extend those of supernovae especially by peak magnitudes, transient behavior, and also by mean decline rates. Further, the rate of OAs may be much larger than the GRBs rate due to different beaming, this factor is unknown but may reach up to four orders of magnitude (e.g. Woods and Loeb, 1998). It is hence obvious that some of detected and poorly in detail investigated supernovae may, in fact, represent unrevealed OAs. The NS mergers are also expected to produce OT events which could be, in principle, found among detected SNe (Li-Xin and Paczynski, 1998).

We present results of cross-correlations (including recently detected triggers) of SNe and GRBs databases to look for any kind of possible correlations in time and position. In this search, we have taken all SNe detected into account including faint and poorly investigated events since such may represent unrevealed OAs. Our approach is to study in detail the coincidences found and SN in question rather than make pure statistical conclusions (which are difficult due to large positional inaccuracies of GRBs and another influences such as incompleteness of SN data bases). We note that the incompleteness of the SN catalogues is large and that only $10^{-3}...10^{-4}$ of all SN down to mag 23 are included.

Since the physical picture is not yet clear, and since there have been very various theoretical predictions for time lags between SNe and GRBs, we have searched for possible correlations for all known SNe occurring both before and after GRBs events, with various time windows. We show that these analyses are heavily affected by the positional inaccuracy of GRBs positions. We have also analysed possible correlations of various physical and observational parameters such as GRB time durations, GRB fluxes, GRB fluences, SN magnitudes, SN redshifts, etc. Although promising temporal and positional coincidences have been found in few cases, we are still unable to provide any kind of direct observational evidence that the two phenomena are physically related. We have also studied whether there may be unrecognized optical afterglows (of GRBs) among detected SNe. No positive case has been revealed, but these results are affected by lack of suitable optical data for faint SNe.

The 4B GRB catalog available at the www has been used as the source of GRBs parameters and positions. This database has been cross-correlated with the list of supernovae compiled by us over the whole time interval of the 4B catalog from reports published in the IAUC as well as with the list of detected SNe by the IAU. All events occurring before May 2000, have been taken into account. Since the OAs are known to peak no more than one day (with only one exception for GRB980508) after the corresponding GRB and then declining gradually, and since the first SN observations are usually delayed after their explosions, we have searched for correlations between GRBs and supernovae detected within the particular GRB error boxes up to 60 days after the gamma ray events.

2   Results

Altogether 2704 GRBs and 958 SN detected between April 1991 and May 2000 have been analyzed. The number of correlations found is consistent with assumption of unrelated samples.

The search was conducted for various time windows between the GRBs and SNe (7, 30 and 60 days). The various observable GRB parameters such as duration, peak flux, fluency etc. have been plotted against the distances of possibly correlated SNe in order to look for possible correlations. The simplified assumption was that, for real correlations, more distant GRB should be of longer duration and of lower fluency. The GRB sample has been investigated as a whole, and also divided into short (<3 s) and long (>3 sec) GRB subgroups, based on the assumption that the long GRB are related to collapses of massive stars. The SNe have been also considered either as a whole, or, alternatively, divided according to their types (if known). No clear correlation has been obtained for any of these groups and subgroups. Elimination of less probable correlations (according to the estimated SN explosion dates etc.) has not influenced these results.

3   Comparison with Previous Searches

Our search is more general than those of Kippen et al. (1998) and Wang and Wheeler (1998): we consider all known SN without any restriction. The study of Kippen et al. (1998) is restricted to brighter events only (brighter than mag 17) but this is in contrast with BeppoSAX OA statistics (only the SN1998bw was brighter than 17 mag, all others 12 OAs were fainter) while Wang and Wheeler restrict to SN of known Ia,Ib types (but the OA may be also among weak and unclassified SN).

For GRBs with beams completely off line of sight, theoretical predictions have been made that the GRB events may be delayed by weeks compared to the SN events, and the GRB should be spatially displaced in the sky by light-weeks (Cen,

1998). The subsequent afterglows will be further delayed and spatially displaced from the SN. The vast majority of GRB/SN events should belong to this class. Combining the time delay and the angular separation between the SN and GRB (probably only for nearby ones in practice) should allow a determination of the angle of the jet relative the line of sight. This is why we have cross-correlated the catalogues also for GRBs following SNe with time windows of up to 60 days. The results are analogous to those obtained for SNe after GRBs i.e. no clear correlation has been confirmed.

4   Discussion

The statistical significance of results of cross-correlations between the GRBs and SNe is affected and limited by the following: (1) the positional uncertainty of a quite large fraction of the GRBs is rather high (2) the dates of SN explosions and/or peaks are unknown in most cases (3) the results of optical SN searches do not represent a full and homogeneous sample so that many (and even a large majority) of the SN may be missed especially at faint magnitudes (4) there is no systematic sky patrol survey at low magnitudes below 15.

For the correlations found, many of the related SNe represent poorly investigated events with poorly known light curves, decline rates, color indices and no or limited spectral information hence the decision whether they could be related to the GRBs in question is difficult. The recent detections of faint OA of GRBs strongly support and justify further more extended and more complete searches for faint new and variable optical objects especially at low magnitudes (18 and less). In principle, such information can be retrieved from deep archival plates (some of the archival plate collections reach the limiting magnitudes of 20 and even 23 such as ROE Edinburg, Palomar Schmidt and TLS Tautenburg) but a systematic deep CCD sky patrol can provide an even more precise and much more complete database. The OAs may be revealed by their characteristic light curves and by their typical colors (for details see Šimon et al., 2001). However, this information is not available for most of the detected SNe. Better optical data are urgently needed for the SNe recently detected which may be related to GRBs.

Acknowledgements. The work has been supported by the project of the Czech Ministry for Education and Youths, No. ES02, by grant of the Grant Agency of the Czech Republic No. 205/99/0145 and by the project CNR-AS CR.

References

1. Cen R., ApJ 507, L131, 1998
2. Li-Xin L. and Paczynski B., ApJ 507, L59, 1998.
3. Kippen R. M. et al., 1998, ApJ 506, L27
4. Šimon V., Pizzichini G. and Hudec R., this volume, 2001.
5. Wang K. and Wheeler J.C., 1998, ApJ 504, L87
6. Woods P. M. and Loeb, ApJ 508, 760, 1998

Optical GRB Analyses: Results from Archival Plates

René Hudec[1] and Wolfgang Wenzel[2]

[1] Astronomical Institute, Academy of Sciences of the Czech Republic,
CZ-251 65 Ondřejov, Czech Republic
[2] Sonneberg Observatory, Sonneberg, Germany

Abstract. The deep archival astronomical plates represent a valuable tool for optical analyses of Gamma-Ray Bursts positions. We report on the results obtained on astronomical plates from the Royal Observatory Edinburgh and Sonneberg Observatory for selected precisely localized GRBs (by BeppoSAX and RXTE).

1 Introduction

The operation of BeppoSAX and RXTE satellites has provided a list of 75 reliably and precisely positioned gamma ray bursts (GRB). For a large fraction of these GRBs, low energy counterparts-afterglows have been found including optical (see McNamara and Harrison 1999 for a review). The optical afterglows (OAs) of GRBs have been found to peak at roughly 18–23 mag and then decreasing slowly according to a power law over a time period of weeks to months. In one case, the direct optical emission peaking at about mag 9 and lasting for about 1 min has been found (Akerlof et al. 1999). The archival plates taken by the UKSTU in Siding Springs, Australia, have typical limiting magnitudes of 19 ... 23 and field of view of 6.4 x 6.4 deg^2 and are hence well suited to search for optical emissions related to GRBs including faint OAs. The archival search conducted has two major prospects: (1) to search for underlying GRB hosts and/or peculiar objects, and (2) to search for possible activity (including recurrence). In addition, two precise GRB positions either with detected or expected known bright optical transient have been analysed on large number of Sonneberg Sky Patrol plates.

2 The Method and the Project Justification

We have analyzed altogether 454 high quality plates covering the positions of 13 BeppoSAX–RXTE precisely positioned GRBs in the UKSTU (United Kingdom Schmidt Telescope Unit) plate archive located at the Royal Observatory Edinburgh, UK. For GRB980425, the error boxes of both X-ray sources as well as the SN1998bw position have been examined. The deepest plates show an extended structure in the spiral arm in the host galaxy of the SN1998bw and consistent with its position, perhaps related to the star-forming region responsible for both events. Since the positional errors of the triggers studied were usually small, of order of 1 arcmin and/or below, the GRBs localizations were investigated both by the Zeiss blinkmicroscope as well as by the plate microscope. This proved to be more effective than the time-consuming scanning of the whole plates. Moreover, the false objects may be much effectively eliminated on original plates if compared with digitized files where the 3rd dimension (along the line of sight) is lost. The goal for the study was to search for any kind of optical activity (including possible burst recurrence and light variability of underlying objects/hosts) at the positions of GRB triggers at times before and after GRBs.

The recent GRB results relate GRBs to faint and distant host galaxies and there are even indications that GRBs are related to star-forming regions (Djorgovski et al. 1998, Bloom et al. 1998). This scenario however do not fully exclude the possibility of events recurrence on a long time basis. The recurrent trigger would then simply represent an another event in the same galaxy.

3   Selected Analyzed GRB in Detail

GRB981016. The plate U18140 was taken nearly 2 months before GRB. For this gamma ray trigger, three possible optical candidates have been suggested by optical follow-up observations. All the three positions have been analyzed on the plate. Their quiet magnitudes are however below plate limits so they cannot be analyzed for light changes.

GRB 990705. This position is covered by very large number of plates (251) taken between 731030 and 981129. One very faint object 0.3 mag above the plate fog has been found at the OT position on the plate R2018, taken on 760126, as a 45 min exposure in R, and 0.2 mag above the plate limit on the plate J1992, a 70 min exposure on 760104. There is another plate taken on 760126 but with worse limiting magnitude. The faintness of both objects makes their final verification as real triggers impossible.

GRB 990908. The GRB 990908 was detected by the BeppoSAX GRBM and WFC instruments on 1999 September 8, around 00:18 UT, to 8 arcmin accuracy (Gandolfi et al. 1999). Only one optical follow-up observation has been reported for this trigger. Only one variable source was detected in the covered area. The stellar object of magnitude approx. 18 was fading by 0.3 magnitudes. The object was found to be present on both images as well as on the Digital Sky Survey (at the position RA = 06:51:10.52, DEC= −75:02:17.3 J2000). No other variable or new sources were seen with residuals brighter than mV = 20.3 (Axelrod et al.1999). The variable object found was declared as to be more likely a short period variable star within our own galaxy, but due to poor seeing conditions, the compact galaxy origin of the host could not be excluded and further investigations have been suggested to prove its nature.

The object position is covered by 52 direct UKSTU plates covering the time interval between 1975 December 7 and 1997 December 20, i.e. more than 22 years. In addition, there are two spectral plates covering the given position, but their limiting magnitudes do not allow to study the variable object mentioned above. The object is seen on all plates available, and is of stellar appearance. The object magnitude has been found nearly constant on all plates evaluated, with light changes less than 0.3 magnitudes. optical candidates (afterglows) of GRBs. With a very high probability, the object is not an optical counterpart of GRB 990908. More detailed spectroscopic and-or photometric CCD observations would be required to confirm the variability of the object as well as its physical nature.

Sonneberg Plate Analyses. Two objects with known and/or suspected bright prompt OT have been analyzed on numerous Sonneberg sky patrol plates. The position of GRB990123 has been analyzed on 1561 blue-sensitive plates with magnitude limit 12–14.5 representing nearly 1 170 hours of exposure in the time interval 1934 to 1998. The position of GRB991208 was analyzed on 1386 blue-sensitive plates with magnitude limit of 12–14.5 taken in the time interval 1957 to 1998 (nearly 1 040 hours of time coverage). No optical activity has been found.

4 Conclusions

We have investigated 454 high quality deep UKSTU Schmidt plates covering the positions of 13 precisely positioned GRBs provided by BeppoSAX and RXTE satellites until November 1998. No optical activity exceeding limiting magnitudes of the corresponding plates which amounts to mag 19 ... 23 has been established except the peculiar variable star BL Cir marginally consistent with the error box of GRB970402 (Hudec et al. 1999). In addition, two precise GRB positions with either known or suspected bright prompt OT (990123 and 991208) have been analyzed with negative results on large number of Sonneberg sky patrol plates.

Acknowledgements. This study has been supported by the grant provided by the Grant Agency of the Czech Republic No. 205/99/0145, by the project KONTAKT ES002 provided by the Ministry of Education and Youth of the Czech Republic and also by the Academic Link between the Astronomical Institute Ondrejov and the University of Westminster, Harrow, UK provided by the British Council in Prague. I also acknowledge the support provided by the UKSTU staff at the ROE.

References

1. Akerlof, C. W. et al., IAUC No. 7100, 1999.
2. Bloom J. S., Djorgovski S. G., Kulkarni, S. R. and Frail, D. A., ApJ 507, L25-L28, 1998.
3. Djorgovski, S. G., Kulkarni, S. R., Bloom, J. S., Goodrich, R., Frail, D. A., Piro, L. and Palazzi, E., ApJ 508, L17-L20, 1998.
4. Hudec R. et al., Proc. 3rd INTEGRAL workshop, Taormina, Italy, in press.
5. McNamara B. J. and Harrison T. E., Nature 396, 233-236, 1999.

6. Hurley K. et al. 1998a, IAU 6966
7. Kouveliotou C. et al. IAUC 6944, 1998b
8. Woods P. M. et al. ApJ 519, 1999, L139-L142.
9. Axelrod T., Mould J. and Schmidt B., GCN Circular No. 408, 1999.
10. Gandolfi G. Et al., GCN Circular No. 406, 1999

EN: Simultaneous Optical Data for GRBs

René Hudec[1], Jan Florian[1], Radomir Smida[1], Ivana Stoklasova[1], Jiři Pálek[1], Nicola Masetti[2], Eliana Palazzi[2], and Graziella Pizzichini[2]

[1] Astronomical Institute, Academy of Sciences of the Czech Republic,
CZ-251 65 Ondřejov, Czech Republic
[2] TESRE, CNR, Bologna, Italy

1 Introduction

The European Fireball Network, EN, is operated on 11 stations in the Czech Republic. The network is based on high quality Fish-Eye lenses with photographic records (sky diameter 8 cm in the focal plane). The main goal of the network is to provide sky monitoring to detect and to study meteors. The use of the sky archive may be however much more general. For example, providing of real-time and pre-burst data (sky monitoring) in visible light is extremely important for investigations of Gamma-Ray Bursts (GRBs) since many recent theories predict related bright optical flashes (e.g. Sari and Piran, 1999, Kobayashi and Sari, 2000, Protheroe and Bednarek, 1999). This kind of information may be provided only by patrol program. No follow-up experiment will be able to provide data for times before GRBs as well as for the first few seconds during the GRB onsets. This is very crucial since some theories predict optical/UV emissions preceding the GRBs (Protheroe and Bednarek, 1999). The system allows a large fraction of GRBs to be observed (so far more than 100) with limiting magnitudes between 4 and 12 at times before, during and after the GRB triggers. This still represents one of the faintest real-time and pre-burst magnitudes of GRBs.

2 Parameters of the Recent System

- Optics: Fish-Eye Objective F-Distagon 3.5/30
- Detector: Planfilm ILFORD FP4 PLUS 125 ASA, alternatively FOMAPAN 400 ASA or 100 ASA (panchromatic emulsion) 90 x 120 mm, sky diameter 80 mm
- Typical exposure time: 3 hrs for guided cameras, whole night for fixed cameras
- 2 stations equipped with guided and fixed cameras
- 9 stations equipped with fixed cameras
- Sensitivity for brief 1 sec triggers 2–4 mag, for stars up to mag 12
- Response limited to the red light above 400 nm

The preferences can be summarized as follows.

- Large sky coverage (full visible hemisphere)

- Large fraction of observation time: 2 400 to 6 000 sr.h for one station/year
- Multiplicity of data to eliminate background triggers easily
- Classification of detected triggers by paralax
- Simultaneous and pre-burst optical data (limits) for GRBs

The recent mechanical and emulsion improvements will result in improved limiting magnitudes (up to mag 6 for stars on trailed and mag 12 on pointed plates). The network will be operated as a fully remote controlled network – without any human assistance in a near future. The future data will be evaluated using a new film scanner, so their evaluation for GRBs related analyses is expected to be less laborious. The above mentioned FE cameras are in the use since 1975. A great number of plates (23 000) were taken by driven Tessar cameras in the years 1955–1978. Another 46 000 plates were taken at two stations equipped with fixed cameras (stars as trails). The Tessar system before 1975 was based on 10 cameras with Tessars 1:4.5 f=18 cm and Agfa ISS 21 DIN emulsion operated at the same time covering nearly 50% of the visible sky hemisphere. The total number of plates in the Ondřejov plate archive is around 110 000.

3   Summary of GRB Related Results

- No optical emission above mag 5 (1 sec duration assumed) or mag 13 (full exposure time) or Lg/Lo> 100 ... 300 has been detected for a few GRBs. The faintest limit (320) exists for GRB 830313 (Hudec, 1993).
- No optical emission above magnitudes 0...3 (1 sec duration assumed) or 4-11 (full exposure time) or Lg/Lo>0.1 ... 10 has been detected for many ($\sim$ 140) GRBs
- Optical Transient (OT) was detected on the plate taken $\sim$ 7 h after the GRB790929 inside its error box (Borovička et al. 1992)
- The pre-burst, simultaneous and quasisimultaneous optical data have been analysed for numerous precisely localized GRBs mostly detected by the BeppoSAX satellite. The knowledge of precise position of the trigger allows to search even for faint optical counterparts just above the detection threshold. The preliminary results are listed in the Table 1.

4   Conclusions

The recently detected optical events related to GRBs are generally considered as optical afterglows (OAs) while the direct optical emission of bursts (optical transients – OTs) has been detected on only one case. The data presented here hence represent valuable limits for direct optical luminosity of GRBs in question: the direct optical emission seems not to exceed V magnitude 10. This is valid also for the optical pre-burst emission. Despite of the recent instrumental developments, the photographic patrol data still represent the unique databases for real-time (simultaneous and pre-bursts) optical observations of GRBs. The automated alert systems were able to provide rapid follow up optical data (with

Table 1. The Ondrejov plates analysed for the recent precisely localized sources

GRB	GRB time (UT)	Plate before (hr)	Limit (mV)	Plate after (hr)	Limit (mV)	OT yes/no	Note
960720	11:37	-10	11.0	+9	11.00	No	
970111	9:44			+7	4.00	No	
970228	2:28	-4.5	4.0	+13	10.00	No	
980326	21:19	-8	10.0	+6	10.00	Yes	Susp. object 23 h before
980329	3:44			+15	9.00	No	
980519	12:20	-11	9.0	+9	9.00	Yes	Susp. object 9 h after
980730	4:24	-51	6.0			No	
981220	3:42	0	6.5	+0	5.50	No	Sim. taken plate
		-26	9.0			No	
		-31	10.0			No	
981223	15:25	-10	9.5			No	
990506	11:54	-10	5.0	+8	4.00	No	
		-12	5.5				
990520	8:35	-9	9.0	+36	7.00	No	
991014	21:52	-18	9.5	+26	8.00	No	
991106	10:54	-6	5.0			No	
		-15	8.0			No	
991208	4:37	-48	6.5	+12	4.50	No	
991216	16:07	-35	5.5			No	
		-40	6.5			No	
		-59	7.5			No	
000115	14:49	-9	12.0			No	Deep plate 9 h before
000126	23:26	-18	6.5	+24	12.00	No	
		-21	12.0			No	
		-23	12.0			Yes	Susp. object 23 h before
000301A	2:34	-22	11.5	+39	12.00	No	
000301C	9:51	-29	12.0	+32	12.00	No	
000408	2:36	-22	12.0	+18	12.00	No	

time delays of order of tens of seconds) for less precisely localized CGRO BATSE triggers but their magnitude limits is such case were generally not substantially better (since they have to cover large FOVs) than those of photograpic sky patrols and further, they were not able to provide data for times during and before bursts. For the BeppoSAX triggers, they are communicated to ground based observers with a time delay of a few hours due to technical reasons, so the real time data provided by the patrol also are important. For HETE, the automated systems are expected to start observations within tens of seconds after the onset of bursts, but even this does not yet represent a true simultaneous observation as can be provided by the patrol experiment. The improved sky patrols can yield valuable real-time and pre-burst optical data for GRBs detected by recent satellite experiments in the magnitude range 12–15.

Acknowledgements. The work has been supported by grant of the Grant Agency of the Czech Republic No. 205/99/0145, by the Ministry of Education of the Czech Republic, project KONTAKT ES002, and by the CNR-AS CR Project "Investigation of GRBs".

References

1. Borovička J., Hudec R., Dedoch A., A&A, 258, 379, 1992.
2. Sari, R. and Piran, T. ApJ 520, 641, 1999.
3. Kobayashi, S. and Sari, R., ApJ 542, 819, 2000.
4. Protheroe, R. J. and Bednarek, W., astro-ph/9904279, 1999.

Near-Infrared Polarimetric Observations of GRB Afterglows

Sylvio Klose[1], Bringfried Stecklum[1], and Olaf Fischer[2]

[1] Thüringer Landessternwarte, Tautenburg, Germany
[2] Universitätssternwarte Jena, Jena, Germany

Abstract. We give a status report on our project of rapid follow-up polarimetric observations of GRB afterglows in the near-infrared.

1 The Project

One of the key issues in GRB research is a determination of the outflow geometry of the underlying explosive events. Since the bursters are at redshifts in the order of 1 indirect methods are required in order to determine whether a GRB explosion was isotropic or beamed. A tool to tackle this observational problem are polarimetric observations. Based on theoretical arguments it has been shown that a collimated relativistic outflow could result in a non-zero and variable degree of linear polarization of the corresponding afterglow light; e.g., [1,2].

In 1999 we established a Target of Opportunity (ToO) program whose goal is the measurement of the degree of linear polarization of GRB afterglows. This project makes use of the Calar Alto 3.5-m telescope equipped with the near-infrared camera Omega Cass. Observations are performed in the K' band. Since the bursters are at redshifts around 1 this corresponds to a wavelength of about 1 μm in the GRB frame. Compared to the optical bands [3,4] such observations are thus less sensitive to a polarization of the afterglow radiation by dust scattering in the corresponding GRB host galaxy.

The Omega Cass camera is equipped with a 1024 $\times$ 1024 pixel array. So far, all polarimetric observations have been performed in the wide field mode (0.3″/pixel). In this mode the field of view is about 5′ $\times$ 5′. We used wire-grid polarizers which provide individual linear polarized images. This means that four single images have to be taken through the four offered analyzers which are mounted at position angles of 0, 45, 90, and 135 degrees. Every observing run is based on 1 h telescope time. This translates into 650 sec total exposure time per polarization angle. The limiting magnitude of the combined four images is then about $K'=19$.

2 First Data

The first phase of the ToO program was running from May 1999 to May 2000. Successful ToOs were carried out for four bursts, GRBs 991014, 991106, 000301C, and 000418. In the cases of GRBs 991014 and 991106 no afterglow was detected

Fig. 1. The 3.2 arcmin *BeppoSAX* error circle of GRB 991106 [9] was imaged 1.4 days after the burst [5]. Shown here is the combined K'-band image of all four polarization angles. North is up, East is left. Observations were performed by R. Lenzen, MPIA Heidelberg.

1.2 and 1.4 days after the burst, respectively (Fig. 1). This could indicate that these afterglows were either declining rapidly or extincted by dust in their host galaxies. The afterglow of GRB 000301C was imaged 1.8 days after the burst at a magnitude of $K' = 17.5$ [6] and a constraint on its degree of linear polarization could be set ($\lesssim 30\%$, [7]). The afterglow of GRB 000418 was discovered based on frames taken with the Omega Cass camera 2.5 days after the burst at a magnitude of $K' \approx 17.5$ [8]. Unfortunately, due to technical reasons its degree of linear polarization could not be measured.

3   Future Work

So far, all polarimetric observations have been performed not earlier than ∼1 day after a burst. Based on our experience with the Omega Cass camera, however, we require a GRB afterglow to be brighter than $K' \approx 16$ in order to constrain its degree of linear polarization with a statistical error of less than ∼10% (Fig. 2). Based on rapid *HETE* 2 alerts this goal should be achievable in the near future. A second year of approved ToOs will start in January 2001. Statistically, the probability that the Omega Cass camera is mounted at the 3.5-m telescope when a burst occurs is about 30%.

We thank the observers and the staff of the German-Spanish Astronomical Centre, Calar Alto, for performing the observations.

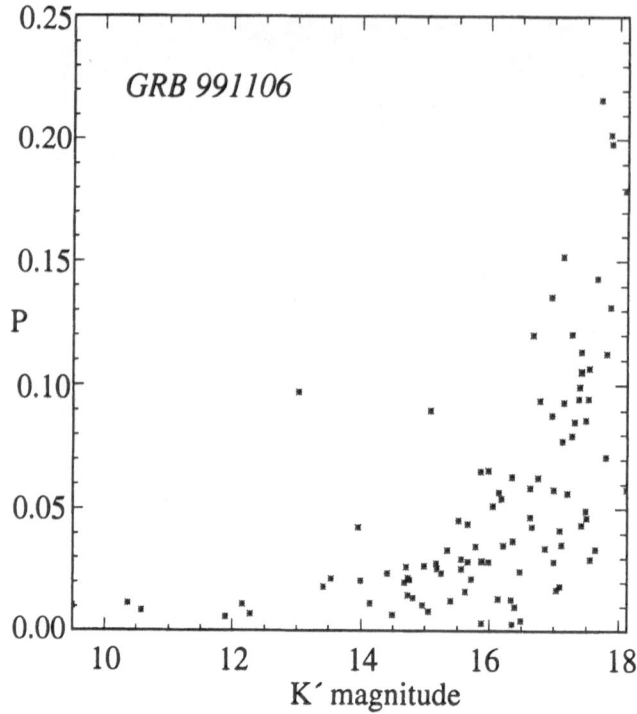

Fig. 2. Results of the polarimetric observations of the error circle of GRB 991106 (Fig. 1). Shown here is the deduced degree of linear polarization P in the K' band for all stellar sources in the GRB error circle as a function of their K'-band magnitude. The data were reduced based on the assumption of a zero net polarization of all stellar sources in the field. No afterglow was detected. The promising object at $(K', P) = (13, 0.1)$ is not the GRB afterglow.

References

1. Ghisellini, G. & Lazzati, D. 1999, MNRAS 309, L7
2. Sari, R. 1999, ApJ 524, L43
3. Wijers, R. A. M. J. et al. 1999, ApJ 523, L33
4. Covino, S. et al. 1999, A&A 348, L1
5. Stecklum, B. et al. 1999, GCN Circ. 446
6. Stecklum, B. et al. 2000, GCN Circ. 572
7. Stecklum, B. et al. 2001, Proc. 20th Texas Symposium, submitted
8. Klose, S. et al. 2000, ApJ 545, 271
9. Piro, L. et al. 1999, GCN Circ. 435

Iron Line Emission in X-Ray Afterglows

Davide Lazzati[1,2], Gabriele Ghisellini[1], Mario Vietri[3],
Fabrizio Fiore[4], and Luigi Stella[4]

[1] Osservatorio Astronomico di Brera, Via Bianchi 46 I-23807 Merate (Lc), Italy
[2] Present Address: Institute of Astronomy, Cambridge CB3 0HA, UK
[3] Università di Roma 3, via della Vasca Navale 84, I-00147 Roma, Italy
[4] Osservatorio Astr. di Roma, via Frascati 33, I-00040 Monteporzio Catone, Italy

Abstract. Recent observations of X-ray afterglows reveal the presence of a redshifted
$K\alpha$ iron line in emission in four bursts. In GRB 991216, the line was detected by the
low energy grating of *Chandra*, which showed the line to be broad, with a full width
of $\sim$15,000 km s^{-1}. These observations indicate the presence of a > 1 $M_\odot$ of iron rich
material in the close vicinity of the burst, most likely a supernova remnant. The fact
that such strong lines are observed less than a day after the trigger strongly limits the
size of the remnant, which must be very compact. If the remnant had the observed
velocity since the supernova explosion, its age would be less than a month. In this case
nickel and cobalt have not yet decayed into iron. We show how to solve this paradox.

1 Introduction

There are now four bursts displaying evidence of an emission line feature during
the X-ray afterglow: GRB 970508 (Piro et al., 1999); GRB 970828 (Yoshida
et al., 1999); GRB 991216 (Piro et al., 2000, hereafter P2000); GRB 000214
(Antonelli et al., 2000). These lines have been observed 8–40 hours after the
burst explosion, have a large equivalenth width (0.5–2 keV) and a flux of about
10^{-13} erg cm^{-2} s^{-1}. Given these properties, each iron atom has to produce at
least 2000 line photons, in order not to exceed 0.1 $M_\odot$ mass of emitting iron,.
Fast recombination and ionization is therefore required. The line of GRB991216
is resolved in the *Chandra* gratings, with a width $0.05c$ (P2000). As discussed
by Lazzati et al. (1999), the detection of the line implies the presence of a sizable
fraction of a solar mass of iron concentrated in the vicinity of the GRB site. This
is naturally accounted for in the SupraNova scenario (Vietri & Stella 1998).

2 General Constraints

The Size Problem. If the line is detected after t_{obs} from the burst, the line
emitting material must be located within a distance R given by:

$$R \leq \frac{ct_{obs}}{1+z} \frac{1}{1-\cos\theta} \simeq \frac{1.1 \times 10^{15}}{1+z} \frac{t_{obs}}{10\,h} \frac{1}{1-\cos\theta} \text{ cm,} \qquad (1)$$

where θ is the angle between the line emitting material and the line of sight at
the GRB site. This limit implies a large scattering optical depth:

$$\tau_{\mathrm{T}} = \frac{\sigma_{\mathrm{T}} M}{4\pi R^2 \mu m_{\mathrm{p}}} \geq 54 \frac{(M/M_\odot)(1+z)^2(1-\cos\theta)^2}{\mu \, (t_{\mathrm{obs}}/10\,\mathrm{h})^2}, \qquad (2)$$

where μ, is the mean atomic weight of the material.

The Kinematic Problem.   For a radial velocity of the remnant of $v = 10^9 v_9$ cm s^{-1} the time elapsed from the supernova (SN) is $t_{\mathrm{SN}} \simeq 12.5(t_{\mathrm{obs}}/10\mathrm{hr})/[(1+z)(1-\cos\theta)v_9]$ days. Such short times implies that most of the ^{56}Co nuclei (and a fraction of the ^{56}Ni nuclei) have not yet decayed to ^{56}Fe (half-life of 77.3 and 6.08 days, respectively, see Vietri et al. 2000).

Line Emission Rate.   We can derive the photon line luminosity by estimating the volume V_{em} effectively contributing to the line emission, and assuming a given iron mass. If the layer contributing to the emission has $\tau_{\mathrm{T}} \sim 1$ (to avoid Compton broadening), and in this layer $\tau_{\mathrm{FeXXVI}} \sim$ a few (to efficiently absorb the continuum), we have $V_{\mathrm{em}} = S/(\sigma_{\mathrm{T}} n_{\mathrm{e}})$, where S is the emitting surface. The line emission rate from V_{em} is then:

$$\dot{N}_{\mathrm{Fe}} = \frac{N_{\mathrm{Fe}}}{t_{\mathrm{rec}}} = \frac{S n_{\mathrm{Fe}}}{1.3 \times 10^{11} T_7^{3/4} \sigma_{\mathrm{T}}} \sim 3 \times 10^{53} \frac{(M_{\mathrm{Fe}}/M_\odot)}{T_7^{3/4} \Delta R_{15}} \; \mathrm{s}^{-1}, \qquad (3)$$

where the *total* volume is $V = S\Delta R$ (slab or shell geometry).

Mass.   Eq. 3 shows that the total iron mass must be a sizable fraction of a solar mass in order to give rise to the observed line photon luminosity of 4×10^{52} s^{-1}. Notice also that Eq. 3 establishes that the line emitting material must be a SNR: no other known astrophysical object contains this iron mass.

3   Models

The Wide Funnel.   Consider a wide funnel excavated in a young plerionic remnant. This solves the *size problem*, since it extends to large radii but can maintain the time-delay contained because it is built close to the polar axis. Fixing the line photon rate (Eq. 3) yields $R = 6 \times 10^{15}$ cm, and thus an opening angle $\theta = 48°$ to fit the time-delay. Assuming a cone geometry for simplicity, we can rewrite Eq. 3 as:

$$\dot{N}_{\mathrm{Fe}} = 3.3 \times 10^{52} \frac{(M_{\mathrm{Fe}}/M_\odot)}{T_7^{3/4}(R_{15}/6)} \tan\theta \; \mathrm{s}^{-1}. \qquad (4)$$

This is a lower limit, since a parabolic funnel has a larger surface and we neglected the (likely) density stratification inside the remnant. Consider now the kinematic properties of the funnel. We expect radiation pressure to exert a force parallel to the surface accelerating the layer with $\tau_{\mathrm{T}} = 1$. The absorbed fluence E_{ion} accelerates the funnel layer to $v_{\mathrm{f}} = (2E_{\mathrm{ion}}/M_{\mathrm{layer}})^{1/2} \sin\phi \simeq 10^4 E_{\mathrm{ion,50}}^{1/2} \sin\phi$ km s^{-1} if $R = 6 \times 10^{15}$ cm. ϕ is the angle between the funnel's normal and the incoming photons. Thus, we expect ablation by radiation pressure to be able to propel the reflecting layer to velocities comparable to those seen in GRB991216.

Back Illuminated Equatorial Material.   The model above assumes that a SN explosion preceded the GRB by some months. We now explore the possibility of a simultaneous GRB–SN explosion. Assume that a GRB ejects and accelerates a small amount of matter in a collimated cone, while a large amount of matter is instead ejected, at sub-relativistic speeds, along the progenitor's equator. Massive star progenitors are inevitably surrounded by dense material produced by strong winds of mass loss rates $\dot{m}_w = 10^{-5}\dot{m}_{w,-5}$ and velocity $v_w = 10^7 v_{w,7}$. This wind scatters back a fraction of the photons produced by the bursts and its afterglow (Thompson & Madau 2000). The scattered luminosity L_{scatt} is constant, since there is an equal number of electrons in a shell of constant width ΔR (for a density profile $\propto R^{-2}$). This luminosity is of order:

$$L_{scatt} \sim m_p c^2 \frac{\dot{m}_w}{m_p v_w/c} = 1.8 \times 10^{45} \frac{\dot{m}_{w,-5}}{v_{w,7}} \text{ erg s}^{-1}. \qquad (5)$$

Scattered photons illuminate the expanding equatorial matter after a time $2R/c$, giving rise to the line emission. Since in this case the SN and GRB explosions are supposed to be simultaneous, the emitting iron must be produced directly by the SN and not through the nickel decay. Iron (^{54}Fe) is directly synthesized for high neutronization of the material at the SN shock.

4   Conclusions

The recently detected features in the X-ray afterglow of GRBs impose strong constraints on models, the most severe being how to arrange a large amount of iron close to the GRB site, while avoiding at the same time a large Thomson scattering opacity. This limit applies to all bursts showing a line feature. An additional limit comes from the *Chandra* observation of a *broad* line in GRB 991216. These observations require a very large amount of iron, known to be contained only in SNe. We have described two models. The "wide funnel" model is in better agreement with observations: its geometry solves the size problem, and the acceleration of the line emitting material by grazing incident photons solves the kinematic problem, allowing the remnant to be a few months old (enough for most cobalt to have decayed into iron). This model implies that the GRB progenitors are massive stars exploded as SNe some months before the burst, inundating the surroundings of the burst with iron rich material. This two-step process and the time-delay between the two steps are exactly what is predicted in the SupraNova scenario of Vietri & Stella (1998).

References

1. Antonelli A. et al., 2000, ApJ, 545, L39
2. Lazzati D., Campana S. & Ghisellini G., 1999, MNRAS, 304, L31
3. Piro L., et al., 1999, ApJ, 514, L73

4. Piro L., et al., 2000, Science, 290, 955
5. Thompson C. & Madau P., 2000, ApJ, 538, 105
6. Vietri M. & Stella L., 1998, Ap.J.Lett., 507, L45
7. Vietri M., Ghisellini G., Lazzati D., Fiore F. & Stella L., 2001, ApJL, in press (astro-ph/0011580)
8. Yoshida, A., et al., 1999, Astr. Ap. Suppl., 138, 433

BeppoSAX Observation of GRB990806: From the Prompt Emission to the X-Ray Afterglow

E. Montanari[1], L. Amati[2], F. Frontera[1,2], C. Guidorzi[1], M. Capalbi[3],
E. Costa[4], M. Feroci[4], L. Piro[4], J. Heise[5], J.J.M. in't Zand[5], E. Pian[6],
E. Palazzi[2], N. Masetti[2], and L. Nicastro[7]

[1] Physics Department, University of Ferrara, Italy
[2] ITeSRE, CNR, Bologna, Italy
[3] BeppoSAX SDC, Roma, Italy
[4] IAS, CNR, Roma, Italy
[5] SPO in the Netherlands, the Netherlands
[6] Osservatorio Astronomico di Trieste, Italy
[7] IFCAI-CNR, Palermo, Italy

Abstract. We present preliminary results of the analysis of BeppoSAX data of GRB 990806. The high energy spectrum of the prompt event is fitted by a single power-law whose spectral index does not vary significantly. In the error box of the GRB, determined by the WFC, the BSAX NFI have detected a previously unknown X-ray source. Based on the power-law temporal decay of its flux in the 2–10 keV range, $f(t) \sim t^{-1.15}$, the source has been identified with the GRB afterglow.

1 Observations

GRB 990806 was detected with the BeppoSAX Gamma-Ray Burst Monitor (GRBM, 40–700 keV [1,2]) and Wide Field Camera (WFC) unit 1 (1.5–26.1 keV [3]) on 6 August at 14:28:07 UT. Its position was determined with an error radius of 3' and was centered at $\alpha_{2000} = 3^h10.6^m$, $\delta_{2000} = -68°07'$ [4].

About eight hours after the burst, the Narrow Field Instruments on-board *BeppoSAX* were pointed at the burst location for a target of opportunity (TOO) observation, from August 6.9268 UT to August 8.6291 UT. A new and variable X-ray source was detected [6] in the GRB error box with the Medium Energy Concentrators/Spectrometers (MECS, 2-10 keV [5]). The net exposure time for MECS was 77.9 ks.

2 Data Analysis and Results

Prompt Emission

In the γ-ray band (40–700 keV), the GRB shows a single peak of about 12 s duration. In the X-ray energy band (2-28 keV), the prompt emission has also a second peak of about 28 s duration; the first one was coincident with that detected by the GRBM. The total duration of the X-ray event is about 40 s (see fig. 1).

The spectral evolution of the GRB data in the 2–700 keV range was studied by subdividing the GRB duration into four temporal slices (A, B, C and D) and performing an analysis of the average spectrum of each slice. We fit the spectra with a power law $(N(E) \propto E^{\alpha})$ photoelectrically absorbed by a neutral hydrogen column density N_H. The N_H value is not constrained well by the data. Its value is the Galactic hydrogen column density along the line of sight to the GRB $(3.53 \ 10^{20} \ cm^{-2})$. The results are shown in Table 1. The photon index does not vary significantly during the GRB.

Table 1.

Slice	Duration (s)	N_H	α	χ^2/ν
A	2	Gal	1.36 ± 0.07	11./7
B	3	Gal	1.37 ± 0.08	3.6/7
C	7	Gal	1.56 ± 0.10	13./7
D*	19	Gal	1.7 ± 0.3	7.6/5

* Only WFC data (2–28 keV).

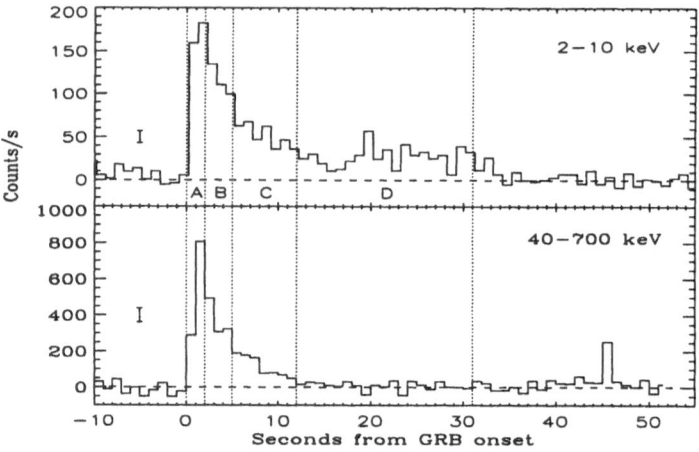

Fig. 1. GRB990806 light curves in different energy bands.

Afterglow Emission

The MECS source count rates and spectra for TOO were extracted using XSE-LECT package, from a ~ 3' radius region around the source centroid, while the background level for the count rates was estimated from a ~ 3' radius region near the source. The spectra from MECS 2 and 3 were equalized and co-added. The use of the XIMAGE package [7] allowed us the choice of a good background region.

During the TOO, a previously unknown X-ray source, 1SAX J0310.6-6806, was detected in the MECS, almost at center of the GRB error box, at coordinates $\alpha_{2000} = 03^h10^m35^s$, $\delta_{2000} = -68°06'35''$ with an error radius of 1 arcmin [6]. Out of the WFC error box, a ROSAT source, 1RXS J031250.0-680915, is visible. This source does not show any statistically significant variation during the entire TOO, contrary to 1SAX J0310.6-6806 which is fading (see below).

We derived the average 1.6–10 keV count spectrum of 1SAX J0310.6-6806 over the whole TOO.We fit it with a power law ($N(E) \propto E^\alpha$) photoelectrically absorbed by the Galactic column density along the GRB direction ($3.53 \times 10^{20} cm^{-2}$). This law provides an acceptable fit of the data: $\chi^2_{red} = 0.32$ (29 degrees of freedom, dof). The best fit power law index is $\alpha = -2.16^{+0.61}_{-0.50}$ (90% confidence level). Assuming no variation of the spectral index, the corresponding fluxes during the first 28 and last 14 hours of the TOO are $(2.3 \pm 0.7) 10^{-13}$ erg cm^{-2} s^{-1}, $(1.5 \pm 0.9) 10^{-13}$ erg cm^{-2} s^{-1} (90% confidence level) respectively.

The 2–10 keV light curve of the prompt emission (WFC data) and the afterglow (MECS2+3 data) is shown in fig. 2. The later fading of 1SAX J0310.6-6806 is apparent, that is described by a power law, $F(t) \propto t^{-\delta}$ with index $\delta = 1.15 \pm 0.03$ (90% confidence level).

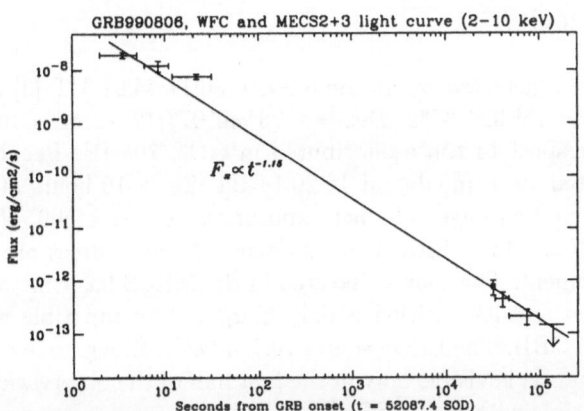

Fig. 2. Light curve of prompt and afterglow emission of GRB990806; the first three points belong to WFC, the others to MECS 2+3.

References

1. F. Frontera et al., *Astron. and Astrophys. Suppl.* **122**, 357 (1997).
2. L. Amati et al., *SPIE Proceedings* **3114**, 176 (1997).
3. R. Jager et al., *Astron. and Astrophys. Suppl.* **125**, 557 (1997).
4. L. Piro, *GCN 392* (1999).
5. G. Boella et al., *Astron. and Astrophys. Suppl.* **122**, 327 (1997).
6. F. Frontera et al., IAU Circ. 7235, 1999
7. P. Giommi et al., *ASP Conf Ser.* **25**, 100 (1991).

GRB000615 in X-Rays

Luciano Nicastro[1], G. Cusumano[1], A. Antonelli[2], L. Amati[3], F. Frontera[3],
E. Palazzi[3], E. Pian[4], E. Costa[5], M. Feroci[5], L. Piro[5], J. in 't Zand[6],
and J. Heise[6]

[1] IFCAI – CNR, Via U. La Malfa 153, 90146 Palermo (I)
[2] Osservatorio Astr. di Roma, Via Frascati 33, 00040 Monteporzio Catone, Roma (I)
[3] Osservatorio Astr. di Trieste, Via G.B. Tiepolo, 11, 34131 Trieste (I)
[4] ITeSRE – CNR, Via P. Gobetti 101, 40129 Bologna (I)
[5] IAS – CNR, Via Fosso del Cavaliere, 00131 Roma (I)
[6] SRON, Sorbonnelaan 2, 3584 CA Utrecht (NL)

Abstract. GRB000615 was detected simultaneously by the *Beppo*SAX GRBM and
WFC 1 with a localization uncertainty of $2'$ (error circle radius). X-ray emission was
detected only in the 0.1–4 keV range during a NFI observation started $\simeq 10$ hours
after the trigger time. The positional and temporal analysis shows the presence of two
sources, one of which may be related to the GRB.

1   Introduction

GRB000615 was detected on 15 June 2000, 06:17:44.91 UT [3] at coordinates
(J2000) R.A. $= 15^h\,32^m\,36^s9$, Dec $= +73°\,49'\,07''$ ($2'$ error radius), which are
revised with respect to those distributed in GCN 705 [1]. *Beppo*SAX NFI observation started the same day at 16:20:14 UT, i.e. $\simeq 10$ hours after the trigger
time, and lasted 1.44 days. The net exposure time was 44609 s for MECS and
31280 s for LECS. Two distinct uncatalogued X-ray sources are detected, one
for each instrument. The source detected in the MECS has constant flux during
the observation and its position is only marginally compatible with the WFC
position of the GRB. The LECS source (0.1–4 keV), though detected at low statistical significance, is visible only in the first half of the observation; its position
is fully compatible with the WFC one.

Optical/IR/radio searches did not detect any new source (see GCN circulars
706, 708, 709, 713, 719, 721, 727).

2   The Burst

The GRBM (40–700 keV) light curve (lasting $\simeq$ **13** s) is shown in the top panel
of Fig. 1. Spectral analysis shows a softening during the burst with a power-law
spectral index ($\mathbf{N_E} \propto \mathbf{E}^{-\alpha}$) from 1.0 ± 0.4 (rise) to 2.4 ± 0.7 (last part of the
decay) with an average of $\alpha = 1.71 \pm 0.46$. The average flux in the 40–700 keV
band is $f_\gamma = (8.2 \pm 0.7) \times 10^{-8}$ erg cm^{-2} s^{-1}, the peak flux is $(1.0 \pm 0.2) \times 10^{-7}$
erg cm^{-2} s^{-1} and the total fluence $F_\gamma = (9.8 \pm 0.9) \times 10^{-7}$ erg cm^{-2}. The
hardness ratio $f_{100-300/50-100} = 2.0 \pm 0.3$.

In the 2–28 keV WFC band, the burst behaviour is quite different. At the time of the GRB onset, the X-ray flux increases only slightly, while a major rise appears some 40 s later and lasts $\simeq 60$ s. Splitting the WFC band in the ranges 2–10 and 10–28 keV we note that the prompt emission comes mainly from the hard band, while the delayed emission from the soft band. This effect, which is under study, can be attributed to scattering of gamma-rays to lower energies rather than to "clean" source photons. The 2–28 keV (delayed) flux is $(2.85 \pm 0.45) \times 10^{-8}$ erg cm^{-2} s^{-1}. In 2–10 keV it is $f_x = (1.74 \pm 0.10) \times 10^{-8}$ erg cm^{-2} s^{-1}.

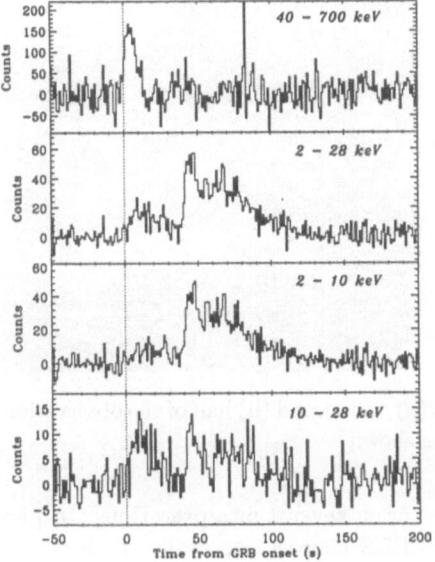

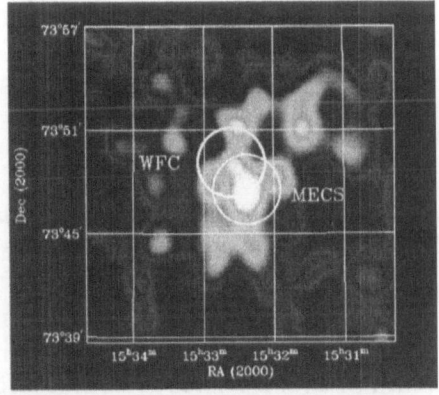

Fig. 1. The 1 s GRBM (*top*) and WFC light curves. A hard X-ray component is visible in the 10–28 keV band at the GRB onset. The soft emission starts after $\simeq 40$ s from the trigger

Fig. 2. MECS 1.6–4 keV smoothed image. The 2′ error circle is shown together with the WFC one (also 2′). The main source is stable during the observation

3 *Beppo*SAX NFIs Observation

The follow-up *Beppo*SAX NFIs observation showed, in the MECS energy range 1.6–4 keV, a soft and stable X-ray source ($\sim 5\sigma$ level). An optimized source counts extraction radius of 3′ was used. Figure 2 shows the smoothed image of the entire observation. The source coordinates are: R.A. = $15^{h}\,32^{m}\,23^{s}.3$, Dec = $+73° 47' 30''$ (1SAX J1532.4+7349). To be *conservative*, we can set an error circle of 2′ (see Fig. 2). This position (note it is $\simeq 1'.5$ away from that reported in GCN 707 [2]) could then be compatible (offset = $1'.8$) with the WFC GRB position. However its flux stability suggests it is not related to the GRB. The average flux is $f_x = (6.7 \pm 1.5) \times 10^{-14}$ erg cm^{-2} s^{-1} in 1.6–4 keV. No significant emission is detected above 4 keV.

In the LECS (0.1–4 keV) a source is detected at 2.5σ level, but *only in the first half* of the observation. The smoothed images of the two halves of the observation are shown in Fig. 3. The position of this source is: R.A. = 15^h 32^m 35, Dec = +73° 48′ 50″ (1SAX J1532.6+7348), fully compatible with the WFC (GRB) position. The low significance of the detection cannot help in discriminating the source characteristics.

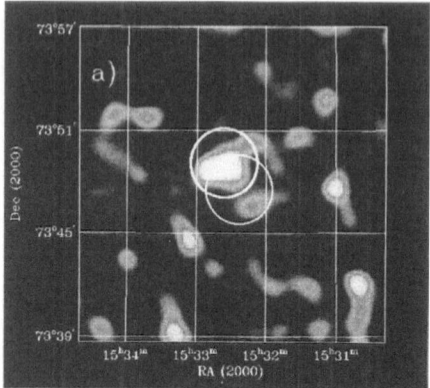 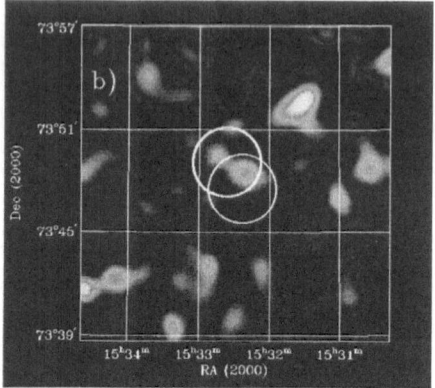

Fig. 3. LECS (0.1–4 keV) images for the first (**a**) and second (**b**) half of the observation. The WFC and MECS source error circles are shown

Unfortunately the low statistics leaves open several interpretations. In particular it is not clear if the MECS and LECS sources are really distinct and which (if any) is related to GRB000615. More sensitive observations of the field may help investigating the two detected sources. Nevertheless, one can speculate that a soft prompt (actually *delayed*) X-ray event could lead to a soft afterglow. In brief, our refined data analysis shows that:

- a stable (probably unrelated to the GRB) source is detected in the MECS. Its spectrum can be fit by a power-law of index $\alpha = 3.2 \pm 0.5$ ($\chi_n^2 = 1.1$).
- after ~ 20 hours from the gamma-ray event the X-ray afterglow is not detected to the limit of the MECS sensitivity suggesting a power-law temporal decay index $\lesssim -1.7$ (respect to the WFC mean flux and $T_{start} = T_{burst}$).
- a source is detected in the LECS band 0.1–4 keV; some hint of decay is present.

References

1. G. Gandolfi et al.: GCN circ. 705 (2000)
2. G. Gandolfi et al.: GCN circ. 707 (2000)
3. L. Piro et al.: GCN circ. 703 (2000)

GRB980613 a Very Faint Burst with a Not So Faint Afterglow Detected by BeppoSAX

Paolo Soffitta[1], Lorenzo Amati[2], Lucio A. Antonelli[3], Enrico Costa[1], Marco Feroci[1], Filippo Frontera[2], John Heise[4], Luciano Nicastro[5], Luigi Piro[1], Giulia Stratta[1], and J.J.M. in't Zand[4]

[1] Istituto di Astrofisica Spaziale del CNR,
    Area di Tor Vergata, Via Fosso del Cavaliere 100, 00133 Roma, Italy
[2] Istituto di Tecnologie e Studio delle Radiazioni Extraterrestri del CNR,
    Via Gobetti 101, 40129 Bologna, Italy
[3] Osservatorio Astronomico di Roma,
    Via Frascati 33, 00040 Monteporzio Catone, Italy
[4] Space Research Organization in the Netherlands,
    Sorbonnelaan 2,3584, CA Utrecht, The Netherlands
[5] Istituto di Fisica Cosmica ed Applicazioni dell'Informatica del CNR,
    Via Ugo La Malfa 153, 90138 Palermo, Italy

Abstract. Here we present the results of the BeppoSAX data analysis of both the prompt event and the afterglow of GRB980613. We summarize, also, the data from optical observation.

1 Detection and History of GRB980613 Observations

GRB980613 was detected [1],[2]on June 13 04:51:06 (13.48) UT by the BeppoSAX Gamma Ray Burst Monitor (GRBM) as a single pulse of 15s and by BATSE/CGRO [3]. The BeppoSAX Wide Field Camera (WFC) no.2 located the burst at the equatorial coordinates R.A. = $10^h17^m44^s$ Decl. = $+71°29'.9$ with a 4' error radius. The BeppoSAX NFI were pointed to the GRB error box about 9 hours after the burst for a total observing time of 100 ks. Two sources [5],[6] were found in the WFC error circle by the BeppoSAX Medium Energy Concentrator Spectrometer (MECS) during the follow-up. Only source A, located at R.A. = $10^h17^m53^s$ Decl. = $+71°27'.4$ showed a rather strong decay while source B did not show substantial variability. An X-ray source, in a position R.A. = $10^h18^m2.4^s$ Decl. = $+71°33'47.6''$, consistent with source B, was already present in a Rosat/PSPC observation of the same field while no source was observed in the position of source A. By comparing two images taken on June 13.9 UT and June 17.9 UT, Hjorth et al. [10] detected, in optical band, a faint point source (not present in the 17.9 UT observation) located (± 0.5") at R.A. = $10^h17^m57^s.64$ Decl. = $+71°27'26^s.4$ consistent with the location of BeppoSAX source A. The magnitude (R) at UT 13.9 was 22.9±0.2 while at UT 17.9 was greater than 24.0 (10-σ limit). Other observers measured consistent upper limits for optical variable objects present in the field [7],[8],[4],[9]. A galaxy, with r = 24.15, coincident within ±0.5" from the Optical Transient (OT) was, found by Djorgovski et al. [12]. The galaxy was resolved [11]. A further OT photometry, by Halpern et al. [13], combined with the flux in [12], implied in the optical R band a power-law decay slope of 1.3. Spectrum of the host galaxy was then found by Djorgovski et al. [14] which resolved the [OII]3727 line placing the distance of the host galaxy at z = 1.0964 for an isotropic gamma-ray energy of the burst of 5.2×10^{51} erg. Djorgovski et al., [17] proposed a model in which a GRB originates from merger-induced starburst system supporting the possible link between GRBs and massive star formation. In fact he noted several extended objects within few arcseconds of the GRB that may be due to tidal interactions.

2   BeppoSAX Observation of the Prompt Event and the Afterglow

The peak flux and fluence for GRB980613 were evaluated both in WFC (2-26 keV) and GRBM (40-700 keV). We measured a WFC peak flux and fluence respectively of $(2.0\pm0.2)10^{-8}$erg cm^{-2} s^{-1} and $(2.7\pm0.4)10^{-7}$erg cm^{-2}. In the GRBM the same quantities are $(1.3\pm0.4)10^{-7}$ erg $cm^{-2}s^{-1}$ and $(1.0\pm0.2)10^{-6}$ erg cm^{-2}. The WFC peak flux ranked this GRB in the last but one position among 31 bursts. The faintest burst is GRB990217. The BATSE peak flux (50-300 keV, integrated over 1s) and fluence (>20 keV) are 0.63 ± 0.05 photons $cm^{-2}s^{-1}$ and $(1.71\pm0.25)\times10^{-6}$erg cm^{-2}, respectively. From the Fourth BATSE catalogue it results that about a fraction of 0.6 and 0.4 of the, respectively, peak flux and fluence are larger than those of GRB980613. The WFC and GRBM data were combined to study the time-resolved spectral evolution. The time history of GRB980613 (see fig. 1) was divided into two segments, interval A (first 4 s; rise) and interval B (second 38 s; fall). We performed a combined spectral analysis which shows a rather flat evolution with a very marginal evidence for a hardening. The photon index of the power-law are 1.33 ± 0.12 (A) and 1.15 ± 0.05 (B) (1-σ error). In interval A the data indicates that the peak maximum is in the gamma energy range. The spectral characteristics of this and other bursts are, also, discussed in [18] while in [19] is presented, among others, the averaged spectrum of GRB980613. The detected afterglow of GRB980613 is relatively speaking not faint. It decays as $(9.0\pm0.2)10^{-13}$ $(t/9hrs)^{(-1.19\pm0.17)}$ erg s^{-1} cm^{-2} (2-10 keV). We note that a strong burst like GRB980329 shows comparable afterglow fluxes with respect to GRB980613 indicating that bursts with very different intensity in the prompt event may evolve in afterglows with comparable intensity (see again fig. 1).

Moreover a refined analysis of GRB980613 with an imaging technique (IROS v.105108) shows a dip between 30 s and 50 s. This gap between the main event and the onset of the afterglow is visible in the images of GRB980613. If we consider the 3-σ upper limit of $1.3 \cdot 10^{-9}$ erg cm^{-2} s^{-1}, this is well below the back extrapolation to the early stage of GRB980613 afterglow, may be indicating that a separation can be present between the main burst and the afterglow luminosity peak. Such a separation could be due to the initial rising of the afterglow luminosity when the dimension of the emitting shell increases. By measuring this gap we can derive the physical properties of the interstellar medium [16]. A very late time rising (7000 s) of the afterglow from below the sensitivity of BeppoSAX NFI was detected by BeppoSAX on GRB981226 [20].

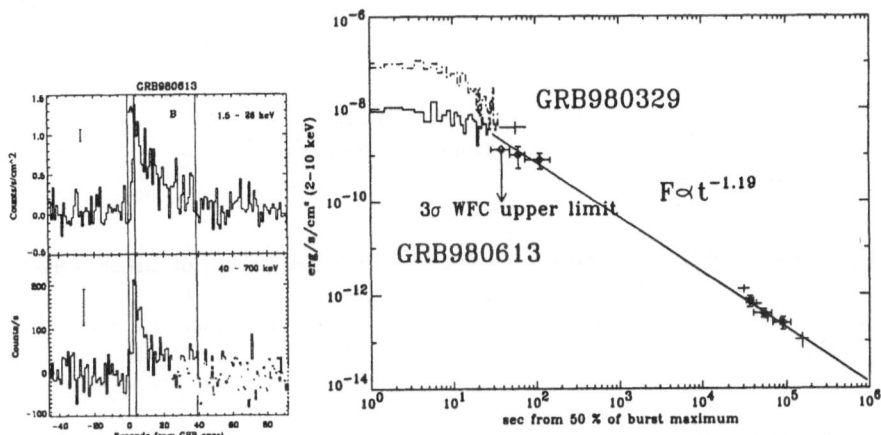

Fig. 1. (Left) Lightcurve of GRBM and WFC. (Right) Afterglow onset of GRB980613.

References

1. Piro L. et al., GCN no. 99.
2. Smith M. et al., IAUC 6938.
3. Woods P. et al., GCN no. 112.
4. Halpern J. et al., GCN no. 106.
5. Piro L. et al., GCN no. 104.
6. Costa et al., IAUC 6939.
7. Castro-Tirado et al., GCN no. 102.
8. Odewahn et al., GCN no. 105.
9. Diercks et al., GCN no. 108.
10. Hjorth et al., GCN no. 109.
11. Djorgovski et al., GCN no. 117.
12. Djorgovski et al., GCN no. 114.
13. Halpern J.P. et al. GCN no. 134.
14. Djorgovski et al. GCN no. 189.
15. Holland et al. GCN no. 777.
16. Sari R., Ap.J. **489**, L37-L40, 1997.
17. Djorgovski S.G. et al. astro-ph 0008029.
18. Frontera F. et al., in preparation.
19. Amati L. et al., these proccedings.
20. Frontera F. et al., Ap.J. **540**, 697-703, 2000

Physical Constraints from Broadband Afterglow Fits: GRB000926 as an Example

Sarah A. Yost[1,2], Re'em Sari[1], Fiona A. Harrison[1], Edo Berger[1],
Alan Diercks[1], Titus Galama[1], Dan Reichart[1], Dale Frail[3], and Paul A. Price[4]

[1] Caltech, Pasadena CA 91125, USA
[2] yost@srl.caltech.edu
[3] National Radio Astronomy Observatory, Socorro NM 87801, USA
[4] Research School of Astronomy & Astrophysics, Australian National University, ACT 2601, Australia

Abstract. We develop a model to fit the broadband afterglows of GRBs from the intrinsic parameters of the fireball's synchrotron emission, and apply it to a few well-studied events, with the goal of constraining the intrinsic variability of GRB parameters. We give an example here of fitting to the recent bright event GRB000926.

1 Introduction

The single successful model of GRB emission to date has been the fireball model. A small amount of matter is accelerated to a large Lorentz factor Γ. Shock expansion produces synchrotron emission of radiation with a well-defined spectrum. The spectral breaks ν_{break} are functions of fireball parameters and depend on the hydrodynamics of the fireball's evolution. The hydrodynamics are strongly affected by the environment and geometry of the fireball, thus the afterglow's broadband lightcurves can in principle constrain fundamental parameters of the burst. For example, collimation of the ejecta produces an achromatic break, but the evolution of ν_{break} past observed frequencies does not.

We consider two possible density profiles for the burst environment, r^0 as in the interstellar medium (ISM) and r^{-2} as from a constant stellar wind (WIND).

We calculate the synchrotron flux as a function of t,ν from the luminosity distance and redshift, and a set of fundamental parameters: isotropic-equivalent energy E, electron powerlaw index p, electron and magnetic energy fractions, ε_e and ε_B, as well as the circumburst density: a constant n in the ISM case or A ($\rho = Ar^{-2}$) in the WIND case. The equations are based on Sari et al [11] and Granot et al [3], [4] for the ISM model and Chevalier & Li [1] for the WIND model. Collimation effects on the evolution are based upon Sari et al [10]. We include the effects of inverse Compton scattering based upon Sari & Esin [9]. Host extinction is parametrized by A_V according to the prescription of Reichart [8].

This calculated flux is compared to observations corrected for Galactic extinction and host flux, and a Powell gradient search optimizes the model parameters.

2 GRB000926: Preliminary Results

The IPN detected this event on 2000 September 26.993 and rapidly disseminated its postion, leading to observations by many. We use the optical observations at ≤ 1 day post-burst by Hjorth et al [1] and Fynbo et al [2], along with the data presented in Price et al [7], with its calibration and host flux subtraction, as well as the x-ray data of Piro & Antonelli [10]. We allow a systematic calibration uncertainty of 4%, account for interstellar scintillations in the radio based on Walker [12], and calculate the best-fit ISM model (I) and WIND model (W).

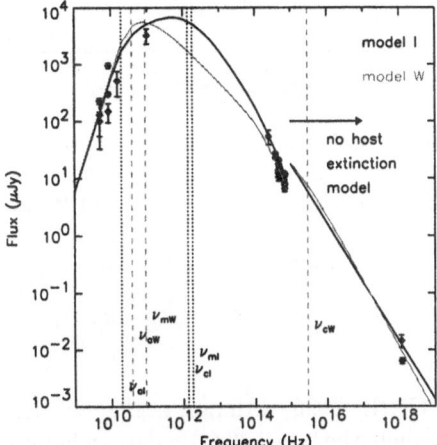

Fig. 1. Broadband spectra of model I (black) and W (grey) at 2 days. Data from 2 ± 1 days is plotted over the curves, interpolated to day 2 by model I. Both models provide a reasonable fit to the broadband data.
* Inverse Compton (IC) cooling constrains the relative evolution of ν_{break}, preventing a high ν_c to better fit the x-ray. A fit to the ISM with no IC gives a notably different fit, with a much higher ν_c. IC effects are not trivial and must be included in model fits.

Table 1. Model Parameters

	I	W	units		I	W	units
E	1.1	39	10^{53} erg	p	2.1	2.2	
n	0.62		cm^{-3}	A		1.5	$5 \times 10^{11} gcm^{-2}$
ε_e	0.27	0.012		ε_B	0.95	0.0025	
θ	0.083	0.044	rad	t_{jet}	1.2	1.8	day
A_V	0.2	0.3	mag				

The fit to model I, including radio scintillation effects, gives a total χ^2 of 197 for 80 degrees of freedom. Model W has $\chi^2 = 171$ for 80 d.o.f.. Both models assume an LMC-like host extinction curve, though an SMC-like curve gave scant difference in the results.

Models I and W both give a reasonable description of the data. Model W gives a better optical fit, but does not seem to fit the late-time radio data. Very late radio data may provide the best discriminant between the ISM and WIND.

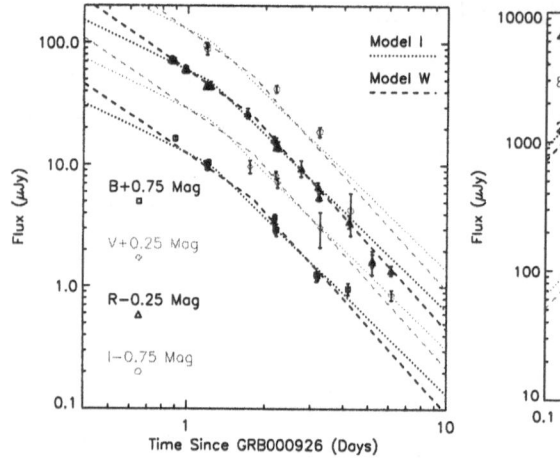

Fig. 2. GRB000926 optical lightcurves. Both I and W models fit reasonably well, though the R band is better fit by W, with its later jet break and steeper slope.

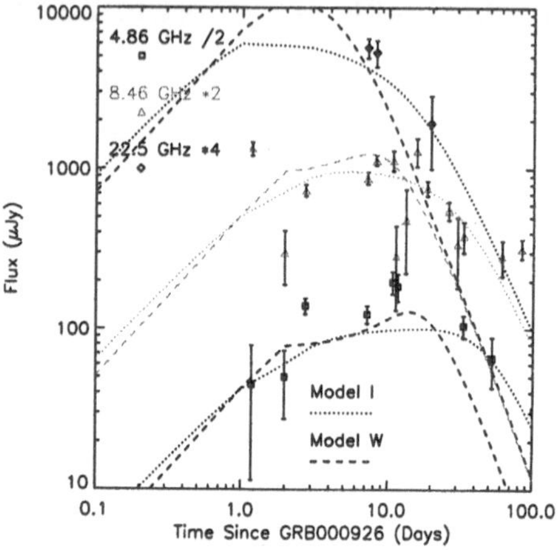

Fig. 3. GRB000926 Radio lightcurves. At 22.5 GHz, early observations could distinguish the models. At lower ν, I and W are distinguishable at late times, and I appears to fit better after about a month.

3   Conclusion

ISM and WIND models fit the afterglow of GRB000926, with non-negligeable Inverse Compton effects. The WIND underpredicts the late 8.46 GHz data, whereas the ISM model is a far better fit to the late radio observations, providing some evidence that this burst occured in a medium of constant density.

References

1. Chevalier, R.A. and Li, Z. ApJ **536**, 195 (2000)
2. Fynbo, J.P.U. et al. GCN 825 (2000)
3. Granot, J., Piran, T. and Sari, R. ApJ **513**, 679 (1999a)
4. Granot, J., Piran, T. and Sari, R. ApJ **527**, 236 (1999b)
5. Hjorth, J. et al GCN 809 (2000)
6. Piro, L. and Antonelli, L.A. GCN 832,833 (2000)
7. Price, P.A. et al. ApJL accepted, astro-ph 0012303 (2000)
8. Reichart, D. astro-ph 9912368 (1999)
9. Sari, R. and Esin, A. A. astro-ph 0005253 (2000)
10. Sari, R., Piran, T. and Halpern, J.R. 1999 ApJ **519** L17
11. Sari, R, Piran, T. and Narayan, R. 1998 ApJ **497** L17
12. Walker, M. A. MNRAS **294**, 307 (1998)

Part III

Host Galaxies and Cosmology with GRBs

The Observed Offset Distribution of GRBs about Their Hosts

Joshua S. Bloom and Shrinivas R. Kulkarni

Palomar Observatory 105-24, California Institute of Technology,
Pasadena, CA 91125 USA

Abstract. We summarize the results of a comprehensive survey to measure the location of 20 cosmological gamma-ray bursts (GRBs) in and around galaxies. We find that all well-localized GRBs to-date fall within 1.2 arcsec (10 kpc in projection) from the nearest detected galaxy. We estimate on statistical grounds that most if not all of these nearby "host" galaxies are indeed physically associated with the respective GRB and not a spurious superposition. The observed distribution of GRBs offsets about their hosts are consistent with a progenitor population that burst near the sites of massive star formation (eg. "collapsars") and inconsistent with the modeled location of degenerate binary systems (eg. neutron-star–neutron-star and neutron-star–black-hole mergers) which travel far from their birthsite.

1 Primary Results

Here we summarize the primary results of our GRB offset census [10]. In total the first 20 long-duration GRBs well-localized by means of their afterglow were included in the sample (up to, and including, GRB 000418). In all but one case (GRB 970828) the GRB field was imaged with HST in the broad STIS/Clear filter. These HST images were registered primarily with early-time Palomar, Keck, HST and VLA positions to obtain the precise location of the GRB afterglow (and hence the GRB itself) with respect to the surrounding field. In addition we found, in archival public HST data, two new probable hosts (GRB 990510 and GRB 990308). With this dataset we note several points.

- *E*very GRB in the sample occurred within 1.2 arcsec (10 kpc in projection) from the nearest galaxy (Fig. 1).
- The GRB positions appear to be preferentially aligned along the optical major-axis of their respective hosts.
- Only a few host assignments could be incorrect given the offsets and nearby galaxy magnitude distribution. We estimate the chance that none of the host assignments are due to a spurious random superposition is $\sim 50\%$. The chance that more than 3 hosts are incorrect is less than 10^{-3}.
- The offset distribution supports the collapsar progenitor model for GRBs. The radial distribution of GRBs about their assumed hosts appear to closely follow the star-formation locations in late-type galaxies (Fig. 2). That is, the radial distribution is consistent with the massive star formation locations in galaxies.

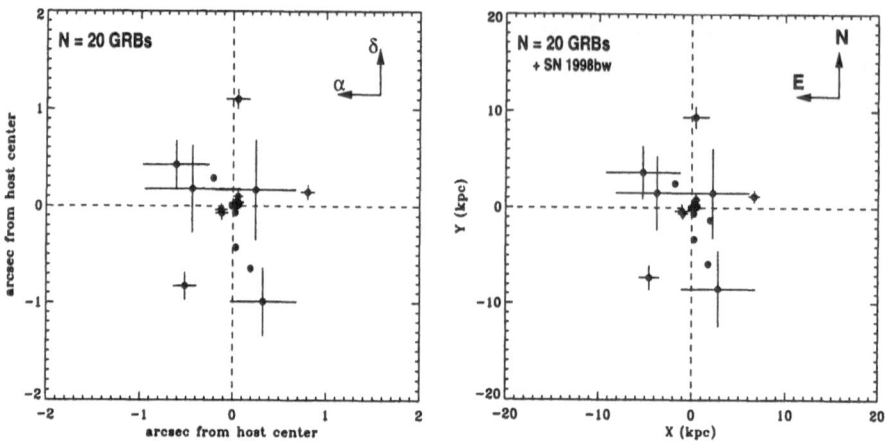

Fig. 1. The angular (left) and projected (right) distribution of 20 GRBs about their presumed host galaxy. The error bars are 1 σ and reflect the total uncertainty in the relative location of the GRB and the apparent host center. The physical offset is assigned assuming $H_0 = 65$ km s^{-1} Mpc^{-1}, $\Lambda = 0.7$, and $\Omega_m = 0.3$ and assuming the GRB and the presumed host are at the same redshift. Where no redshift has been directly measured a redshift is assigned equal to the median redshift ($z = 0.966$) of all GRBs with measured redshifts. The qualitative similarity between the physical and sky distributions demonstrates the relative insensitivity of the physical distribution to assigned redshifts for those GRB without known redshifts.

- Insofar as the modeled positions of "delayed merging binary" progenitors are correct, the offset distribution disfavors the progenitor scenarios of merging NS–NS or NS–BH binaries (Fig. 2). Such models predict that roughly one third to one half of all GRBs should occur beyond 10 kpc in projection, depending on the particular host galaxy potential.

In our analysis we included the uncertainties in the measured offset and host half-light radii of the GRBs in the sample; through Monte Carlo modeling we found the conclusions based on the simplistic KS probabilities to be insensitive to these uncertainties.

There are two important biases which would preclude some long-duration GRBs from entering in our sample. First, heavily obscured GRBs might escape precise localization since optical localizations are currently more sensitive than in the X-ray and radio (which do not suffer as severely from dust extinction). Even so, such "dark GRBs" which escape localization would likely occur preferentially in molecular clouds and HII regions; thus our claim that GRBs appear to occur where stars are formed would be *strengthened* by such a bias. Second, GRBs which occur in low-density environments such as in the intergalactic medium (IGM) could have significantly fainter afterglow than in the interstellar medium (ISM) and thus escape detection. However, Piro (Amsterdam, June 2001) noted that the first 29 out of 34 GRBs followed-up by BeppoSAX had detectable X-ray afterglow. Since the intensity of the prompt GRB emission should be relatively decoupled from the ambient environment (in the internal shock model of GRBs),

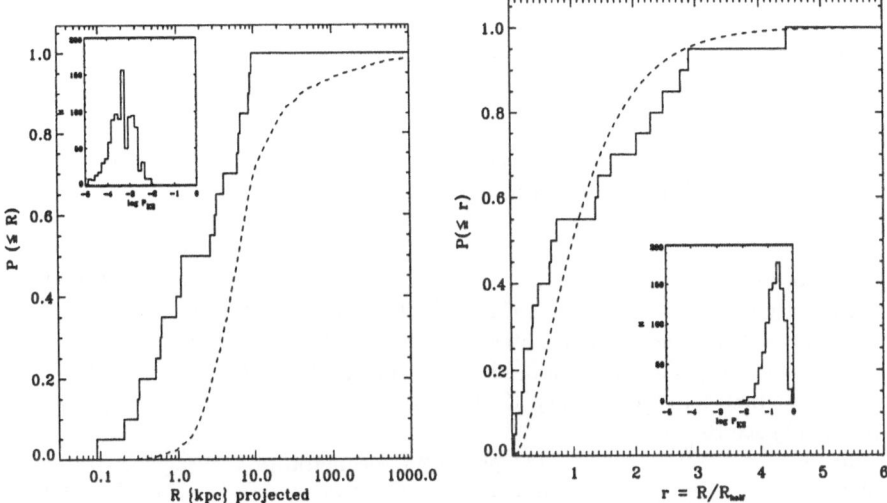

Fig. 2. (left) Offset distribution of GRBs compared with delayed merging remnant binaries (NS–NS and BH–NS) prediction and (right) compared with host galaxy star formation model. Inset in both are the distribution of KS statistics determined from a Monte Carlo which accounts for the varying uncertainty in the observed offsets. The predicted NS–NS merger distribution (*dashed line*) is inconsistent with the observed distribution with a KS probability less than 10^{-3} despite the conservative assumption that all GRB hosts, many of which are dwarf starburst galaxies, are as massive as the Milky Way. At right, an exponential disk profile is assumed (*dashed line*) as an approximation to the distribution of the location of collapsars and promptly bursting remnant binaries (BH–He). The observed GRB distribution provides a remarkable fit to the model considering we make few assumptions to perform the comparison.

at most $\sim 10\%$ of long-duration GRBs could suffer from systematically fainter afterglow relative the prompt emission due to locations in the IGM.

Thus, we are confident that the observed offset distribution is not significantly biased against large offsets and hence the conclusion about the inconsistency with NS–NS and NS–BH remnants stands. The dust–"dark GRB" bias only strengthens the case that GRBs are closely connected to the locations of star formation.

References

1. J. S. Bloom, S. R. Kulkarni, S. G. Djorgovski: Submitted to Astronomical J.; astro-ph/0010176 (2001).

A Deep, High-Resolution Imaging Survey of GRB Host Galaxies

N.R. Tanvir[1], S. Holland[2], M.I. Andersen[3], G. Björnsson[4], J.U. Fynbo[5],
J. Gorosabel[6], J. Hjorth[7], A. Jaunsen[5], P. Møller[5], P. Natarajan[8],
and B. Thomsen[9]

[1] Dept. of Physical Sciences, University of Hertfordshire, Hatfield, AL10 9AB. UK.
[2] University of Notre Dame, Notre Dame, IN46556-5670. USA
[3] Division of Astronomy, PO Box 3000, FIN-90014 University of Oulu, Finland.
[4] University of Iceland, Dunhaga 3, IS-107 Reykjavik, Iceland.
[5] European Southern Observatory, Karl Schwarzschild Strasse 2, Garching, Germany.
[6] Danish Space Research Inst.,
    Juliane Maries Vej 30, DK-2100 Copenhagen, Denmark.
[7] Astronomical Obs., University of Copenhagen, DK-2100 Copenhagen, Denmark.
[8] Department of Astronomy, Yale University, New Haven, CT, USA.
[9] Inst. of Physics and Astronomy, University of Århus, DK-8000, Århus, Denmark.

Abstract. We have begun a survey of GRB host galaxies using HST/STIS to obtain very deep, high-resolution images. The primary aims are to constrain progenitor models by revealing any relationship between GRBs and particular regions or stellar populations within the host, and to study the morphologies of the hosts themselves. The data from the survey has all been made public via the web.

1 Introduction

Vital clues to the nature of GRBs may be obtained by studies of their host galaxies. The type of the galaxy and position of the GRB in relation to the various stellar populations constrains possible progenitors. In particular, the two most-often discussed scenarios: core collapse of massive stars, and coalescence of compact binaries; predict a close association with sites of star formation in the former case, and a much more extended distribution around the host galaxy in the latter case.

However, because most GRBs have so far been found at high redshift, the host galaxies are typically small and faint, making analysis of the GRB environment uncertain. We have therefore embarked on a deep, high-resolution survey of a sample of GRB host galaxies [2].

2 Description of Survey

So far HST/STIS images have been obtained for eleven GRB hosts, selected primarily for their accurate localisations (see table 1). Exposure times are typically several orbits, reaching $R \sim 28$ or fainter. In some cases both clear (unfiltered) and "long-pass" (approximately a broad RI-band filter) images have been obtained to provide some colour information. Because the images are obtained over a year after the GRB event, analysis is not confused by significant light from the afterglows themselves.

Table 1. GRBs so far observed as part of the survey.

GRB	Exposure times (seconds) and pass-bands
990705	8851 clear, 8203 LP
990506	7856 clear, 8000 LP
990308	7842 clear
990123	8224 LP
981226	7865 clear, 7910 LP
980703	5118 clear, 5264 LP
980613	5851 clear, 5936 LP
980519	8983 clear
980425	1240 clear, 1185 LP
980329	8072 clear, 5416 LP
971214	8599 clear

3   Data Reduction and Web Access to Database

Images were passed through the standard STScI pipeline, which performs basic calibration, and then coadded with the DRIZZLE software [4] with a final pixel scale of 0.0254 arcsec per pixel.

An important aspect of our survey is that the reduced images have all been made available on the web within a day of the data being obtained. Thus they are available for other researchers immediately, and indeed various groups have already made use of this data set (including some in this proceedings!). For information and access to the data, go to:

http://www.ifa.au.dk/~hst/grb_hosts/intro.html

The web pages also include links to other images of these galaxies available in the HST archive.

4   Discussion

Results of this survey have been reported in a series of GCN circulars (specifically 778, 777, 753, 749, 731, 726, 715, 704 and 698). They are broadly consistent with the emerging picture for "long-soft" bursts, that at least most are located in high surface brightness regions within their hosts (cf. [5]). Although not a statistically complete sample, these results, combined with the evidence that GRB host galaxies have unusually high star formation rates (eg. [6]), support models in which GRBs are associated with massive, short-lived stars.

The survey includes the so-far unique low-redshift event GRB980425 (=SN 1998bw). The proximity of this GRB means that we can continue to follow the fading afterglow remnant more than 2 years on [7]. Other important results are tentative evidence for a supernova underlying GRB 980703 [3] , and measurement of the highest specific star formation rate yet found in a GRB host in the case of GRB 980613 [1].

We hope to extend this survey in the future to a larger, statistically useful sample, to both test this association with star formation and compare hosts to galaxy samples selected in other ways (eg. Lyman break galaxies). In the longer term, a close association with star formation could open the door to using GRBs to select galaxies and constrain star formation activity at very high redshifts or in dusty galaxies, where direct observation is difficult (eg. [8]).

References

1. J. Hjorth *et al.*, in prep.
2. S. Holland *et al.* 2000, GCN 698.
3. S. Holland *et al.* A&A, submitted.
4. R. Hook & A. Fruchter 2000, ADASS IX, ASP conf. ser. 216, eds. N. Manset *et al,* p 521.
5. J. Bloom *et al* submitted, astroph/0010176.
6. L. Hanlon *et al* 2000 A&A, 359, 941.
7. J. Fynbo *et al* 2000, ApJ, 542, L89.
8. A. W. Blain & P. Natarajan 2000, MNRAS, 312, L35.

Host Galaxies as Gamma-Ray Burst Distance Indicators

David Band[1], Raul Jimenez[2], and Tsvi Piran[3]

[1] X-2, Los Alamos National Laboratory, Los Alamos, NM 87505, USA
[2] Dept. of Physics and Astronomy, Rutgers University, Piscataway, NJ 08854, USA
[3] Hebrew University, Jerusalem, Israel

Abstract. We calculate the distributions of the total burst energy, the peak luminosity and the X-ray afterglow energy using burst observations and distances to the associated host galaxies. To expand the sample, we include redshift estimates for host galaxies without spectroscopic redshifts. The methodology requires a model of the host galaxy population; we find that in the best model the burst rate is proportional to the host galaxy luminosity at the time of the burst.

1 Introduction

The first few bursts with afterglows, host galaxies and spectroscopic redshifts demonstrated that most, if not all, classical bursts occur at cosmological distances. As the sample of such bursts grows, we can address fundamental characteristics of the burst population. Here we calculate the distributions of the total gamma-ray energy emitted, the peak luminosity, and the energy of the X-ray afterglow. The average values of these quantities set fundamental physical constraints on burst models, while the widths of the distributions indicate the variability these models must permit.

To estimate the distribution of these different intrinsic intensities requires both the apparent intensity of the burst and the distance to the burst. We fit spectra to BATSE data, where available, to calculate the energy fluence and the peak luminosity. The distance is obtained from the spectroscopic redshift, if observed, or from a redshift probability distribution derived from the host galaxy brightness. We include bursts without spectroscopic redshifts out of a desire to use all available data. The number of bursts with spectroscopic redshifts has increased since we began this project, and the inclusion of bursts without such redshifts is not as compelling, although the methodology will again be useful if the determination of spectroscopic redshifts does not keep up with the discovery of host galaxies. The redshift probability distribution requires a model of the host galaxy population, which we choose using a sample of bursts with both a spectroscopic redshift and a host galaxy brightness. This host galaxy population model is intrinsically interesting since it depends on the type of progenitor.

2 Methodology

We present the methodology for determining the distribution of the gamma-ray energy, but the approach is easily generalized to the distributions of the peak gamma-ray luminosity and the X-ray afterglow energy. We assume that the gamma-ray energy is characterized by a lognormal burst distribution. We convert the energy to an observed fluence using the redshift and the cosmological model with $H_0 = 65\,\mathrm{km\ s^{-1}\,Mpc^{-1}}$, $\Omega_M = 0.3$ and $\Omega_\Lambda = 0.7$. As discussed above, we use the spectroscopic redshift, if available; otherwise we use an estimated redshift probability distribution $p_B(z\,|\,D_0)$ based on the host galaxy's R magnitude, as described below. The result is a probability distribution for the fluence for each burst parameterized by the average burst energy and the logarithmic width of the energy distribution.

The likelihood function is the product of the probability of observing each fluence. This probability is the probability distribution for the burst fluence above the fluence threshold, the minimum fluence for which the burst would be included in our sample. Using the fluence threshold mitigates the selection effect where at high redshift only intrinsically bright bursts can be detected. We then maximize this likelihood by varying the parameters of the lognormal distribution – the average energy and the logarithmic width of the distribution; confidence ranges are also determined.

Our sample consists of bursts with spectroscopic redshifts or host galaxy detections which were observed by BATSE between 1997 and CGRO's untimely demise in June, 2000. We fit BATSE spectra with the four parameter "GRB" function [1]. The fit to the spectrum accumulated over an entire burst was integrated over the 20–2000 keV energy range and the burst duration to give that burst's bolometric fluence, while the spectrum accumulated over the shortest available period which included the burst's peak count rate was integrated over 50–300 keV in the burst's frame to provide the peak luminosity. We find that fluences derived by integrating over spectral fits are a factor of ~ 2 smaller than the values in the online BATSE catalog [2], but given the large range of fluence values, this discrepancy is insignificant. In many bursts most of the energy is above 300 keV where the fits are uncertain.

The X-ray afterglow was first extrapolated back to 100 s after the burst using the observed X-ray flux and power-law temporal decay. The X-ray afterglow was then integrated over time and frequency to provide the total X-ray afterglow energy. Since the temporal decay is usually close to t^{-1}, the calculated energy depends only logarithmically on the uncertain beginning of the afterglow.

For the bursts without a spectroscopic redshift we need $p_B(z\,|\,R)$, the probability distribution that a host galaxy with magnitude R is at redshift z. Empirically, the Hubble Deep Field (HDF) provides $p_{HDF}(z\,|\,R)$, the probability that a galaxy with magnitude R is at redshift z. However, this counts each galaxy equally – a nearby dwarf is equal to a distant luminous galaxy. Thus, we must weight p_{HDF} by the probability that the galaxy is a host, which requires a model of the host galaxy population. We consider four models. In the first, bursts are equally likely to occur in any galaxy and thus p_{HDF} is unweighted. In the second, the burst rate is proportional to the galaxy mass; p_{HDF} is weighted by the luminosity with k- and e-corrections. In the third, the burst rate is proportional to the galaxy luminosity at the time of the burst; p_{HDF} is weighted by the luminosity with only k-corrections. Finally, in the fourth, the burst rate is proportional to the galaxy mass times the cosmic star formation rate; the distribution from model 2 is multiplied by this star formation rate.

We choose the host galaxy model which best fits the set of 10 bursts with both spectroscopic redshifts and host galaxy R magnitudes. For each model we form a likelihood which is the product of p_B for each burst in this dataset; each model has a different p_B. Technically p_B should be modified to account for the redshift ranges in which the observing telescope would have been unable to determine the redshift. However, accounting for the capabilities of all the different telescopes is beyond the scope of this work. We find the following likelihoods for the different models: model $1 - 4 \times 10^{-3}$; model $2 - 2 \times 10^{-4}$; model $3 - 4 \times 10^{-2}$; and model 4 $- 10^{-4}$. Thus for all our calculations we use model 3, where the burst rate is proportional to the galaxy's luminosity at the time of the burst.

3 Results and Discussion

By applying the methodology described above to a sample of 9 bursts with spectroscopic redshifts and 3 bursts with only host galaxy magnitudes we find that the distribution for the total energy is characterized by $\langle E_\gamma \rangle = 1.6^{+1.7}_{-1.2} \times 10^{53}$ erg with a logarithmic width of $\sigma_\gamma = 1.7^{+0.8}_{-0.4}$. We emphasize that this is the total energy if the burst radiates isotropically. We use natural logarithms in defining the width. The uncertainties are 2σ. Using the same dataset we find $\langle E_L \rangle = 4.6^{+5.4}_{-2.5} \times 10^{51}$ erg s^{-1} and $\sigma_L = 1.4^{+0.7}_{-0.35}$ for the peak gamma-ray luminosity. For the afterglow X-ray energy we have $\langle E_X \rangle = 4.5^{+6.7}_{-2.8} \times 10^{51}$ erg and $\sigma_X = 1.6^{+0.85}_{-0.55}$; this was calculated using 6 bursts with redshifts and 1 with only a host R magnitude. The total energy radiated is enormous, even with beaming factors which may reduce the actual energy release by one or two orders of magnitude. Of course, the actual energy release must be even greater because of the imperfect transfer of energy into radiating particles and then into emission. The large energies required severely constrain the physical models.

This methodology can be used wherever burst distances and energies are required. For example, a burst sample with known distances is required to calibrate proposed correlations between the burst energy and the frequency-dependent lags in pulses [3] or lightcurve variability [4].

References

1. D. Band, et al.: Astrophys. J. **413**, 281 (1993)
2. BATSE GRB team: http://www.batse.msfc.nasa.gov/batse/grb/catalog/current/ (2001)
3. E. Fenimore, E. Ramirez-Ruiz: Astrophys. J., submitted, astro-ph/0004176 (2001)
4. J. Norris, G. Marani, and J. Bonnel: Astrophys. J. **534**, 248 (2000)

The GRB Host Galaxies and Redshifts*

S.G. Djorgovski[1], S.R. Kulkarni[1], J.S. Bloom[1], D.A. Frail[2], F.A. Harrison[3], T.J. Galama[1], D. Reichart[1], S.M. Castro[1], D. Fox[1], R. Sari[4], E. Berger[1], P. Price[3], S. Yost[3], R. Goodrich[5], and F. Chaffee[5]

[1] Palomar Observatory, Caltech, Pasadena, CA 91125, USA
[2] NRAO/VLA, Socorro, NM 87801, USA
[3] Space Radiation Laboratory, Caltech, Pasadena, CA 91125, USA
[4] Theoretical Astrophysics, Caltech, Pasadena, CA 91125, USA
[5] W.M. Keck Observatory, CARA, Kamuela, HI 96743, USA

Abstract. Observations of GRB host galaxies and their environments in general can provide valuable clues about the nature of progenitors. Bursts are associated with faint, $\langle R \rangle \sim 25$ mag, galaxies at cosmological redshifts, $\langle z \rangle \sim 1$. The host galaxies span a range of luminosities and morphologies, and appear to be broadly typical for the normal, evolving, actively star-forming galaxy populations at comparable redshifts and magnitudes, but may have somewhat elevated SFR per unit luminosity. There are also spectroscopic hints of massive star formation, from the ratios of [Ne III] and [O II] lines. The observed, unobscured star formation rates are typically a few $M_\odot$/yr, but a considerable fraction of the total star formation in the hosts may be obscured by dust. A census of detected optical afterglows provides a powerful new handle on the obscured fraction of star formation in the universe; the current results suggest that at most a half of the massive star formation was hidden by dust.

1 Introduction

Host galaxies of GRBs serve a dual purpose: determination of the redshifts, which are necessary for a complete physical modeling of the bursts, and to provide some insights about the possible nature of the progenitors, e.g., their relation to massive star formation, etc.

Table 1 summarizes the host galaxy magnitudes and redshifts known to us as of mid-June 2001. The median apparent magnitude is $R = 24.8$ mag, with tentative detections or upper limits reaching down to $R \approx 29$ mag. Down to $R \sim 25$ mag, the observed distribution is consistent with deep field galaxy counts [56], but fainter than that, selection effects may be playing a role. We note also that the observations in the visible probe the UV in the restframe, and are thus especially susceptible to extinction.

The majority of redshifts so far are from the spectroscopy of host galaxies, and some are based on the absorption-line systems seen in the spectra of the afterglows (which are otherwise featureless power-law continua). Reassuring

---

* Based in part on the observations obtained at the W.M. Keck Observatory, operated by the California Association for Research in Astronomy, a scientific partnership among Caltech, the Univ. of California and NASA; and with the NASA/ESA Hubble Space Telescope, operated by the AURA, Inc., under a contract with NASA.

Table 1. GRB Host Galaxies and Redshifts (June 2001)

GRB	R mag	Redshift	Type [a]	References [b]
970228	25.2	0.695	e	[25,9]
970508	25.7	0.835	a,e	[48,7,26]
970828	24.5	0.9579	e	[17]
971214	25.6	3.418	e	[39,49]
980326	29.2	~1?		[6], GCN 1029
980329	27.7	<3.9	(b)	GCN 481, 778
980425 [c]	14	0.0085	a,e	[27]
980519	26.2			[29]
980613	24.0	1.097	e	[16]
980703	22.6	0.966	a,e	[15]
981226	24.8			[21]
990123	23.9	1.600	a,e	[40,5]
990308 [d]	>28.5			GCN 726
990506	24.8	1.30	e	[59,8]
990510	28.5	1.619	a	[62], GCN 757
990705	22.8	0.86	x	[47,52]
990712	21.8	0.4331	a,e	[30,31,62]
991208	24.4	0.7055	e	[14], GCN 475, 481
991216	24.85	1.02	a,x	[52], GCN 496, 751
000131	>25.7	4.50	b	[2]
000214		0.37–0.47	x	[3]
000301C	28.0	2.0335	a	[35], GCN 603, 605, 1063
000418	23.9	1.1185	e	GCN 661, 733
000630	26.7			GCN 1069
000911	25.0	1.0585	e	[53]
000926	23.9	2.0369	a	GCN 851, 871
010222	>24	1.477	a	[36], GCN 965, 989, 1002

NOTES:
[a] e = line emission, a = absorption, b = continuum break, x = x-ray
[b] GCN circulars available at http://gcn.gsfc.nasa.gov/gcn/gcn3_archive.html
[c] Association of this galaxy/SN/GRB is somewhat controversial
[d] Association of the OT with this GRB may be uncertain

overlap exists in some cases; invariably, the highest-z absorption system corresponds to that of the host galaxy, and has the strongest lines. In some cases no optical transient (OT) is detected, but a combination of the X-ray (XT) and radio transient (RT) unambiguously pinpoints the host galaxy. A new method for obtaining redshifts may come from the X-ray spectroscopy of afterglows, using the Fe K line at ~ 6.55 keV [51,3,52], or the Fe absorption edge at ~ 9.28 keV [65,64,1]. Rapid X-ray spectroscopy of GRB afterglows may become a powerful tool to understand their physics and origins.

Are the GRB host galaxies special in some way? If GRBs are somehow related to the massive star formation (e.g., [60,50], etc.), it may be worthwhile to examine their absolute luminosities and star formation rates (SFR), or spectroscopic properties in general. This is hard to answer [38,34,55] from their visible ($\sim$ restframe UV) luminosities alone: the observed light traces an indeterminate mix of recently formed stars and an older population, cannot be unambiguously interpreted in terms of either the total baryonic mass, or the instantaneous SFR.

The magnitude and redshift distributions of GRB host galaxies are typical for the normal, faint field galaxies, as are their morphologies [32] when observed with the HST: often compact, sometimes suggestive of a merging system [16], but that is not unusual for galaxies at comparable redshifts. Their redshift distribution is about what is expected for an evolving, normal field galaxy population at these magnitude levels. There is an excellent qualitative correspondence between the observations and simple galaxy evolution models [46]. This is further illustrated in Fig. 1, which compares the observed GRB host magnitudes as a function of redshift with stellar population synthesis models which are used to describe the evolution of normal field galaxies. The observed spread of luminosities is reasonable for a normal, evolving galaxy luminosity function.

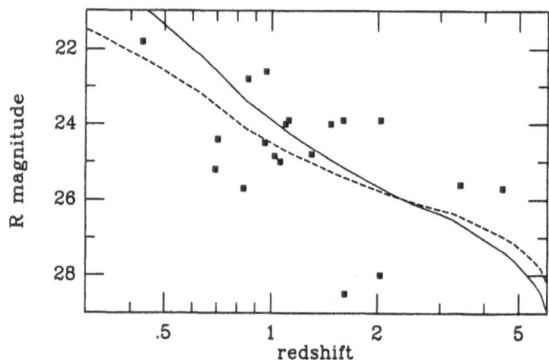

Fig. 1. The R-band Hubble diagram for GRB host galaxies with measured magnitudes and redshifts, as of June 2001. The points represent the hosts; typical error bars are a few tenths of a magnitude. The lines represent Bruzual-Charlot galaxy evolution models [13], normalised to an L_* galaxy today. Both models assume Salpeter IMF, no extinction, galaxy formation redshift $z_{gf} = 6$, and a standard Friedmann cosmology with $h = 0.65$, $\Omega_0 = 0.2$ and $\Lambda_0 = 0$. The dotted line corresponds to a model with a constant SFR, and the dashed line to a model with a SFR declining exponentially with an e-folding time of 9.5 Gyr. Allowing for extinction would lower the model curves at higher redshifts; assuming more star formation in the past, and/or a top-heavy IMF would move them higher.

Fig. 2 shows the distribution of absolute B-band luminosities of GRB hosts identified to date. If GRB's follow the luminous mass, then the expected distribution would be approximated by the luminosity-weighted galaxy luminosity function for the appropriate redshifts. The hosts span a wide range of luminosities, with the sample median very close to a present-day average (L_*) galaxy. The observed distribution can indeed be reasonably described by an evolved Schechter luminosity function, with an overall brightening and a steepening on the low-luminosity end. The interpretation of this result is complex: the observed light reflects an unknown combination of the unobscured fraction of recent star formation (especially in the high-z galaxies, where we observe the restframe UV continuum) and the stellar populations created up to that point.

This is in a qualitative agreement with studies of field galaxy evolution [43,20], which indicate both a brightening of M_*, and a steepening of α, relative to $z \sim 0$ galaxy samples. The apparent frequent occurence of "subluminous" hosts is not surprising, since most star forming activity at $z \sim 1$ is in starbursting dwarfs (corresponding to the steepening of the power-law tail of the GLF). Thus the GRB hosts seem to be representative of the normal, star-forming field galaxy population at comparable redshifts, and so far there is no evidence for any significant systematic differences between them.

One could also speculate that lower luminosity galaxies may on average have lower metallicities, where the mean extinction may be lower, making them easier to detect, or whose stellar IMF may be biased towards more massive stars (this is *highly* speculative).

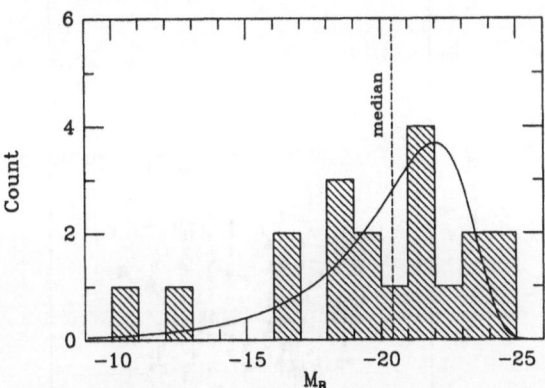

Fig. 2. Distribution of estimated absolute B-band magnitudes for GRB host galaxies with known redshifts, as of June 2001. These restframe magnitudes were computed from the observed R-band magnitudes by approximating the galaxy spectra as $f_\nu \sim \nu^{-1}$ and no additional extinction correction. Standard Friedmann cosmology with $H_0 = 65$ km s^{-1} Mpc^{-1}, $\Omega_0 = 0.2$, and $\Lambda_0 = 0$ was used. The dashed line indicates the sample median, $M_B = -20.43$ mag. The solid curve is a heuristic model, representing a luminosity-weighted Schechter function, with $M_* = -23$ mag and $\alpha = -1.6$. The coice of parameters is meant to be illustrative, rather than a best fit.

Within the host galaxies, the distribution of GRB-host offsets follows closely the light distribution [9,11], which is roughly proportional to the density of star formation (especially for the high-z galaxies). It is thus fully consistent with a progenitor population associated with the sites of massive star formation.

Spectroscopic measurements provide direct estimates of the recent, massive SFR in GRB hosts. Most of them are based on the luminosity of the [O II] 3727 doublet [37], the luminosity of the UV continuum at $\lambda_{rest} = 2800$ Å [4], in one case so far [39] from the luminosity of Lyα 1216 line [39], and in one case [15] from the luminosity of Balmer lines [37]. All of these estimators are susceptible to the internal extinction and its geometry, and have an intrinsic scatter of at least 30%. The observed SFR's range from a few tenths to a few $M_\odot$ yr^{-1}, again typical for the normal field galaxy population at comparable redshifts.

Equivalent widths of the [O II] 3727 doublet in GRB hosts, which may provide a crude measure of the SFR per unit luminosity (and a worse measure of the SFR per unit mass), are on average somewhat higher [19] than those observed in magnitude-limited field galaxy samples at comparable redshifts [33]. This is illustrated in Fig. 3. A larger sample of GRB hosts, and a good comparison sample, matched both in redshift and magnitude range, are necessary before any solid conclusions can be drawn from this apparent difference.

One intriguing hint comes from the flux ratios of [Ne III] 3869 to [O II] 3727 lines: they are on average a factor of 4 to 5 higher in GRB hosts than in star forming galaxies at low redshifts. These strong [Ne III] require photoionization

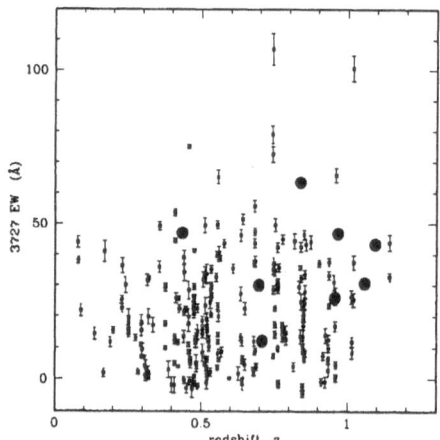

Fig. 3. Restframe equivalent widths of [O II] 3727 line emission for a magnitude-limited sample of field galaxies from Hogg et al. [33] (small squares), and a sample of GRB hosts (large circles). There is a hint that the GRB hosts tend to have higher equivalent widths at comparable redshifts, relative to this field galaxy sample. However, the GRB hosts tend to be fainter on average, as this field sample is magnitude-limited at $R = 23$ mag. The apparent difference in mean equivalent widths may be a combination of sample selection and evolution effects, rather than an evidence for enhanced SFR per unit mass for GRB hosts.

by massive stars in hot H II regions, and may represent an indirect evidence linking GRBs with massive star formation.

The interpretation of the luminosities and observed star formation rates is vastly complicated by the unknown amount and geometry of extinction. The observed quantities (in the visible) trace only the unobscured stellar component, or the components seen through optically thin dust. Any stellar and star formation components hidden by optically thick dust cannot be estimated at all from these data, and require radio and sub-mm observations. Thus, for example, optical colors of GRB hosts cannot be used to make any meaningful statements about their net star formation activity. The broad-band optical colors of GRB hosts are not distinguishable from those of normal field galaxies at comparable magnitudes and redshifts [9,57].

Already within months of the first detections of GRB afterglows, no OT's were found associated with some well-localised bursts despite deep and rapid searches; the prototype "dark burst" was GRB 970828 [17]. Perhaps the most likely explanation for the non-detections of OT's when sufficiently deep and prompt searches are made is that they are obscured by dust in their host galaxies. This is an obvious culprit if indeed GRBs are associated with massive star formation.

Support for this idea also comes from detections of RTs without OTs, including GRB 970828, 990506, and possibly also 981226 (see [59,22]). Dust reddening has been detected directly in some OTs (e.g., [54,4,15] etc.); however, this only covers OTs seen through optically thin dust, and there must be others, hidden by optically thick dust. An especially dramatic case was the RT [58] and IR transient [42] associated with GRB 980329. We thus know that at least some GRB OTs must be obscured by dust.

The census of OT detections for well-localised bursts can thus provide a completely new and independent estimate of the mean obscured star formation fraction in the universe. Recall that GRBs are now detected out to $z \sim 4.5$ and that there is no correlation of the observed fluence with the redshift [18], so that they are, at least to a first approximation, good probes of the star formation over the observable universe.

As of mid-June 2001, there have been $\sim 52 \pm 5$ adequately deep and rapid searches for OTs from well-localised GRBs. We define "adequate searches" as reaching at least to $R \sim 20$ mag within less than a day from the burst, and/or to at least to $R \sim 23 - 24$ mag within 2 or 3 days; this is a purely heuristic, operational definition. The uncertainty comes from the subjective judgement of whether the searches really did go as deep and as fast, and whether the field was at a sufficiently low Galactic latitude to cause concerns about the foreground extinction and confusion by Galactic stars. Out of those, $\sim 27 \pm 2$ OTs were found (the uncertainty being due to the questionable nature of some candidates). Some OTs may have been missed due to an intrinsically low flux, an unusually rapid decline rate, or very high redshifts (so that the brightness in the commonly used BVR bands would be affected by the intergalactic absorption). Thus the *maximum* fraction of all OTs (and therefore massive star formation) hidden by the dust is $(48 \pm 8)\%$.

This is a remarkable result. It broadly agrees with the estimates that there is roughly an equal amount of energy in the diffuse optical and FIR backgrounds (see, e.g., [45]). This is contrary to some claims in the literature which suggest that the fraction of the obscured star formation was much higher at high redshifts. Recall also that the fractions of the obscured and unobscured star formation in the local universe are comparable. GRBs can therefore provide a valuable new constraint on the history of star formation in the universe.

There is one possible loophole in this argument: GRBs may be able to destroy the dust in their immediate vicinity (up to ~ 10 pc?) [8,28], and if the rest of the optical path through their hosts ($\sim$ kpc scale?) was dust-free, OTs would become visible. Such a geometrical arrangement may be unlikely in most cases, and our argument probably still applies. Further support for the use of GRBs as probes of obscured star formation in distant galaxies comes from radio and sub-mm detections of the hosts [24,12].

Acknowledgments: We wish to acknowledge the efforts of numerous collaborators worldwide, and the expert help of the staff of Palomar, Keck, and VLA observatories and STScI. This work was supported in part by grants from the NSF, NASA, and private foundations to SRK, SGD, FAH, and RS, Fairchild Fellowships to RS and TJG, Hubble Fellowship to DER, and Hertz Fellowship to JSB.

References

1. Amati, L., *et al.* 2000, Science, 290, 953
2. Andersen, M.I., *et al.* 2000, A&A, 364, L54
3. Antonelli, L.A., *et al.* 2000, ApJ, 545, L39
4. Bloom, J.S., *et al.* 1998, ApJ, 508, L21
5. Bloom, J.S. *et al.* 1999, ApJ, 518, L1
6. Bloom, J.S. *et al.* 1999, Nature, 401, 453
7. Bloom, J.S., Djorgovski, S.G., Kulkarni, S.R., & Frail, D.A. 1998, ApJ, 507, L25
8. Bloom, J.S., Frail, D.A., & Sari, R. 2001, AJ, 121, 2879
9. Bloom, J.S., Djorgovski, S.G. & Kulkarni, S.R. 2001, ApJ, 554, 678
10. Bloom, J.S., Kulkarni, S.R., & Djorgovski, S.G. 2001, [astro-ph/0010176]
11. Bloom, J.S., & Kulkarni, S.R. 2001, this volume
12. Berger, E., Kulkarni, S.R., & Frail, D.A. 2001, ApJ in press [astro-ph/0105081]
13. Bruzual, G., & Charlot, S. 1991, ApJ, 367, 126
14. Castro-Tirado, A.J., *et al.* 2001, A&A, 370, 398
15. Djorgovski, S.G., *et al.* 1998, ApJ, 508, L17
16. Djorgovski, S.G., Bloom, J.S., & Kulkarni, S.R. 2001, ApJ in press [astro-ph/0008029]
17. Djorgovski, S.G., *et al.* 2001, ApJ in press
18. Djorgovski, S.G., *et al.* 2001, in Proc. IX Marcel Grossmann Meeting, eds. V. Gurzadyan *et al.* Singapore: World Scientific, in press [astro-ph/0106574]
19. Djorgovski, S.G., *et al.* 2001, in prep.
20. Ellis, R. 1997, ARAA, 35, 389

21. Frail, D.A., *et al.* 1999, ApJ, 525, L81
22. Frail, D.A., *et al.* 2000, ApJ, 538, L129
23. Frail, D.A., *et al.* 2001, submitted [astro-ph/0102282]
24. Frail, D.A., *et al.* 2001, in prep.
25. Fruchter, A.S., *et al.* 1999, ApJ, 516, 683
26. Fruchter, A.S., *et al.* 2000, ApJ, 545, 664
27. Galama, T.J. *et al.* 1998, Nature, 395, 670
28. Galama, T.J., & Wijers, R. 2000, ApJ, 549, L209
29. Hjorth, J., *et al.* 1999, A&ASup, 138, 461
30. Hjorth, J., *et al.* 2000a, ApJ, 534, L147
31. Hjorth, J., *et al.* 2000b, ApJ, 539, L75
32. Holland, S., 2001, [astro-ph/0102413]
33. Hogg, D., Cohen, J., Blandford, R., & Pahre, M. 1998, ApJ, 504, 622
34. Hogg, D., & Fruchter, A. 1999, ApJ, 520, 54
35. Jensen, B.L., *et al.* 2001, A&A, 370, 909
36. Jha, S., *et al.* 2001, ApJ, 554, L155
37. Kennicut, R. 1998, ARAA, 36, 131
38. Krumholtz, M., Thorsett, S., & Harrison, F. 1998, ApJ, 506, L81
39. Kulkarni, S.R., *et al.* 1998, Nature, 393, 35
40. Kulkarni, S.R., *et al.* 1999, Nature, 398, 389
41. Kulkarni, S.R., *et al.* 2000, Proc. SPIE, 4005, 9 [astro-ph/0002168]
42. Larkin, J. *et al.* 1998, GCN Circ. 44
43. Lilly, S., *et al.* 1995, ApJ, 455, 108
44. Madau, P., Pozzetti, L.,& Dickinson, M. 1998, ApJ, 498, 106
45. Madau, P. 1999, ASPCS, 193, 475 [cf. also astro-ph/9902228 and 9907268]
46. Mao, S., & Mo, H.J. 1998, A&A, 339, L1
47. Masetti, N., *et al.* 2000a, A&A, 354, 473
48. Metzger, M., *et al.* 1997, Nature, 387, 879
49. Odewahn, S.C., *et al.* 1998, ApJ, 509, L5
50. Paczyński, B. 1998, ApJ, 494, L45
51. Piro, L., *et al.* 1999, A&ASup, 138, 431
52. Piro, L., *et al.* 2000, Science, 290, 955
53. Price, P.A., *et al.* 2001, in prep.
54. Ramaprakash, A., *et al.* 1998, Nature, 393, 43
55. Schaefer, B. 2000, ApJ, 532, L21
56. Smail, I., Hogg, D., Yan, L., & Cohen, J. 1995, ApJ, 449, L105
57. Sokolov, V.V. *et al.* 2001, A&A, in press [astro-ph/0104102]
58. Taylor, G.B., *et al.* 1998, ApJ, 502, L115
59. Taylor, G.B., *et al.* 2000, ApJ, 537, L17
60. Totani, T. 1997, ApJ, 486, L71
61. van Paradijs, J., Kouveliotou, C., & Wijers, R. 2000, ARAA, 38, 379
62. Vreeswijk, P.M., *et al.* 2001, ApJ, 546, 672
63. Waxman, E., & Draine, B. 2000, ApJ, 537, 796
64. Weth, C., Mészáros, P., Kallman, T., & Rees, M.J. 2000, ApJ, 534, 581
65. Yoshida, A., *et al.* 1999, A&ASup, 138, 433

Gamma-Ray Bursts as a Probe of Cosmology

Donald Q. Lamb[1] and Daniel E. Reichart[2]

[1] Department of Astronomy & Astrophysics, University of Chicago,
   5640 South Ellis Avenue, Chicago, IL 60637
[2] Department of Astronomy, California Institute of Technology,
   Mail Code 105-24, 1201 East California Boulevard, Pasadena, CA 91125

Abstract. We show that, if the long GRBs are produced by the collapse of massive stars, GRBs and their afterglows may provide a powerful probe of cosmology and the early universe.

1   Introduction

There is increasingly strong evidence that gamma-ray bursts (GRBs) are associated with star-forming galaxies [1,2,3,4] and occur near or in the star-forming regions of these galaxies [2,3,4,5,6]. These associations provide indirect evidence that at least the long GRBs detected by BeppoSAX are a result of the collapse of massive stars. The discovery of what appear to be supernova components in the afterglows of GRBs 970228 [7,8] and 980326 [9] provides tantalizing direct evidence that at least some GRBs are related to the deaths of massive stars, as predicted by the widely-discussed collapsar model of GRBs [10,11,12,13,14]. If GRBs are indeed related to the collapse of massive stars, one expects the GRB rate to be approximately proportional to the star-formation rate (SFR).

2   Detectability of GRBs and Their Afterglows

We have calculated the limiting redshifts detectable by BATSE and HETE-2, and by *Swift*, for the sixteen GRBs with well-established redshifts and published peak photon number fluxes. In doing so, we have used the peak photon number fluxes given in Table 1 of [15], taken a detection threshold of 0.2 ph s^{-1} for BATSE and HETE-2 and 0.04 ph s^{-1} for *Swift*, and set $H_0 = 65$ km s^{-1} Mpc^{-1}, $\Omega_m = 0.3$, and $\Omega_\Lambda = 0.7$ (other cosmologies give similar results). Figure 1 displays the results. This figure shows that BATSE and HETE-2 would be able to detect half of these GRBs out to a redshift $z = 20$ and 20% of them out to a redshift $z = 50$. *Swift* would be able to detect half of them out to redshifts $z = 70$, and 20% of them out to a redshift $z = 200$, although it is unlikely that GRBs occur at such extreme redshifts. Consequently, if GRBs occur at very high ($z > 5$) redshifts (VHRs), BATSE has probably already detected GRBs at these redshifts, and HETE-2 and *Swift* should detect them as well.

The soft X-ray, optical and infrared afterglows of GRBs are also detectable out to VHRs. The effects of distance and redshift tend to reduce the spectral flux in GRB afterglows in a given frequency band, but time dilation tends to increase

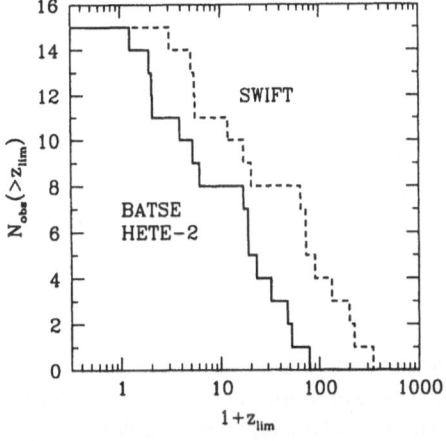

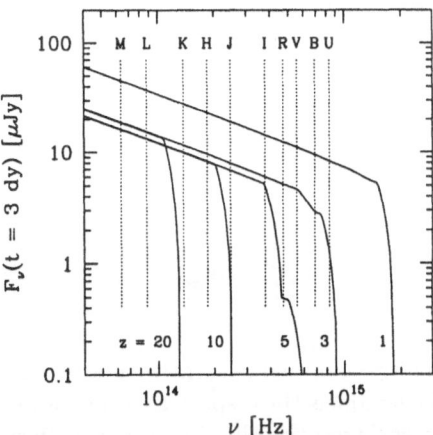

Fig. 1. Cumulative distributions of the limiting redshifts at which the 15 GRBs with well-determined redshifts and published peak photon number fluxes would be detectable by BATSE and HETE-2, and by *Swift*.

Fig. 2. The best-fit spectral flux distribution of the early afterglow of GRB 000131, as observed one day after the burst, after transforming it to various redshifts, and extinguishing it with a model of the Lyα forest.

it at a fixed time of observation after the GRB, since afterglow intensities tend to decrease with time. These effects combine to produce little or no decrease in the spectral energy flux F_ν of GRB afterglows in a given frequency band and at a fixed time of observation after the GRB with increasing redshift:

$$F_\nu(\nu, t) = \frac{L_\nu(\nu, t)}{4\pi D^2(z)(1 + z)^{1-a+b}}, \tag{1}$$

where $L_\nu \propto \nu^a t^b$ is the intrinsic spectral luminosity of the GRB afterglow, which we assume applies even at early times, and $D(z)$ is the comoving distance to the burst. Many afterglows fade like $b \approx -4/3$, which implies that $F_\nu(\nu, t) \propto D(z)^{-2}(1 + z)^{-5/9}$ in the simplest afterglow model, where $a = 2b/3$ [16]. In addition, $D(z)$ increases very slowly with redshift at redshifts greater than a few. Consequently, there is little or no decrease in the spectral flux of GRB afterglows with increasing redshift beyond $z \approx 3$.

In fact, in the simplest afterglow model where $a = 2b/3$, if the afterglow declines more rapidly than $b \approx 1.7$, the spectral flux actually *increases* as one moves the burst to higher redshifts! An example of this is the afterglow of GRB 000131. Its peak flux F_{peak} was in the top 5% of all BATSE bursts and the break energy E_{break} in its spectrum was 164 keV, yet it occurred at a redshift $z = 4.50$. We have calculated the best-fit spectral flux distribution of the afterglow of GRB 000131 from [17], as observed three days after the burst, transformed to various redshifts. The transformation involves (1) dimming the afterglow, (2) redshifting its spectrum, (3) time dilating its light curve, and (4) extinguishing the spectrum using a model of the Lyα forest (for details, see [15]). Finally, we have convolved

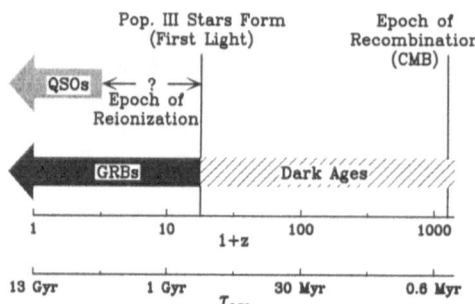

Fig. 3. Cosmological context of VHR GRBs. Shown are the epochs of recombination, first light, and re-ionization. Also shown are the ranges of redshifts corresponding to the "dark ages," and probed by QSOs and GRBs.

the transformed spectra with a top hat smearing function of width $\Delta\nu = 0.2\nu$. This models these spectra as they would be sampled photometrically, as opposed to spectroscopically; i.e., this transforms the model spectra into model spectral flux distributions.

Figure 2 shows the resulting spectral flux distribution. The spectral flux distribution of the afterglow is cut off by the Lyα forest at progressively lower frequencies as one moves out in redshift. Thus high redshift afterglows are characterized by an optical "dropout" [4], and VHR afterglows by a near infrared "dropout." We conclude that, if GRBs occur at very high redshifts, both they and their afterglows can be easily detected.

3   GRBs as a Probe of Cosmology and the Early Universe

Theoretical calculations show that the birth rate of Pop III stars produces a peak in the SFR in the universe at redshifts $16 \lesssim z \lesssim 20$, while the birth rate of Pop II stars produces a much larger and broader peak at redshifts $2 \lesssim z \lesssim 10$ [18,19,20]. Therefore one expects GRBs to occur out to at least $z \approx 10$ and possibly $z \approx 15 - 20$, redshifts that are far larger than those expected for the most distant quasars.

Figure 3 places GRBs in a cosmological context. At recombination, which occurs at redshift $z = 1100$, the universe becomes transparent. The cosmic background radiation originates at this redshift. Shortly afterwards, the temperature of the cosmic background radiation falls below 3000 K and the universe enters the "dark ages" during which there is no visible light in the universe. "First light," which occurs at $z \approx 20$, corresponds to the epoch when the first stars form. Ultraviolet radiation from these first stars and/or from the first active galactic nuclei re-ionizes the universe. Afterward, the universe is transparent in the ultraviolet.

QSOs are currently the most powerful probes of the high redshift universe. GRBs have several advantages relative to QSOs as probes of cosmology. First, GRBs are expected to occur out to $z \approx 20$, whereas QSOs occur out to only $z \approx 5$. Second, very high redshift GRB afterglows can be 100–1000 times brighter at early times than are high redshift QSOs. This makes possible very sensitive high dispersion spectroscopy of the metal absorption lines and the Lyman α forest

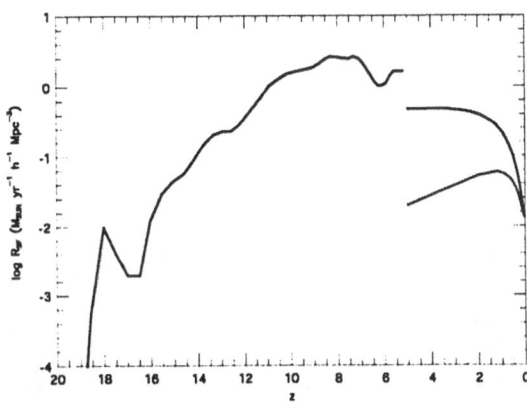

Fig. 4. The cosmic SFR R_{SF} as a function of redshift z. The solid curve at $z < 5$ is the SFR derived by [25]; the solid curve at $z \geq 5$ is the SFR calculated by [18] (the dip in this curve at $z \approx 6$ is an artifact of their numerical simulation). The dotted curve is the SFR derived by [24].

in the spectrum of the afterglows. Third, no "proximity effect" on intergalactic distances scales is expected for GRBs and their afterglows, in contrast to QSOs. Thus GRBs may be relatively "clean" probes of the intergalactic medium, the Lyman α forest, and damped Lyman α clouds, even in the vicinity of the GRBs.

The important cosmological questions that observations of GRBs and their afterglows may be able to address include the following:

• Information about the epoch of "first light" and the earliest generations of stars from merely the detection of GRBs at very high redshifts;

• Information about the growth of metallicity in the universe in the star-forming entities in which the bursts occur, in damped Lyman α clouds, and in the Lyman α forest from observations of the metal absorption line systems in the spectra of their afterglows;

• Information about the large-scale structure of the universe at VHRs from the clustering of the Lyman α forest lines and the metal absorption-line systems in the spectra of their afterglows; and

• Information about the epoch of re-ionization from the depth of the Lyman α break in the spectra of their afterglows.

Below we consider the first of these questions: the epoch of "first light" and the earliest generations of stars.

4 GRBs as a Probe of Star Formation

Observational estimates [21,22,23,24] indicate that the SFR in the universe was about 15 times larger at a redshift $z \approx 1$ than it is today. The data at higher redshifts from the Hubble Deep Field (HDF) in the north suggests a peak in the SFR at $z \approx 1 - 2$ [24], but the actual situation is highly uncertain.

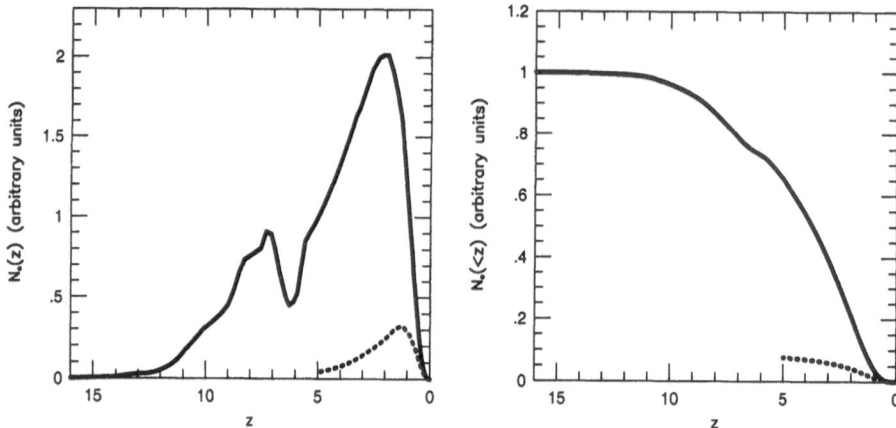

Fig. 5. Left panel: The number N_* of stars expected as a function of redshift z (i.e., the SFR from Figure 4, weighted by the differential comoving volume, and time-dilated) assuming that $\Omega_M = 0.3$ and $\Omega_\Lambda = 0.7$. Right panel: The cumulative distribution of the number N_* of stars expected as a function of redshift z. Note that $\approx 40\%$ of all stars have redshifts $z > 5$. The solid and dashed curves in both panels have the same meanings as in Figure 4.

In Figure 4, we have plotted the SFR versus redshift from a phenomenological fit [25] to the SFR derived from submillimeter, infrared, and UV data at redshifts $z < 5$, and from a numerical simulation by [18] at redshifts $z \geq 5$. The simulations done by [18] indicate that the SFR increases with increasing redshift until $z \approx 10$, at which point it levels off. The smaller peak in the SFR at $z \approx 18$ corresponds to the formation of Population III stars, brought on by cooling by molecular hydrogen. Since GRBs are detectable at these VHRs and their redshifts may be measurable from the absorption-line systems and the Lyα break in the afterglows [4], if the GRB rate is proportional to the SFR, then GRBs could provide unique information about the star-formation history of the VHR universe.

We have calculated the expected number N_* of stars as a function of z assuming (1) that the GRB rate is proportional to the SFR[1], and (2) that the SFR is that given in Figure 4 (see [15] for details). The left panel of Figure 5 shows our results for $N_*(z)$ for an assumed cosmology $\Omega_M = 0.3$ and $\Omega_\Lambda = 0.7$ (other cosmologies give similar results). The solid curve corresponds to the star-formation rate in Figure 4; the dashed curve corresponds to the star-formation rate derived by [24]. Figure 5 shows that $N_*(z)$ peaks sharply at $z \approx 2$ and then drops off fairly rapidly at higher z, with a tail that extends out to $z \approx 12$. The rapid rise in $N_*(z)$ out to $z \approx 2$ is due to the rapidly increasing volume of space. The rapid decline beyond $z \approx 2$ is due almost completely to the "edge" in the spatial distribution produced by the cosmology. In essence, the sharp peak

---

[1] This may underestimate the GRB rate at VHRs since it is generally thought that the initial mass function will be tilted toward a greater fraction of massive stars at VHRs because of less efficient cooling due to the lower metallicity of the universe at these early times.

in $N_*(z)$ at $z \approx 2$ reflects the fact that the SFR we have taken is fairly broad in z, and consequently, the behavior of $N_*(z)$ is dominated by the behavior of the co-moving volume $dV(z)/dz$; i.e., the shape of $N_*(z)$ is due almost entirely to cosmology. The right panel in Figure 5 shows the cumulative distribution $N_*(> z)$ of the number of stars expected as a function of redshift z. The solid and dashed curves have the same meaning as in the upper panel. Figure 5 shows that for the particular SFR we have assumed, $\approx 40\%$ of all stars (and therefore of all GRBs) have redshifts $z > 5$.

5 Conclusions

If the long GRBs are indeed produced by the collapse of massive stars, one expects GRBs to occur out to $z \approx 15 - 20$, redshifts that are far larger than those expected for the most distant QSOs. We have shown that both GRBs and their afterglows are easily detected out to these VHRs. GRBs can therefore give us information about the star-formation history of the universe, including the earliest generations of stars. The absorption-line systems and the Lyα forest visible in the spectra of GRB afterglows can be used to trace the evolution of metallicity in the universe, and to probe the large-scale structure of the universe at VHRs. Finally, measurement of the Lyα break in the spectra of GRB afterglows can be used to constrain, or possibly measure, the epoch at which re-ionization of the universe occurred.

References

1. Castander, F. J., & Lamb, D. Q. 1999, ApJ, **523**, 593
2. Fruchter, A. S., et al. 1999, ApJ, **516**, 683
3. Kulkarni, S. R., et al. 1998, Nature, **395**, 663
4. Fruchter, A. S. 1999, ApJ, **516**, 683
5. Sahu, K. C., et al. 1997, Nature, **387**, 476
6. Kulkarni, S. R., et al. 1999, Nature, **398**, 389
7. Reichart, D. E., 1999, ApJ, **521**, L111
8. Galama, T. J., et al. 2000, ApJ, **536**, 185
9. Bloom, J. S., et al. 1999, Nature, **401**, 453
10. Woosley, S. E. 1993, ApJ, **405**, 273
11. Woosley, S. E. 1996, in Gamma-Ray Bursts, eds. C. A. Meegan, R. D. Preece, & T. M. Koshut (New York: AIP), 520
12. Paczyński, B. 1998, ApJ, **494**, L45
13. MacFadyen, A. I., & Woosley, S. E. 1999, ApJ, **524**, 262
14. Wheeler, J. C., et al. 2000, ApJ, **537**, 810
15. Lamb, D. Q., & Reichart, D. E., 2000, ApJ, **536**, 1
16. Wijers, R. A. M. J., Rees, M. J., & Mészáros, P. 1997, MNRAS, **288**, L51
17. Andersen, M. I., et al. 2000, A&A, **364**, L54
18. Ostriker, J. P., & Gnedin, N. Y. 1996, ApJ, **472**, L63
19. Gnedin, N. Y., & Ostriker, J. P. 1997, ApJ, **486**, 581
20. Valageas, P., & Silk, J. 1999, A&A, **347**, 1

21. Gallego, J. 1995, ApJ, **455**, L1
22. Lilly, S. J., et al. 1996, ApJ, **460**, L1
23. Connolly, A. J. 1997, ApJ, **486**, L11
24. Madau, P., Pozzetti, L., & Dickinson, M. 1998, ApJ, **498**, 106
25. Rowan-Robinson, M. 1999, Ap&SS, **266**, 291

Construction
of the Variability → Luminosity Estimator

D.E. Reichart[1,2] and D.Q. Lamb[3]

[1] Department of Astronomy, California Institute of Technology,
Mail Code 105-24, 1201 East California Boulevard, Pasadena, CA 91125
[2] Hubble Fellow
[3] Department of Astronomy and Astrophysics, University of Chicago,
5640 South Ellis Avenue, Chicago, IL 60637

1 Introduction

In this paper, we present a possible luminosity estimator for the long-duration bursts, the construction of which was motivated by the work of [1] and [2]. We term the luminosity estimator "Cepheid-like" in that it can be used to infer luminosities L and luminosity distances for the long-duration bursts from the variabilities V of their light curves alone. A preliminary application of this luminosity estimator to 907 long-duration bursts appears in [3].

We discuss the construction of our measure of V in §2. In §3, we discuss our expansion of the original [1] sample of 7 bursts to include a total of 20 bursts, including 13 BATSE bursts, 5 *Wind*/KONUS bursts, 1 *Ulysses*/GRB burst, and 1 NEAR/XGRS burst. Also in §3 we discuss the construction of our luminosity estimator.

2 The Variability Measure

Qualitatively, V is computed by taking the difference of the light curve and a smoothed version of the light curve, squaring this difference, summing the squared difference over time intervals, and appropriately normalizing the result. We rigorously construct V in [4]. We require it to have the following properties: (1) we define it in terms of physical, source-frame quantities, as opposed to measured, observer-frame quantities; (2) when converted to observer-frame quantities, all strong dependences on redshift and other difficult or impossible to measure quantities cancel out; (3) it is not biased by instrumental binning of the light curve, despite cosmological time dilation and the narrowing of the light curve's temporal substructure at higher energies [5]; (4) it is not biased by Poisson noise, and consequently can be applied to faint bursts; and (5) it is robust; i.e., similar light curves always yield similar variabilities. Also in [4], we derive an expression for the statistical uncertainty in a light curve's measured variability, and we describe how we combine variability measurements of light curves acquired in different energy bands into a single measurement of a burst's variability. We plot the > 25 keV light curves of the most and least variable cosmological BATSE bursts in our sample in Figure 1.

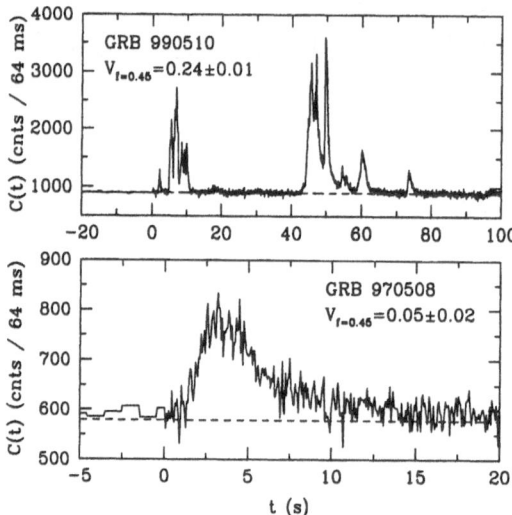

Fig. 1. The > 25 keV light curves of the most (GRB 990510) and least (GRB 970508) variable cosmological BATSE bursts in our sample. In the case of GRB 990510 ($z = 1.619$), we find that $V = 0.24 \pm 0.01$. In the case of GRB 970508 ($z = 0.835$), we find that $V = 0.05 \pm 0.02$.

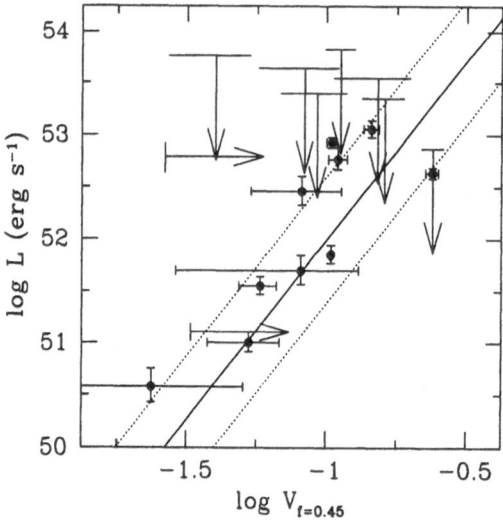

Fig. 2. The variabilities V and isotropic-equivalent peak photon luminosities L between source-frame energies 100 and 1000 keV (see [4]) of the bursts in our sample, excluding GRB 980425. The solid and dotted lines mark the center and 1 σ widths of the best-fit model distribution of these bursts in the $\log L$-$\log V$ plane.

3   The Luminosity Estimator

We list our sample of 20 bursts in Table 1 of [4]; it consists of every burst for which redshift information is currently available. Spectroscopic redshifts, peak fluxes, and high resolution light curves are available for 11 of these bursts; partial information is available for the remaining 9 bursts. We rigorously construct the luminosity estimator in [4], applying the Bayesian inference formalism developed by [6]. We plot the data and best-fit model of the distribution of these data in the $\log L$-$\log V$ plane in Figure 2.

References

1. Ramirez-Ruiz, E., & Fenimore, E. E. 1999, contributed oral presentation at the 5th Huntsville Gamma-Ray Burst Symposium
2. Fenimore, E. E., & Ramirez-Ruiz, E. 2000, ApJ, submitted (astro-ph/0004176)
3. Reichart, D. E., & Lamb, D. Q., 2001, in Procs. of the 20th Texas Symposium on Relativistic Astrophysics, in press
4. Reichart, D. E., et al. 2001, ApJ, in press (astro-ph/0004302)
5. Fenimore, E. E., et al. 1995, ApJ, 448, L101
6. Reichart, D. E. 2001, ApJ, in press (astro-ph/9912368)

Early Afterglows as Probes
for the Reionization Epoch

Davide Lazzati[1,2], Gabriele Ghisellini[1], Francesco Haardt[3],
and Alberto Fernández-Soto[1]

[1] Osservatorio Astronomico di Brera, Via Bianchi 46 I-23807 Merate (Lc), Italy
[2] Present Address: University of Cambridge, Institute of Astronomy,
   Cambridge CB3 0HA, UK
[3] Università dell'Insubria, via Valleggio 11, I-22100 Como, Italy

Abstract. The nature of Gamma-Ray Burst (GRB) progenitors is still a debated issue, but consensus is growing on the association of GRBs with massive stars. Furthermore, current models for the reionization of the universe consider massive Pop-III stars as the sources of the ionizing photons. There could then be a natural link between GRBs and reionization. The reionization epoch can be measured through prompt IR spectral observations of high redshift GRBs. For this, GRBs are better than quasars: they produce a smaller HII region even if they are much brighter than quasars (but for a much shorter time) and then, contrary to quasars, they do not modify the absorption properties of the surrounding IGM.

1   Introduction

The importance of Gamma-Ray bursts (GRB) for cosmological studies has been realized immediately after the measure of the first high-redshift bursts (GRB971214: Kulkarni et al., 1998 and GRB 000131: Andersen et al., 2000). Their importance for cosmological studies is based on observed properties and on the likely association with very massive stars:

- For a few hours, the early afterglow flux in the NIR, optical and X-ray bands is larger than the flux from any other cosmological object;
- The afterglow brightness depends very weakly on redshift, given its time evolution and spectral properties (Ciardi & Loeb, 2000);
- GRBs are probably connected with the death of massive stars and can hence be used to investigate the star formation rate at very high redshift, (see, e.g., Lamb & Reichart, 2000);
- GRBs are likely linked to the Pop-III objects and hence to the reionization of the universe (see, e.g. Rees 2000);

We here concentrate on the use of GRBs for the measurement of the redshift of the reionization of the intergalactic medium (IGM).

This measurement is made difficult by the fact that the opacity of the IGM is given by (see, e.g., Madau & Rees, 2000):

$$\tau(z) = 1.5 \times 10^5 \, h^{-1} \, \Omega_M^{-1/2} \left(\frac{\Omega_b \, h^2}{0.019} \right) \left(\frac{1+z}{8} \right)^{3/2} \tag{1}$$

implying that at high redshift ($z > 2$), even if only a hydrogen atom over 10^4 is neutral, yet the universe has a huge ($\tau \sim 10$) opacity to UV photons.

Since all observed quasars have a strong Lyman-α line in emission in their spectra, it has been speculated that the discovery of a quasar without an emission Lyman-α line proves its location at a redshift larger than reionization. On the other hand, Madau & Rees (2000) have shown that a quasar is always surrounded by a large HII region. This implies that the Lyman-α photons produced by the central source will be redshifted at a lower frequency before reaching the edge of the HII region, and will not be scattered by neutral hydrogen any more. The presence/absence of Lyman-α emission line in QSO spectra is then not a probe of the reionization epoch.

We here show a possible solution: a GRB is more luminous than a quasar but, since its duty cycle is much shorter, the total number of ionizing photons is smaller and the size of the HII region smaller.

2 Strömgren Spheres

An important advantage of GRBs with respect to QSOs for cosmological use is the dimension of the Strömgren sphere surrounding them.

Consider a source which radiates ultraviolet photons. The recombination time of the hydrogen with interstellar medium densities, even at redshift of order 10 and in presence of moderate clumping, is longer than 1 Gy (Madau & Rees, 2000). In these conditions, the radius of the HII region surrounding the photon source is obtained by equating the number of hydrogen atoms within a certain volume with the number of ionizing photons emitted during the life of the source itself. We obtain:

$$R = \left(\frac{3 \dot{N}_{\text{ion}} t}{4\pi n_H} \right)^{1/3}, \qquad (2)$$

where $\dot{N}_{\text{ion}}$ is the ionizing photon rate, t is the lifetime of the photon source and n_H the hydrogen number density.

By adopting fiducial values for the luminosity and the duty cycle of a quasar and a GRB, we obtain[1]:

$$R_{\text{QSO}} = 1.5 \times 10^5 \, L_{46}^{1/3} \, t_{15}^{1/3} \, n_H^{-1/3} \, \text{pc}; \qquad R_{\text{GRB}} = 70 \, L_{46}^{1/3} \, t_{15}^{1/3} \, n_H^{-1/3} \, \text{pc} \qquad (3)$$

There is then a factor greater than 1000 between the size of the HII region of a GRB, (which remains well within a galaxy), and the size of the HII region of a QSO. In particular, the size of the Strömgren sphere of the quasar is so large that a Lyman-α line emitted by the central object can travel through an opaque universe and be observed at infinity. This happens because the line photons would be redshifted outside the Lyman-α resonance before reaching the edge of the HII region (Madau & Rees, 2000; Cen & Haiman, 2000).

---

[1] Here and in the following we parameterize a quantity Q as: $Q = 10^x \, Q_x$

3 Discussion

We have shown that the early afterglows of GRBs are better suited for the study of the properties of the IGM at cosmological distance than any other known class of objects. GRBs, in fact, have at the same time a high luminosity, which makes them easy to detect and to study, and a small number of emitted photons, so that their presence do not influence the properties of the surrounding medium. They are therefore ideal probes, with the smallest possible impact on what we want to measure. Yet, they are very bright. It is fair to say that it is still unknown whether or not GRBs emit Lyman-α line radiation. What we have shown here is that, should such a line be observed in a GRB spectrum, this would imply $z_{GRB} < z_{Reion}$, while this is not true for QSOs.

This set of good properties is possible thanks to the fact that GRBs are transient phenomena. For this reason, we must be able to detect and follow up them in real time in order to fully exploit all the information they carry. In particular, it is important to select the high redshift bursts as soon as possible in order not to waste too much telescope time on nearby objects. This can be done through prompt NIR imaging, by the Lyman drop-out technique. For this reason, robotic IR telescopes are planned to complement the foreseen SWIFT mission (see, in particular, Zerbi et al. 2001).

References

1. Andersen, M., I., et al., 2000, A&A, 364, L54
2. Cen, R. & Haiman, Z. 2000, ApJ, 542, L75
3. Ciardi, B. & Loeb, A. 2000, ApJ, 540, 687
4. Kulkarni, S. R., et al., 1998, Nature, 393, 35
5. Lamb, D. Q. & Reichart, D. E. 2000, ApJ, 536, 1
6. Madau, P. & Rees, M. J. 2000, ApJ, 542, L69
7. Rees, M. J., 2000, Physics Reports, in press (astro-ph/9912345)
8. Zerbi, F., et al., 2001, this volume

γ-Ray Burst Remnants: How Can We Find Them?

Rosalba Perna

Harvard-Smithsonian Center for Astrophysics,
60 Garden street, Cambridge, MA 01238, USA

1 Introduction

By now there is substantial evidence that Gamma-Ray Bursts (GRBs) originate at cosmological distances from unusually powerful explosions. The interaction between a GRB and its surrounding environment has dramatic consequences on the environment itself. At early times, the strong X-ray UV flux photoionizes the medium on distance scales on the order of 100 pc [7], destroys dust [10], creates photoionization edges [2]. These are short term effects, that occur while the afterglow is propagating in the medium.

In this contribution, I discuss the long term effects resulting from the interaction between a GRB and its environment. In particular, in §2, I discuss the emission spectrum which results as the heated and ionized gas slowly cools and recombines. Besides photoionizing the medium with its afterglow, a GRB explosion drives a blast wave into the medium. In §3, I discuss possible candidates for such GRB remnants in our own and in nearby galaxies, and ways to distinguish them from remnants due to other phenomena, such as multiple supernova (SN) explosions.

2 Spectral Signatures of a Cooling GRB Remnant

Let us consider a GRB afterglow source which turns on at time $t = 0$ and illuminates a stationary ambient medium of uniform density n, initially neutral and in thermodynamic equilibrium. The afterglow is propagated while computing, as a function of position and time, the temperature of the plasma, the ionization state of all the elements, and the emission in the most important lines. The gas is heated up to $T \sim 10^5$ K, and its cooling time $\sim 10^5 (T/10^5 \text{K})/(n_e/1 \text{ cm}^{-3})$ yr, combined with the GRB rate $\sim (10^{6-7} f_b \text{ yr})^{-1}$ per galaxy ($f_b \leq 1$ is the beaming factor), implies that in every galaxy there is a non-negligible probability of finding an ionized GRB remnant at any given time. In order to identify the emission due to a cooling GRB remnant, and distinguish it from other emitting regions, such as SN remnants, HII regions, etc., line diagnostics are extremely useful. Figure 1 (from [8]) shows an example of line ratios which are very sensitive to the type of mechanism by which the gas has been ionized. The ratios [O III] $\lambda 5007/\text{H}\beta$, [O III] $\lambda 4686/\text{H}\beta$, [O II] $\lambda 3727/\text{H}\beta$ reach much higher values than in shock heated gas and HII regions. This is a consequence of having a

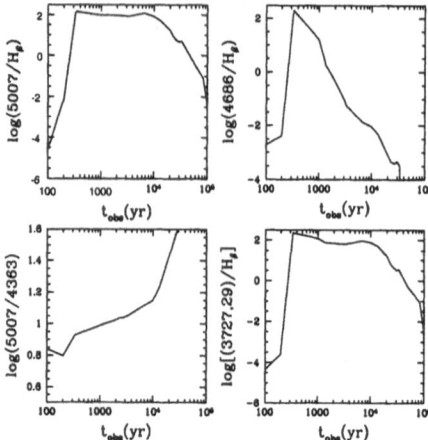

Fig. 1. Line diagnostics of cooling GRB remnants.

huge mass of gas which is suddenly photoionized and then let cool. Besides being characterized by unusual values of line ratios, cooling GRB remnants can also be identified as they are center-filled with high ionization lines, and limb-brightened with low-ionization lines. The non-relativistic blast wave might be visible separately, since it does not reach the outer edge of these young photo-ionized remnants. Furthermore, the remnants should show evidence for ionization cones if the prompt or afterglow UV emission from GRBs is beamed. Therefore, their identification could help constrain the degree of beaming of the GRB emission.

3   GRB Remnants: Hydrodynamic Effects

Besides photoionizing the medium with its afterglow, a GRB explosion drives a blast wave into the medium. This will be washed out by interstellar turbulence only after it has slowed down to a velocity of ~ 10 km s^{-1}. For a uniform medium of density n_1 cm^{-3} and an energy of 10^{54} ergs deposited in the gas, this will happen at a radius $R_{\rm kpc} = 0.7\ E_{54}^{0.32} n_1^{-0.36}$ after a time of $t = 2.1\ \times\ 10^7 E_{54}^{0.32} n_1^{-0.36}$ yr. Combining this timescale with the GRB rate per galaxy, one estimates that a few such GRB remnants should be present in every galaxy at any given time. Have we already identified them? Maybe. For several decades, 21 cm surveys of spiral galaxies have revealed the puzzling existence of expanding giant HI supershells [2] The radii of these shells are much larger than those of ordinary supernova remnants and often exceed ~ 1 kpc. Their estimated ages are in the range of 10^6–10^8 years. The Milky Way galaxy contains probably several tens of them and in one case the estimated kinetic energy is as high as $\sim 10^{54}$ ergs. Whereas smaller shells of radii ~ 200–400 pc and energies $\leq 3 \times 10^{52}$ ergs are often explained as a consequence of the collective action of stellar winds and supernova explosions originating from OB star associations [5], the energy source of the largest supershells is still a subjecy of debate. If due to

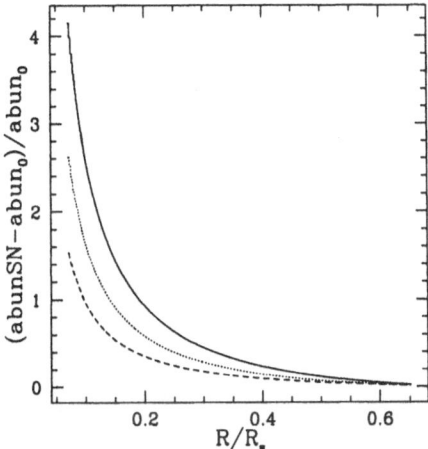

Fig. 2. Metal enhancements for a giant supershell powered by multiple SN explosions.

multiple SNe, some of these supershells would require about 10^4 SNe to power them; such large OB associations, even though not impossible, have never been observed in a survey of thousands of HII regions in nearby galaxies [3]. The similarity between the expected properties of GRB remnants and those of HI supershells prompted the suggestion that the energy sources of HI supershells might actually be GRBs [1,4]. This hypothesis could be tested based on the fact that SNe inject metals in the ISM in which they explode. As a result, if a supershell has been powered by multiple SNe, the abundances of some specific metals in its interior should be enhanced with respect to the typical values in the ISM surrounding the shell [9]. Figure 2 shows the expected metal enhancements in the abundances of O, Ne, and Si for a supershell of energy $E_K = 5 \times 10^{53}$ ergs and age $t = 5 \times 10^7$ yr that has been powered by multiple SNe. The enhancements in the abundances are more pronounced in the inner regions of the supershell, as a consequence of the fact that most of the extra mass is injected at early times, due to the shorter lifetimes of the most massive stars. Such peculiar abundances can be probed by measuring ratios between X-ray emission lines of two elements, one of which is more enhanced than the other [9]. Being able to discriminate between the multiple SNe and the GRB scenario for the production of HI supershells would help constrain GRB rates and energetics, as well as their location within a galaxy.

References

1. Efremov, Y. N., Elmegreen, B. & Hodge, P. W. 1998, ApJ, 501, L163
2. Heiles, C. 1979, ApJ, 229, 533
3. Kennicutt, R. C., Edgar, B. K. & Hodge, P. W. 1989, ApJ, 337, 761
4. Loeb, A., & Perna, R. 1998, ApJ, 503, L35
5. McCray, R. & Kafatos, M. 1987, ApJ, 317, 190

6. Meszaros, P.& Rees, M. 1998, MNRAS, 299, L10
7. Perna, R., & Loeb, A. 1998, ApJ, 501, 467
8. Perna, R., Raymond, J., & Loeb, A. 2000, ApJ, 533, 658
9. Perna, R., & Raymond, J. 2000, ApJ, 539, 706
10. Waxman, E., & Draine, B. T. 2000, ApJ, 537, 796

Progenitors of GRBs Originated
in the Dense Star Clusters

Yuri N. Efremov

Sternberg Astronomical Institute, 119899 Moscow, RF

The suggestion formulated in the title of this paper arised first to explain the origin of the giant (170–300 pc in radii) stellar arcs in the region of the supershell LMC4 in the LMC [4,5,7]. The formation of stellar arcs in the swept up gas shells formed by the central sources of pressure needs some 10^{52} ergs, yet neither multiple SNe in suggested central clusters [6] nor cloud impacts were able to form these arcs. I believe the first suggestion failed to explain why are all the arcs so rare features, all the arcs in the LMC being close to each other, and the second one – why are the ages of arcs different.

Yet another source of the energy imput to ISM to produce star-forming HI shells was suggested to be the GRB events [8,15] Along with the only known in the LMC Soft Gamma Repeater SGR 0526-66, within the same region of $\sim$ 1 kpc in diameter there are HI supershell and three or four arc-shaped star complexes. There should be in this region the common source for the progenitors of all these objects! And there is indeed an unique star cluster in the same region: the NGC 1978 cluster, 2 Gyr old and 10^6 suns massive. There are no clusters of similar mass and age in the Milky Way galaxy, and only a handful of these is in the LMC, NGC 1978 being the most massive and the most elongated one.

The binaries of compact objects, the progenitors of GRB, might not be results of two SN outbursts in the primeval binaries of massive stars. Instead of this, the single stellar remnants could form the hard binaries in result of dynamical interactions in dense cluster cores. Many of them might then have been dynamically ejected from the cluster, to merge after escaping in GRB events, and to form the giant arcs not centered on NGC 1978 itself [4,7].

This conclusion was supported by observation that X-ray binaries concentrate near NGC 1978. The recent data for the LMC X-ray sources [19] suggest that in 10 × 10 degrees square there are nine X-ray binaries, whereas within 0.6 × 0.6 degrees square including NGC 1978, there are four of these nine stars. It is tempting to explain this with the origin of these four X-ray binaries in NGC 1978, facing the high rate of occurence of X-ray binaries in globular clusters and the high chance for them to be dynamically ejected from the cluster (see refs. in [4,7]).

The recent studies of dynamical evolution of star clusters suggested the high rate of the binary formation and star coalescences and ejections. Portegies Zwart and MacMillan [18] have argued that the BH/BH binaries are formed in (and many of these then ejected from) the massive clusters in a few Gyr after a number of close encounters. The ejected pairs are close enough to merge in a few Myr and the GRB events should be then rather close to the parent cluster. The objects escaped from NGC 1978 must have had quite reasonable spans of time and velocities to merge in the centers of the present day stellar arcs [5].

The possibility of the formation of compact binaries – progenitors of GRB in stellar encounters within the globular clusters was suggested also by authors [14], whereas authors [2] considered the generation of GRB due to the stellar encounters in dense stellar clusters of evolved galactic nuclei. Other arguments for the origin of GRBs in star clusters were given by Kulkarni (this conference).

Bloom et al. [1] argued that the GRB afterglows observed mostly in the regions of star formation (SFR) and therefore the GRB events are connected with Hypernovae. Even so, the progenitors of the latters are suggested to be the very massive fast rotating stars, and the most plausible chanell of their formation is the coalescences of massive stars inside dense young clusters. There are indeed observational evidences that hypernovae from these stars did occur inside or near such clusters in NGC 6946 [5,11] and M82 galaxies [16].

Anyway, the occurence of an afterglow near a SFR might just imply that this SFR is the result of the previous GRB event near the region, from the same parent cluster, which might be rather old, like NGC 1978. The afterglow distribution in Z suggests that most GRBs have arised 8–12 Gyrs ago, being a few Gyrs younger than the systems of classical globulars in galaxies like ours. The delay of about 2 Gyrs is just the age of the NGC 1978 cluster. It is even probable that this delay, which is close to the age difference between the globular clusters and the oldest objects of the Galactic disk, implies that the formation of the massive stars in the disk was triggered by the first GRB occured when the present day globulars were some 2 Gyr old [5].

Contrary to authors [1] conclusions, the host-normalized offset distribution of the GRBs seems to be similar nor to distribution of the star formation regions neither to distribution of SNe [21]. In fact, it looks rather like the distribution of the classical (old) globular clusters, with the clear concentration to the center of the composite galaxy. It may be also compatible with the origin of GRB progenitors in the very dense star clusters in galactic nuclei, suggested by authors [2]. Note also, that many GRBs had nor the observed afterglows neither the host galaxies and some GRB afterglows were observed near the cores of elliptic-like galaxies.

There is the direct evidence for the occurence of GRB 980425, identified with SN1998bw, in a cluster [13]; moreover, the authors [13] noted near the SN the arc-shaped feature. This GRB being by far the nearest one ($\sim$ 40 Mpc), only there the liner resolution is high enough. Note also that both well observed SGRs in the Galaxy, SGR 1906-20 and SGR 1900+14, are found to be in the dense young clusters [17].

At any rate, assuming the giant stellar arcs were formed by GRB-connected events, it is possible to get some conlusions on the geometry of jets. The shapes and orientations of the LMC arcs suggest they are the partiall shells and cannot be results of isotropic bursts in non-uniform ISM [9]. The arcs might be formed by the jets with the corresponding opening angle (60–90 degrees), what is compatible with Usov [22] model of GRB. Otherwise, there might exist the multiprecessing and long-standing narrow jets (Fargion: [12] and this conference). Their working surface might fill up the partiall shells, triggering star formation [10]. Note that the long-standing precessing jet of SS433 has being formed the HI bubble during some 10000 years [3]. Otherwise, the jet instability might result in the bow shock with the wide working surface, like it is the case for the star-forming jets from AGN.

Note anyway, that the visible shapes of the triggered star complexes which are intrisically the partiall shells, depend on their orientation to the line of sight, whereas their intrinsic parameters depend on the properties of the target clouds to be swept up and compressed into star-forming ones. Also, there was evidently no gas to form the opposite-side stellar arcs in the LMC, considering their sizes.

References

1. J.S.Bloom J.S., S.R.Kulkarni, and S.G.Djorgovski: astro-ph/0010176 (2000)
2. V.I.Dokuchaev, Yu.N.Eroshenko, and L.M.Ozernoy: Astroph. J. **502**, 192 (1998)
3. G.M.Dubner et al.: Astron. J. **116**, 1842 (1998)
4. Yu.N.Efremov: Astron. Lett. **25**, 74 (1999)
5. Yu.N.Efremov: Astron. Lett. **26**, 558 (2000)
6. Yu.N.Efremov, B.G.Elmegreen, Mon. Not. RAS **299**, 643 (1998)
7. Yu.N.Efremov and B.G.Elmegreen: 'Spontaneous and Induced Star Formation in the LMC'. In: *New Views of the Magellanic Clouds, IAU Symposimu No. 190, July 12–17 1998*, ed. by You-Hua Chu et al. (Sheridan Books, Chelsea 1999), pp. 422–427
8. Yu.N.Efremov, B.G.Elmegreen, and P.W.Hodge: Astroph. J. **501**, L163 (1998)
9. Yu.N.Efremov, S.Ehlerova, and J.Palous: Astron Astroph. **350**, 457 (1999)
10. Efremov Yu. N. and Fargion D., astro-ph/9912562 (1999)
11. B.G.Elmegreen, Yu.N.Efremov, and S.S.Larsen: Astroph. J. **535**, 748 (2000)
12. D.Fargion: Astron. Astroph. Suppl. **138**, 507 (1999)
13. J.M.Fynbo et al. astro-ph/0009014
14. B.M.S.Hanson and C.Murali: Astroph. J. **505**, L15 (1998)
15. A.Loeb and R.Perna: Astroph. J. **503**, L35 (1998)
16. S.Matshushita et al.: astro-ph/0011071 (2000)
17. I.F.Mirabel, Y.Fuchs, and S.Chaty: astro-ph/9912446 (1999)
18. S.Portegies Zwart and S.L.M.McMillan: Astroph. J. **528**, L17 (2000)
19. M.Sasaki M., F.Haberl, and W.Pietsch: astro-ph/0002318 (2000)
20. H.C.Spruit: Astro. Astroph. **341**, L1 (1999)
21. D.Yu.Tsvetkov D.Yu., S.I.Blinnikov, and N.N.Pavlyuk N.N.: astro-ph/0101362 (2001)
22. V.V.Usov: Nature, **357**, 344 (1992)

The Multiband Photometry of GRB Host Galaxies: Comparison with the Spectral Energy Distributions of Nearby Galaxies and Theoretical Modeling

V.V. Sokolov[1,3], T.A. Fatkhullin[1], V.N. Komarova[1,3], E.R. Kasimova[2], and V.I. Korchagin[2]

[1] Special Astrophysical Observatory of RAS,
   Karachai-Cherkessia, Nizhnij Arkhyz, 369167 Russia
[2] Institute of Physics, Rostov University, Stachki 194, Rostov-on-Don, 344090, Russia
[3] Isaac Newton Institute of Chile, SAO Branch

Abstract. We present one of the results of $BVRI$ photometry of the hosts of GRB for the host galaxy of GRB 970508 and the theoretical modeling of its continuum spectral energy distribution (SED) to show that it is important to take into account internal extinction in the host galaxies. We compared the $BVRI$ broad-band flux spectrum of the host to template SEDs of local starburst galaxies and found that there is a significant internal extintion in this host. Moreover, this comparison allows us to derive the absolute magnitude ($M_{B_{rest}}$) and rouhgly estimate reddening (A_V). Population synthesis modeling of the continuum SED for different reddening laws ([4] and [5]) demostrates that the observational data of the host galaxy of GRB 970508 are best fitted by the spectral properties of a model SED with extinction of $A_V \approx 2$.

The multiband observations of GRB host galaxies were performed with the 6-m telescope of SAO RAS in 1998–2000. In details the data reduction and photometry of the host galaxies are described in [12]. As *a first approximation* for an estimate of internal extinction, we compared our broad-band flux spectrum of the host galaxy of GRB 970508 (as an example) to SEDs of local starburst galaxies. We used S1, S2, S3, S4, S5, S6 averaged template SEDs for the local starburst galaxies from [3]. The spectra of local starburst were grouped according to increasing values of the color excess $E(B-V)$. Fig. 1 demonstrates the best fitting (minimum of $\chi^2/d.o.f$) of our broad-band flux spectrum by the starburst template galaxies (in Fig. 1 and 2 FWHM of each band is shown, taking into account z, by dashed horizontal lines with bars). Using $E(B-V)$ for the S5 template and reddening laws from [5] and [4] we can estimate the value of A_V, which is in ranges $A_V = 1.58 \div 1.86$ and $A_V = 2.07 \div 2.43$, respectively. The same comparisons of the $BVRI$ broad-band spectra with the local template SEDs allows us to derive K-correction and absolute magnitudes. In Table 1 we present the observed R-band magnitudes and the estimates of $M_{B_{rest}}$ derived by us [12] and other authors for eight GRB host galaxies.

As *a second approximation* for the estimate of internal extinction, we constructed a set of model theoretical templates for the host galaxy of GRB 970508 using population synthesis modeling. We used the PEGASE package ([7]) and

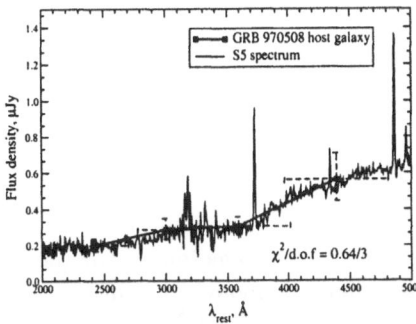 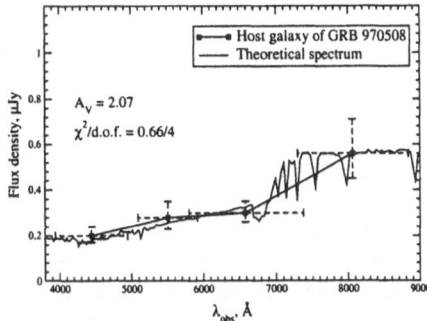

Fig. 1. A comparison of the GRB 970508 host galaxy broad-band rest-frame ($z = 0.835$) flux spectrum with the SED of S5 template galaxies (see [6]). Fluxes of S5 template were scaled for the best fitting.

Fig. 2. The best fit for the model SED to the $BVRI$ photometry of the host galaxy GRB 970508, assuming the extinction law from [4]. Wavelengths are in the observed frame.

Table 1. Observed and absolute magnitudes of GRB host galaxies

Host	observed magn. R	absolute magn. $M_{B_{rest}}$	reference
GRB 970228	24.6±0.2	−18.6	R: [8], $M_{B_{rest}}$: [2]
GRB 970508	24.99±0.17	−18.6	[12]
GRB 991208	24.36±0.15	−18.8	[12]
GRB 990712	21.80±0.06	−19.9	[9]
GRB 980613	23.58±0.1	−20.8	[12]
GRB 990123	24.47±0.14	−20.9	[12]
GRB 971214	25.69±0.3	−21.1	R: [12], $M_{B_{rest}}$: [10]
GRB 980703	22.30±0.08	−21.3	[12]

the following assumptions: the Sun metalicity, instantaneous burst of star formation, Salpeter initial mass function with the low and high mass cut-offs to be $0.1\,M_\odot\,\mathrm{yr}^{-1}$ and $120\,M_\odot\,\mathrm{yr}^{-1}$, respectively; cosmology with $H_0=60\,\mathrm{km\,s}^{-1}\,\mathrm{Mpc}^{-1}$, $\Omega_M=0.3$ and $\Omega_\Lambda=0.7$. For the calculations of the resulting SED we applied a two-component model: the first component is just a burst ("burst" component) of star formation and the second one is an old ("old" component) stellar population. Both give corresponding contributions into the resulting continuous SED. The "burst" component is responsible for emission lines and nebular continuum. For this reasons, we roughly fixed the "burst" component parameters using the luminosity of the forbidden emission line [O II] by fitting to its observed flux from [1] and taking into account the assumed reddening laws. With the constructed set of the theoretical templates we found the minimum of $\chi^2/d.o.f$ for the $BVRI$ broad-band flux spectrum of the GRB host in two ranges of A_V obtained from the comparison with local starburst templates in *a first approxi-*

Table 2. The best fit model parameters

reddening curve	A_V	$\chi^2/d.o.f.$
Cardelli et al, [5]	1.6	1.03/4
Calzetti et al, [4]	2.07	0.66/4

mation: $A_V = 2.07 \div 2.43$ and $A_V = 1.58 \div 1.86$ for two reddening laws [4] and [5]. According to our method, we derived the best fit parameters of the model SEDs which are given in Table 2. In Fig. 2 we plot the model SED in the case of the reddening curve from [4]. As it can be seen, the $BVRI$ broad-band flux spectrum of the host galaxy of GRB 970508 is best fitted by the theoretical template with sufficiently high internal extinction. We notice that the best fit parameters correspond to the A_V, which indeed, lies within the range of A_V derived from the comparison with the S5 template SED. Taking into account the reddening curve from [4], A_V from Table 2 and the lower limit of SFR from [1], we can estimate the extinction-corrected SFR (star formation rate) as follows: $SFR_{corr} \approx 17\,M_\odot\,yr^{-1}$.

We emphasize that only the simplest model assumptions were made and we did not include in the modeling other possibilities (e.g. exponentially decreasing star formation scenario, subsolar metallicity). For comparison of our value of A_V, we draw attention that in the case of the host galaxy of GRB 990712 the extinction was obtained to be $A_V = 3.4^{+2.4}_{-1.7}$ for the extinction law from [5] (see [13]), which is about 2 times higher than our one ($A_V = 1.6$ for the same law). It should be noted in conclusion that the comparison of $BVRI$ broad-band flux spectrum of the host galaxy of GRB 970508 with local starburst templates and theoretical templates shows that it is likely to be of great importance to take into account internal extinction in GRB host galaxies.

Acknowledgements: This work was supported by INTAS N96-0315, "Astronomy" Foundation (grant 97/1.2.6.4), RFBR N98-02-16542, RFBR N00-02-17689.

References

1. Bloom J. S., Djorgovski S. G., Kulkarni S. R., Frail D. A. 1998, ApJ, 507, L25
2. Bloom J. S., Djorgovski S. G., Kulkarni S. R. 2000, astro-ph/0007244, http://xxx.lanl.gov
3. Calzetti D., Kinney A. L., Storchi-Bergmann T. 1994, ApJ, 429, 582
4. Calzetti D., Armus L., et al. 2000, ApJ, 533, 682
5. Cardelli J. A., Clayton G. C. & Mathis J.S. 1989, ApJ, 345, 245
6. Connoly A. J., Szalay A. S., et al. 1995, AJ, 110, 1071
7. Fioc M. and Rocca-Volmerange B. 1997, A&A, 326, 950
8. Galama T. J. et al. 2000, ApJ, 536, 185
9. Hjorth J., Holland S., et al. ApJL, 534, 147
10. Kulkarni S. R., Djorgovski S. G. et al. 1998, Nature, 393, 35
11. Landolt A. U. 1992, AJ, 104, 340
12. Sokolov V. V., Fatkhullin T. A., Komarova V. N. 2000, astro-ph/0006207
13. Vreeswijk P. M., Fruchter A. et al. 2000, accepted for publication in The ApJ, astro-ph/0009025, http://xxx.lanl.gov

Multiscale Statistical Methods and the Angular Distribution of Gamma-Ray Bursts

Roland Vavrek[1], Lajos G. Balázs[1], Attila Mészáros[2], István Horváth[3], and Zsolt Bagoly[4]

[1] Konkoly Observatory, Box 67, H-1505 Budapest, Hungary
[2] Astronomical Institute of the Charles University,
   V Holešovičkách 2, CZ-180 00 Prague 8, Czech Republic
[3] Dept. of Physics, Bolyai Military University, Box 12, H-1456 Budapest, Hungary
[4] Laboratory for Information Technology, Eötvös University,
   Pázmány Péter sétány 1/A, H-1518 Budapest, Hungary

Abstract. The spherical variants of multiscale methods – Voronoi tesselation (VT), minimal spanning tree (MST), and multifractal (MFR) analysis – are used to study the angular distributions of three subgroups of gamma-ray bursts (GRBs) collected in Current BATSE Gamma-Ray Burst Catalog [4] until the end of August 2000 to test the isotropy of the sky distribution of GRBs. The "short/fainter" ($T_{90} < 2s$ and $0.65 < P_{256} < 2$ photons/(cm^{2}s)) subclass has different distribution of the MFR spectrum compared with the "long" subclass and also with the 200 Monte Carlo samples on the confidence level > 99.9%. Contrary this, VT and MST methods do not show this intrinsic anisotropy on a satisfactorily high confidence level. All this suggests, but not confirm, the anisotropy for the $T_{90} < 2s$ subgroup.

1 Studied Samples

Table 1 summarizes the selection criteria of GRBs for the three samples. In fact, the usual $T_{90} < 2$ s subgroup is divided into two further subclasses with respect to peak flux P_{256}; the third, *intermediate subgroup* [3] is not considered in this study, because it is incoveniently less abundant.

We made analysis on subsets of subclasses having equal numbers of GRBs; i.e. we assort 200 times $N = 260$ subsets from the "Long", "Short1" and "Short2" samples. Pseudo-random samples were also generated by Monte Carlo (MC) simulations taking into account BATSE's non-uniform exposure function in order to determine the confidence levels.

2 Survey of Methods

Voronoi Tesselation (VT). The Voronoi cell [10] of an object is the region of sphere being closer to the given point than to any other point of the sphere. The set of Voronoi cells for a point field, called Voronoi diagram, provides a partition of a point pattern according to its spatial structure. The behaviour of the tesselation methods on the sphere is quite different like on the infinite plane. The polygon's area will not be independent from each other, since the 4π surface

Table 1. Selection criteria of studied GRB subsets and simulations.

Sample	Duration (s)	Peak flux (photons/(cm²s))	No. of events	No. of sets
Short1	$T_{90} < 2$ s	$0.65 < P_{256} < 2$	$261 \rightarrow 260$	200
Short2	$T_{90} < 2$ s	$0.65 < P_{256}$	$406 \rightarrow 260$	200
Long	$T_{90} > 2$ s	$0.65 < P_{256} < 2$	$676 \rightarrow 260$	200
MC Simulations	–	–	260	200

of the sphere is given. Hence the derived statistics should be more sensitive to the clustering on the sphere. Confidence levels are derived from the fundamental cell parameters: area (A), perimeter (P), and the number of vertices (N_v).

Minimal Spanning Tree. For a given point set, the minimal spanning tree (MST) is defined as the unique graph connecting all the points, with no closed loops and having minimal length. For this reason the lengths of its edges (L_{MST}) form the minimal covering of the set.

Multifractal Spectrum (MFS). A MFR on a point process can be defined as unification of subsets of different (fractal) dimensions [9]. The contribution of this subsets to the whole pattern is not necessarily equally weighted, practically it depends on the relative abundances of subsets. The functional relationship between the subsets and the corresponding fractal dimension is called the *MFR or Hausdorff spectrum*, $f(\alpha)$. In the vicinity of point i one can measure from the neighbourhood structure a local dimension α_i or *pointwise dimension*. This measure approximates the dimension of the embedding subset, giving a possibility to construct the MFR spectrum which characterizes the whole pattern. If the maximum of this convex spectrum is equal to the Euclidean dimension of the space, then in classical sense the pattern *is not a fractal*, but the spectrum remains sensitive to the inhomogeneities and anisotropies of the point set. This is the case in the analysed patterns of selected GRB subgroups.

3   Results and Conclusions

For all three subgroups VT and MST methods show no difference in angular distribution compared with the simulated patterns on a satisfactorily high confidence level ($> 99.0\%$); for sample "Short2" this is not so obvious, and the results are not coherent (Table 2).

On angular scales $\leq (20° - 40°)$ the "Short1" subsample gives a MFR spectrum, which is wider than any other spectra (for values $\alpha > 2$) over the 4σ error boxes at confidence level $> 99.9\%$. Error boxes are evaluated regarding the variances of derived measures for each pattern of the 200 samples.

Within the 4σ error boxes the MFR spectra, the brighter "Short2" sample is similar to the MC simulated patterns and the to the "Long" samples.

All this means that the Short1 sample, and hence the whole $T_{90} < 2$s subgroup, seems to be anisotropic from MFR method. The remaining two methods do not confirm this. This is in accordance with the earlier authors' tests of isotropy of the sky distribution of GRB [1], [2], [5], [6], [7], [8], where also ambiguous results were obtained for the short subgroup.

R.V. acknowledges the Research Fellowship of the Hungarian Academy of Sciences. This research was supported by Domus Hungarica Scientiarum et Artium Grant (A.M.), by OTKA grants T024027 (L.G.B.), F029461 (I.H.), T034549.

Table 2. Results of VT, MST and MFR tests. High confidence levels from the two-sample KS statistics means that the sample significantly differs from simulated random patterns.

Statistics		Long	Short1	Short2
Round. factor average	$\overline{4\pi A/P}$	No	No	> 98%
Round. factor homogeneity	$1 - \sigma(RFav)/RFav$	> 99.4%	No	No
AD factor	$1 - \frac{1}{1-\sigma(A)/\langle A \rangle}$	No	No	No
Shape factor average	$\overline{A/P^2}$	No	> 95%	> 98%
Edge variance	$\sigma(N_v)$	No	No	> 95%
SE factor	$\overline{N_v(A/P^2)}$	No	No	> 95%
MST average	$\overline{L_{MST}}$	No	No	> 99.1%
MST variance	$\sigma(L_{MST})$	No	No	No
MFR spectra	$f(\alpha)$	No	> 99.9%	No

References

1. L.G. Balázs, A. Mészáros, I. Horváth: A&A **339** 1 (1998)
2. L.G. Balázs, A. Mészáros, I. Horváth, R. Vavrek: A&A Suppl. **138** 417 (1999)
3. I. Horváth: ApJ **508** 757 (1998)
4. C.A. Meegan, et al.: Current BATSE Gamma-Ray Bursts Catalog, http://gammaray.msfc.nasa.gov/batse/grb/catalog/current/ (2000)
5. A. Mészáros, Z. Bagoly, R. Vavrek: A&A **354** 1 (2000)
6. A. Mészáros, Z. Bagoly, I. Horváth, L.G. Balázs, R. Vavrek: ApJ **539** 98 (2000)
7. A. Mészáros, Z. Bagoly, I. Horváth, L.G. Balázs, R. Vavrek: In: *Gamma-Ray Bursts: 5th Huntsville Symposium*, eds. R.M. Kippen et al. (AIP Conference Proceedings, Vol. 526, Melville, New York, 2000) pp. 102-106
8. A. Mészáros, Z. Bagoly, L.G. Balázs, I. Horváth, R. Vavrek: these Proceedings
9. G. Paladin, A. Vulpiani: Physics Reports, **156** N4 (1987)
10. D. Stoyan, H. Stoyan: *Fractals, Random Shapes and Point Fields*, John Wiley & Son Ltd; ISBN: 0471937576 (1994)

Determining the Gamma-Ray Burst Rate as a Function of Redshift

Nevin Weinberg[1], Carlo Graziani[1], Donald Q. Lamb[1], and Daniel E. Reichart[2]

[1] Department of Astronomy & Astrophysics, University of Chicago,
  5640 South Ellis Avenue, Chicago, IL 60637
[2] Department of Astronomy, California Institute of Technology,
  Mail Code 105-24, 1201 East California Boulevard, Pasadena, CA 91125

Abstract. We exploit the 14 gamma-ray bursts (GRBs) with known redshifts z and the 7 GRBs for which there are constraints on z to determine the GRB rate $R_{GRB}(z)$, using a method based on Bayesian inference. We find that, despite the qualitative differences between the observed GRB rate and estimates of the SFR in the universe, current data are consistent with $R_{GRB}(z)$ being proportional to the SFR.

1 Introduction

There is increasing evidence that GRBs are due to the collapse of massive stars (see, e.g., [1] for a discussion of this evidence). If GRBs are indeed related to the collapse of massive stars, one expects the GRB rate to be roughly proportional to the SFR. However, the observed redshift distribution of GRBs differs noticeably from that of the SFR: the observed GRB redshift distribution peaks at $z \approx 1$ and few bursts are observed beyond $z \sim 1.5$, while the SFR peaks at $z \approx 2$ and 10-40% of stars are thought to form beyond $z = 5$ (see, e.g., [2,3]).

However, observational selection effects play an important role in determining the observed redshift distribution of GRBs. The important question is therefore whether or not the discrepancy between the observed GRB redshift distribution and the redshift dependence of the SFR is entirely due to selection effects; i.e., is the GRB rate roughly proportional to the SFR after taking observational selection effects into account? We address this question in this paper.

2 Method

We adopt a Bayesian approach. We calculate the likelihood of the data given the model, and convert it to a posterior distribution on the model parameters. We assume a very general model for the GRB rate, and a power-law model for the intrinsic GRB photon luminosity distribution (we assume that the amplitude and the power-law index of the photon luminosity distribution does not evolve; we relax this assumption in future work). We determine the efficiency with which BeppoSAX and the IPN detect GRBs as a function of peak photon flux P by comparing the BeppoSAX and IPN peak photon flux distributions to that of BATSE. We fit the model jointly to the peak fluxes and redshifts of the 14 GRBs with known z, and the 7 GRBs for which there are constraints on z.

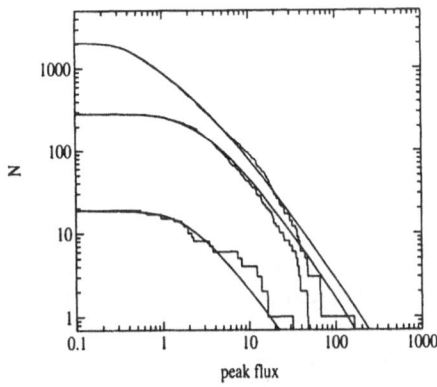

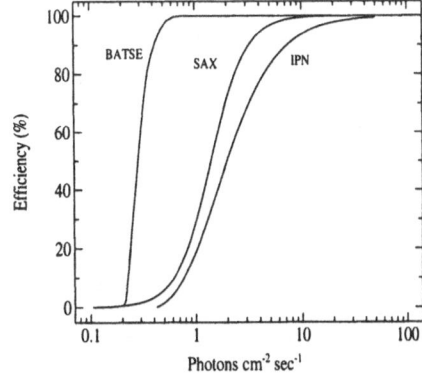

Fig. 1. Comparison of the best-fit models to the differential distribution of peak fluxes of BATSE, IPN and BeppoSAX bursts, and the cumulative peak flux distributions for the three experiments.

Fig. 2. Efficiency $\epsilon(P)$ with which BATSE, IPN and BeppoSAX detect GRBs as given by the best-fit models to the differential distribution of peak fluxes of BATSE, IPN and BeppoSAX bursts.

We write the rate of GRBs that occur per unit redshift and luminosity as

$$dN/dz\,dL_N = \rho(z)f(L_N)\,, \tag{1}$$

where

$$\rho(z) = R_{\mathrm{GRB}}(z; P, Q) \times (1+z)^{-1} \times 4\pi r(z)^2 (dr/dz) \tag{2}$$

is the rate of GRBs that occur at redshift z,

$$R_{\mathrm{GRB}}(z; P, Q) = \left[\frac{t(z)}{t(0)}\right]^{P} \exp\left[Q\left(1 - \frac{t(z)}{t(0)}\right)\right] \tag{3}$$

is the rate of GRBs that occur at redshift z per unit comoving volume (see [5]), $t(z)$ is the elapsed time since the Big Bang, and

$$f(L_N) = L_N^{-\beta} \times \Theta(L_N - L_{\min})\Theta(L_{\max} - L_N) \tag{4}$$

is the intrinsic photon luminosity distribution of GRBs. Thus the model has five parameters: P, Q, β, $L_{\min}$, and $L_{\max}$.

We write the efficiency with which GRBs with known redshifts are found as $\epsilon(z, P) = \epsilon_z(z) \times \epsilon_{\mathrm{ST}}\epsilon_P(P)$, where $\epsilon_z(z)$ is the efficiency with which the redshifts of GRBs are determined from optical observations once they are detected by a γ-ray burst instrument. We take $\epsilon_z(z) = 1$ for GRBs whose redshifts were determined by detection of an absorption-line system in the optical afterglow of the burst and $\epsilon_z(z) = \Theta(1 - z) + \Theta(z - 2.5)$ for GRBs whose redshifts were determined by measuring emission lines in the spectra of the host galaxy. The latter expression accounts qualitatively for the difficulty in measuring redshifts when the H_α and O[II] emission lines from host galaxies do not lie in the visible spectrum. The quantity ϵ_{ST} is the "stereo-temporal" efficiency that accounts for limitations of exposure in time and solid angle and $\epsilon_P(P)$ is the efficiency with which BATSE, the IPN and BeppoSAX detect GRBs as a function of peak flux

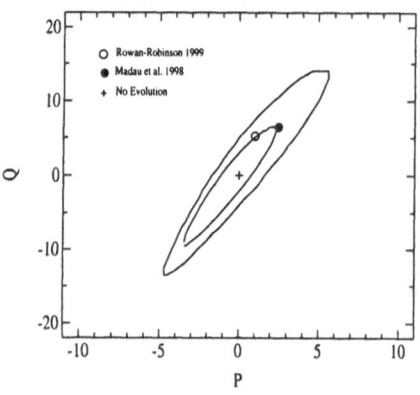

Fig. 3. Probability contours in the (P,Q)-plane for the GRB rate parameters (P, Q) found from fitting to the 14 GRBs with known z and the 7 GRBs with constraints on z. The solid curves correspond to the 68% and 95% probability contours. Also shown are the (P,Q)-values corresponding to no space density evolution (+), the Madau et al. SFR [4], and the Rowan-Robinson phenomenological model fit to IR, optical and UV data [5].

P. Figure 1 compare the best-fit models of ϵ_P and the cumulative peak flux distributions of BATSE, the IPN, and BeppoSAX, respectively. Figure 2 shows the best-fit ϵ_P for each of the three experiments.

The likelihood function is then given by

$$\mathcal{L} = \exp\left\{-\int dz\, dP\, \mu(z, P)\epsilon(z, P)\right\} \prod_{i=1}^{N} \mu(z_i, P_i), \qquad (5)$$

where

$$\mu(z, P) = \int_0^\infty dL_N\, \rho(z) f(L_N) \delta\left(P - \frac{L_N}{4\pi r(z)^2 (1+z)^\alpha}\right)$$
$$= \rho(z) \times f\left(4\pi r(z)^2 (1+z)^\alpha P\right) \times 4\pi r(z)^2 (1+z)^\alpha \qquad (6)$$

is the expected number of events observed within $dz\, dP$ of (z, P). The quantity α is the burst spectral index, which we set equal to one in this work. By an application of Bayes' Theorem, we now regard $\mathcal{L}$ as an (unnormalized) probability distribution on the model parameters.

3   Results

Figure 3 shows 68% and 95% probability contours for the GRB rate parameters P and Q. Also shown on the plots are the best-fit SFR models of Madau et al. [4] and Rowan-Robinson [5]. The SFR models lie at about a 68% excursion from the best-fit GRB rate model. Thus we find that, despite the qualitative differences that exist between the observed GRB rate and estimates of the SFR in the universe, current data are consistent with the actual GRB rate being approximately proportional to the SFR.

References

1. Lamb, D. Q. 2000, Phys. Reports, **333**, 505
2. Gnedin, N. Y., & Ostriker, J. P. 1997, ApJ, **486**, 581
3. Valageas, P., & Silk, J. 1999, A&A, **347**, 1
4. Madau, P., Pozzetti, L., & Dickinson, M. 1998, ApJ, **498**, 106
5. Rowan-Robinson, M. 1999, Ap&SS, **266**, 291

Part IV

Theories for GRBs and Their Afterglows

Gamma-Ray Burst Models: The Central Engine

S.E. Woosley

Department of Astronomy and Astrophysics
Santa Cruz CA 95064

Abstract. Twinkle, twinkle superstar. How I wonder what you are. [12]

1 A Zoo of Models

Observations summarized in these proceedings and elsewhere (e.g., [5]) show that those bursts whose counterparts have been localized happen, for the most part, in star-forming regions in distant galaxies. So far, this sample includes only bursts of the "long-soft" variety (duration ~ 10 s) and we must keep in mind that the "short-hard" bursts (duration ~ 0.3 s) are not so constrained. Given the twin requirements of enormous energy and association with star formation, the currently favored models all involve massive stars and their byproducts, especially black holes. At this meeting, Frail presented evidence, based upon radio measurements of the afterglow, that the total energy in the form of relativistic matter in most GRBs (excluding 980425) is approximately 10^{51} erg (e.g., [4]). That this is so much less than the equivalent isotropic energy in gamma-rays for these same bursts reinforces our belief that GRBs are strongly beamed in the direction of the observer.

There turn out to be a surprisingly large number of ways to create a relativistic jet of 10^{51}–10^{52} erg. Hyper-accreting black holes (0.01 to 10 $M_\odot$ s^{-1}) can arise from collapsars of various types [32][19][11]; white dwarf–black hole mergers [10]; helium core–black hole mergers [8][34]; and merging neutron stars and black holes [17][23][24]. Those models that involve merging neutron stars and black holes mostly transpire outside star forming regions [10] and are not currently thought to be appropriate for long-soft GRBs. Such models may also have difficulty producing adequate energy, collimation, and duration, but each of these problems is model specific and might be circumvented by different choices of disk viscosity and jet acceleration mechanism.

Collapsars are discussed extensively in §2. Of the other accretion-based models, black holes merging with both helium cores and white dwarfs have the advantage that they could occur in star forming regions, produce adequate energy, and have sufficient angular momentum to form a disk. However, if the burst is to be beamed to less than 1% of the sky, the white dwarf model may provide an inadequate number of events [10]. The helium core model requires that a hydrogen envelope initially be present in order that the black hole (or neutron star) experience a common envelope evolution an merge with the core. Yet that same envelope must be absent when the burst occurs or prohibitive baryon loading will occur [20]. It may be that the merger peels off the envelope as the compact remnant goes in, but calculations to support this hypothesis are needed. The duration of the rapid accretion phase for the white dwarf and lower mass helium core models is also overly long, unless the jet orientation varies with time (§2.3). However, it should be noted that the merger of a black hole with a massive helium core produces essentially the same conditions as in ordinary collapsars, except for a larger angular momentum. The beaming and energetics should be comparable.

An entirely different sort of model also occurs in massive stars, but makes use of the rotational energy of a young millisecond magnetar [31]. There is adequate energy to produce a jet of 10^{51} to 10^{52} erg, but the requisite ordered *dipole* (i.e., not just surface) field is very large, $\sim 10^{16}$ gauss. It is not clear what supports the mantle of the star against collapse while the magnetar is forming, nor why the rotational energy is converted, with high efficiency, to a narrowly focused jet. The model's proponents suggest that a "LeBlanc-Wilson jet" initially weakens the star's inertial confinement along the rotational axis, but more work is needed.

Finally, there are models that do not specify the energy source, but assume that a strong shock accelerates down the density gradient of a stripped helium core. Hydrodynamic acceleration produces a small amount of relativistic matter and the interaction of this matter with the circumstellar wind makes the GRB [33][21][29]. This model can explain soft, low-energy bursts such as 980425, especially for asymmetric explosions, but has difficulty as a model for common GRBs. The Lorentz factors estimated by [29] as adequate to produce cosmological GRBs, even for hypothetical 10^{54} erg explosions, are low compared to [13]. However, it may be that shock break out in *highly asymmetric* explosions does produce events like 980425 as well as hard x-ray precursors to common GRBs [33]. This break-out interaction may also be relevant to the hard x-ray transients announced at the meeting (§3.2).

2   The Collapsar Model – Pro and Con

A collapsar forms when the evolved core of a massive star collapses to a black hole and the remaining star has too much angular momentum to fall in directly. One can divide collapsars into those where the black hole is formed promptly (Type I, [19]) when iron core collapse fails to produce an outgoing shock; those where the black hole forms by fallback (Type II, [20]) after the launch of a successful low energy explosion; and those that may have existed in the early universe (near zero metallicity) by collapse due to the pair instability in rotating stars of a few hundred solar masses (Type III, [11]). Type II produces longer bursts that could only be common GRBs if the jet varies its orientation (§2.3). Type III produces very energetic bursts, but only in regions of extremely low metallicity. For clarity, we will concentrate here on Type I.

Collapsars are currently a favored model because of a) their clear association with massive stars; b) the large energy potentially available when several solar masses accretes into a black hole; c) the unavoidable production of a stellar explosion accompanying the GRB which may produce enough ^{56}Ni (e.g., from a disk wind) to be a bright supernova; d) the existence of detailed models showing strong collimation and relativistic motion for the jet; and e) a possible high event rate – up to $\sim 1\%$ that of supernovae. It may or may not be an advantage that the star also produces a dense circumstellar medium with a density that declines $\sim r^{-2}$ (§2.2). The physics of jet formation in collapsars is uncertain, just as in all models where MHD processes may be involved, and no simulation so far has reached a Lorentz factor of even 100. However, Type I collapsars (and neutron star black hole mergers) do have the advantage of a reasonable efficiency for jet formation by neutrino energy deposition.

2.1 Progenitors

The two necessary ingredients for the collapsar are a failed or weak initial supernova explosion that produces a black hole and sufficient angular momentum to form a disk. In fact, high angular momentum, $\gtrsim 10^{16}$ erg s, is a common requirement in *all* current GRB models that involve massive stars, including those that use a pulsar power source Yet the actual angular momentum in a presupernova star is unknown. Calculations by Heger [15] that ignore magnetic fields provide the necessary angular momentum – and more, but Spruit & Phinney [25] have suggested that the magnetic interaction of the rapidly rotating helium core with its stationary envelope during the red supergiant stage will halt the core, making it unfit as a collapsar progenitor. More recently, calculations by Spruit [28] that include magnetic instabilities have greatly reduced the braking action of the hydrogen envelope. Working with Spruit, Heger and I [16] have implemented an approximate representation of magnetic torques in our stellar evolution code and recomputed the angular momentum distribution in presupernova stars. The good news is that the results are consistent with a pulsar rotation period at birth near 7 ms, not too bad for the Crab pulsar (though 20 ms would be better). However, this is too slow to form a disk. Either the stellar model is inaccurate or some other route than single star evolution must be involved in making GRBs. The natural alternative is a close binary or merger. But if the hydrogen envelope is removed too early, mass loss from the WR star will also reduce the angular momentum. We are currently studying whether there remain viable paths to a rapidly rotating helium core at collapse. Low metallicity may be essential.

2.2 Winds

Considerable discussion was given at the meeting to whether the expected r^{-2} density structure for a stellar wind is compatible with analysis of the afterglow light curves (see papers by Chevalier and by Kulkarni). Panaitescu and Kumar have estimated the density in the circum-source region where the afterglow is produced to be $\lesssim 10^{-3}$ g cm^{-3}.

It should be noted that the afterglows sample a region $\sim 10^{17}$ cm in size. Depending upon the density of the interstellar medium, this may be comparable to the wind termination radius. Moreover, the wind history of a WR star during its last few centuries could be quite complicated as the star enters advanced burning stages unlike those in any WR star observed so far. Unlike a red supergiant, the core and surface of a WR star remain in communication at late times. That is, pulsations or flashes in the center during carbon or oxygen burning are communicated to the surface where they are amplified by the steep density gradient. As noted above, the WR star mass loss may also be influenced by a close binary companion. The complex density structure we see in SN 1987A could be a hint of what exists in some GRB progenitors. Finally the jet is characterized by a distribution of Lorentz factors with angle, not a single value.

Still, there is no obvious way for the density at 10^{17} cm to be much lower than about 1 cm^{-3} (assuming $\dot{M} = 10^{-6}$ M$_\odot$ y^{-1}; v $\sim 10^8$ cm s^{-1}). The low densities inferred by Panaitescu and Kumar are thus problematic for the collapsar model.

2.3   Short Bursts from Collapsars?

It is frequently stated (e.g., [19],[22]) that collapsars are only capable of explaining GRBs of the "long-soft" variety and not those having duration shorter than a few seconds. This is based upon the reasonable assumption that the disk will continue to be fed by the collapsing star at an approximately constant rate for a time comparable to the free fall time for the inner helium core. Further, it is often stated that, for large enough viscosity, the duration of merging black holes and neutron stars is better matched to "short hard" bursts [22],[17]. If so, and if the mergers are occurring chiefly in intergalactic space [9], one may understand why Beppo-Sax has yet to discover an afterglow from a short burst. Lacking an interstellar medium to make an external shock, the afterglow would be much fainter, at least when viewed the following day [18]. But what if HETE-2 discovers short hard bursts in star forming regions?

It is possible for a collapsar to make a shorter burst provided that the beaming is tight and the direction of the jet varies. Suggestions along these lines have been made previously [7] [14] [6]. However, I would like to mention an alternate reason the jet might vary its orientation. During the time the relativistic jet is at its maximum energy, it is still passing through a massive stellar core that is in the process of exploding. The jet interacts hydrodynamically with this helium core and is, in part, focused by the interaction [1]. The input of energy at the base of the jet is likely to be variable, both in rate and orientation because of black hole precession, disk instabilities, and kicks. Given the hydrodynamic instabilities the jet may experience on the way out, it is reasonable that the centroid of the jet erupting from the surface of the star may wander in polar angle. Special relativistic calculations (of necessity 3D) to explore this possibility are in progress. At a minimum, time structure may be introduced into the burst.

3   Other Exotica

3.1   Iron Lines

Another topic receiving much attention at the meeting was the discovery by BeppoSax and Chandra of iron features in the x-ray afterglow of several GRBs

[2][3][26]. One model proposed to explain these features is the "supranova" [30][26], a two stage explosion in which a supernova occurs many days, or even years prior to the GRB. The GRB then occurs inside the previous supernova and powered by black hole accretion. The black hole and its disk are in turn a consequence of fall back or the delayed collapse of a rapidly rotating neutron star. The fall back scenario resembles a Type II collapsar [20] except that the time scale would be very long. It seems unlikely that a typical 10 s GRB could be produced in this fashion without unrealistic collimation. It is also whether the collapse of a rapidly rotating neutron star to a black hole will leave behind an appreciable disk. This could be addressed with computation. To leave a rotating supermassive neutron star hovering upon the brink of collapse for so long probably takes fine tuning.

A different sort of model has been suggested by Rees and Meszaros [27] who argue that the line features can be attributed to the interaction of a continuing (but decaying) post-burst relativistic outflow. The densities in the region where the line is produced are higher and the necessary iron mass is consequently reduced from $\sim$0.1 $M_\odot$ to 10^{-8} $M_\odot$. They discuss a magnetar as the possible source of this late time outflow, but, in the collapsar model, accretion of $\sim$$10^{-6}$ $M_\odot$ s^{-1} continues a day after the event [20]. This could provide the necessary 10^{47} erg s^{-1} needed in the Rees-Meszaros model. Alternatively one may consider the possibility that the accreting material and the jet may themselves be iron rich since the material falling back has been mixed with ejecta from the inner core.

3.2 Hard X-Ray Bursts

At the meeting, Heise also announced the discovery of hard x-ray transients (2–10 keV) that otherwise seem to share many properties (duration, distribution, frequency) with long GRBs. Hard x-ray transients are expected from the collapsar model because the jet that produces the GRB is only the central portion ($\vartheta \lesssim 0.1$ radian) of a much broader eruption with declining $\Gamma(\vartheta)$, with ϑ the polar angle. Even on axis, mildly relativistic material is ejected while the jet is developing its full energy [1]. Mildly relativistic material is also present as a consequence of shock break out [33][29]. The interaction of this material with circumstellar matter – the wind of the pre-explosive star, creates hard x-ray emission. The hard bursts reported by Heise would thus be extreme cases (in terms of soft spectra) of GRB 980425. They would originate relatively nearby, would be very common in the universe, and might even display an asymmetric sky distribution (though they are certainly extra-galactic). They would also have systematically longer durations than GRBs.

This work has been supported by NASA (NAG5-8128 and MIT 292701).

References

1. M. A. Aloy, E. Müller, J. M. Ibanez, J. B. Marti, & A. MacFadyen, ApJL, **531**,L119, (1999)
2. L. Amati, F. Frontera, M. Vietri, J.J.M. in't Zand, et al., Science, **290**, 953 (2000).
3. L.A. Antonelli, L. Piro, M. Vietri, E. Costa, et al., ApJ, in press, (2001), astro-ph/0010221
4. E. Berger, R. Sari, D. A. Frail, S. R. Kulkarni, et al., ApJ, **545**, 56, (2000)
5. J. S. Bloom, S. R. Kulkarni, S. G. Djorgovski, ApJ, submitted, (2001), astro-ph/0010176
6. A. Dar & A. De Rjula, submitted to A&A, (2001), astro-ph/0012227
7. D. Fargion, A&ASup, **138**, 507, (1999)
8. C.L. Fryer & S.E. Woosley, ApJL, **502**, L9, (1998)
9. C.L. Fryer, S.E. Woosley, M. Herant, M. B. Davies, ApJ, **520**, 650 (1999)
10. C. L. Fryer, S. E. Woosley, & D. Hartmann, ApJ, **526**, 152, (1999).
11. C. L. Fryer, S. E. Woosley, and A. Heger, ApJ, in press, (2001), astro-ph/0007176
12. adapted from G. Gamow and his poem "Quasar", (1964)
13. D. Guetta, M. Spada, & E. Waxman, ApJ, submitted, (2001), astro-ph/0011170
14. D. H. Hartmann & S. E. Woosley, Proc. COSPAR meeting, *Advances in Space Research*, Vol 15, No. 5, 143 (1995)
15. A. Heger, N. Langer, & S. E. Woosley, ApJ, **528**, 368
16. A. Heger, H. C. Spruit, & S. E. Woosley, in preparation, (2001)
17. H.-Th. Janka & M. Ruffert, in Proc. Conf. on Stellar Collisions and Mergers, ed. M. Shara, ASP Conf. Series, in press, (2001), astro-ph/0101357
18. P. Kumar & A. Panaitescu, ApJL, submitted (2001), astro-ph/0006317
19. A. MacFadyen & S. E. Woosley, ApJ, **524**, 262, (1999)
20. A. MacFadyen, S. E. Woosley, & A. Heger, ApJ, in press, (2001), astro-ph/9910034
21. C. D. Matzner & C. F. McKee, ApJ, **510**, 379, (1999)
22. R. Popham, S. E. Woosley, & C. Fryer, ApJ, **518**, 356, (1999)
23. S. Rosswog, M.B. Davies, F.-K. Thielemann, & T. Piran, A&A, **360**, 171, (2000).
24. H.-T. Janka, T. Eberl, M. Ruffert, & C. L. Fryer, ApJL, **527**, 39, (1999)
25. H. C. Spruit & E. S. Phinney Nature, **393**, 139, (1998)
26. L. Piro, G. Garmire, M. Garcia, G. Stratta, et al., Science, **290**, 955, (2000)
27. M. J. Rees & P. Meszaros, ApJL, **545**, 73, (2000), astro-ph/0010258
28. H. C. Spruit, A&A, **349**, 189, (1999)
29. J. Tan, C. Matzner, & C. McKee, ApJ, in press, (2001), astro-ph/0012003
30. M. Vietri, & L. Stella, ApJ, **507**, L45, (1998)
31. J. C. Wheeler, I. Yi, L. Wang, & P. Höflich, ApJ, **537**, 810, (2000)
32. S. E. Woosley, ApJ, **405**, 273, (1993)
33. S. E. Woosley, R. G. Eastman, & B. Schmidt, 1998, ApJ, **516**, 788, (1999)
34. W. Zhang & C. Fryer ApJ, in press, March (2001), astro-ph/0011236

High-Energy Particles from γ-Ray Bursts

Eli Waxman

Weizmann Institute of Science, Rehovot 76100, Israel

The widely accepted interpretation of the phenomenology of γ-ray bursts (GRBs), bursts of 0.1 MeV–1 MeV photons lasting for a few seconds [1], is that the observable effects are due to the dissipation of the kinetic energy of a relativistically expanding wind, a "fireball," whose primal cause is not yet known [2]. The recent detection of "afterglows," delayed low energy (X-ray to radio) emission of GRBs [3], confirmed the cosmological origin of the bursts, through the redshift determination of several GRB host-galaxies, and confirmed standard model predictions of afterglows that result from the collision of an expanding fireball with its surrounding medium [4].

In §1 we discuss the association of GRBs and ultra-high energy cosmic-rays (UHECRs) [5]. We show that recent afterglow observations strengthen the evidence for GRB and UHECR association, which is based on two key points. First, the constraints imposed on fireball model parameters by recent observations imply that acceleration of protons is possible to higher energy than previously assumed, $\sim 10^{21}$ eV. Second, the inferred local ($z = 0$) GRB energy generation rate of γ-rays, $\sim 10^{44}$erg/Mpc^{3}yr, is remarkably similar to the local generation rate of UHECRs implied by cosmic-ray observations.

The GRB model for UHECR production makes unique predictions [5], that may be tested with planned large area UHECR detectors [6]. In this review we focus, however, on more recent predictions for the emission of high energy neutrinos. We discuss in §2 high energy neutrino production in fireballs and its implications for planned high energy neutrino telescopes (the ICECUBE extension of AMANDA, ANTARES, NESTOR; see [7] for review).

1 UHECRs from GRB Fireballs

In the fireball model of GRBs, a compact source, of linear scale $r_0 \sim 10^7$ cm, produces a wind characterized by an average luminosity $L \sim 10^{52}$erg s^{-1} and mass loss rate $\dot{M} = L/\eta c^2$. At small radius, the wind bulk Lorentz factor, Γ, grows linearly with radius, until most of the wind energy is converted to kinetic energy and Γ saturates at $\Gamma \sim \eta \sim 300$. Variability of the source on time scale Δt, resulting in fluctuations in the wind bulk Lorentz factor Γ on similar time scale, then leads to internal shocks in the expanding fireball at a radius $r_i \approx \Gamma^2 c \Delta t$. If the Lorentz factor variability within the wind is significant, internal shocks reconvert a substantial part of the kinetic energy to internal energy. It is assumed that this energy is then radiated as γ-rays by synchrotron and inverse-Compton emission of shock-accelerated electrons.

1.1 Fermi Acceleration in GRBs

In the region where electrons are accelerated, protons are also expected to be shock accelerated. Since the internal shocks are mildly relativistic, with Lorentz factors $\Gamma_i - 1 \sim 1$ in the wind frame, the predicted energy distribution of accelerated protons is $dn_p/d\epsilon_p \propto \epsilon_p^{-2}$ [8], similar to the electron energy spectrum inferred from the observed photon spectrum. Two constraints must be satisfied by fireball wind parameters in order to allow proton acceleration to $\epsilon_p > 10^{20}$ eV in internal shocks [9,10]:

$$\xi_B/\xi_e > 0.02\Gamma_{2.5}^2 \epsilon_{p,20}^2 L_{\gamma,52}^{-1}, \tag{1}$$

and [9]

$$\Gamma > 130\epsilon_{p,20}^{3/4}\Delta t_{-2}^{-1/4}. \tag{2}$$

Here, $\epsilon_p = 10^{20}\epsilon_{p,20}$ eV, $\Delta t = 10^{-2}\Delta t_{-2}$ s, $\Gamma = 10^{2.5}\Gamma_{2.5}$ is the plasma expansion Lorentz factor, $L_\gamma = 10^{52}L_{\gamma,52}$erg/s is the γ-ray luminosity, ξ_B is the fraction of the wind energy density which is carried by magnetic field, $4\pi r^2 c\Gamma^2(B^2/8\pi) = \xi_B L$, and ξ_e is the fraction of wind energy carried by shock accelerated electrons. Since the electron energy is lost radiatively, $L_\gamma \approx \xi_e L$. The first constraint must be satisfied in order for the proton acceleration time t_a to be smaller than the wind expansion time. The second constraint must be satisfied in order for the synchrotron energy loss time of the proton to be larger than t_a.

The constraints that must be satisfied to allow acceleration of protons to energy $> 10^{20}$ eV are remarkably similar to those inferred from γ-ray observations. $\Gamma > 100$ is implied by observed γ-ray spectra, and magnetic field close to equipartition, $\xi_B \sim 1$, is required in order for electron synchrotron emission to account for the observed radiation.

We have assumed in the discussion so far that the fireball is spherically symmetric. However, since a jet-like fireball behaves as if it were a conical section of a spherical fireball as long as the jet opening angle is larger than Γ^{-1}, our results apply also for a jet-like fireball (we are interested only in processes that occur when the wind is ultra-relativistic, $\Gamma \sim 300$, prior to significant fireball deceleration). For a jet-like wind, L in our equations should be understood as the luminosity the fireball would have carried had it been spherically symmetric.

1.2 Energy Generation Rate

The observed GRB redshift distribution implies a typical GRB γ-ray energy of $E \approx 10^{53}$ erg, and a GRB rate of $R_{\text{GRB}} \sim 3/\text{Gpc}^3\text{yr}$ at $z \sim 1$. The present, $z = 0$, rate is less well constrained, since most observed GRBs originate at redshifts $1 \leq z \leq 2.5$ [3]. Present data are consistent with both no evolution of GRB rate with redshift, and with strong evolution (following, e.g., star formation rate), in which $R_{\text{GRB}}(z = 1)/R_{\text{GRB}}(z = 0) \sim 10$. Detailed analyses, assuming R_{GRB} is proportional to star formation rate, lead to $R_{\text{GRB}}(z = 0) \sim 0.5/\text{Gpc}^3\text{yr}$ [3]. The energy observed in γ-rays reflect the fireball energy in accelerated electrons. If accelerated electrons and protons carry similar energy (as indicated by afterglow observations) then the GRB cosmic-ray production rate is

$$\epsilon_p^2(d\dot{n}_p/d\epsilon_p)_{z=0} \approx 10^{44} \text{erg/Mpc}^3\text{yr}. \tag{3}$$

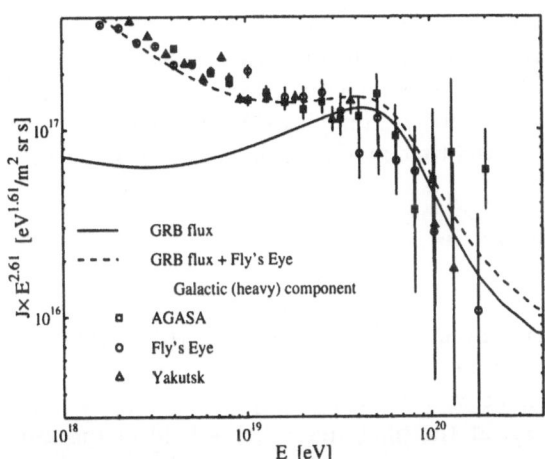

Fig. 1. The UHECR flux expected in a cosmological model, where high-energy protons are produced at a rate $(\epsilon_p^2 d\dot{n}_p/d\epsilon_p)_{z=0} = 0.5 \times 10^{44}\text{erg/Mpc}^3\text{yr}$ as predicted in the GRB model [Eq. (3)], solid line, compared to the Fly's Eye [11], Yakutsk [12] and AGASA [13] data. 1σ flux error bars are shown. The highest energy points are derived assuming the detected events represent a uniform flux over the energy range 10^{20} eV–3×10^{20} eV. The dashed line is the sum of the GRB model flux and the Fly's Eye fit to the Galactic heavy nuclei component, $J \propto E^{-3.5}$ [11], with normalization increased by 25%.

In Fig. 1 we compare the observed UHECR spectrum with that predicted by the GRB model (assuming GRB rate follows star formation rate). The generation rate (3) of high energy protons is remarkably similar to that required to account for the flux of $> 10^{19}$ eV cosmic-rays. The flux at lower energies is most likely dominated by heavy nuclei of Galactic origin [11], as indicated by the flattening of the spectrum at $\approx 10^{19}$ eV and by the evidence for a change in composition at this energy [14].

The suppression of model flux above $10^{19.7}$ eV is due to energy loss of high energy protons in interaction with the microwave background, i.e. to the "GZK cutoff" [15]. The available data do not allow to determine the existence (or absence) of the "cutoff" with high confidence. The AGASA results show an excess (at a $\sim 2.5\sigma$ confidence level) of events compared to model predictions above 10^{20}eV. This excess is not confirmed, however, by the other experiments. Moreover, since the 10^{20}eV flux is dominated by sources at distances < 100 Mpc, over which the distribution of known astrophysical systems (e.g. galaxies, clusters of galaxies) is inhomogeneous, significant deviations from model predictions presented in Fig. 1 for a uniform source distribution are expected at this energy [16]. Clustering of cosmic-ray sources leads to a standard deviation, σ, in the expected number, N, of events above 10^{20} eV, given by $\sigma/N = 0.9(d_0/10\text{Mpc})^{0.9}$ [17], where d_0 is the unknown scale length of the source correlation function and $d_0 \sim 10$ Mpc for field galaxies.

Although the rate of GRBs out to a distance of 100 Mpc from Earth, the maximum distance traveled by $> 10^{20}$ eV protons, is in the range of 10^{-2} to 10^{-3}yr^{-1}, the number of different GRBs contributing to the flux of $> 10^{20}$ eV protons at any given time may be large. This is due to the dispersion τ in proton arrival time, which is expected due to deflection by inter-galactic magnetic fields and may be as large as 10^5 yr, implying that the number of sources contributing to the flux at any given time may be as large as $\tau \times 10^{-2}\text{yr}^{-1} = 10^3$ [9].

2   GRB Neutrinos

A burst of $\sim 10^{14}$eV neutrinos, accompanying observed γ-rays, is a natural consequence of the conventional fireball scenario [18]. The neutrinos are produced by π^+ created in interactions between fireball γ-rays and accelerated protons. The key relation is between the observed photon energy, ϵ_γ, and the accelerated proton's energy, ϵ_p, at the photo-meson threshold of the Δ-resonance. In the observer frame,

$$\epsilon_\gamma \, \epsilon_p = 0.2 \, \text{GeV}^2 \, \Gamma^2 \,. \tag{4}$$

For $\Gamma \approx 300$ and $\epsilon_\gamma = 1$ MeV, we see that characteristic proton energies $\sim 10^{16}$ eV are required to produce pions, leading to $\sim 10^{14}$ eV neutrinos.

The fraction $f_\pi(\epsilon_p)$ of proton energy lost to pion production is determined by the number density of photons at the dissipation region, and is given by [18]

$$f_\pi \approx 0.2 \min(1, \epsilon_p/10^{16}\text{eV}) \frac{L_{\gamma,52}}{\Gamma_{2.5}^4 \Delta t_{-2}} \,. \tag{5}$$

Assuming that GRBs are the sources of observed ultra-high energy cosmic rays, i.e. that GRBs produce high energy protons at a rate given by Eq. (3), the intensity of high energy neutrinos implied by Eq. (5) is [18]

$$\epsilon_\nu^2 \Phi_{\nu_x} \approx 10^{-9} \left(\frac{f_\pi}{0.2} \right) \min \left(1, \frac{\epsilon_\nu}{10^{14}\text{eV}} \right) \frac{\text{GeV}}{\text{cm}^2 \, \text{sr} \, \text{s}} \,. \tag{6}$$

ν_x stands for ν_μ, $\bar{\nu}_\mu$ and ν_e. The neutrino flux (6) is suppressed at high energy, $> 10^{16}$ eV, due to synchrotron energy loss of pions and muons [19,18].

High energy protons may also interact with 10 eV–1 keV photons produced on a time scale ~ 10 s following the GRB, due to interaction of the fireball with its surrounding medium. Inserting in Eq. (4) typical photon energy of ~ 100 eV, neutrinos of energy $\sim 10^{18}$ eV may be expected. The expected neutrino intensity due to this process is [20]

$$\epsilon_\nu^2 \Phi_{\nu_x} \approx 10^{-10} \left(\frac{\epsilon_\nu}{10^{17}\text{eV}} \right)^\alpha \frac{\text{GeV}}{\text{cm}^2 \, \text{s} \, \text{sr}} \,, \tag{7}$$

where $\alpha = 1/2$ for $\epsilon_\nu > 10^{17}$eV and $\alpha = 1$ for $\epsilon_\nu < 10^{17}$eV. The neutrino flux is expected to be strongly suppressed at energy $> 10^{19}$ eV, since protons are not expected to be accelerated to energy $\gg 10^{20}$ eV.

The prediction of Eq. (7) is based on the assumption that the fireball expands into inter-stellar medium gas, with typical density ~ 1cm^{-3}. This is expected, e.g., if the underlying progenitor is a binary neutron star merger. Some GRBs may result, however, from the collapse of a massive star, in which case the fireball is expected to expand into a pre-existing wind. The relevant plasma density is much higher in this case, $\sim 10^4$cm^{-3}, implying a lower expansion Lorenz factor, higher proper photon density, and hence a larger fraction of proton energy lost to pion production. Protons of energy $\epsilon_p \geq 10^{18}$ eV lose all their energy to pion production in this case, and the expected neutrino intensity is [20,21]

$$\epsilon_\nu^2 \Phi_{\nu_x} \approx 10^{-8} \min\left(1, \frac{\epsilon_\nu}{10^{17}\text{eV}}\right) \frac{\text{GeV}}{\text{cm}^2\,\text{s}\,\text{sr}}. \tag{8}$$

The predicted intensity of 10^{14} eV neutrinos produced by photo-meson interactions with observed 1 MeV photons, Eq. (6), implies a detection of ~ 10 neutrino induced muons per year in planned 1km^3 Cerenkov neutrino detectors, correlated in time and direction with GRBs. The predicted intensity of 10^{17} eV neutrinos, produced by photo-meson interactions during the onset of fireball interaction with its surrounding medium in the case of fireball expansion into a pre-existing wind, Eq. (8), implies a detection of several neutrino induced muons per year in a 1km^3 detector. In this case, the predicted flux of 10^{19} neutrinos may also be detectable by planned large air-shower detectors [6,23].

Inelastic p-n collisions may produce ~ 10 GeV neutrinos with a fluence of $\sim 10^{-4}$cm^{-2} per burst, due to either p-n decoupling in a wind with high neutron fraction and high, > 400, Lorentz factor, or to neutron diffusion in a wind with, e.g., strong deviation from spherical symmetry [22]. The predicted number of events in a 1km^3 detector is ~ 10yr^{-1}. Their detection will, however, be difficult, since at ~ 10 GeV the effective volume of planned detectors is much smaller than 1km^3.

Detection of high energy neutrinos will test the shock acceleration mechanism and the suggestion that GRBs are the sources of ultra-high energy protons, since $\geq 10^{14}$ eV ($\geq 10^{18}$ eV) neutrino production requires protons of energy $\geq 10^{16}$ eV ($\geq 10^{19}$ eV). It should be pointed out, that despite the apparent strong dependence of f_π on wind Lorentz factor, Eq. (5), the range of allowed f_π values is $\sim 10\%$ to $\sim 30\%$. This is due to the fact that GRB observations imply that wind model parameters (Γ, L, Δt) are correlated, and that Γ is restricted to values in a relatively narrow range [24].

The model discussed above predicts the production of high energy muon and electron neutrinos. However, for the neutrino parameters inferred from recent atmospheric experiments [25], we expect flavor oscillations to produce equal fluxes of muon and tau neutrinos upon their arrival to Earth, thus allowing for a "τ appearance" experiment. Furthermore, since the neutrino signal should coincide with the γ-ray signal, checking the simultaneity of photon and neutrino arrival times will allow to test for deviations from Lorentz invariance and from the weak equivalence principle. With 1 s accuracy ($\sim$ 1 ms for short bursts), a burst at 100 Mpc would reveal a fractional difference in limiting speed of 10^{-16}, and a fractional difference in gravitational time delay of order 10^{-6} (considering the Galactic potential alone), many orders of magnitude better than present limits based on supernova 1987A [26].

References

1. For review see: Fishman, G. J. & Meegan, C. A., ARA&A **33**, 415 (1995).
2. For review see: Piran, T., Phys. Rep. **333**, 529 (2000).
3. For review see: Kulkarni, S. R. *et al.*, To appear in Proc. of the 5th Huntsville Gamma-Ray Burst Symposium (astro-ph/0002168)
4. For review see: Mészáros, P., A&AS **138**, 533 (1999).
5. For review see: Waxman, E., Nucl. Phys. B (Proc. Suppl.) **87**, 345 (2000).
6. Cronin, J. W., Nucl. Phys. B (Proc. Suppl.) **28B**, 213 (1992); Teshima, M. *et al.*, Nuc. Phys. **28B** (Proc. Suppl.), 169 (1992).
7. Halzen, F., in Proc. 17th International Workshop on Weak Interactions and Neutrinos (Cape Town, South Africa, January 1999) (astro-ph/9904216).
8. For review see: Blandford, R., & Eichler, D., Phys. Rep. **154**, 1 (1987).
9. Waxman, E., Phys. Rev. Lett. **75**, 386 (1995).
10. Vietri, M., Ap. J. **453**, 883 (1995); Milgrom, M. & Usov, V., Ap. J. **449**, L37 (1995).
11. Bird, D. J., *et al.*, Ap. J. **424**, 491 (1994).
12. Efimov, N. N. *et al.*, in *Proceedings of the International Symposium on Astrophysical Aspects of the Most Energetic Cosmic-Rays*, ed. M. Nagano and F. Takahara (World Scientific, Singapore, 1991), p. 20.
13. Takeda, M. *et al.*, Phys. Rev. Lett. **81**, 1163 (1998).
14. Dawson, B. R., Meyhandan, R., Simpson, K.M., Astropart. Phys. **9**, 331 (1998).
15. Greisen, K., Phys. Rev. Lett. **16**, 748 (1966); Zatsepin, G. T., & Kuzmin, V. A., JETP lett., **4**, 78 (1966).
16. Waxman, E., Ap. J. **452**, L1 (1995).
17. Waxman, E., & Bahcall, J. N., Ap. J. **542**, 542 (2000).
18. Waxman, E., & Bahcall, J. N., Phys. Rev. Lett. **78**, 2292 (1997); Phys. Rev. **D59**, 023002 (1999).
19. Rachen, J. P., & Mészáros, P., Phys. Rev. D **58**, 123005 (1998).
20. Waxman, E., & Bahcall, J. N., Ap. J. **541**, 707 (2000).
21. Dai, Z. G., & Lu, T., submitted to Ap. J. (astro-ph/0002430).
22. Derishev, E. V., Kocharovsky, V. V., & Kocharovsky, Vl. V, Ap. J. **521**, 640 (1999); Bahcall, J. N., & Mészáros, Phys. Rev. Lett., 85, 1362 (2000); Mészáros, P., & Rees, M., Ap. J. **541**, L5 (2000).
23. J. Linsley, MASS/AIRWATCH Huntsville workshop report, pp. 34–74 (1995).
24. Guetta, D., Spada, M. & Waxman, E., these proceedings (submitted to apJ).
25. Y. Fukuda *et al.*, Phys. Lett. **B335**, 237 (1994); D. Casper *et al.*, Phys. Rev. Lett. **66**, 2561 (1991); G. L. Fogli, and E. Lisi, Phys. Rev. **D52**, 2775 (1995).
26. J. N. Bahcall, *Neutrino Astrophysics*, Cambridge University Press (NY 1989), pp. 438–460.

Ultra-high Energy Cosmic Rays and Neutron-Decay Halos from Gamma Ray Bursts

C.D. Dermer

Code 7653, Naval Research Laboratory,
4555 Overlook Ave., SW, Washington, DC 20375-5352 USA

Abstract. Simple arguments concerning power and acceleration efficiency show that ultra-high energy cosmic rays (UHECRS) with energies $\gtrsim 10^{19}$ eV could originate from GRBs. Neutrons formed through photo-pion production processes in GRB blast waves leave the acceleration site and travel through intergalactic space, where they decay and inject a very energetic proton and electron component into intergalactic space. The neutron-decay protons form a component of the UHECRs, whereas the neutron-decay electrons produce optical/X-ray synchrotron and gamma radiation from Compton-scattered background radiation. A significant fraction of galaxies with GRB activity should be surrounded by neutron-decay halos of characteristic size ~ 100 kpc.

1 Introduction

Gamma-ray bursts produce enough power within the Greisen-Zatsepin-Kuzmin (GZK) photopion production radius to power the UHECRs [1,2,3]. Stochastic gyroresonant acceleration of protons and ions by turbulence generated in relativistic blast waves can accelerate particles to ultra-high energies [4]. Energetic neutrons are formed by photopion interactions of accelerated hadrons with nonthermal synchrotron radiation in GRB blast waves. The neutrons travel through intergalactic space and decay, and the neutron-decay electrons form synchrotron and Compton halos around galaxies with GRB activity. The discovery of neutron-decay halos around galaxies with vigorous star-forming activity will provide strong evidence for a GRB origin of the UHECRs [3].

2 GRB Origin of UHECRs: Power and Acceleration

As a consequence of Beppo-SAX results, we now know that GRBs are extragalactic and originate from sources with a broad distribution of redshifts and mean redshift $\bar{z} \approx 1$. Beppo-SAX has a much smaller field-of-view than BATSE, but triggers on nearly the same sample of long-duration ($t_{50} \gtrsim 1$ s) GRBs. If UHECRs originate from GRBs, then the product of the UHECR energy density u_{UH} and the characteristic source volume V is equal to the product of the GRB power L_{GRB}, the loss time from the source volume, and the efficiency ϵ to convert GRB energy into UHECRs. For protons with energies $\gtrsim 10^{20}$ eV, the GZK radius is ~ 140 Mpc [5]. Thus the loss time $t_{p\gamma} \cong 140$ Mpc/c $\cong 1.4 \times 10^{16}$ s. Hence $u_{UH} \cong \epsilon f L_{GRB} t_{p\gamma}/V$, where f is a factor that takes into account present day star-formation activity compared with that occurring at $\bar{z}$. If $\bar{z} = 1$, then $f \cong 1/6$.

Let $\bar{d} = 10^{28} d_{28}$ cm represent the average luminosity distance to observed GRBs, so that $V \cong 4\pi\bar{d}^3/3$. The power of GRBs into the volume V is given by the typical GRB energy E_{GRB} multiplied by the GRB rate. BATSE is sensitive to GRBs with peak fluxes $\varphi \gtrsim 10^{-7}\varphi_{-7}$ ergs cm^{-2} s^{-1}. The observed mean duration of the long duration GRBs is $t_{dur} = 30 t_{30}$ s. Thus $E_{GRB} \approx 4\pi\bar{d}^2 \cdot 10^{-7}\varphi_{-7} \cdot 30 t_{30}(1+\bar{z})/(0.1\eta_{-1}) \approx 4 \times 10^{52} d_{28}^2 \varphi_{-7} t_{30}(1+\bar{z})/\eta_{-1}$ ergs, where $\eta = 0.1\eta_{-1}$ is the efficiency for transforming the GRB explosion energy into γ rays in the Beppo-SAX and BATSE energy bands. The long-duration GRBs occur at a rate of $\approx 1/(t_{day}$ day), with $t_{day} \approx 1$, so that $L_{GRB} \approx 4 \times 10^{47} d_{28}^2 \varphi_{-7} t_{30}(1+\bar{z})^2/\eta_{-1}$ ergs s^{-1}. We therefore find that

$$u_{UH} \text{ (ergs cm}^{-3}) \approx 1.5 \times 10^{-21} \frac{k\epsilon f \varphi_{-7} t_{30}(1+\bar{z})^2}{\eta_{-1} t_{day} d_{28}}. \tag{1}$$

The factor k represents the energy released by the dirty and clean fireballs which do not trigger the BATSE detector. Detailed calculations within the context of the external shock model show that $k \approx 3$ [5].

Observations show that $u_{UH} \cong 10^{-20}$ and 2×10^{-21} ergs cm^{-3} for cosmic rays with $E \gtrsim 10^{19}$ eV and 10^{20} eV, respectively. (For protons with energies $\gtrsim 10^{19}$ eV, $t_{p\gamma} \cong 1000$ Mpc/c.) If an efficient mechanism for converting the energy of the relativistic outflows into UHECRs exists, then there is sufficient power in the sources of GRBs to power the UHECRs. Detailed calculations [3,6,7] verify this result.

Particle acceleration in GRB blast waves must satisfy the Hillas [8] condition for UHECR production, which requires that the Larmor radius be smaller than the characteristic size of the acceleration region. For GRB blast waves, this size is the blast-wave width. Hence the particle Larmor radius $r_L = (Am_pc^2/ZeB)(\gamma_{max}/\Gamma) < \Delta' = f_\Delta x/\Gamma$, where γ_{max} is the maximum particle Lorentz factor measured in the explosion frame, $\Gamma = 300\Gamma_{300}$ is the blast wave Lorentz factor, Δ' is the comoving blast wave width, $f_\Delta \cong 1/12$ from hydrodynamics, and $x = 10^{16} x_{16}$ cm is the location of the blast wave from the explosion center. The blast-wave magnetic field $B \cong \sqrt{32\pi e_B n_{ISM} m_p c^2}\Gamma \cong 0.4\sqrt{e_B n_{ISM}}\Gamma$ G is defined by a magnetic-field parameter $e_B(< 1)$, and the term n_{ISM} is the particle density of the surrounding medium. Thus

$$E_{max} = Am_pc^2\gamma_{max} = ZeBf_\Delta x \simeq 3 \times 10^{19} Z\sqrt{e_B n_{ISM}}\left(\frac{f_\Delta}{1/12}\right) x_{16}\Gamma_{300} \text{ eV.} \tag{2}$$

Provided that the magnetic field is near its equipartition value $e_B \cong 1$, a wide range of parameter values can satisfy the Hillas condition for accelerating UHE-CRs by stochastic acceleration through gyroresonant interactions with MHD turbulence in the blast wave fluid [4,9]. The Alfvén speed v_A in the relativistic shocked fluid is also relativistic (naively using the nonrelativistic expression gives $v_A/c \cong \sqrt{2e_B\Gamma}$), resulting in an acceleration rate that is much more rapid for second-order than for first-order processes [9].

3   Neutron-Decay Halos

Photopion processes involving accelerated protons and ions interacting with non-thermal synchrotron radiation in the blast wave will produce neutrons through the process $p + \gamma \to n + \pi^+$. The neutrons, unbound by the magnetic field in the blast wave, leave the acceleration site with Lorentz factors $\gamma_n = 10^{10}\gamma_{10}$, with $0.1 \lesssim \gamma_{10} \lesssim 100$. The neutrons decay on a timescale $\gamma_n t_n \simeq 3 \times 10^5 \gamma_{10}$ yr, where the neutron β-decay lifetime $t_n \cong 900$ s. The neutrons travel a characteristic distance $\lambda_n \simeq 90\gamma_{10}$ kpc before they decay and inject highly relativistic electrons and protons into intergalactic space. Approximately 1% of the energy of a GRB explosion with $10^{54}E_{54}$ ergs is deposited into highly relativistic neutrons when $E_{54} \approx 1$ [3]. The neutron-decay electron halo surrounding a galaxy from a single GRB reaches a maximum power of $L_{halo} \approx 0.01 \times \mathcal{F}10^{54}E_{54}(m_e/m_p)/(\gamma_n t_n) \approx 10^{36}E_{54}\mathcal{F}/\gamma_{10}$ ergs s^{-1}. Detailed calculations show that $\mathcal{F} \approx 0.1$ [3]. The neutron-decay protons become part of the UHECRs. GRB explosions with $E_{54} \gtrsim 0.2$ occur at a rate of about once every 5 Myrs per L* galaxy, implying that $\sim (5\text{-}10)\gamma_{10}\%$ of L* galaxies should display a neutron-decay halo at maximum power.

The beta-decay electrons radiate nonthermal synchrotron emission and Compton scatter CMB photons to high energies. The maximum synchrotron frequency is $\nu \sim 3 \times 10^{20}B(\mu\text{G})\gamma_{10}^2$ Hz, where $B(\mu\text{G})$ is the mean magnetic field in the region surrounding the galaxy in μG. The halo will display a cooling synchrotron spectrum at optical and soft X-ray energies. The electromagnetic cascade formed by the Compton-scattered γ rays terminates when the γ rays are no longer energetic enough to pair produce with the diffuse radiation fields. The relative intensity of the synchrotron and Compton components depends on the magnitude of $B(\mu\text{G})$. The best prospect for discovering neutron-decay halos is by optical observations of field galaxies that display active star formation [3].

We also note that the emission of nonthermal synchrotron and Compton radiation from photopion processes by UHECRs traveling through intergalactic space will produce a nonthermal component of the diffuse radiation background, irrespective of the sources of UHECRs.

References

1. E. Waxman: *Phys. Rev. Lett.* **75**, 386 (1995)
2. M. Vietri: *Astrophys. J. Lett.* **453**, 883 (1995)
3. C.D. Dermer: *Astrophys. J.*, submitted (2000), astro-ph/0005440
4. C.D. Dermer, M. Humi: *Astrophys. J.*, submitted (2000), astro-ph/0012272
5. T. Stanev, R. Engel, A. Mücke, R.J. Protheroe, R. J., and J.P. Rachen: *Phys. Rev. D*, **62**, 093005
6. M. Böttcher and C.D. Dermer: *Astrophys. J.* **529**, 635 (2000)
7. C.D. Dermer: In *Heidelberg 2000 High-Energy Gamma-Ray Workshop*, ed. F.A. Aharonian and H. Völk (AIP, New York, 2000), astro-ph/0010564
8. A.M. Hillas: *Ann. Rev. Astron. Astrophys.* **22**, 425 (1984)
9. R. Schlickeiser and C.D. Dermer: *Astron. Astrophys.* **360**, 789 (2000)
10. C.D. Dermer: *Astrophys. J. Lett.*, submitted (2000), astro-ph/0012490

On the Neutrino Flux from Gamma-Ray Bursts

Dafne Guetta[1], Maddalena Spada[1], and Eli Waxman[2]

[1] Osservatorio Astrofisico di Arcetri, L. E. Fermi 5, Firenze, Italy
[2] Weizmann Institute of Science, Rehovot 76100, Israel

Abstract. Observations imply that gamma-ray bursts (GRBs) are produced by the dissipation of the kinetic energy of a highly relativistic fireball. Photo-meson interactions of protons with γ-rays within the fireball dissipation region are expected to convert a significant fraction of fireball energy to $> 10^{14}$ eV neutrinos. We show that the fraction of fireball energy converted in this process to high energy neutrinos is not very sensitive to uncertainties in fireball model parameters, such as the expansion Lorentz factor and characteristic variability time. This is due in part to the constraints imposed on fireball parameters by observed GRB characteristics, and, more important, to the fact that for parameter values for which the photo-meson optical depth is high (implying high proton energy loss to pion production) neutrino production is suppressed by pion and muon synchrotron losses and by competition between two-photon annihilation and photo-meson interactions. The neutrino flux is therefore expected to be correlated mainly with the observed γ-ray flux.

1 Introduction

Within the fireball model framework, observed γ-rays are produced by synchrotron emission of electrons accelerated to high energy by internal shocks within the expanding wind. In the region where electrons are accelerated, protons are also expected to be shock accelerated: Plasma parameters in the dissipation region allow proton acceleration to $> 10^{20}$ eV. A natural consequence of proton acceleration to high energy is the production of a burst of $> 10^{14}$ eV neutrinos [1], produced by the decay of charged pions created in interactions between fireball photons and high energy protons. The predicted flux of neutrinos is determined by the fraction f_π of fireball proton energy lost to pion production which is given by [1].

$$f_\pi(E_p) \approx 0.2 \min(1, E_p/E_p^b) \frac{L_{\gamma,52}}{\Gamma_{2.5}^4 \Delta t_{-2} E_{\gamma,\mathrm{MeV}}^b}. \tag{1}$$

This value is strongly dependent on Γ. It has recently been pointed out by [2] that if the Lorentz factor Γ varies significantly between bursts, then we can have large burst to burst fluctuations in the neutrino bursts.

The main goal of the present work is to determine the allowed range of variation in the fraction of fireball energy converted to high energy neutrinos, under the assumption that GRBs are produced by internal dissipation shocks in an ultra-relativistic wind [3].

In this work, we use the model described in [1] The wind flow with an average luminosity $L_w \sim 10^{52} \text{erg s}^{-1}$, produced by a source of $R_0 \sim 10^6$ cm, is approximated as a set of discrete shells. Each shell is characterized by four parameters: ejection time t_j, where the subscript j denotes the j-th shell, Lorentz factor Γ_j, mass M_j, and width Δ_j. We consider shells of initial thickness $\Delta_j = ct_d = R_0$, ejected from the source at an average rate t_v^{-1}. In [1] we have shown that in order to obtain γ-ray flux and spectrum consistent with observations, large variance is required in wind Lorentz factor distribution. We therefore restrict the following discussion to the bimodal case, where Lorentz factors are drawn from a bimodal distribution, $\Gamma_j = \Gamma_m$ or $\Gamma_j = \Gamma_M \approx \eta_* \gg \Gamma_m$, with equal probability and the shell masses are taken equal. Once shell parameters are determined, we calculate the radii where collisions occur and determine the photon and neutrino emission from each collision following the procedure given in [1].

The fraction $f_\pi(E_p)$ of proton energy lost to pion production is estimated as $f_\pi = \min(1, \Delta t / t_\pi)$ where Δt is the comoving shell expansion time and t_π is the proton photo-pion energy loss time calculated as [1]. Photo-meson production of low energy protons, well below E_p^b, requires interaction with high energy photons, well above E_γ^b. Such photons may be depleted by pair production. For each collision we find the photon energy $E_\gamma^\pm$, for which the pair production optical thickness, $\tau_{\gamma\gamma}$, is unity. A large fraction of photons of energy exceeding $E_\gamma^\pm$ will be converted to pairs, and hence will not be available for photo-meson interaction, leading to a suppression of the neutrino flux at low energies. In order to take this effect into account, we use $f_\pi = \min[1, \Delta(t/t_\pi) \min(1, \tau_{\gamma\gamma}^{-1})]$. Neutrino production is suppressed at high energy, where neutrinos are produced by the decay of muons and pions whose lifetime τ exceeds the characteristic time for energy loss due to synchrotron emission. We therefore define an effective f_π, $f_{\pi,\text{eff.}}$, as $4f_\pi$ times the fraction of pions' energy converted to muon neutrinos. In the absence of pion and muon energy loss, $\approx 1/4$ of the pions' energy is converted to muon neutrinos, since $\approx 1/2$ the energy of charged pions is converted to $\nu_\mu + \bar{\nu}_\mu$. Thus, $f_{\pi,\text{eff.}}$ is the fraction of proton energy that, in the absence of synchrotron losses, leads to approximately the same muon neutrino flux as that obtained when synchrotron losses are taken into account.

2 Results and Discussion

In the framework of the model described above, we determine the dependence of $f_{\pi,\text{eff.}}$ on wind model parameters which is showed in Figure 1. Over the range of wind model parameters, which produce photon spectrum consistent with observations (bounded by the dashed line in the figures), the value of $f_{\pi,\text{eff.}}$ at high neutrino energy, $E_\nu > E_\nu^b$, is within the range of $\approx 5\%$ to $\approx 15\%$, close to the value of $\approx 20\%$ estimated in [1]. The weak dependence of $f_{\pi,\text{eff.}}$ on wind model parameters, in contrast with the strong dependence implied by Eq. (1), is due to two reasons. First, for low values of Γ and Δt, where large values of f_π are implied by Eq. (1), f_π is suppressed at low proton energy by competition between photo-meson and two-photon pair production interactions, and only a

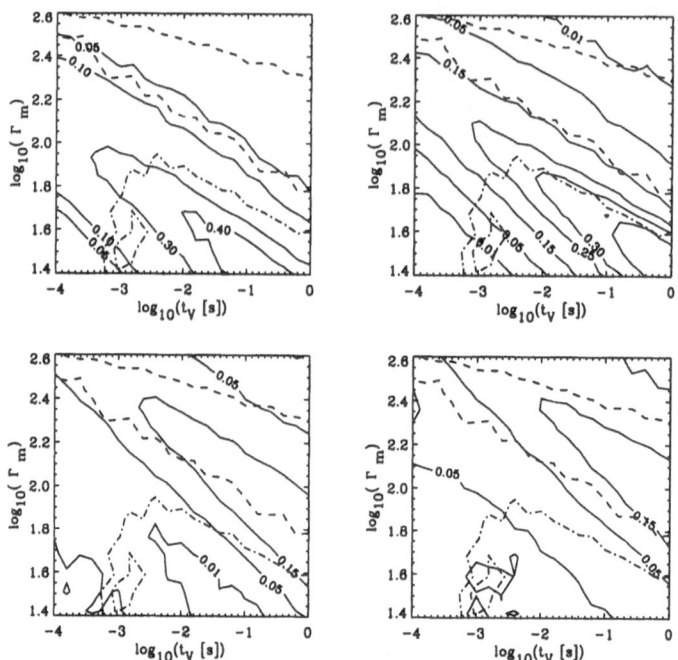

Fig. 1. Contour plots of $f_{\pi,\text{eff.}}$, as function of wind variability time t_v and minimum Lorentz factor Γ_m, for wind luminosity $L_w = 10^{52}\,\text{erg/s}$. The four panels correspond to four observed neutrino energy bins, clockwise from top left: $10^{14}\text{eV} < E_\nu < 10^{15}\text{eV}$, $10^{15}\text{eV} < E_\nu < 10^{16}\text{eV}$, $10^{17}\text{eV} < E_\nu < 10^{18}\text{eV}$, $10^{18}\text{eV} < E_\nu < 10^{19}\text{eV}$. The region in Γ_m–t_v plane where $E_\gamma^b > 0.1$ MeV is bound by the dashed lines. The dash-dotted lines outline the region in which the fraction of wind energy converted to radiation exceeds 2% (higher fraction is obtained at larger Γ_m values).

small fraction of the pions' energy is converted to neutrinos at high proton energy due to pion and muon synchrotron losses. Second, the observational constraints imposed by γ-ray observations imply that wind model parameters (Γ, L, Δt) are correlated.

We note, in particular, that the value of $f_{\pi,\text{eff.}}$ does not significantly exceed 20% also in wind model parameter regions which are outside the parameter region implied by observations. Our results imply that GRB neutrino flux of individual bursts should correlate mainly with the burst γ-ray flux.

References

1. Waxman, E., & Bahcall 1997, J. N., Phys. Rev. Lett. **78**, 2292.
2. Halzen, F. & Hooper, D. W. 1999, ApJ, **527**, L93.
3. Guetta, D., Spada, M. & Waxman, E., in preparation.
4. Guetta, D., Spada, M. & Waxman, E., astro-ph/0011170.

Efficiency and Spectrum
of Internal γ-Ray Burst Shocks

Maddalena Spada[1], Dafne Guetta[1], and Eli Waxman[2]

[1] Osservatorio Astrofisico di Arcetri, L. E. Fermi 5, Firenze, Italy
[2] Weizmann Institute of Science, Rehovot 76100, Israel

Abstract. We present an analysis of the Internal Shock Model of GRBs, where gamma-rays are produced by internal shocks within a relativistic wind. We show that observed GRB characteristics impose stringent constraints on wind and source parameters. We find that a significant fraction, of order 20%, of the wind kinetic energy can be converted to radiation, provided the distribution of Lorentz factors within the wind has a large variance and provided the minimum Lorentz factor is > 100. For a high efficiency wind, a radiation spectral energy break in the 0.1–1 MeV range is obtained due to pair-production optical depth for source sizes $R/c < 1$ ms, suggesting a possible explanation for the observed clustering of spectral break energies in the 0.1–1 MeV range. Natural consequences of the model are absence of bursts with peak emission energy significantly exceeding 1 MeV, and existence of low luminosity bursts with low, 1 keV to 10 keV, break energies.

1 Introduction

The main goal of this work is to address the questions of whether, and under which conditions, high radiative efficiency can be obtained in the framework of the internal shock model, and whether the clustering of peak emission energies between 50 keV and 300 keV can be naturally explained by the model [1].

We have analyzed a model of GRBs, in which a compact source of linear scale R_0 produces a wind characterized by an average luminosity L_w and mass loss rate $\dot{M} = L_w/\eta c^2$ [2]. Variability of the source results in fluctuations in the wind saturation Lorentz factor Γ, leading to internal shocks in the expanding wind. These shocks reconvert a fraction of the kinetic energy back to internal energy, which is assumed to be radiated as γ-rays by synchrotron (and inverse-Compton) emission of shock-accelerated electrons. Since the wind duration, $t_w \sim 10$ s, is much larger than the dynamical time of the source, $t_d \sim R_0/c$, variability of the wind on a wide range of time scales, $t_d < t_v < t_w$, is possible. For simplicity, we have assumed that the wind is characterized by a single time scale $t_v > t_d$. We approximate the wind flow as a set of discrete shells. Each shell is characterized by four parameters: ejection time t, Lorentz factor Γ, mass M, and width Δ. Once the injection features are determined, we calculate the radii where collisions occur and determine the emission from each collision. We assume that following each collision the two colliding shells merge and continue to expand as a single shell. The energy released in each shock is distributed among electrons, magnetic field and protons with fractions ϵ_e, ϵ_B and $1 - (\epsilon_e + \epsilon_B)$ respectively.

In the following we adopt electron and magnetic field energy fractions close to equipartition, ϵ_e=0.45, $\epsilon_B = 0.1$. The particles are accelerated by the shocks to a power law distribution with index -2.

The photons radiated can be scattered by the electrons within the shell. The optical depth to Thomson scattering may be increased significantly beyond the value derived taking into account cooled shell electrons, due to the production of $e^{\pm}$ pairs. This contribution has not been taken into account in previous analyses of the internal shock model, since for a uniform distribution of Lorentz factors, and under the hypothesis of equipartition, the break energy, $h\nu_{sy}$, in the shell co-moving frame is well below the pair production threshold. However, since the photon spectrum extends to high energy as a power law with spectral index $-p/2 \approx -1$, there is an equal number of photons per logarithmic energy intervals, and there may exist a large number of photons beyond the pair production threshold. The pairs produced by these photons would have low energy (compared to the electron rest mass) since most pair-production interactions occur near the threshold, and the produced pairs may therefore contribute significantly to the Thomson optical depth.

2   Results and Discussion

In the framework of the internal shock model described above, we determine the dependence of the wind radiative efficiency ϵ_γ, defined as the fraction of wind energy converted to radiation, and peak emission energy ε_p, the photon energy at which the maximum of νf_ν is obtained, on the wind model parameters.

In order to obtain high radiative efficiency and peak emission energy $\sim$ 1 MeV, the minimum radius R_i at which internal collisions in the expanding wind occur is required to be similar to $R_{\pm}$, the radius where the Thomson optical depth due to $e^{\pm}$ pairs produced by shock synchrotron emission equals unity:

$$R_{\pm} \approx 10^{13} \mathrm{cm}\, \epsilon_e^{1/2}\, L_{w,52}^{1/2}\, t_{v,-3}^{1/2}\, (\Gamma_{\mathrm{M},3}\Gamma_{\mathrm{m},2})^{-1/4}. \qquad (1)$$

and a large variance (compared to the mean) of the colliding shells' Lorenz factor (LF) distribution is required. The constraint $R_i \sim R_{\pm}$ is equivalent to a constraint on the minimum LF, Γ_m, of expanding shells, $\Gamma_m \sim \Gamma_{m\pm} \approx 10^{2.5}(L_{w,52}/t_{v,-3})^{2/9}$, where $L_w = 10^{52}L_{w,52}\mathrm{erg\,s}^{-1}$ and $t_v = 10^{-3}t_{v,-3}$ s. Large variance in the LF distribution of colliding shells than requires a non-uniform LF distribution: as a bimodal distribution of shell LFs ($\Gamma_j = \Gamma_m$ or $\Gamma_j = \Gamma_M$ with equal probability) or a truncated log-normal distribution, with Γ_M, the maximum LF of wind shells, close to the upper limit set by the shell acceleration process, $(\Gamma < \eta_* \approx (\sigma_T L_w/4\pi m_p c^3 R_0)^{1/4} = 2 \times 10^3(L_{w,52}/R_{0,6})^{1/4}$, where $R_0 = 10^6 R_{0,6}$ cm).

Fig. 1 shows the contour plots of ϵ_γ and ε_p in the case of a bimodal distribution of shell LFs. For high value of Γ_m the shocks are optically thin, decreasing Γ_m leads to a decrease in the initial radius of the collisions, increasing the efficiency up to $\approx 15\%$ and the peak emission energy up to 0.1–1 MeV. At lower values of $\Gamma_m < \Gamma_{m\pm}$, the shocks become optically thick due to pair production

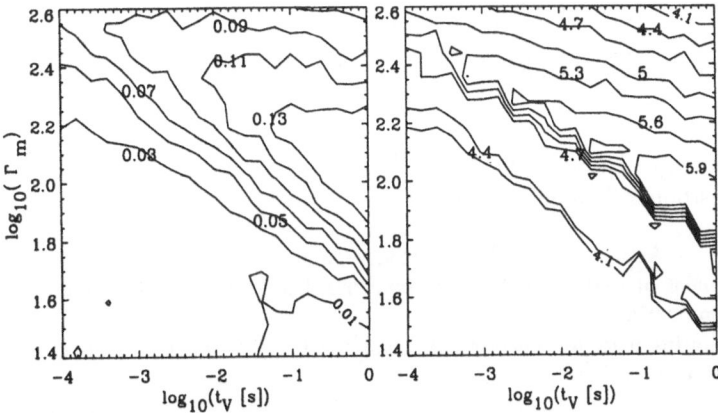

Fig. 1. Contour plots of the radiative efficiency (left panel) and peak emission energy (right panel) as function of t_v and Γ_m. The LF distribution is bimodal and the shell have equal masses and time indipendent shell tickness $\Delta = R_0$.

($\Gamma_m = 50$ to $\Gamma_m = 130$ for t_v varying between 1 s and 10^{-4} s respectively). A small decrease in Γ_m results in collisions below leading to a steep decrease in peak emission energy, and radiative efficiency.

High radiative efficiency and peak emission energy consistent with observations are therefore obtained for $\Gamma_m \sim \Gamma_{m\pm}$. This does not necessarily imply that fine tuning of this model parameter is required. For $\Gamma_m < \Gamma_{m\pm}$, most efficient collisions occur at radii where the optical depth is high, leading to low efficiency, and hence low luminosity, bursts with peak emission energy ~ 1 keV (see figure 1), which would not have been detected by BATSE. For Γ_m significantly higher than $\Gamma_{m\pm}$, LF variance is small, leading to low efficiency, low luminosity bursts with peak emission energy ~ 10 keV, which may have been difficult to detect with BATSE. Natural consequences of the model considered here are therefore absence of bursts with peak emission energy significantly exceeding ~ 1 MeV, and existence of low luminosity bursts with low, ~ 1 keV to ~ 10 keV, peak emission energy. The frequency of such bursts depends on the distribution of Γ_m in different winds.

References

1. Guetta, D., Spada, M. & Waxman, E., astro-ph/0011170
2. Mészáros, P., & Rees, M.J. 2000, ApJ, **530**, 292

Observational Consequences of $e^{\pm}$ Pair Creation in γ-Ray Bursts

Enrico Ramirez-Ruiz[1], Piero Madau[1,2], Martin J. Rees[1], and Christopher Thompson[3]

[1] Institute of Astronomy, Madingley Road, Cambridge, CB3 0HA.
[2] Department of Astronomy and Astrophysics, University of California, Santa Cruz, CA 95064, USA.
[3] Canadian Institute for Theoretical Physics (CITA), 60 St. George St., Toronto, M 5S 3H 8, Canada

Abstract. Physical models for γ-ray emission from a relativistic fireball and the ensuing synchrotron emission from the decelerating shock have generally neglected the feedback of the intense γ-ray flux on the dynamics of the fireball, or on the prompt and delayed emission. There are, however, several reasons to believe that this interaction can have an important influence on the observed multi-wavelength emission from γ-ray bursts. Pair creation by high energy photons will raise the radiative efficiency of a shock and, if the ambient medium is sufficiently dense, it will significantly modify or even thermalize the original spectrum. The resulting spectral modifications could be indicative of birth in a dense environment from a massive progenitor.

1 An Overview of the Generic Model

A generic scheme for a cosmological GRB model has emerged in the last few years. According to this scheme the observed γ-rays are emitted when an ultra-relativistic energy flow is converted to radiation. Possible forms of the energy flow are kinetic energy of ultra-relativistic particles or electromagnetic Poynting flux. This energy must be converted to radiation in an optically thin region, as the observed bursts are not thermal. However, the observed spectra are hard, with a significant fraction of the energy above the $\gamma\gamma \rightarrow e^{\pm}$ formation energy threshold, and a high compactness parameter can result in new pairs being formed outside the originally optically thin shocks responsible for the primary radiation. In this paper, we present a simplified discussion of the generic effects of pair formation arising when γ-rays are generated in internal shocks (Dermer & M. Böttcher 2000 considered the effect of pair formation for an external shock model of GRB), which can have high compactness parameters resulting in pair cascades and a large pair multiplication factor [1].

The simplest model of relativistic energy flow is in the form of the kinetic energy of a shell of relativistic particles with a width Δ. The kinetic energy is converted to "thermal" energy of relativistic particles via shocks. These particles then release this energy and produce the observed radiation. There are two

---

[1] Further details of this work may be found in Ramirez-Ruiz et al. (2001) and in Mészáros et al. (2001).

modes of energy conversion (i) External shocks, which are due to interaction with an external medium and (ii) internal shocks that arise due to shocks within the flow when fast moving particles catch up with slower ones. External shocks arise from the interaction of the shell with external matter. The typical length scale is the Sedov length, $l \equiv (E/n_{ext}m_pc^2)^{1/3}$. The rest mass energy within a sphere of radius l equals the energy of the shell. Typically $l \sim 10^{18}$cm. The relativistic external shocks convert a significant fraction of their kinetic energy at $r_\Delta = l/\gamma^{2/3} \approx 10^{15} - 10^{16}$cm. Internal shocks occur when one shell overtakes another. If the initial separation between the shells is δ and the difference in Lorentz factors is of the order γ, these shocks take place at: $\delta\gamma^2$. A typical value is $10^{12} - 10^{14}$cm.

2 $e^{\pm}$ Pair Formation in the Generic Model

As shown by Madau & Thompson (2000), Thompson & Madau (2000) and Madau, Blandford & Rees (2000), a strong burst of radiation will have important dynamical effects on the surrounding ISM. Optically thin material overtaken by an expanding photon shell at radius r will develop a large bulk Lorentz factor $\gamma_{\pm}$ when the energy deposited by Compton scattering exceeds the rest-mass energy of the scatterers, i.e. when $S = \frac{E\bar{\sigma}}{\Omega_b r^2 \mu_e m_p c^2} \gg 1$, where Ω_b is the burst beaming angle, μ_e is the molecular weight per electron, m_p the proton mass, and $\bar{\sigma}$ is the spectrum-weighted total cross section. The pair cascades saturate after the pair screen reaches a maximum Lorentz factor below that of the fireball producing the original burst, $\gamma_{\pm} \sim \cosh(\ln(1 + S)) \ll \gamma$. Both the energy deposited by Compton scattering S and the maximum $\gamma_{\pm}$ decrease with distance from the source, and the $e^{\pm}$ pair formation process shuts off for $S \leq 1$ at a radius, $r_c(S \sim 1) \approx 6 \times 10^{14}$ cm $\left(\frac{4\pi E}{10^{53}\,\Omega_b\,\text{ergs}}\right)^{1/2}$ where the maximum $\gamma_{\pm} \sim 1$ ($\mu_e = 2$). When pairs are produced in sufficient numbers, i.e. when $2m_e n_{e+} \gg m_p n_p$, the mean mass per scattering charge drops to $\sim m_e$ and runaway pair production may occur well beyond the radius r_c by as much as a factor $(\mu_e m_p/m_e)^{1/2}$. Because of the reduced inertia per particle, and also because pair-producing collisions impart direct momentum to the gas, such a pair-loaded plasma may, under some circumstances, be more efficiently accelerated to relativistic bulk velocities than a baryonic gas.

The pair production process depends only on the "seed" γ-ray photon, while its manifestations depend on the external density and on the bulk Lorentz factor γ. The external baryon density n_{ext} determines the optical depth that can be built up through back-scattering and pair multiplication. For a fixed Lorentz factor γ, the external density determines when the outer shock and the reverse shock become important and whether this happens within the radius already polluted with pairs. There are two rather different cases depending on whether or not $l/\gamma^{2/3}$ is less than r_c. In the former case the external shock responsible for the afterglow occurs beyond the region "polluted" by new pairs (see Case II in Fig. 1), while in the second case the afterglow shock may experience, after starting out in the canonical manner, a *resurgence* as its radiative efficiency is boosted by running into an pair-enriched gas (see Case I in Fig. 1). Additional effects are expected when $\tau_\pm \rightarrow 1$. This requires very high external baryon densities. Such high densities would only be expected if the burst is associated with a massive star in which prior mass loss led to a dense circumstellar envelope. The pair optical depth saturates to $\tau_\pm \sim 1$ and in addition to an increased efficiency and softer spectrum of the afterglow reverse shock, the original gamma-ray spectrum of the GRB will be modified: (i) a quasi-thermal pulse of X-rays is expected to accompany the burst and (ii) the high energy photons will be absorbed due to pair creation, creating a cut-off in the spectrum. Detailed Monte Carlo simulations (Ramirez-Ruiz et al., 2001) should provide a more detailed assessment of the self-consistent spectrum of a GRB in the presence of self-induced pair formation.

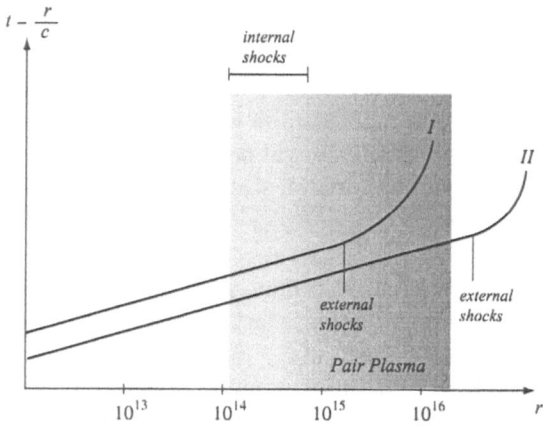

Fig. 1. Two illustrative cases are depicted. In case (II), the external medium has a low density, and the blast wave, with high γ, sweeps up all the pair-enriched medium before it has been much decelerated. In case (I), with higher external density the blast wave is still moving through pair-enriched material during the afterglow [4].

References

1. C. Dermer & M. Böttcher: ApJ, **534**, L155 (2000)
2. P. Madau & C. Thompson: ApJ, **534**, 239 (2000)
3. P. Madau, R. Blandford & M. J. Rees: ApJ, **541**, 712 (2000)
4. P. Mészáros, E. Ramirez-Ruiz & M. J. Rees: ApJ submitted (2001)
5. E. Ramirez-Ruiz et al.: in preparation (2001)
6. C. Thompson & P. Madau: ApJ, **538**, 105 (2000)

The Close Environment of GRB

Peter Mészáros

Pennsylvania State University, University Park, PA 16802, USA

Abstract. We discuss three aspects of the interaction between GRB and their immediate surroundings. Pair production induced by the initial gamma-rays in the nearby
environment will modify the initial spectrum and the afterglow light curve, and the
magnitude of these changes provides a diagnostic for the external density. The presence
of very large dust column densities, capable of obscuring the GRB optical afterglow,
will lead to characteristic X-ray and far-IR light curve signatures. The illumination
of the progenitor remnant and/or the surroundings by the X-ray afterglow continuum
can produce substantial Fe K-alpha line and edge emission, with implications for the
progenitor model.

1 Pair Production Effects in the GRB Environment

Gamma-ray burst sources with a high luminosity can produce $e^{\pm}$ pair cascades in
their environment as a result of back-scattering of a seed fraction of their original
hard spectrum. New pairs can be made as some of the initial energetic photons
are backscattered and interact with other incoming photons. Previous work on
this investigated the acceleration of new pairs for a particular fireball model
[1,4], the effect of pair formation for a low compactness parameter external shock
model of GRB [3], and Compton echos produced by pairs [2]. Here we discuss a
simplified analytical treatment [5] of pair effects from γ-rays arising in internal
shocks in a wind; the remaining wind energy drives a blast wave which decelerates
as it sweeps up the external medium, and gives rise to the afterglow emission.
The γ-rays would propagate ahead of the blast wave, leading to pair production
(and an associated deposition of momentum) into the external medium. The pair
cascades saturate after the external (pair-enriched) medium reaches a critical
bulk Lorentz factor, which is generally below that of the original relativistic wind.
For external baryonic densities similar to those in molecular clouds the pairs can
achieve scattering optical depths $\tau_{\pm} \lesssim 1$. Even for less extreme external densities
the effect of the additional pairs can be substantial, increasing the radiative
efficiency of the blast wave and leading to distortions of the original spectrum.
This provides a potential tool for diagnosing the compactness parameter of the
bursts and thus the radial distance at which shocks can occur. It also provides a
tool for diagnosing the baryonic density of the external environment, and testing
the association with star-forming regions.

Considering the maximum Lorentz factor to which an $e^{\pm}$ can be accelerated
by scattering, and the maximum Lorentz factor at which back-scattered photons
can still make new pairs, one finds two regimes defined by the effective duration

of the light pulse seen by the screen of accelerated pairs. These are, at low radii, the wind regime where the effective duration is the burst duration t_w, and for large radii the impulsive regime, when the effective pulse duration is $\Delta t \sim r/c\Gamma_{\pm}^2$. For an incident photon number index $\beta = 2$, in the former $\Gamma_{\pm} \propto r^{-1/3}$ and in the latter $\Gamma_{\pm} \propto r^{-2}$. The critical radius and Lorentz factor for the transition between the wind and the impulsive dominated regimes are [5]

$$r_c = 5 \times 10^{14} L_{w50}^{2/5} t_{w1}^{3/5} \quad , \quad \Gamma_c = 3 \times 10^1 L_{w50}^{1/5} t_{w1}^{-1/5}, \tag{1}$$

The maximum radius at which pair cascades cut off is

$$r_\ell \sim (4r_* ct_w/3)^{1/2} \sim 4 \times 10^{15} L_{w50}^{1/2} t_{w1}^{1/2} \text{ cm}. \tag{2}$$

Before the pairs start accelerating, assuming they are held back by the environmental protons through magnetic fields, an initial cascade amplification factor $k_p \sim (m_p/m_e)$ is achieved. After the mean mass per scatterer drops to a value comparable to the electron mass, before reaching r_ℓ a further amplification $k_a \sim 2^s \sim 50$ (where $s \sim log\Gamma_c/\log 2$) is possible, so the total pair amplification factor is [5] $k_c = k_p k_a(r_c) \sim (m_p/m_e)\Gamma_c \sim 5 \times 10^4 L_{w50}^{1/5} t_{w1}^{-1/5}$. The maximum pair optical depth at r_c, which is prevented from exceeding $\tau_{\pm} \sim 1$ by self-shielding, is achieved for external densities $n_p \gtrsim n_{p,c}$, where

$$n_{p,c} \simeq 10^5 L_{w50}^{-3/5} t_{w1}^{-2/5} \text{cm}^{-3} . \tag{3}$$

The external density (along with the initial Lorentz factor η) determines when the outer shock and the reverse shock become important and whether this happens within the radius already polluted with pairs (and pre-accelerated by radiation pressure before the shock hits). There are two rather different cases depending on whether or not η^2 is less than r_ℓ/ct_w). In the former case the external shock responsible for the afterglow occurs beyond the region "polluted" by new pairs, while in the second case the afterglow shock may experience, after starting out in the canonical manner, a "resurgence" or second kick as its radiative efficiency is boosted by running into an $e^{\pm}$-enriched gas [5]. Internal shocks in the pair-dominated external plasma can lead to self-absorbed radiation at $\sim 10^9 - 10^{10}$ Hz, while the swept-up pairs can also contribute a $10^{11} - 10^{12}$ Hz 'prompt' signal, which precedes the onset of the standard deceleration afterglow phase.

Additional effects are expected when $\tau_{\pm} \to 1$. This requires external baryon densities at radii $r < r_\ell$ of $n_p \gtrsim n_{c,p} \sim 10^5 L_{w50}^{-2/5} t_{w1}^{-3/5} \text{cm}^{-3}$. Such high densities would only be expected if the burst is associated with a massive star in which prior mass loss led to a dense circumstellar envelope The pair optical depth saturates to $\tau_{\pm} \sim 1$ and in addition to an increased efficiency and softer spectrum of the afterglow reverse shock, the original gamma-ray spectrum of the GRB will be modified as well. The specific nature of this spectral modification depends on the value of the luminosity, which influences the bulk Lorentz factor of the reprocessing pair cloud before it has been swept up by the ejecta, and also on the extent to which the outflow is beamed. One of the consequences of such a critical external density leading to $\tau_{\pm} \sim 1$ would be the presence of an X-ray quasi-thermal pulse, whose total energy may be a few percent of the total burst energy [5]. While quasi-thermal X-ray pulses might also arise due to other reasons, e.g. from an underlying optically thick central engine, if the X-ray luminosity scales as expected here and is accompanied by radio or far-IR signals such as discussed, this could be indicative of birth in a dense environment from a massive progenitor.

2   Delayed X-Ray and IR Afterglows from Obscured GRB in Star-Forming Regions

For GRB in large star forming regions, a significant fraction of the prompt X-ray emission will be scattered by dust grains. Since dust grains scatter X-rays by a small angle, time delays of the scattered x-rays will be small (minutes to days, depending on the X-ray energy and the grain size). If the dust column density is substantial, the softer part of the X-ray afterglow on the above timescales will be dominated by the dust scattering, the direct X-ray emission from the blast wave being weaker. This intermediate time, soft(er) X-ray light curve will generally be steeper than the original unscattered X-ray afterglow.

As a specific example [6], consider a typical GRB whose unscattered X-ray light curve is parametrized as $F_0(t) = [1 + (t/100s)]/[1 + (t/100s)^{2.3}]$, with an arbitrary normalization depending on the X-ray energy band. This is represented by the thin line in Fig.1. We assume that the GRB occurs in a large star forming region, of typical radius R about 100pc, where the dust grain populations and optical depths are close to what is observed in our Galactic center region. Thus for numerical estimates we assume that (1) visual extinction is ~ 10, (2) X-rays are scattered preferentially by those dust grains whose size is in the range $a \sim 0.06\mu m$, (3) the optical depth to dust scattering at the X-ray energy ϵ is

$$\tau(\epsilon) = 3 \left(\frac{\epsilon}{1keV} \right)^{-2}. \tag{4}$$

At X-ray optical depths less than few, dust grains of size a will scatter X-rays of energy ϵ by an angle $\vartheta \sim 0.2\lambda/a$, where λ is the X-ray wavelength, $\vartheta(\epsilon) \simeq 4 \times 10^{-3} \left(\frac{a}{0.06\mu m} \right)^{-1} \left(\frac{\epsilon}{1keV} \right)^{-1}$. The corresponding time lag is $t \sim R\vartheta^2/2c$, or

$$t(\epsilon) \sim 9 \times 10^4 s \left(\frac{a}{0.06\mu m} \right)^{-2} \left(\frac{\epsilon}{1keV} \right)^{-2} \left(\frac{R}{100pc} \right). \tag{5}$$

At 2 keV, the optical depth is $\tau \sim 1$. The time lag is $t \sim 2 \times 10^4 s$. The scattered flux is $F_s \sim \tau f/t \sim 0.03$. The unscattered flux at $2 \times 10^4 s$ is $F_0 \sim 10^{-3}$. In the time interval from hours to weeks, the dust scattering dominates the afterglow, and, as shown in Fig.1 , the afterglow is approximately a power law $F \propto t^{-1.75}$ [6]. This is because dust grains of radius $a < 0.06\mu m$ will scatter the prompt emission with longer time lags, $t \propto a^{-2}$, and with smaller optical depths τ. To calculate τ, we take a standard dust grain size distribution where the number of grains of size of order a is $\propto a^{-2.5}$ For a scattering cross section $\propto a^4$ the optical depth is $\tau \propto a^{1.5} \propto t^{-0.75}$, so the flux $F \propto t^{-1.75}$.

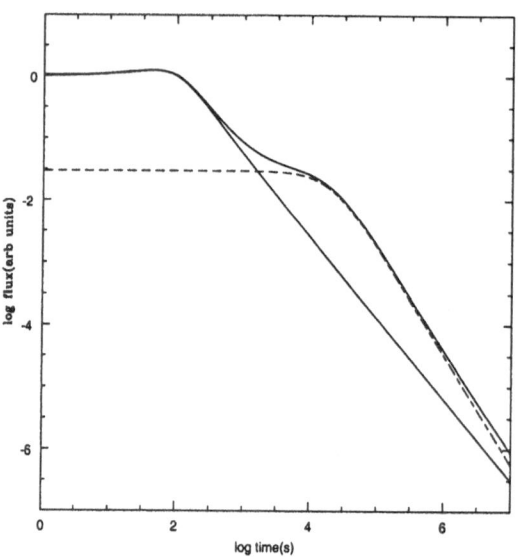

Fig. 1. Dust-scattered X-ray afterglow. Thin line: unscattered X-ray flux. Thin dashed line: scattered X-ray flux. Thick line: total flux. The flux normalization is arbitrary, while the relative fluxes correspond to the example discussed in the text for an energy of 2 keV [6].

A GRB in such a highly obscured star-forming region should lead to specific signatures in the X-ray afterglow light curve, consisting of a secondary flattening or bump in X-ray light curve at energies $\epsilon \sim 2 - 3$ keV [6]. This X-ray signature is expected for bursts which do not produce a detectable optical transient (OT). Since optical observations first require an X-ray position, which so far has been possible only after hours of delay and with arc-minutes accuracy, there is currently a possible bias against finding OTs. Stronger constraints may be however available with the faster coordinate alerts and smaller X-ray error circles expected from dedicated GRB afterglow missions such as HETE-2 and Swift. Such OT-less, X-ray peculiar GRBs will also lead to thermal reemission and scattering of the O/UV flux causing a delayed IR emission, as is the case also for partially absorbed bursts [8,9]. For an isotropic equivalent total burst energy $E \sim 10^{53}$ erg at a redshift $z \sim 1$ the normalization of the X-ray flux for the burst of Figure 1 would be $F_x \sim 10^{-9}$ erg cm^{-2} s^{-1} keV^{-1} for $t \lesssim 100$ s, in the usual range of X-ray afterglow fluxes detected by Beppo-SAX. The dust reradiation occurs beyond the sublimation radius $R_s \sim 10\, L_{49}^{1/2}$ pc at wavelengths $\lambda \gtrsim 2(1+z)\mu$m, where

$10^{49}L_{49}$ erg/s is the early UV component of burst afterglow [8]. The time delay associated with the reradiated flux is $t_{IR} \sim (R_s/2c)\vartheta_j^2$ where $\vartheta_j = 10^{-1}\vartheta_{-1}$ is a typical collimation half-angle of the burst radiation. At $z \sim 1$ the corresponding infrared flux at 2.2 μm would be $F_{2.2\mu m} \sim L_{49}\vartheta_j^2/[4\pi D_L^2(R_s/2c)\vartheta_j^2\nu] \sim 0.3L_{49}^{1/2}$ μJy, independent of ϑ_j, or $m_K \sim 23.3$ compared to Vega [6], approximately constant for a time $t_{IR} \sim 5 \times 10^6\vartheta_{-1}^2 L_{49}^{1/2}$ s. The IR flux of the host galaxy at that redshift could exceed this value, but 8-m class telescopes in good seeing conditions or with adaptive optics would resolve the galaxy, facilitating detection of the point-like IR afterglow. If the X-ray to IR fluxes can be calibrated for a sample of sources at $z \sim 1$, such γ-ray detected GRBs with anomalous X-ray afterglow behavior and no OT may be used as tracers of massive stellar collapses. It may thus be possible to detect star-forming regions out to redshifts larger than those detectable with optical or infrared techniques, since typical GRB γ-ray and X-ray fluxes can in principle be measured out to $z \sim 10 - 15$.

3   Fe K-α Lines as GRB Progenitor Diagnostics: Supranovae or Decaying Long-Lived Jets?

Important clues for identifying the nature of the progenitors of the long $(t \gtrsim 2$ s) GRBs may be available from the recent report at a 4.7σ level of X-ray Fe line features in the afterglow after 1.5 days of the gamma-ray burst GRB 991216 [10], as well as similar detections at the 3σ level in 5 other bursts with Beppo-SAX and ASCA (these proceedings). X-ray atomic edges and resonance absorption lines are theoretically expected to be detectable from the gas in the immediate environment of the GRB, and in particular from the remnants of a massive progenitor stellar system [15,14,16].

A straightforward interpretation [10] of the GRB 991216 observation would imply a mass $\gtrsim 0.1 - 1M_\odot$ of Fe at a distance of about 1-2 light-days, possibly due to a remnant of an explosive event or supernova which occurred days or weeks prior to the gamma-ray burst itself (a 'supranova', [10,12]). The long time delay is necessary both to get the relatively massive, slow moving ejecta out to few light-day distances (to explain the line appearance at a few days with light travel arguments), and in order to get the initial Ni and Co to decay to Fe (half-life 55 days). This requires a two-step process, in which an initial supernova leads to a temporarily stabilized neutron star remnant, which after weeks collapses to a black hole leading to a canonical burst ([11,12]). It is unclear whether fall-back from the supernova leading to the second collapse to a BH could occur with such a ($\sim$ weeks) long delay (e.g. [18]). Another possibility is that a massive progenitor has previously emitted a copious wind ($\dot{M} \gtrsim 10^{-4}M_\odot/$yr), which would need to be unusually Fe-rich and highly inhomogeneous ([14]; c.f. [10]).

An alternative, and perhaps less restrictive scenario for such Fe lines [7] involves an extended, possibly magnetically dominated wind from a GRB impacting the expanding envelope of a massive progenitor star. This could be due either to a spinning-down millisecond super-pulsar or to a highly-magnetised torus around a black hole (e.g. [13]), which could produce a luminosity that was still, one day after the original explosion, as high as $L_m \sim 10^{47} t_{day}^{-1.3}$ ergs. An outflow with such a dependence can also be powered by accretion of fall-back material onto a central black hole [18]. This jet luminosity may not dominate the continuum afterglow; but its impact on the outer portions of the expanding stellar envelope at distances $\lesssim 10^{13}$ cm, even with just solar abundances, can be efficiently reprocessed into an Fe line luminosity comparable to the observed value, together with a contribution to the X-ray continuum. Under this interpretation, the dominant continuum flux in the afterglow, even in the X-ray band, is still attributable to a standard decelerating blast wave.

The magnetised wind from the compact remnant (which we assume to be relativistic) would develop a stand-off shock before encountering the envelope material, and shocked relativistic plasma would be deflected along the funnel walls. Non-thermal electrons are expected to be accelerated behind the standoff shock in the jet material; the transverse magnetic field strength (which decreases as $1/r$ in an outflowing wind) would be of order 10^4 G at 10^{13} cm – strong enough to ensure that the shock-accelerated electrons cool promptly, yielding a power-law continuum extending into the X-ray band. Some of these X-rays would escape along the funnel, but at least half (the exact proportion depending on the geometry and flow pattern) would irradiate the material in the stellar envelope. Pressure balance in the shoked envelope wall implies densities of $n_e = \alpha L_m / 6\pi r^2 ckT \sim 10^{17} \alpha L_{47} r_{13}^{-2} T_8^{-1}$ cm^{-3}, where $\alpha \sim 1$ is a geometric factor, and the recombination time for hydrogenic Fe in the funnel walls photoionized by the non-thermal continuum is $t_{rec} = 6 \times 10^{-6} T_8^{1/2} n_{17}^{-1} \sim 6 \times 10^{-6} \alpha L_{m47}^{-1} r_{13}^2 T_8^{3/2}$ s. Standard calculations of photoionization of optically-thin slabs (e.g. [7]) show that the equivalent width of the Fe K-alpha line, for solar abundances, is about 0.5 kev, or twice as strong if the Fe has ten times solar abundances. These results are applicable provided that the ionizing photons encounter a Fe ion before being scattered by free electrons i.e. provided that $\tau_T = \sigma_T d_i n_e \lesssim 1$. Under these conditions the Fe K-α photon flux is about 0.1 of the X-ray continuum [7],

$$\dot{N}_{LFe} \sim 10^{54} L_{47} \beta \text{ ph/s}, \tag{6}$$

where $\beta < 1$ is the ratio of ionizing to MHD luminosity. This line luminosity compares well with Fe line luminosity 6×10^{52} ph/s observed $t \sim 1.5$ day after the GRB 991216 burst by [10].

The total amount of Fe needed to explain the observed K-α line flux, arising in a thin layer of the funnel walls of a collapsar model, amounts to a very modest mass of $M_{Fe} \sim 10^{-8} M_\odot$, which could be Fe synthesized in the core. The Fe-enriched core material can easily reach a distance comparable to $r \sim 10^{13}$ cm in 1 day for an expansion velocity below the limit $v \sim 10^9$ cm s^{-1} inferred by [10] from the line widths. Even without this, a solar abundance ($10^{-5} M_\odot$ of Fe) in the envelope is sufficient to explain the observations. The initial, energetic portion of the relativistic jet, with a typical burst duration of $1 - 10$ s, will rapidly expand beyond the stellar envelope, leading in the usual way to shocks and a decelerating blast wave. A continually decreasing fraction of energy, such as put out by a decaying magnetar, may continue being emitted for periods of a day or longer, and its reprocessing by the stellar envelope can be responsible for the observed Fe line emission in GRB 991216. Since the energy in this tail can decay faster than t^{-1}, the usual standard shock gamma-ray and afterglow scenario need not be affected, being determined by the first 1-10 s worth of the energy input.

Acknowledgements: Research supported by NASA NAG5-9192, the Guggenheim Foundation and the Sackler Foundation. I am grateful to M.J. Rees, A. Gruzinov and E. Ramirez-Ruiz for valuable discussions on these topics.

References

1. Madau, P & Thompson, C, 2000 ApJ, 534, 239
2. Madau, P, Blandford, R & Rees, M.J., 2000, ApJ, 541, 712.
3. Dermer, C & Böttcher, M., 2000 ApJ 534, L155
4. Thompson, C & Madau, P, 2000 ApJ, 538, 105
5. Mészáros, P, Ramirez-Ruiz, E & Rees, M, 2001, ApJ subm(astro-ph/0011284)
6. Mészáros, P & Gruzinov, A, 2000, ApJL, 543, L35 (astro-ph/0007255)
7. Young, A.J., 1999, Ph.D. thesis, Cambridge University
8. Waxman, E & Draine, B., 2000, ApJ, in press (astro-ph/9909020)
9. Esin, A & Blandford, R.D., 2000, ApJL in press (astro-ph/0003415)
10. Piro, L, et al., 2000, Science, 290, 955)
11. Vietri, M & Stella, L.A., 1998, ApJ 507, L45
12. Vietri, M, Perola, G, Piro, L & Stella, L, 2000, MNRAS 308, L29.
13. Wheeler, J.C, et al., 2000, ApJ in press (astro-ph/9909293)
14. Weth, C, Mészáros, P, Kallman, T & Rees, M.J, 2000, ApJ 534, 581
15. Mészáros, P & Rees, M.J. 1998, MNRAS, 299, L10
16. Böttcher, M & Fryer, C.L, 2000, ApJ, subm. (astro-ph/0006076)
17. Rees, M.J. & Mészáros, P, 2000, ApJ, 545, L73
18. MacFadyen, A, Woosley, S & Heger, A, 2000, ApJ(astro-ph/9910034)

Failed Optical Afterglows

Gabriele Ghisellini[1], Davide Lazzati[1,2], and Stefano Covino[1]

[1] Osservatorio Astronomico di Brera, Via Bianchi 46 I-23807 Merate (Lc), Italy
[2] Present Address: University of Cambridge, Institute of Astronomy,
    Cambridge CB3 0HA, UK

Abstract. While all but one Gamma-Ray Bursts observed in the X-ray band showed an X-ray afterglow, about 60 per cent of them have not been detected in the optical band. We show that this is not due to adverse observing conditions. We then investigate the hypothesis that the failure of detecting the optical afterglow is due to absorption at the source location. We find that this is a marginally viable interpretation, but only if the X-ray burst and afterglow emission and the possible optical/UV flash do not destroy the dust responsible for absorption in the optical band. If dust is efficiently destroyed, we are led to conclude that bursts with no detected optical afterglow are intrinsically different.

1 Observations

Figure 1 shows magnitudes of the detected bursts and upper limits of failed optical afterglows (FOAs), all in the R band, versus the time of observation. Filled and empty circles correspond to *Beppo*SAX and non-*Beppo*SAX bursts with detected optical afterglows, while arrows are upper limits.

The visual inspection of Figure 1 reveals a clear segregation of arrows from dots, the former being systematically fainter than the latter at comparable times. This impression is confirmed by the application of a bidimensional KS test (Press et al. 1992). The probability for the circles (empty + filled) and the arrows being derived from the same parent distribution is $P \sim 0.2$ per cent.

This result shows that in most cases we failed to detect the optical afterglow not because the search was conducted without the necessary depth, but instead because the FOAs are indeed fainter than the detected ones. Yet, it is possible that FOAs are optically fainter because intrinsically less energetics at all wavelengths, or because they are more distant. In order to check this, we compared the X-ray and R band flux densities of bursts with and without optical detection 12 hours after the burst event, finding that the X-ray fluxes of FOAs are not systematically fainter than the fluxes of afterglows with optical detection, indicating that FOAs are indeed optically poor and define a different population with respect to optically detected afterglows.

We have checked that local Galactic extinction does not play a crucial role by comparing the hydrogen column densities in the direction of detected afterglows with those in the direction of FOAs.

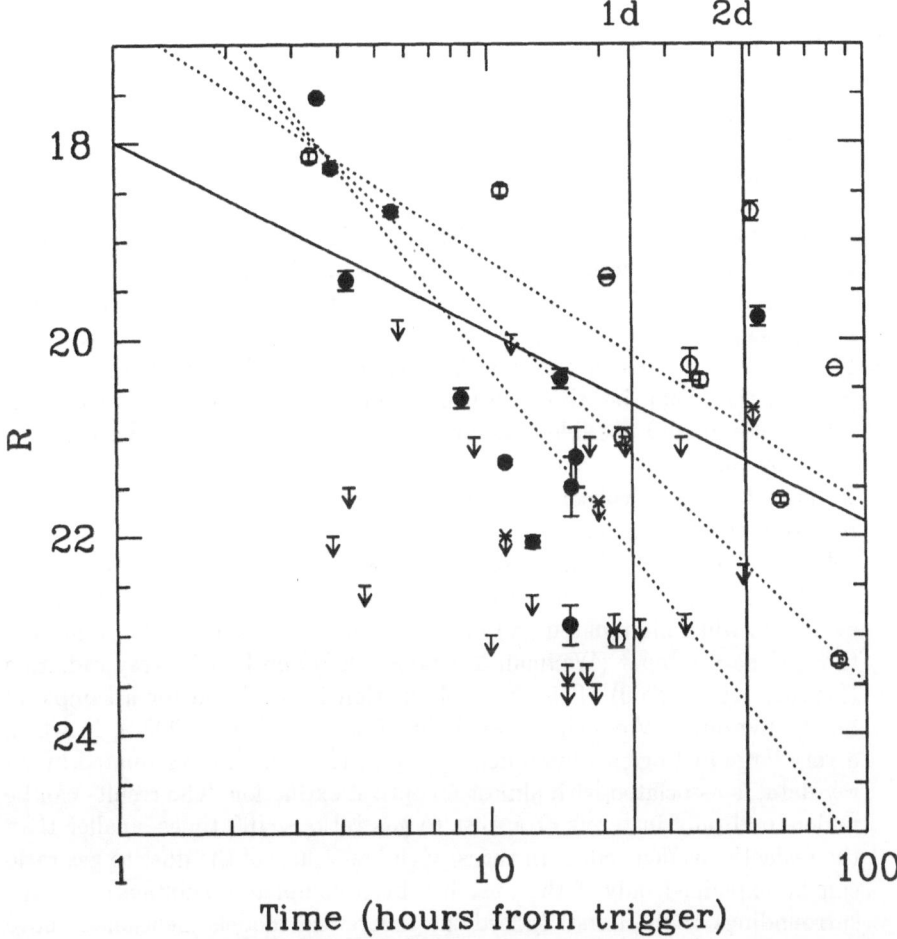

Fig. 1. Detection R magnitude (or upper limits) versus the time of observation for a set of afterglows. Filled circles show optical detections of *Beppo*SAX afterglows while empty circles show detections of non *Beppo*SAX afterglows. Arrows show upper limits for *Beppo*SAX failed optical afterglows. Arrows with crosses refer to the upper limits on γ-ray poor X-ray transients detected by *Beppo*SAX(their inclusion/exclusion from the sample does not alter any of the presented result). The dark solid line is the best fit for the magnitudes of detections vs. time. Dotted lines show the $F_\nu(t) \propto t^{-1}$, $t^{-1.5}$ and t^{-2} relations. From Lazzati et al., 2001.

2 Intrinsic Absorption?

We have investigated the possibility that the difference between the two groups is due to absorption local to the burst. We can quantify in roughly 2 magnitudes the amount of average absorption in the R band needed for more than half of the bursts to go undetected in the optical.

Can a typical molecular cloud produce such an absorption in more than half of the bursts? In order to answer this question we (Lazzati et al. 2001) have computed the *average* (i.e. over many line of sights) and the *maximum* absorption of known molecular clouds in our galaxy, taking into account that observations of bursts in the R band actually correspond to light emitted at shorter wavelengths, where extinction is more effective. Since the redshifts of FOAs is obviously unknown, we have assumed $z = 1$ for all of them.

Our results can be summarized as follows:

- If the burst is located at random within a molecular cloud, it will on average be absorbed with the mean value of the cloud absorption. In this case we found that only a few percent of the burst afterglows could be missed for this reason.

- If bursts are located in star forming regions, then they lie in the densest parts of the cloud, i.e. those with maximum absorption. In this case it is (albeit marginally) possible that up to 60 per cent of the bursts have optical afterglows sufficiently absorbed to have avoided detection. But consider that we have been very conservative in our procedure, because our results are based on considering *upper limits* on the optical flux, and *peak* absorption columns expected in giant molecular clouds.

- The latter assumptions may well be too conservative, if the dust is bound to evaporate when illuminated and heated by the powerful optical/UV flash of the gamma-ray burst (Waxman & Draine 2000) and by its X-ray radiation (Fruchter et al. 2000). This dust sublimation is suggested for a sample of burst afterglows (Vreeswijk et al. 1999, Galama & Wijers 2000), in which a very large hydrogen column density $N_H \gtrsim 10^{22}$ cm^{-2}, as estimated by X-ray data, is associated with almost no optical extinction. The results can be understood only in terms of a dust to gas ratio ~ 100 times smaller than the Galactic average value. In turns, such low values of the dust to gas ratio can be explained only if the dust has been completely sublimated in the surroundings of the burst. Indeed the theoretical models mentioned above predict that dust can be destroyed by the burst emission out to a radius comparable to the dimension of a typical molecular cloud (up to a few tens of parsecs). If this is the case, the material responsible for absorption in FOAs is not the overdense cocoon surrounding the star forming region, but the cloud as a whole (or even less), and the discrepancy between the observed and measured value becomes extremely compelling.

References

1. Fruchter A.S., Krolik J. & Rhoads J., 2000, ApJ submitted
2. Galama T. J. & Wijers R. A. M. J., 2000, ApJ submitted (astro-ph/0009367)
3. Lazzati D., Covino S. & Ghisellini G., 2001, submitted to MNRAS
4. Press W.H., Teukolsky S.A., Vetterling W.T. & Flannery B.P., 1992, Numerical recipes (second edition), Cambridge University Press
5. Vreeswijk P.M. et al., 1999, GCN Circ No 398
6. Waxman E. & Draine B. T., 2000, ApJ, 537, 796

Winds from Massive Stars and the Afterglows of γ-Ray Bursts

Enrico Ramirez-Ruiz[1], Lynnette M. Dray[1], Piero Madau[1,2], and Christopher A. Tout[1]

[1] Institute of Astronomy, Madingley Road, Cambridge, CB3 0HA.
[2] Department of Astronomy and Astrophysics, University of California, Santa Cruz, CA 95064, USA.

Abstract. The observed distribution of optical afterglows with respect to their host galaxies may suggest that some gamma-ray bursts (GRBs) are associated with star forming regions, and therefore with the explosions of massive stars rather than with merging neutron stars. We construct a simple computational scheme to explore the expected contribution of the presupernova ejecta of single Wolf-Rayet (WR) stars to the circumstellar environment. We make specific predictions for the interaction of the relativistic blast wave with the density bumps that arise when the progenitor star rapidly loses a large fraction of its initial mass or when the ejected wind interacts with the external medium and decelerates. A re-brightening of the afterglow with a spectrum redder than the typical synchrotron spectrum (as seen in GRB 970508 and GRB 980326) is predicted.

1 Introduction

Recent observations suggest that long-duration γ-ray bursts (GRBs) and their afterglows are produced by highly relativistic jets emitted in core-collapse explosions. MacFadyen & Woosley (1999) have explored the evolution of rotating helium stars whose iron core collapse does not produce a successful traditional neutrino-powered explosion. In the collapsar model for GRBs, or in any other model involving a massive star, the key to obtaining relativistic motion is the escape of an energy-loaded fireball from the stellar environment. This is aided if the progenitor undergoes a Wolf-Rayet (WR) phase, which is characterized by a strong stellar wind that causes the star to lose enough of its outer layers for the surface hydrogen abundance to become minimal. The radius of a WR star is sufficiently small for the explosion energy to break out before the engine ceases to operate. Because of the intrinsic variations of mass-loss rates in WR evolution, the GRB blast wave expands into shells of varying gas density. The effects of the WR ejecta interacting with the interstellar medium (ISM) can be observed in the wind bubbles around some of these objects. In this paper, we use detailed stellar evolution of WR stars ($M_0 > 35 M_\odot$) to show how the progenitor initial mass (M_0) and metallicity Z affect the density profile of the ambient material and therefore the evolution of the relativistic blast wave [1].

---

[1] Further details of this work may be found in Ramirez-Ruiz et al. (2001).

2   Evolutionary Models of Massive Stars

To explore the expected contribution of the presupernova ejecta of single Wolf-Rayet (WR) stars to the circumstellar environment, we use detailed stellar evolutionary tracks for stars of initial main sequence mass, M_0, from 10–150 $M_\odot$ and for solar metallicity, $Z_\odot = 0.02$, with the enhanced mass-loss rates of Meynet et al. (1994). We carry out the computational simulations with the stellar evolution code first developed by Eggleton (1971). The evolutionary tracks are part of an ongoing series of WR simulations performed by Dray & Tout (2001). In describing the effects of stellar winds on the surrounding medium we used the basic theory for the wind-ISM interaction develop by Castor, McCray & Weaver (1975). During the evolution of a wind-driven circumstellar shell the system has a four-zone structure. From the inside to the outside these are: (I) a supersonic stellar wind with density $\rho(r) = \dot{M}/4\pi r^2 v_\infty$; (II) a hot, almost isobaric region consisting of shocked stellar wind mixed with a small fraction of the swept-up interstellar gas; (III) a thin, dense, cold shell containing most of the swept-up interstellar gas; (IV) ambient interstellar gas of number density n_0. For a typical WR wind expanding into an homogeneous ISM, the free expansion ends when the wind has reached a radius of $8.8 \times 10^{17} \dot{M}_{-5} n_{0,1}^{1/2} v_3^{-1/2}$ cm [2]. Note that if a neighbouring star is present, the density of the surrounding medium can be as large as $10^6 \mathrm{cm}^{-3}$, causing the ejected wind to slow rapidly at a much smaller radius.

3 γ-Ray Burst Afterglows

One can understand the dynamics of the afterglows of GRB in a fairly simple manner, independent of any uncertainties about the progenitor systems, from the relativistic generalisation of the method used to describe supernova remnants. The basic model for GRB afterglow hydrodynamics is a relativistic blast wave expanding into the surrounding medium. In most early discussions [6,2,8], the progenitor star was expected to be surrounded by a substantial $n(r) = Ar^{-s}$ medium at the end of its life, where $s = 2$ for a wind ejected at a constant speed. For the stellar models computed here, we found no significant deviations from an r^{-2} density gradient as a result of the small mass-loss variations from the progenitor star prior to core collapse. The highest deviation from the extrapolated density profile is mainly provoked by the interaction between the winds and the interstellar medium. As time progresses the amount of material swept up increases and the momentum from the wind cannot drive the shell to such high speeds. The ejected mass then accumulates at a corresponding radius, creating an overdense region. The presence of these density bumps in the nearby ambient environment are inherent to the evolution of WR stars. Interestingly, the density ring in the wind profile lies closer to the progenitor for WR stars with low initial mass and low metallicity. This characteristic offers a direct observational test of which stars are likely to produce a GRB.

---

[2] $\dot{M}_{-5} = \dot{M}/10^{-5} M_\odot \mathrm{yr}^{-1}$, $n_{0,1} = n_0/1\mathrm{cm}^{-3}$ and $v_3 = v/10^3 \mathrm{km\ s}^{-1}$

The impact between the forward shock and these high-density regions should be observed as a re-brightening of the afterglow with a typically redder spectrum. For WR stars surrounded by an $s \approx 2$ medium at the end of their life we found $3.0 \times 10^{34} < A < 3.0 \times 10^{37} \text{cm}^{-1}$. For these stellar parameters, a re-brightening of the afterglow as a consequence of the collision of the shock front with the high density spherical shell is observed at a time $\approx 0.25 E_{52}^{-1} r_{\text{shell},17}^{2} A_{35}$ after the burst. Here t_{day} is the observer time measured in days, $E_{52} = E/10^{52}$ ergs, $r_{\text{shell},17} = r/10^{17}$ cm and $A_{35} = A/3.0 \times 10^{35} \text{cm}^{-1}$.

4 Conclusions

Quantitative insight into the formation of GRBs is hindered by the lack of detailed core-collapse calculations. The ground-breaking work of MacFadyen & Woosley (1999) suggests that GRBs are more likely to occur in stars that have lost their hydrogen envelope. Stars with less radiative mass-loss retain a hydrogen envelope in which a poorly collimated jet that loses its energy before breaking through the stellar surface is likely to arise. Highly relativistic jets will not escape red super-giants with radii $> 10^{13}$ cm. By contrast, a focused low-entropy jet that has broken free of its stellar cocoon is likely to arise from a WR progenitor. The total energy observed in γ-rays from GRBs whose redshift has been determined is diverse. One appealing aspect of a massive star progenitor is that the great variety of stellar parameters can probably explain this diversity. Given the need for a large helium core mass in progenitors, the burst formation may be favoured not only by rapid rotation but also by low metallicity. Larger mass helium cores might have more energetic jets, but it is unclear whether they can be expected to be accelerated to large Lorentz factors. Many massive stars may produce supernovae by forming neutron stars in spherically symmetric explosions, but some may fail neutrino energy deposition, forming a black hole in the centre of the star and possibly a GRB. One expects various outcomes ranging from GRBs with large energies and durations, to asymmetric, energetic supernovae with weak GRBs. The medium surrounding a GRB would provide a natural test to distinguish between different stellar explosions.

References

1. J. Castor, R. Weaver, R. McCray: ApJ,200, L107 (1975)
2. R. A. Chevalier, Z.-Y. Li: ApJ, 520, L29 (1999)
3. L. M. Dray, C. A. Tout: in preparation
4. P. P. Eggleton: MNRAS, 151, 351 (1971)
5. A. I. MacFadyen & S. Woosley: ApJ, 524, 262 (1999)
6. P. Mészáros, M. J. Rees, M. Wijers: ApJ, 499, 301 (1998)
7. G. Meynet, A. Maeder, G. Schaller, D. Schaerer, C. Charbonnel: A&AS, 103, 97 (1994)
8. A. Panaitescu, P. Kumar: ApJ 543, 66 (2000)
9. E. Ramirez-Ruiz, L. M. Dray, P. Madau, C. A. Tout: submitted to MNRAS, astro-ph/0012396 (2001)

The Effects of a Gamma-Ray Burst
on Nearby Preplanetary Systems

P. Duggan[1], B. McBreen[1], L. Hanlon[1], L. Metcalfe[2], A. Kvick[3],
and G. Vaughan[3]

[1] Physics Department, University College, Dublin 4.
[2] Astrophysical Division, European Space Agency, Villafranca, Spain.
[3] European Synchrotron Research Facility (ESRF), Grenoble, France.

Abstract. Intense irradiation from a GRB can melt dust balls in nearby preplanetary disks. Iron in the dust is the major absorber of x-rays and gamma-rays. In the case of the preplanetary solar system, a nearby GRB would have melted about 30 Earth masses of dust to form chondrules. It is now possible to create astrophysical conditions of this type in the laboratory using synchrotron radiation. The results of the experiment performed at ESRF are presented. A GRB in a nearby galaxy will reveal preplanetary disks by transient infrared emission from the dust melting process.

1   Introduction

One GRB is known to have occurred in a luminous infrared galaxy [3], and some GRBs are close to star forming regions suggesting a connection with massive stars. Models of 'failed supernova' and 'hypernova' have been proposed [10,7] in which the inner core of a massive rotating star collapses to a black hole while the outer core forms a massive disk or torus that somehow generates a relativistic fireball and GRB. In these models GRBs represent the violent end to massive stars. The role of nearby supernovae that preceded the formation of the solar system have been considered [2] along with the serious consequences for life on Earth of nearby GRBs [8]. The effects of a nearby GRB on preplanetary systems have been considered and chondrules can be formed by this process [6].

Chondrules are millimetre sized, spherical to irregular shaped objects (Fig. 1 (A)) that constitute the major component of most chondrite meteorites that originate in the region between Mars and Jupiter and which fall to the Earth. They appear to have crystallised rapidly from molten or partially molten drops. The properties of the chondrules and chondrites have been exquisitely deduced from an extensive series of experiments and a recent conference has been devoted completely to chondrules [4]. The mineralogy of chondrules is dominated by olivine ($(FeMg)_2SiO_4$) and pyroxene ($(FeMg)SiO_3$) and there is a wide range of compositions for all elements. This diversity is consistent with the melting of heterogeneous precursor solids or dust balls.

The heat source that melted the chondrules remains uncertain and a critical summary of the heating mechanisms was given by Boss[1]. The only model for producing chondrules from outside the solar nebula is flash heating by a nearby GRB [6]. The distance to the source would have to be about 300 light years (or

100 pc) for a GRB output of 10^{53} ergs assuming a minimum value of 2×10^{10} erg g^{-1} required to heat and melt the precursor grains [9].

2 Absorption of X-Rays, γ-Rays and Melting

The absorption of the GRB energy by the gas and dust in the preplanetary nebula occurs by the photoelectric effect and Compton scattering. In gas and dust with solar composition the absorption by Fe is dominant between the K edge at 7.2 keV and about 30 keV. Fe makes the major contribution to the absorption by dust and is the key to chondrule formation [6]. The elements O, Si, Mg and Fe dominate the composition of the chondrules but the composition of the precursor grains has been the subject of much study and speculation. There are a number of chondrule classification systems and McSween[5] recognised two main types i.e., type I or FeO poor and type II or FeO rich.

The thickness of the dust layer converted to chondrules depends on the GRB spectrum which must have significant emission below 30 keV where dust absorption dominates. In the case of solar abundance the thickness of the chondrule layer created is 0.18 g cm^{-2} corresponding to one optical depth for 30 keV x-rays. The layer thickness increases to about 0.8 g cm^{-2} and 2.0 g cm^{-2} for optical depths to 40 keV and 55 keV x-rays with H and He abundances reduced by factors of 3 and 10 respectively. The thickness of the chondrule layer is therefore controlled by the degree of gas depletion from the nebula.

3 Astrophysical Conditions in the Laboratory

There are a few synchrotrons in the world that produce sufficient power in x-ray and γ-rays to melt milligram samples of precursor dust material. We were granted time on the ID11 white beam at the ESRF to test the theory of GRB melting of materials. Precursor pellets of size 3 mm and composition type I, type II and solar abundance were placed in the beam. The samples were rapidly heated to above the liquidus temperature of about 1400 C. The pressure in the container was typical of the nebula at about 10^{-5} of atmospheric pressure. In a few cases the residual air in the container was replaced with hydrogen. The samples were cooled over a period of about 15 minutes from 1400 C to 1000 C by reducing the synchrotron radiation intensity.

The major result is that the samples melted and did not disintegrate because of charging or gas escape. Usually the samples boiled and often had several eruptions as pieces were ejected. The cool remnants generally have the characteristics of chondrules (Fig. 1(B)). The melted samples are being analysed using a scanning electron microscope (SEM), EDX and x-ray diffraction (XRD). The XRD results clearly reveal olivine crystals in many samples. An example of one of the samples with large dark-coloured olivine crystals is given in Fig. 1(B). This experiment is the first ever report on x-ray melting of materials.

4   Infrared Flashes from Chondrule Formation

A GRB in a nearby galaxy ($\approx$ 30 Mpc) could be used to reveal protoplanetary disks because of the transient infrared emission from chondrule formation. In K band, the transient source would be about 1 μJy and good angular resolution is required to separate the transient emission from the galactic background. The transient sources occur over a period of hundreds of years after the GRB.

5   Conclusions

Astrophysical conditions near a GRB source have been created in the laboratory for the first time. The experiment performed at ESRF has successfully melted millimetre-sized samples with composition similar to that of the solar nebula chondrules. The faint transient infrared emission from the dust melting process after a GRB can reveal preplanetary systems in nearby galaxies.

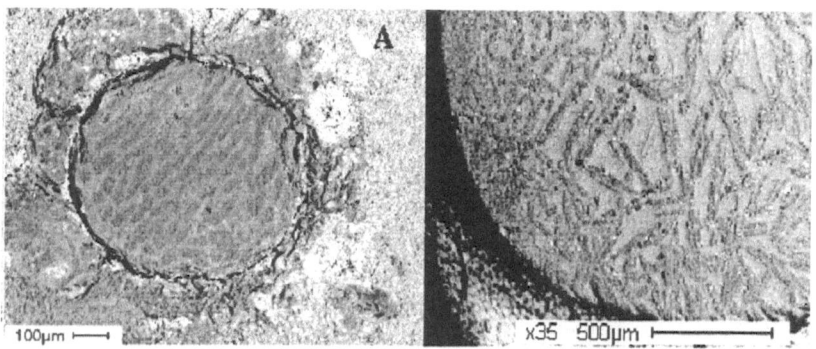

Fig. 1. (A) Backscatter image of Barred Olivine chondrule found in the Allende meteorite. (B) Backscatter image of type I sample melted in a low pressure H_2 atmosphere.

References

1. A.P. Boss: *Chondrules and the Protoplanetary Disk* (Cambridge University Press, Cambridge, 257, Eds. R. H. Hewins et al.,)
2. A.G.W. Cameron, P. Hoeflich, P.C. Myers, D.D. Clayton: ApJ **447**, L53 (1995)
3. L. Hanlon, R.J. Laureijs et al.: A&A **359**, 941 (2000)
4. R.H. Hewins, R.H. Jones, E.R.D. Scott: *Chondrules and the Protoplanetary Disk* (Cambridge University Press, Cambridge)
5. H.Y. McSween: Geochim. Cosmochim. Acta **41**, 1843 (1977)
6. B. McBreen, L. Hanlon: A&A **351**, 759 (1999)
7. B. Paczynski: ApJ **494**, L45 (1998)
8. S.E. Thorsett: ApJ **444**, L53 (1995)
9. J.T. Wasson: Meteoritics **28**, 14 (1993)
10. S.E. Woosley: ApJ **405**, 273 (1993)

The Role of Dust in GRB Afterglows

Donald Q. Lamb

Department of Astronomy & Astrophysics, University of Chicago,
5640 South Ellis Avenue, Chicago, IL 60637

Abstract. We show that the clumpy structure of star-forming regions can naturally explain the fact that 50–70% of GRB afterglows are optical "dark." We also show that dust echos from the GRB and its afterglow, produced by the clumpy structure of the star-forming region in which the GRB occurs, can lead to temporal variability and peaks in the NIR, optical, and UV lightcurves of GRB afterglows.

1 Introduction

There is increasing evidence that the long gamma-ray bursts (GRBs) are associated with galaxies undergoing copious star formation, and occur near or in the star-forming regions of these galaxies (see, e.g., [1] for a discussion of this evidence). Star-forming regions contain large amounts of dust that can extinguish the optical and UV light of GRB afterglows. Indeed, no optical afterglows have been detected for 60-70% of the long GRBs. Some of these failures may be due to the relatively large size of the GRB error box, or to a delay in observing the error box. Some may be because the GRB lies at a very high redshift, and the Lyman limit lies longward of the optical band [2,3]. However, the majority of the failures are most likely because the optical afterglow is faint or absent due to its extinction by dust in the host galaxy of the GRB.

We show that the clumpy structure of star-forming regions can naturally explain the statistics of optically "dark" GRB afterglows. We also show that dust echos from the GRB and its afterglow, produced by the clumpy structure of the star-forming region in which the GRB occurs, can lead to temporal variability and peaks in the NIR, optical, and UV lightcurves of GRB afterglows.

2 Structure of Star-Forming Regions

Star-forming regions are thought to be clumpy, with dense dust clouds embedded in a much less dense intercloud medium. A simple model of star-forming regions can therefore be characterized by three parameters: (1) the amount of dust equivalent to a radial optical depth τ_H of a homogeneous uniform density medium, the volume filling factor f of the dust clumps, and the density contrast k_1/k_2 of the clumps relative to the interclump medium [4,5]. In this model, the dense dust clumps form connected structures, as is true in real star-forming regions.

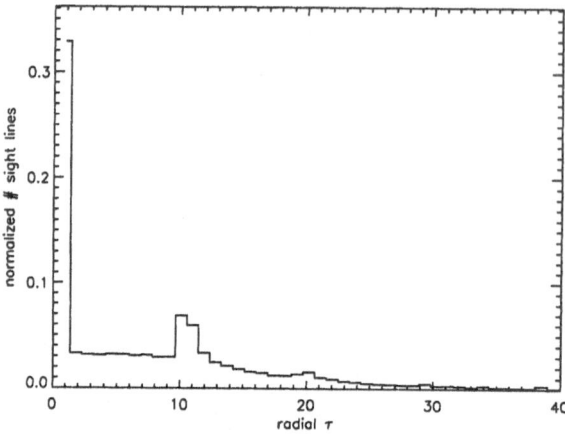

Fig. 1. Distribution of optical depths, as expereinced by random photons emitted isotropically by a central source in a clumpy medium with an amount of dust equivalent to a radial optical depth $\tau_H = 10$ in a homogeneous uniform density medium, a volume filling factor $f = 0.10$ of the clumps, and a density contrast $k_1/k_2 = 100$ between the clumps and the interclump medium. From [4].

Figure 1 shows the of a Monte Carlo calculation of the distribution of optical depths for random lines of sight (LOS) from the center of a clumpy star-forming region to an external observer. The parameters of this particular calculation are $\tau_H = 10$, $f = 0.10$, $k_1/k_2 = 100$, and the number N^3 of spatial bins is 20^3 [4]. These parameter values lead to results that are consistent with observations of star-forming regions in the Milky Way. Figure 1 shows that 35% of LOS have optical depths $\tau_{obs} = 1$ (the minimum value possible in this particular model), while the remaining 65% of LOS have $\tau_{obs} \gg 1$. The distribution of optical depths can be understood as follows. A substantial fraction of the photons emitted by the central source do not encounter a dense clump; these photons correspond to the LOS with $\tau_{obs} = 1$. The remaining photons emitted by the central source encounter one or more clumps and experience $\tau_{obs} \gg 1$. The distribution of τ_{obs} in this model is consistent with the statistics of GRB optical afterglows, 50-70% of which are optically "dark."

3   Dust Echos

Dust echos from the clumpy structure of the star-forming region can also produce variability and peaks $1^d - 30^d$ after the GRB in the NIR, optical, and UV lightcurves of GRB afterglows. As an illustrative example, we have calculated the optical light curve produced by the dust echo from a single clump at a substantial distance from the burst source but along the LOS from the burst source to the observer. The delay time δt between the GRB and the echo is characterized by $\delta t = (1 + z)R_{\perp,min}^2/cD$, where D is the distance between the GRB and the dust cloud and $R_{\perp,min}$ is the minimum of the perpendicular extent of the dust cloud $R_{\perp,cloud}$, $\vartheta_{jet}(t)D$, and $\vartheta_{forward}D$. Here $\vartheta_{jet}(t)$ is the half opening angle of the afterglow jet and $\vartheta_{forward}$ is the half angular width ($\approx 10° - 20°$) of the forward scattering peak for scattering by dust grains.

The amplitude A of the dust echo relative to the direct light from the afterglow is a function of the albedo a of the dust, the scattering phase function

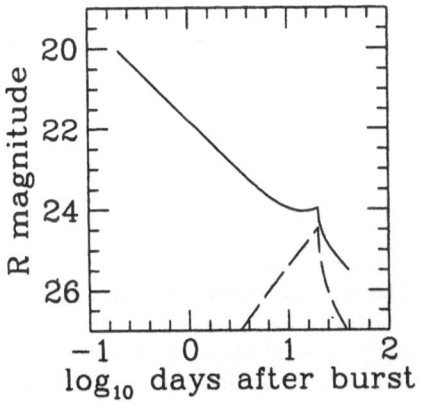
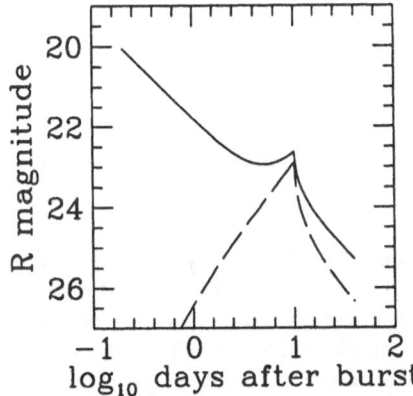

Fig. 2. Contribution to the observed R-magnitude from the light forward-scattered by the dust cloud (dashed line), and the sum of this light and the light seen directly from the point-like source (solid curve) for two sets of parameters of the simple model described in the text. Left: $\delta t = 20$ days $A = 0.2$, $\epsilon = 0.1$ days. Right: $\delta t = 10$ days $A = 0.5$, $\epsilon = 0.1$ days.

$\Phi(\vartheta)$, and the optical depth τ_{clump} of the dust clump, all of which are wavelength dependent. For forward scattering, which we assume here, $\Phi(\vartheta) \approx 1$ [6]. Then in the limiting cases of small and large optical depths, $A \approx a\tau_{\text{clump}}$ and $A \approx a$, respectively. The prominence of the dust echo also depends on the time ϵ after the GRB that the afterglow begins.

Figure 2 shows the dust echo from a single clump along the LOS from the burst source to the observer, assuming $\tau_{\text{clump}} < 1$. In the two cases shown, the rate of temporal decline of the GRB afterglow was taken to be a power law with $b = 1$. Most GRB afterglows decline more rapidly with time (i.e., $b = 1.3 - 2.25$). The dust echo is more prominent for afterglows that decline more rapidly, since the contrast between the direct light from the afterglow and the echo – which reflects the brightness of the afterglow at an earlier time – is then larger. Thus the examples we have shown are conservative.

Studies of the temporal variability of the NIR, optical, and UV lightcurves of the afterglows of GRBs may allow "reverberation mapping" of the structure of the star-forming regions in which GRBs occur.

References

1. Lamb, D. Q. 2000, Phys. Reports, **333**, 505
2. Fruchter, A. S. 1999, ApJ, **512**, L1
3. Lamb, D. Q. & Reichart, D. E. 2000, ApJ, **536**, 1
4. Witt, A. N. & Gordon, K. D. 1996, ApJ, **463**, 681
5. Witt, A. N. & Gordon, K. D. 2000, ApJ, **528**, 799
6. van de Hulst, H. C. 1957, Light Scattering by Small Particles (New York: Wiley), Chapter 9

Theory of GRB Afterglow

T. Piran and J. Granot

Racah Institute of Physics, Hebrew University, Jerusalem 91904, Israel

Abstract. The most interesting current open question in the theory of GRB afterglow is the propagation of jetted afterglows during the sideway expansion phase. Recent numerical simulations show hydrodynamic behavior that differs from the one suggested by simple analytic models. Still, somewhat surprisingly, the calculated light curves show a 'jet break' at about the expected time. These results suggest that the expected rate of orphan optical afterglows should be smaller than previously estimated.

1 Introduction

Our understanding of GRBs has been revolutionized by the BeppoSAX discovery of GRB afterglow. While GRBs last seconds or minutes the afterglow lasts days, weeks months or even years. This makes afterglow observations much richer. These observations provide us with multi-wavelength and multi-timescales data. At the same time the afterglow, which is a blast wave propagating into the surrounding matter is a much simpler phenomena than the GRB and it is possible to construct a simple theory that can be compared directly with the observations.

In this short review we describe the theory of GRB afterglow. We begin with the simplest idealized model and continue with various levels of complications. The final level is full numerical simulations. We present preliminary results of such simulations and compare them with analytic models. At present there is no simple analytic explanation for the features seen in the numerical results.

2 Spherical Hydrodynamics

The theory of relativistic blast waves has been worked out in a classical paper by Blandford & McKee (BM) already in 1976 [1]. The BM model is a self-similar spherical solution describing an adiabatic ultra relativistic blast wave in the limit $\Gamma \gg 1$. The basic solution is a blast wave propagating into a constant density medium. However, Blandford and McKee also describe in the same paper a generalization for varying ambient mass density, $\rho = AR^{-k}$, R being the distance from the center. The latter case would be particularly relevant for $k = 2$, as expected in the case of wind from a progenitor, prior to the GRB explosion.

The BM solution describes a narrow shell of width $\sim R/\Gamma^2$, in which the shocked material is concentrated, where Γ is the typical Lorentz factor. The conditions in this shell can be approximated if we assume that the shell is homogeneous. Then the adiabatic energy conservation yields:

$$E = \frac{\Omega}{3-k} AR^{3-k} \Gamma^2 c^2 \,, \tag{1}$$

where E is the energy of the blast wave and Ω is the solid angle of the afterglow. For a full sphere $\Omega = 4\pi$, but it can be smaller if the expansion is conical with an opening angle ϑ: $\Omega = 2\pi\vartheta^2$ (assuming a double sided jet).

A natural length scale, $l = \left[(3-k)E/\Omega Ac^2\right]^{1/(3-k)}$, appears in equation 1. For a spherical blast wave Ω does not change with time, and when the blast wave reaches $R = l$ it collects ambient rest mass that equals its initial energy, the Lorentz factor Γ drops to 1 and the blast wave becomes Newtonian. The BM solution is self-similar and assumes $\Gamma \gg 1$. Obviously, it breaks down when $R \sim l$. We therefore expect that a Relativistic-Newtonian transition should take place around $t_{NR} = l/c \approx 1.2 \, \mathrm{yr}(E_{\mathrm{iso},52}/n_1)^{1/3}$, where the scaling is for $k = 0$, E_{52} is the isotropic equivalent energy, $E_{\mathrm{iso}} = 4\pi E/\Omega$, in units of 10^{52}ergs and n_1 is the external density in cm^{-3}. After this transition the solution will turn into the Newtonian Sedov-Taylor solution. Clearly this produces an achromatic break in the light curve.

The adiabatic approximation is valid for most of the duration of the fireball. However, during the first hour or so (or even for the first day, for $k = 2$), the system could be radiative (provided that $\epsilon_e \approx 1$). During a radiative phase the evolution can be approximated as:

$$E = \frac{\Omega}{3-k} AR^{3-k}\Gamma\Gamma_0 c^2 \, , \qquad (2)$$

where Γ_0 is the initial Lorentz factor. Cohen, Piran & Sari [2] derived an analytic self-similar solution describing this phase. Cohen & Piran [3] describe a solution for the case when energy is continuously added to the blast wave by the central engine, even during the afterglow phase. A self-similar solution arises if the additional energy deposition behaves like a power law. This would arise naturally in some models, e.g. in the pulsar like model [4].

3   Spherical Afterglow Models

A good model for the observed emission from spherical blast waves can be obtained by adding synchrotron radiation to these hydrodynamic models. Sari, Piran & Narayan [5] used the simple adiabatic scaling (1) together with synchrotron radiation model and the relation between the observer time t, and R:

$$t = R/C_1 c\Gamma^2 \, , \qquad (3)$$

where C_1 is a constant that may vary from 2 to 16 [6].

Assuming a powerlaw energy distribution of the shocked relativistic electrons: $N(E_e) \propto E_e^{-p}$, and that the electrons and the magnetic field energy densities are ϵ_e and ϵ_B times the total energy density, Sari, Piran & Narayan [5] estimate the observed emission as a series of power law segments (PLSs), where

$$F_\nu \propto t^{-\alpha}\nu^{-\beta} \, , \qquad (4)$$

that are separated by break frequencies, across which the exponents of these power laws change: the cooling frequency, ν_c, the typical synchrotron frequency

ν_m and the self absorption frequency ν_{sa}. The analytic calculations were done for a homogeneous shell and for emission from a single representative point. At a specific frequency one will observe a break in the light curve when one of these break frequencies passes the observed frequency. An intriguing feature of this model is that for a given PLS, say for emission above the cooling frequency, there is a unique relation between α, β and p. The power law index p is expected to be a universal quantity as it depends on the, presumably common, acceleration processes and it is expected to be between 2 and 2.5 [7]. The consistency of those observed parameters could be a simple check of the theory.

The simple solution, that is based on a homogeneous shell approximation, can modified by using the full BM solution and integrating over the entire volume of shocked fluid [8]. Such an integration can be done only numerically. It yields a smoother spectrum and light curve near the break frequencies, but the asymptotic slopes, away from the break frequencies and the transition times, remain the same as in the simpler theory.

Chevalier & Lee [9] estimated the emission from a blast wave propagating into a wind profile $n(R) \propto R^{-2}$. They use equation (1) and calculate the synchrotron emission from a single representative point. This leads to different temporal scalings α of the PLSs, while the spectral indices β remain the same, since they are independent of the hydrodynamic solution. This results in different relations between α, β and p, providing in principle a way to distinguish between different neighborhoods of GRBs and between different progenitor models.

Another modification to the "standard" model arises from a variation of the emission process. Sari & Esin [10] considered the influence of Inverse Compton on the observed spectrum. They find that in some cases the additional cooling channel might have a significant effect on the observed spectrum and light curves.

4   Jets

The afterglow theory becomes much more complicated if the relativistic ejecta is not spherical. To model jetted afterglows we consider relativistic matter ejected into a cone of opening angle ϑ. Initially, as long as $\Gamma \gg \vartheta^{-1}$ [11] the motion would be almost conical. There isn't enough time, in the blast wave's rest frame, for the matter to be affected by the non spherical geometry, and the blast wave will behave as if it was a part of a sphere. When $\Gamma = C_2 \vartheta^{-1}$, namely at[1]:

$$t_{\text{jet}} = \frac{1}{C_1} \left(\frac{l}{c} \right) \left(\frac{\vartheta}{C_2} \right)^{\frac{2(4-k)}{(3-k)}} = \frac{1 \, \text{day}}{C_1 C_2^{8/3}} \left(\frac{E_{\text{iso},52}}{n_1} \right)^{1/3} \left(\frac{\vartheta}{0.1} \right)^{8/3} , \qquad (5)$$

rapid sideway propagation begins. The last equality holds, of course for $k = 0$.

The sideways expansion continues with $\vartheta \sim \Gamma^{-1}$. Plugging this relations in equation (1) we find that $R \approx$ const. This is obviously impossible. A more

---

[1] The exact values of the uncertain constants C_2 and C_1 are extremely important as they determine the jet opening angle (and hence the total energy of the GRB) from the observed breaks, interpreted as t_{jet}, in the afterglow light curves.

detailed analysis [12–14] reveals that according to the simple one dimensional analytic models Γ decreases exponentially with R on a very short length scale.[2]

The sideways expansion causes a change in the hydrodynamic behavior and hence a break in the light curve. Additionally, when $\Gamma \sim \vartheta^{-1}$ relativistic beaming of light will become less effective. This would cause an extra spreading of the emission (that was previously focused into a narrow angle ϑ and is now focused into a larger cone of opening angle Γ^{-1}). If the sideways expansion is at the speed of light than both transitions would take place at the same time [15]. If the sideways expansion is at the sound speed then the beaming transition would take place first and only later the hydrodynamic transition would occur [16]. This would cause a slower and wider transition with two distinct breaks, the first and steeper break when the edge of the jet becomes visible and later a shallower break when sideways expansion becomes important.

The analytic or semi-analytic calculations of synchrotron radiation from jet-ted afterglows [12,15–17,14] have led to different estimates of the jet break time t_{jet} and of the duration of the transition. Rhoads [12] calculated the light curves assuming emission from one representative point, and obtained a smooth 'jet break', extending $\sim 3 - 4$ decades in time, after which $F_{\nu > \nu_m} \propto t^{-p}$. Sari Piran & Halpern [15] assume that the sideway expansion is at the speed of light, and not at the speed of sound $(c/\sqrt{3})$ as others assume, and find a smaller value for t_{jet}. Panaitescu and Mészáros [16] included the effects of geometrical curvature and finite width of the emitting shell, along with electron cooling, and obtained a relatively sharp break, extending $\sim 1 - 2$ decades in time, in the optical light curve. Moderski, Sikora and Bulik [17] used a slightly different dynamical model, and a different formalism for the evolution of the electron distribution, and obtained that the change in the temporal index α ($F_\nu \propto t^{-\alpha}$) across the break is smaller than in analytic estimates ($\alpha = 2$ after the break for $\nu > \nu_m$, $p = 2.4$), while the break extends over two decades in time. Kumar and Panaitescu [14] find that for a homogeneous (or stellar wind) environment there is a steepening of $\Delta\alpha \sim 0.7$ (0.4) when the edge of the jet becomes visible, while the steepening due to sideways expansion extends over 2 (4) decades in time. They conclude that a jet running into a stellar wind will not leave a prominent detectable signature in the light curve.

The different analytic or semi-analytic models have different predictions for the sharpness of the 'jet break', the change in the temporal decay index α across the break and its asymptotic value after the break, or even the very existence a 'jet break' [18]. All these models rely on some common basic assumptions, which have a significant effect on the dynamics of the jet: (i) the shocked matter is homogeneous (ii) the shock front is spherical (within a finite opening angle) even at $t > t_{\mathrm{jet}}$ (iii) the velocity vector is almost radial even after the jet break.

However, recent 2D hydrodynamic simulations [19] show that these assumptions are not a good approximation of a realistic jet. Figure 1 shows the jet at

---

[2] Note that the exponential behavior is obtained after converting equation 1 to a differential equation and integrating over it. Different approximations used in deriving the differential equation lead to slightly different exponential behavior, see [13].

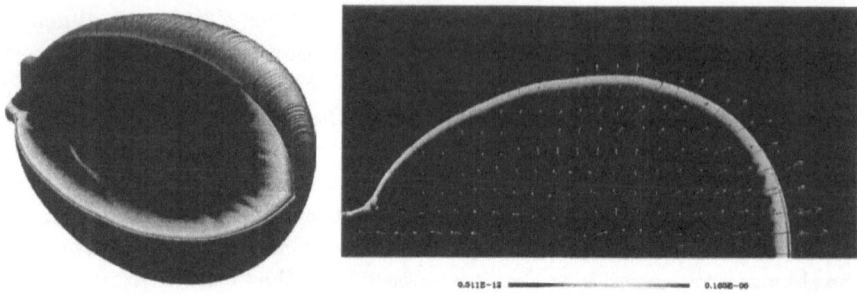

Fig. 1. A relativistic jet at the last time step of the simulation [19]. (**left**) A 3D view of the jet. The outer surface represents the shock front while the two inner faces show the proper number density (*lower face*) and proper emissivity (*upper face*) in a logarithmic color scale. (**right**) A 2D 'slice' along the jet axis, showing the velocity field on top of a linear color-map of the lab frame density.

the last time step of the simulation. The matter at the sides of the jet is propagating sideways (rather than in the radial direction) and is slower and much less luminous compared to the front of the jet. The shock front is egg-shaped, and quite far from being spherical. Figure 2 shows the radius R, Lorentz factor Γ, and opening angle ϑ of the jet, as a function of the lab frame time. The rate of increase of ϑ with $R \approx ct_{\text{lab}}$, is much lower than the exponential behavior predicted by simple models [12]. The value of ϑ averaged over the emissivity is practically constant, and most of the radiation is emitted within the initial opening angle of the jet. The radius R weighed over the emissivity is very close to the maximal value of R within the jet, indicating that most of the emission originates at the front of the jet[3], where the radius is largest, while R averaged over the density is significantly lower, indicating that a large fraction of the shocked matter resides at the sides of the jet, where the radius is smaller. The Lorentz factor Γ averaged over the emissivity is close to its maximal value, (again since most of the emission occurs near the jet axis where Γ is the largest) while Γ averaged over the density is significantly lower, since the matter at the sides of the jet has a much lower Γ than at the front of the jet. The large differences between the assumptions of simple dynamical models of a jet and the results of 2D simulations, suggest that great care should be taken when using these models for predicting the light curves of jetted afterglows. Since the light curves depend strongly on the hydrodynamics of the jet, it is very important to use a realistic hydrodynamic model when calculating the light curves.

Granot et al. [19] used 2D numerical simulations of a jet running into a constant density medium to calculate the resulting light curves, taking into account the emission from the volume of the shocked fluid with the appropriate time delay in the arrival of photons to different observers. They obtained an achromatic jet break for $\nu > \nu_m(t_{\text{jet}})$ (which typically includes the optical and near IR), while at lower frequencies (which typically include the radio) there is a more

---

[3] This implies that the expected rate of orphan optical afterglows should be smaller than estimated assuming significant sideways expansion!

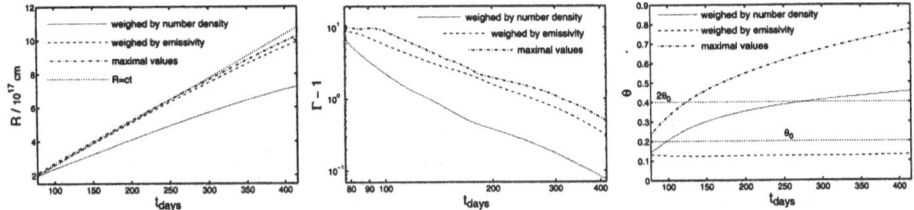

Fig. 2. The radius R (*left frame*), Lorentz factor $\Gamma - 1$ (*middle frame*) and opening angle ϑ of the jet (*right frame*), as a function of the lab frame time in days [19].

moderate and gradual increase in the temporal index α at t_{jet}, and a much more prominent steepening in the light curve at a latter time when ν_m sweeps past the observed frequency. The jet break appears sharper and occurs at a slightly earlier time for an observer along the jet axis, compared to an observer off the jet axis (but within the initial opening angle of the jet). The value of α after the jet break, for $\nu > \nu_m$, is found to be slightly larger than p ($\alpha = 2.85$ for $p = 2.5$).

Somewhat surprisingly we find that in spite of the different hydrodynamic behavior the numerical simulations show a jet break at roughly the same time as the analytic estimates. This encourages us to trust the current estimates of the jet opening angles. However, we should search for an intuitive explanation for the nature of the hydrodynamic behavior and for a simple analytic model that would predict it.

References

1. R.D. Blandford, C.F. McKee: Phys. of Fluids, **19**, 1130 (1976).
2. E. Cohen, T. Piran, R. Sari: Ap. J., **509**, 717 (1998).
3. E. Cohen, T. Piran: Ap. J., **518**, 346 (1999).
4. V.V. Usov: MNRAS, **267**,1035 (1994)
5. R. Sari, T. Piran, R. Narayan: Ap. J. Lett., **497**, L17 (1998).
6. R. Sari: Ap. J. Lett., **489**, L37 (1997).
7. R. Sari, T. Piran: MNRAS, **287**, 110 (1997).
8. J. Granot, T. Piran, R. Sari: Ap. J., **513**, 679 (1999).
9. R.A. Chevalier, Z.-Y. Li: Ap. J. Lett., **520**, L29 (1999).
10. R. Sari, A.A. Esin: Ap. J., **548**, 787 (2001).
11. T. Piran: in AIP Conference Proceedings **307**, *Gamma-Ray Bursts, Second Workshop, Huntsville, Alabama, 1993*, Fishman, G.J., Brainerd, J.J., & Hurley, K., Eds., (New York: AIP), p. 495. (1994)
12. J.E. Rhoads: Ap. J., **525**, 737 (1999)
13. T. Piran: Phys. Rep., **333**, 529 (2000)
14. P. Kumar & A. Panaitescu: Ap. J., **541**, L9 (2000)
15. R. Sari, T. Piran, T. Halpern: Ap. J., **519**, L17 (1999).
16. A. Panaitescu & P. Mészáros: Ap. J., **526**, 707 (1999).
17. R. Moderski, M. Sikora, T Bulik: Ap. J., **529**, 151 (2000)
18. Y. Huang, Z. Dai & T. Lu: A&A **355**, L43 (2000)
19. J. Granot, et al.: These proceedings (astro-ph/0103038) (2001)

Strange Afterglows from Embedded GRBs: Reconciling Hypernovae with Slow Decays

Ralph A.M.J. Wijers

Department of Physics and Astronomy, SUNY at Stony Brook,
Stony Brook, NY 11794-3800, USA; rwijers@mail.astro.sunysb.edu

Abstract. The hypothesis that GRBs originate in massive stars has gained popularity in recent times. With this comes the expectation that the afterglow should bear the rapid decay characteristic of a $1/r^2$ density fall-off in the ambient stellar wind, but few – if any – such afterglows have been seen. I show here that inclusion of the wind termination shock and the region behind it may resolve this paradox if GRB are embedded in molecular clouds, and suggest HETE-era timing measurements that may reveal whether real afterglows are of this type.

1 Introduction

Much progress has been made in our understanding of GRB afterglows. Some likely progenitor properties have emerged from detailed analysis of several afterglows: first, the association of some GRBs with supernovae has pointed a finger at deaths of massive stars as the cause of GRBs, or at least of a subset thereof. Further evidence comes from the location of GRBs in star forming galaxies (see contributions of Fruchter and Djorgovski in this volume). Also, the extinctions and column densities toward GRBs provide evidence, in that they associate GRBs specifically with molecular clouds (see Galama, this volume, and [3]).

There are, however, a few other lines of evidence that appear to contradict the association of GRBs with dying massive stars. First, massive stars have strong winds via which they lose many solar masses of material over their lives. Since a fraction of that mass suffices to slow the blast wave down to non-relativistic speed, we should expect the blast wave evolution as we see it to take place in that wind. Since that wind according to naive expectation has a $1/r^2$ density distribution, we expect a steeply declining light curve characteristic of that decreasing density. In most cases, however, observations show slowly declining light curves consistent with a uniform density around the burst. Second, if GRBs are in molecular clouds, the density in which the blast wave propagates should be of order $1000 \, \mathrm{cm}^{-3}$. We see, however, that where the density can be estimated it is usually quite low, $10^{-3} - 1 \, \mathrm{cm}^{-3}$.

The key to resolving these two paradoxes is one and the same: wind bubbles. Where the wind of the star meets the ISM, shocks form. One shock propagates back into the wind, and at the end of the star's life, it has gone far enough that most of the stellar wind is not unshocked $1/r^2$ wind, but shocked, hot, tenuous wind with roughly uniform density. Within hours to days after the burst, the blast wave is propagating into this shocked wind, making it the same as a blast wave in normal ISM, except for the lower density. The calculations below demonstrate that this only works for massive stars embedded in dense molecular clouds, and show how the measurable properties of such afterglows depend on the nature of the progenitor star and the medium around it.

2 Theoretical Light Curves

2.1 Blast Waves

Whatever the origin of the energy, after an initial explosion phase, the fireball of relativistic temperature is converted to a cold, expanding shell with a large Lorentz factor, η, and energy, E. As it starts to sweep up ambient matter, its kinetic energy is used to shock and heat that matter. Deceleration due to this starts in earnest when about half the initial energy is transferred to the shocked matter, i.e. when it has swept up $1/\eta$ times its own rest mass [6]. The characteristic mass where this happens is

$$M_{\rm dec} = E/(\eta^2 c^2) = 6 \times 10^{-7} E_{53} \eta_{300}^{-2} \; M_\odot, \qquad (1)$$

where $E_{53} = E/(10^{53}\,{\rm erg})^1$, and $\eta_{300} = \eta/300$. For this discussion I shall assume the blast wave is adiabatic, i.e. its energy is constant in time, and effectively spherical[1]. The relativistic expansion is then gradually slowed down, and the blast wave evolves in a self-similar manner with a power-law light curve, as has now been seen some twenty times. This phase ends when so much mass shares the energy that the Lorentz factor drops to 1. Obviously, this happens when a mass E/c^2 has been swept up. This sets the non-relativistic mass scale

$$M_{\rm NR} = E/c^2 = 0.06\,E_{53} \; M_\odot. \qquad (2)$$

Beyond that point, the event slowly changes into a classical Sedov-Taylor supernova remnant evolution, leading to a steeper decline of the light curve (e.g., [9,6]).

2.2 Expected Environments of Progenitors

Usually in the study of afterglows, one considers either the uniform ambient medium case or the $1/r^2$ wind case on its own. However, since the wind of a star meets the interstellar medium at some point, the density structure is more complex.

---

[1] Our discussion is limited to the pre-beaming break phase of the blast wave, where the correct scaling is with the energy per unit solid angle (see, e.g., [8]). This means the E here is the isotropic equivalent energy as, e.g., derived from gamma-ray output, rather than the smaller true energy.

A Stationary Windy Star. A massive star blows a wind bubble around itself. Most of this wind bubble, however, is not filled with a $1/r^2$ wind, but with a roughly uniform shocked wind region, as we now show. (For details, see, e.g. [2].) The wind material and the ISM are separated by a contact discontinuity. For a fast wind from a massive hot star (the only case we consider, as GRB progenitors are usually deemed to be very massive, thus hot even in advanced evolution stages) the material on either side approaches the discontinuity supersonically, so a shock forms. Specifically, the shock in the stellar wind is a strong shock because the wind velocity is very supersonic, and the post-shock material is therefore so hot ($T \sim 10^7$ K) that it can hardly cool. It thus forms a pressurized cavity which propels the discontinuity into the surroundings. The pressure in this cavity is about equal to the total injected wind energy divided by the volume. From this it follows that the outer radius evolves as $t^{3/5}$ and grows to many tens of parsecs within the ~ 3 Myr life of the star. Since the gas in the cavity moves subsonically its pressure keeps it approximately at uniform density. The radius of the wind termination shock at the inner edge of the wind bubble can be found by balancing the wind ram pressure with the post-shock cavity pressure. For a star that loses mass at a rate $10^{-6}\dot{M}_{-6}M_\odot$/yr with a wind velocity $1000v_{w,3}$ km/s, in interstellar gas with density $1000n_3$ cm^{-3}, we have a termination shock radius

$$R_t = 0.4\,\dot{M}_{-6}^{3/10}v_{w,3}^{1/10}n_3^{-3/10}t_6^{2/5} \text{ pc}, \qquad (3)$$

where t_6 is the time in Myr. We can find the mass in the $1/r^2$ part of the wind by integrating the density profile, to get

$$M_t = \dot{M}R_t/v_w = 3.1 \times 10^{-4}\dot{M}_{-6}^{13/10}v_{w,3}^{-9/10}n_3^{-3/10}t_6^{2/5} \text{ M}_\odot. \qquad (4)$$

Comparison with the previous section shows that for typical massive-star parameters, this falls within the range of the relativistic expansion, which means that we should see a transition in the afterglow from the steeper wind decay to the shallower uniform-ISM decay. The time of the transition we can estimate from $t_t = R_t/(2\Gamma_t^2 c)$. Noting that $\Gamma = \eta(M/M_{\rm dec})^{-1/2}$, we find that $t_t = R_t M_t c/(2E)$, which is 0.4 days for our standard molecular-cloud case, but much longer in normal ISM. At the transition, apart from the change in slope, there is a modest re-flare of the event (by about a factor 2 bolometrically), because the density jumps by a factor 4 across the wind termination shock. The density in the uniform shocked wind region at late times is

$$n_{\rm sw} = \frac{3\dot{M}t}{4\pi R_{\rm out}^3 m_p} = 0.06\dot{M}_{-6}^{4/5}n_3^{3/5}v_{w,3}^{6/5}t_6^{-4/5} \text{ cm}^{-3}, \qquad (5)$$

which shows that even if the progenitor is embedded in a dense molecular cloud the observed blast wave can propagate in a low-density, uniform medium. A possible observed example is the afterglow of GRB 970508, which is well modeled by a blast wave in a uniform medium of density $n \simeq 0.03$ cm^{-2} beyond 2 days after the burst [8], but the X-ray absorption may indicate that it is embedded in a molecular cloud [3]. The rise in its light curve during the first day might then mark the time when the blast wave crossed the wind termination shock (see Figure 1, left panel).

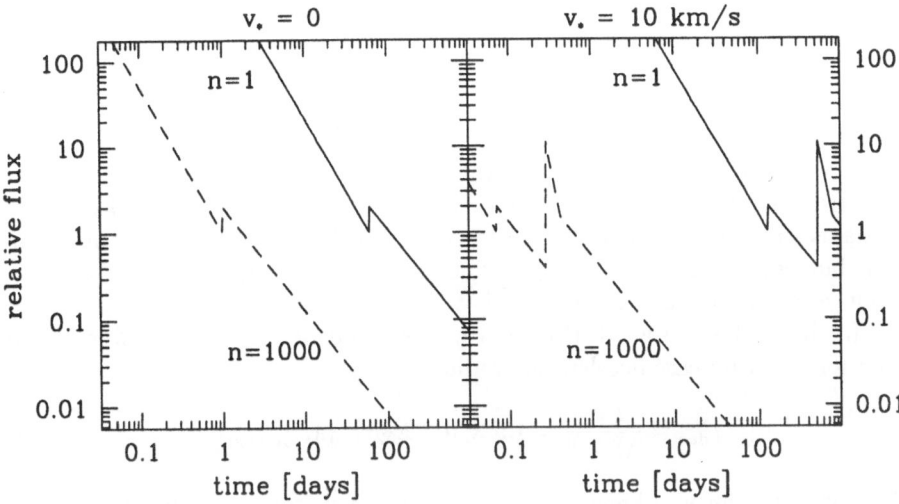

Fig. 1. Schematic light curves of blast waves, from stationary (left) and moving (right) stars with a wind. Parameters used are $\dot{M} = 10^{-6} M_\odot/\text{yr}$, $v_w = 1000\,\text{km/s}$, $t = 3\,\text{Myr}$, $\eta = 300$. The curves are simplified considerably by assuming piecewise self-similar evolution.

A Moving Windy Star. This case is more complex, and its full discussion outside the scope of this paper. However, it is useful to see roughly what might happen. First, the motion will be inconsequential as long as the star moves with less than the speed of the outer shock into the ISM:

$$v_{\text{s,out}} = 3.3 \dot{M}_{-6}^{1/5} n^{-1/5} v_{\text{w},3}^{2/5} t_6^{-2/5} \,\text{km s}^{-1}. \tag{6}$$

Therefore, by the time the star is ready to explode at age $t_6 \sim 2 - 3$, even a $10\,\text{km/s}$ speed may modify the structure of the medium in which the blast wave will propagate. For stellar speeds like this, most of the lost wind will be left behind in a trail. In the forward direction, the contact discontinuity location will be determined by ram pressure balance between stellar wind and ambient medium. In this case, the $1/r^2$ bubble is even smaller, and the GRB blast wave emerges from it within hours. Also, because now most of the shocked stellar wind is swept back, there is little enough mass in it in the forward direction that the blast wave may emerge from it, into the ISM, while still relativistic. Therefore, after having gone through the same two phases as before, the blast wave may enter the highly compressed shocked ISM, giving a sharp non-selfsimilar spike in the light curve. It may then even enter into the uniform unshocked ISM. This depends rather strongly on collimation: these events occur in the mildly relativistic part of a burst, which is very different for a collimated outflow so that we do not expect well-collimated outflows to display these later phases. A schematic illustration of the resulting light curve is in the right panel of Fig. 1, and full discussion is given in [7].

2.3 Timing

It has been commented that wind solutions can be ruled out, because the observed sizes in excess of 10^{17} cm of some afterglows, and also the late times of many days after trigger at which rapid decays are seen are inconsistent with blast waves in winds. Supposedly, the high densities in the inner wind cause rapid deceleration at a small radius, and therefore the afterglow size always remains small. I show here that this is incorrect: deceleration does happen earlier, but because a blast wave in a wind expands more rapidly, it overtakes the uniform-medium one after about a day.

In the unshocked wind, the mass within radius r is $\dot{M}r/v_w$; with eq. 1 we find for the wind case deceleration radius:

$$r_{\mathrm{dec}}^{\mathrm{w}} = \frac{E v_w}{\dot{M} c^2 \eta^2} = 1.9 \times 10^{14} E_{52} v_{w,3} \dot{M}_{-6}^{-1} \eta_{300}^{-2} \text{ cm}. \tag{7}$$

By contrast, the well-known expression for the uniform-medium deceleration radius is

$$r_{\mathrm{dec}}^{\mathrm{u}} = 2.6 \times 10^{16} (E_{52}/n)^{1/3} \eta_{300}^{-8/3} \text{ cm}, \tag{8}$$

and so indeed the wind case decelerates at much smaller radius. The deceleration time is given by $t_{\mathrm{dec}} = r_{\mathrm{dec}}/(2\eta^2 c)$ in both cases, and thus is correspondingly smaller for the wind case. However, because the Lorentz factor decreases as $M^{-1/2}$ (with M the swept-up mass) beyond this point, the mass-starved wind blast wave decelerates much more slowly, and therefore begins to catch up. Specifically, for $r \gg r_{\mathrm{dec}}$, we get

$$r^{\mathrm{w}} = 4.1 \times 10^{17} E_{52}^{1/2} v_{w,3}^{1/2} \dot{M}_{-6}^{-1/2} (t/1\,\mathrm{d})^{1/2} \text{ cm} \tag{9}$$

$$r^{\mathrm{u}} = 4.3 \times 10^{17} E_{52}^{1/2} n^{-1/4} (t/1\,\mathrm{d})^{1/4} \text{ cm}. \tag{10}$$

(For the wind case, this is of course only true for $r^{\mathrm{w}} < R_t$.) Therefore, as stated in the beginning, there is not much difference in size between wind and uniform-ISM blast waves in the most commonly observed interval from 0.3 to 10 days after the burst; the wind blast waves can even be bigger. Conversely, if we find that the $1/r^2$ wind terminates a fraction of a parsec away from the star, this is consistent with the blast wave propagating into a uniform shocked-wind bubble within a day if the burst takes place in a molecular cloud.

Table 1. Properties of the two stationary-star cases shown in left panel of Figure 1. t_t is the time at which we see the light curve change from wind-like to the more slowly decaying uniform-like mode.

		$n = 1\,\mathrm{cm}^{-3}$	$n = 1000\,\mathrm{cm}^{-3}$
stationary	$R_t =$	5 pc	0.6 pc
	$M_t =$	0.004$M_\odot$	0.0005$M_\odot$
	$t_t =$	61 d	1.0 d
	$n_{sw} =$	0.001 cm^{-3}	0.02 cm^{-3}

Note that as we get observations closer to the trigger time, and radius determinations can be made from optically thick parts of the spectrum, the early small radii (combined with later 'normal' ones) may be one of the best diagnostics to distinguish wind blast waves from uniform ones.

3 Discussion and Conclusion

It is evident from the discussion above that the environment of a massive star at the time of its death is a very rich one. Even in the simplest case of a wind whose properties do not vary over the life of the star, rich behavior with multiple possible transitions in the observable part of the afterglow lifetime is seen. The eventual resulting light curve depends fairly strongly on the properties of the system, especially the mass loss rate of the star and the ambient density. There is a good and a bad side to this. On the bad side, it implies we cannot be too specific about the times at which we expect to see transitions. On the good side, if and when we do see these transitions, they can be fairly constraining on the properties of the progenitor.

If we continue to see the population of afterglows dominated by slow decays with low inferred densities, this is support for blast waves in wind bubbles, i.e., support for the origin of bursts from windy stars embedded in molecular clouds. We chose the mass loss rates to be $10^{-6}M_{\odot}/\text{yr}$, which is low for metal-rich helium stars; this partly explains our different conclusions from earlier work [1]. However, it is well known that radiatively driven winds are much weaker in metal-poor environments, e.g., from comparing measured winds in our Galaxy with those in the Magellanic Clouds. So since we are looking at high-redshift events, it may be quite reasonable to suppose that winds of our progenitors are weaker than is usually assumed for helium stars.

Thus far, the only candidate for having shown a shock transition in an afterglow is the flare-up of GRB 970508 one day after trigger. However, due to the paucity of early data, this is only one of many allowable interpretations of the flare-up. We expect that the shock transitions can be quite early, especially for moving stars, so that it will take the rapid alerts provided by HETE to explore what fraction of afterglows may come from shock-terminated stellar winds.

References

1. R. A. Chevalier, Z.-Y. Li: ApJ, **520**, L29 (1999)
2. J. E. Dyson and D. A. Williams: *The Physics of the Interstellar Medium.* IOP, Bristol (1997)
3. T. J. Galama, R. A. M. J. Wijers: ApJ Lett, in press (2001)
4. P. Mészáros, M. J. Rees, R. A. M. J. Wijers: ApJ, **499**, 301 (1998)
5. M. J. Rees and P. Mészáros: MNRAS, **258**, L41 (1992)
6. E. Waxman, S. R. Kulkarni, D. A. Frail: ApJ, **497**, 288 (1998)
7. R. A. M. J. Wijers: in preparation (2001)
8. R. A. M. J. Wijers, T. J. Galama: ApJ, **523**, 177 (1999)
9. R. A. M. J. Wijers, M. J. Rees, P. Mészáros: MNRAS, **288**, L51 (1997)

Light Curves from an Expanding Relativistic Jet

J. Granot[1], M. Miller[2], T. Piran[1], W.M. Suen[2], and P.A. Hughes[3]

[1] Racah Institute of Physics, Hebrew University, Jerusalem 91904, Israel
[2] Department of Physics, Washington University, St. Luis, MO 63130, USA
[3] Department of Astronomy, University of Michigan, Ann Arbor, MI 48109, USA

Abstract. We perform fully relativistic hydrodynamic simulations of the deceleration and lateral expansion of a relativistic jet as it expands into an ambient medium. The hydrodynamic calculations use a 2D adaptive mesh refinement (AMR) code, which provides adequate resolution of the thin shell of matter behind the shock. We find that the sideways propagation is different than predicted by simple analytic models. The physical conditions at the sides of the jet are found to be significantly different than at the front of the jet, and most of the emission occurs within the initial opening angle of the jet. The light curves, as seen by observers at different viewing angles with respect to the jet axis, are then calculated assuming synchrotron emission. For an observer along the jet axis, we find a sharp achromatic 'jet break' in the light curve at frequencies above the typical synchrotron frequency, at $t_{jet} \approx 5.8(E_{52}/n_1)^{1/3}(\vartheta_0/0.2)^{8/3}$ days, while the temporal decay index α ($F_\nu \propto t^\alpha$) after the break is steeper than $-p$ ($\alpha = -2.85$ for $p = 2.5$). At larger viewing angles t_{jet} increases and the jet break becomes smoother.

1 The Hydrodynamics

The hydrodynamic simulations are performed using a fully relativistic 2D AMR code. The jet propagates into a cold and homogeneous ambient medium of number density n_1 cm^{-3}. We use $n_1 = 1$, typical of the ISM. We assume an adiabatic evolution (i.e. that radiation losses do not influence the dynamics). For the initial conditions, we use a wedge of opening angle $\vartheta_0 = 0.2$ taken out of the Blandford McKee [1] self similar spherical solution, with an isotropic equivalent energy of $E_{52}10^{52}$ ergs, where we use $E_{52} = 1$. Since the lateral expansion is expected to become significant only when the Lorentz factor of the flow drops to $\gamma \sim 1/\vartheta_0 = 5$, the initial Lorentz factor of the shock was chosen to be $\Gamma = \sqrt{2}\gamma = 23.7$.

Figure (1) shows a 3D view of the jet at the last time step of the simulation, with color-maps of the proper number density, n', and proper emissivity. Primed quantities are measured in the local rest frame of the fluid. While the number density does not change significantly between the front and the sides of the jet, the emissivity is significant only at the front of the jet, and drops sharply at angles larger than the initial opening angle, ϑ_0. The overall egg-shaped structure is very different from the quasi-spherical structure assumed in 1D analytic models.

2 The Light Curves

The synchrotron spectral emissivity, $P'_{\nu'}$, is calculated using the physical conditions determined by the hydrodynamic simulation. The electrons and the magnetic field are assumed to hold fractions ϵ_e and ϵ_B, respectively, of the internal

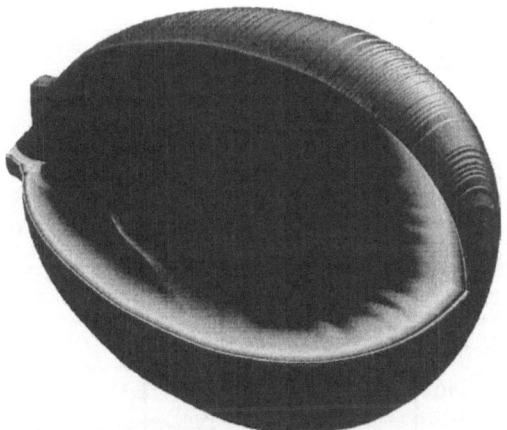

Fig. 1. A 3D view of the jet at the last time step of the simulation. The outer surface represents the shock front while the two inner faces show the proper number density (*lower face*) and proper emissivity (*upper face*) in a logarithmic color scale

energy density, e', while the electrons posses a power law energy distribution, $N(\gamma_e) \propto \gamma_e^{-p}$ for $\gamma_e > \gamma_m = [(p-2)/(p-1)]\,(\epsilon_e e'/n'm_e c^2)$. For simplicity, we ignore the effects of electron cooling and self absorption, so that $P'_{\nu'} \propto \nu'^{1/3}$ for $\nu' < \nu'_m$ and $P'_{\nu'} \propto \nu'^{(1-p)/2}$ for $\nu' > \nu'_m$, where ν'_m is the local synchrotron frequency of an electron with $\gamma_e = \gamma_m$; F_ν is calculated using the formalism of [2], summing over the contributions from the finite 4-volume of the simulation.

Figure (2) shows the radio light curves seen by an observer along the jet axis ($\vartheta_{obs} = 0$). For simplicity, cosmological corrections are not included. The insert shows an optical light curve as seen by observers at three different viewing angles with respect to the jet axis: $\vartheta_{obs}/\vartheta_0 = 0, 1, 2$. We obtain an achromatic 'jet break' in the light curve for $\nu > \nu_m(t_{jet})$ (as predicted by simple semi-analytic models [3,4]) at $t_{jet}(\vartheta_{obs})$, where $0.66 t_{jet}(\vartheta_0) \approx t_{jet}(0) \approx 5.8(E_{52}/n_1)^{1/3}(\vartheta_0/0.2)^{8/3}$ days. Defining α, β by $F_\nu \propto \nu^\beta t^\alpha$, the shape of the break may be approximated by

$$F_\nu = F_0 \nu^\beta \left[(t/t_{jet})^{-s\alpha_1} + (t/t_{jet})^{-s\alpha_2} \right]^{-1/s} , \qquad (1)$$

where $\alpha_1 = \alpha(t \ll t_{jet})$, $\alpha_2 = \alpha(t \gg t_{jet})$ and the parameter $s(\vartheta_{obs})$ determines the sharpness of the break, and ranges between $s(\vartheta_0) \approx 1$ to $s(0) \approx 4.5$, indicating that the break is sharper at smaller ϑ_{obs}. For $\nu < \nu_m(t_{jet})$, there is only a moderate and more gradual change in α, until the time when $\nu_m = \nu$. For $\nu > \nu_m$ we find that α_2 is slightly smaller than $-p$ (the value predicted by most simple models) for $\vartheta_{obs} = 0$ ($\alpha_2 = -2.85$ for $p = 2.5$).

3   Discussion

We find that the physical conditions at the sides of the jet are significantly different than at the front of the jet, and most of the radiation is emitted

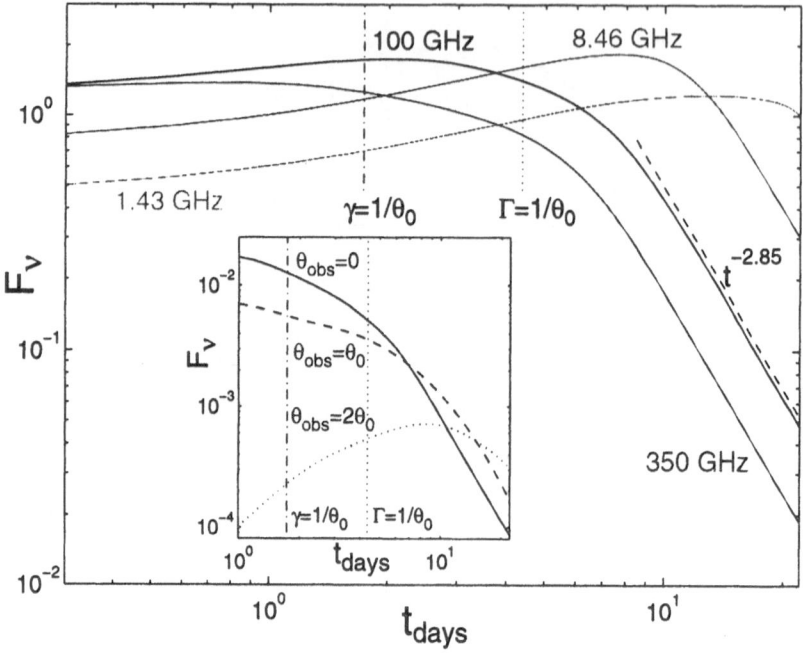

Fig. 2. Radio light curves (flux density, F_ν, in arbitrary units, as a function of the observed time in days) for an observer along the jet axis. We use $\epsilon_e = \epsilon_B = 0.1, p = 2.5$. The observed times, for an observer along the jet axis, when the Lorentz factors γ of the shocked fluid (*vertical dash-dotted line*) or Γ of the shock (*vertical dotted line*) drop to $1/\vartheta_0$, for an extrapolated spherical evolution, are indicated. (**insert**) Optical light curves for observers at viewing angles $\vartheta_{obs}/\vartheta_0 = 0, 1, 2$ with respect to the jet axis.

within the initial opening angle of the jet $[\vartheta < \vartheta_0$; see Figure (1)]. Therefore, the frequently used assumption of a homogeneous jet seems inadequate. For $\nu > \nu_m(t_{jet})$ we find a sharp achromatic break in the light curve at $t_{jet} = 5.8(E_{52}/n_1)^{1/3}(\vartheta_0/0.2)^{8/3}$ days, for $\vartheta_{obs} = 0$, while at larger viewing angles t_{jet} increases and the break becomes smoother. For $\nu < \nu_m(t_{jet})$ the change in the temporal index α near t_{jet} is more moderate and gradual. The value of α for $t > t_{jet}$, $\nu > \nu_m$, $\vartheta_{obs} = 0$ is slightly smaller than $-p$ ($\alpha = -2.85$ for $p = 2.5$). Finally, we note that $t_{jet}(\vartheta_{obs} = \vartheta_0) \approx 1.5 t_{jet}(\vartheta_{obs} = 0)$. Since, in order to detect a burst in γ-rays we require that $\vartheta_{obs} \lesssim \vartheta_0$, this may induce an uncertainty of up to 15% when deducing the value of ϑ_0 from t_{jet}, unless ϑ_{obs} is well constrained.

References

1. R.D. Blandford, C. F. McKee: Phys. of fluids, **19**, 1130 (1976)
2. J. Granot, T. Piran, R. Sari: ApJ, **513**, 679 (1999)
3. J.E. Rhoads: ApJ, **525**, 737 (1999)
4. R. Sari, T. Piran, T. Halpern: ApJ, **519**, L17 (1999)

GRBs, Fireballs and Precessing Gamma Jets

Daniele Fargion[1,2]

Universitá degli Studi di Roma I, *La Sapienza*[1] and, $INFN$[2],
Piazzale Aldo Moro 5, 00185, Rome, Italy

Abstract. Fireballs are huge isotropic explosions models widely believed to explain Gamma Ray Burst, GRBs (Piran,1999); ever-new versions consider wide beamed (10°) Jet explosions hitting external shells. On the contrary, since 1994-1998, we argued (Fargion 1995-2000; see also Blackman et all.1996) that GRBs (as well as Soft Gamma Repeaters SGR) are spinning and precessing Gamma Jets, produced by collimated e^+,e^- Jet via Inverse Compton Scattering, in a very narrow (0.1°) angles, blazing and flashing the observer. The Jet arises in Super-Nova (SN) explosions; its energy decays slowly from earliest SN powers (corresponding to GRB) toward lower stable power as Soft Gamma Repeaters (SGR) regimes. GRBs and SGRs shared (sometimes) same spectra and time structure: then SGRs are low-power GRBs, but without SN relics (or GRB afterglows, signatures of Jets in SN-GRBs). Moreover weak isolated X-ray precursor signals,(such as $GRB980519$, $GRB981226$, $GRB000131$), corresponding to huge isotropic $\sim 10^{47}$ erg s^{-1}, followed by the extreme GRB $\sim 10^{52}$ erg s^{-1} powers, disagree with any Fireball explosive scenarios. We naturally interpret these X-Ray precursors as rare earliest marginal blazes of outlying X conical precessing Jet tails, surrounding the γ Jet, later hitting in-axis as a GRB.

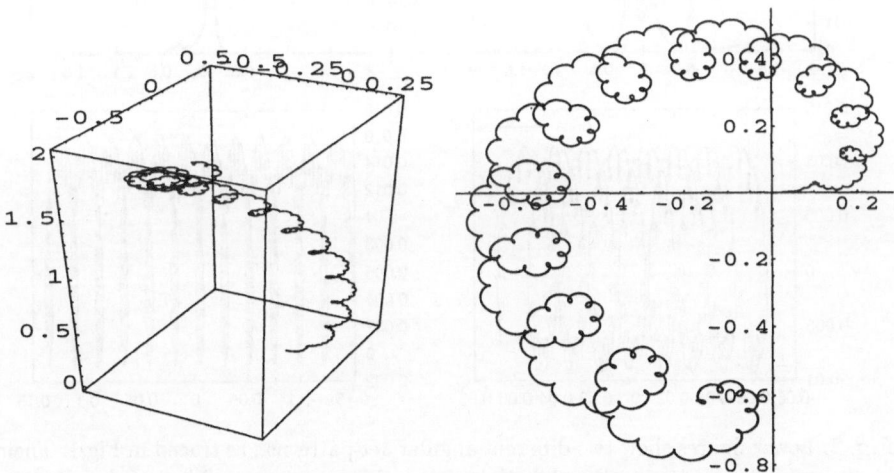

Fig. 1. Left: Three dimensional space evolution of a Spinning while Precessing X-γ Jet, leading by its blazing to X precursor and to the main γ GRB. Right: The same jet pattern observed from above along the vertical axis , on a two dimensional plane.

Since GRB980425 we argued (Fargion 1998-2000) that GRBs and SGRs can
be explained by a comprehensive theory where a thin (tens of seconds) γ beam
Jet, spinning in multi-precession, is sprayed by a Neutron Star, NS, or a Black
Hole, BH, flashing and blazing the observer. Indeed the extreme energy released
in GRB990123 and GRB000131, ($\gg 10^{54}$ erg), (or even twice as much, keeping into account neutrino budget) leads to a conflict with any isotropic GRB
model: Schwarzschild scale times (corresponding to the needed solar masses),
above milliseconds, disagree with the observed GRBs fine time structures (below a fraction of millisecond). GRBs and SGRs share, in a few cases, the same
spectra (Fargion 1998-1999-2000;Wood et all 1999) and time structure, suggesting an unique model. The γ Jet for GRB and SGR is produced, trough Inverse
Compton Scattering (ICS), by GeVs e^+ e^- (secondaries of penetrating GeVs
μ^+ μ^-), scattering on infrared photons, (Fargion,Salis 1995-1998), leading to a
collimated, spinning and precessing γ (MeVs) precessing Jet.

The peak γ Jets has power of a Supernova ($10^{44}ergs^{-1}$) appearing beamed
as ($10^{52}ergs^{-1}$) decaying by power law $\sim t^{-1}$ in 3-6 hour scale times, to ancient,
lower power SGRs stages. SGRs are powered by X-ray pulsars Jet ($10^{35}ergs^{-1}$)
whose collimated beam is amplified up to ($10^{43}ergs^{-1}$). Both of GRB and SGR

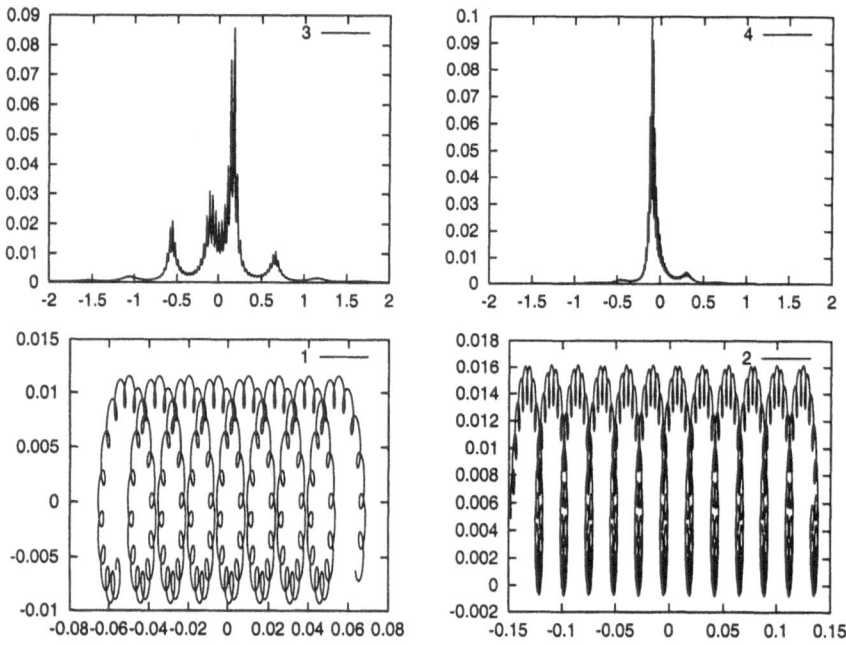

Fig. 2. Lower figures show two different angular Jet patterns , as traced in Fig.1. Their
bi-dimensional opening angles while Spinning and Precessing, are blazing the observer
at the center (origin $(0,0)$) leading, by ICS, to the consequent GRB signal described
above. Upper figures show the the consequent X, GRB intensity evolution (time in secs)
derived by the ICS formula and the corresponding geometrical Jet patterns evolutions
below. The X ray precursor may naturally arise in some pattern configurations.

show an apparent luminosity amplified by the inverse of the beamed solid angle $(10^7 - 10^9)$. The earliest and puzzling X-Ray precursors in few GRBs (as well as SGRs) is an obvious peripherical off-axis flashing followed by main in-axis GRB blaze, in some geometrical configurations, as shown in simulations in Fig.2. Data on X-Ray precursor and GRB are shown for comparison in Fig3.

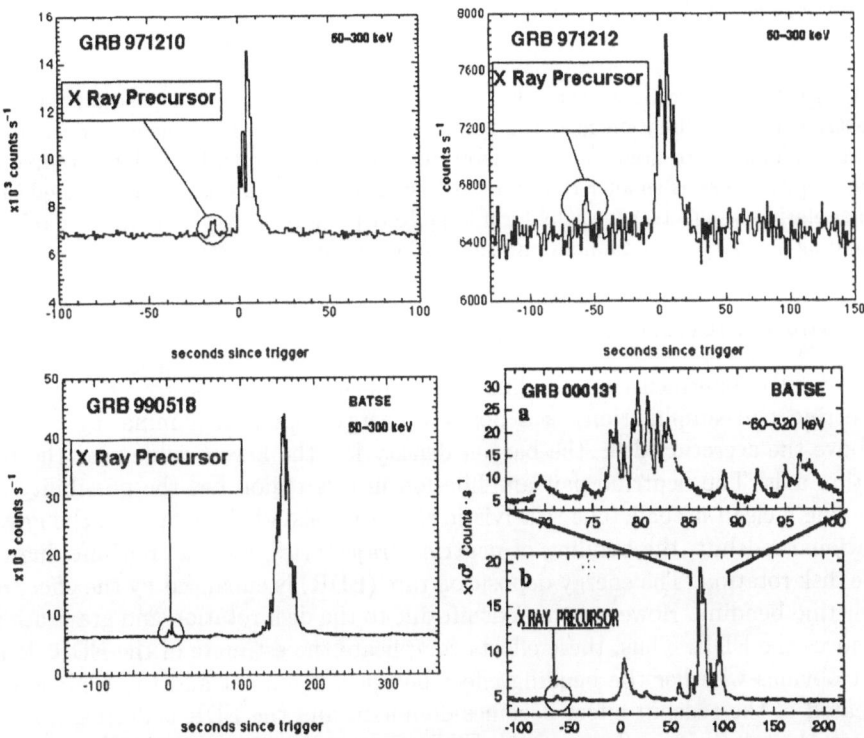

Fig. 3. Up: Time evolution and X precursors in GRB 971210 and GRB 971212. Down: The same evolution in GRB 990518 and in most distant (red-shift 4.5) GRB 000131; note the (surprising for Fireball) tiny X-Ray precursor a minute before the main GRB.

References

1. Blackman, E: G., Yi, I., Field G. B.: 1996, ApJ 479, L79-L82
2. Fargion, D., Salis, A.: 1995, Nuc. Phys B, 43, 269-273; As.& Spac.Sc.231,191-194.
3. Fargion, D., Salis, A.: 1998,Physics-Uspekhi, 41(8), 823-829
4. Fargion, D. 1999, A & A, 138, 507; astro-ph/0011403.
5. Woods P.M.et all. astro-ph/9909276

Neutrino Pair Annihilation above a Kerr Black Hole with the Accretion Disk

Katsuaki Asano

Osaka University, Toyonaka 560-0043, Japan

Abstract. We investigate the relativistic effects on the energy deposition rate via neutrino pair annihilation near the rotation axis of a Kerr black hole. For the disk with a temperature gradient, the energy deposition rate for a small inner radius of the accretion disk is smaller than that estimated by neglecting the relativistic effects. The relativistic effects, especially for a large Kerr parameter, a, play a negative role in avoiding the baryon contamination problem in gamma-ray bursts.

1 Introduction

The neutrino-antineutrino annihilation into electrons and positrons (hereafter neutrino pair annihilation) is a possible energy source of gamma-ray bursts. Above the accretion disk, the baryon density has the lowest value near the rotation axis. The neutrino pair annihilation in this region has the possibility of making a clean fireball. The relativistic effects consist of three factors: the gravitational redshift, the bending of neutrino trajectories, and the redshift due to the disk rotation. The energy deposition rate (EDR) is enhanced by the effect of neutrino bending. However, the redshift due to the disk rotation and gravitation reduces the EDR. Thus, these effects complicate the estimate of the EDR. It is not obvious whether the bending effect becomes dominant and the EDR is increased or the redshift effect becomes dominant and the EDR is decreased. We study the relativistic effects on the EDR above a Kerr black hole. We consider the most idealized situations and compare the results under the same conditions, except for the presence or absence of gravitation.

2 Relativistic Effects

The formulation in this study is obtained by modifying a work of Asano & Fukuyama (2000, hereafter AF) for the Schwarzschild black hole. In the present case we consider the following situations: Around a rotating black hole a hot thin accretion disk is formed and emits a vast number of neutrinos. The inner (outer) edge of this accretion disk is denoted by $R_{\rm in}$ ($R_{\rm out}$). The geometrically thin disk is assumed to be sufficiently opaque to neutrinos over the whole region. We calculate the EDR along the rotation axis and take it as a rough standard in the vicinity of the rotation axis. As was derived in AF, the EDR via neutrino pair annihilation for a distant observer is expressed as

$$\frac{dE_0(r)}{dtdV} = \frac{21\pi^4}{4}\zeta(5)\frac{KG_{\rm F}^2}{h^6c^5}k^9T_{\rm eff}^9(3r_g)F(r), \tag{1}$$

where $F(r)$ is the function of the distance from the center, r, including the relativistic effects; the bending of neutrino trajectories and the redshift due to the disk rotation and gravitation.

3 Results

Integration of equation (1) over the volume $dV = (r^2 + a^2 \cos^2 \vartheta) \sin \vartheta dr d\vartheta d\varphi$ gives the energy deposition per unit world time for a distant observer. However, we estimate the EDR within the infinitesimal angle $d\vartheta$ along the rotation axis over $r = 1.5\text{-}10 r_g$. We do not consider the EDR at $r < 1.5 r_g$ since in this region the baryon contamination occurs severely and the reabsorption rate of the deposited energy to the black hole is large. Then the EDR is proportional to the dimensionless integral of $G(r) \equiv F(r)(r^2 + a^2)/r_g^2$ over $\hat{r} \equiv r/r_g$;

$$\frac{dE_0}{dt} \simeq 4.41 \times 10^{48} \left(\frac{d\vartheta}{10°}\right)^2 \left(\frac{kT_{\text{eff}}(3r_g)}{10\text{MeV}}\right)^9 \left(\frac{r_g}{10\text{km}}\right)^3 \int_{1.5}^{10} G(\hat{r}) d\hat{r} \quad \text{ergs s}^{-1},$$
(2)

for ν_e. We discuss the cases for the various dimensionless Kerr parameter $a_* \equiv a/0.5 r_g$. We adopt the innermost stable orbit as the inner edge of the disk, R_{in}. As a_* takes a larger value, R_{in} becomes smaller. The value R_{out} is fixed at $10 r_g$ in every case. First of all, we adopt $T_{\text{eff}}(R) = T_{\text{eff}}(3r_g) \cdot 3r_g/R \propto R^{-1}$ as the simplest and acceptable temperature profile of the accretion disk. In this model the temperature profiles for different a_* remain common outside of $R = 3r_g$. The relativistic effects are estimated by comparing it to the EDR for the same R_{in} without the relativistic effects.

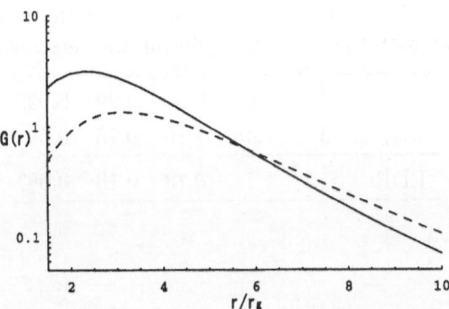

Fig. 1. Plot of $G(r)$ (solid line) against r/r_g for the case of $a_* = 0$ and $T_{\text{eff}} \propto R^{-1}$. The dashed line corresponds to the case for the same R_{in} ignoring the relativistic effects.

We plot $G(r)$ in Fig. 1 for the Schwarzschild black hole. The EDR increases at small r because of the gravitational lensing. Since the neutrino energy decreases because of the gravitational redshift, the EDR is less efficient than that without the relativistic effects at r larger than $r = 6r_g$. As is shown in Table 1, the relativistic effects for $a_* = 0$ enhance the EDR by factor 1.7. As a_* takes a larger value, the EDR itself increases. For $a_* \geq 0.9$, however, the energy reduction by the relativistic effects becomes larger as a_* increases (see Table 1). In Fig. 2 $G(r)$ for $a_* = 0.99$ is much smaller than the $G(r)$ ignoring the relativistic effects in all regions. These results are disadvantageous in avoiding the baryon contamination problem in gamma-ray bursts.

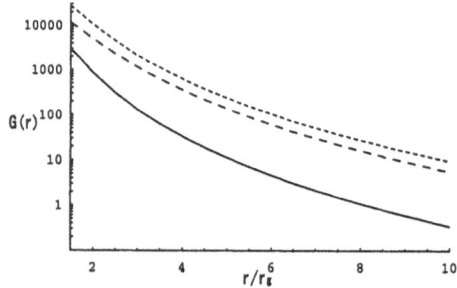

Fig. 2. Same as Fig. 1 but for $a_* = 0.99$. The dashed line and dotted line correspond to the cases of $R_{in} = 0.73$ and $0.59 r_g$ ignoring the relativistic effects, respectively.

The calculations for $T_{eff} \propto R^{-1}$ show the following things: If the deposited energy is mainly contributed by the neutrinos coming from the central region, the bending effect is dominant for large R_{in}, and the redshift effect is dominant for small R_{in}. Namely, as R_{in} approaches the horizon, the redshift effect overwhelms the bending effect. This qualitative result is unchanged, even if we assume a model with a flatter temperature gradient than $T_{eff} \propto R^{-1}$.

Table 1. EDR for the disk with $T_{eff} \propto R^{-1}$ including the relativistic effects. Each value is divided by that with the same R_{in} ignoring the relativistic effects.

a_*	0	0.8	0.9	0.99	0.999
R_{in}/r_g	3	1.45	1.16	0.73	0.59
EDR	1.7	1.1	0.73	0.18	0.087

References

1. K. Asano, T. Fukuyama: ApJ. **531**, 949 (2000)
2. K. Asano, T. Fukuyama: ApJ. **546**, 1019 (2001)

Dissipation Efficiency of Internal Shocks in GRB

Andrei M. Beloborodov

Stockholm Observatory, Saltsjöbaden, SE-133 36, Sweden

Abstract. Previous simulations of internal shocks in GRB ejecta (e.g. [2-5]) indicated that a small fraction $\epsilon \sim 0.1$ of the ejecta kinetic energy is dissipated in this way. The low ϵ can lead to an energy crisis of the model: prompt GRB should then be strongly dominated by the early afterglow from the external shock, in conflict with observations. More detailed analysis shows that ϵ depends on the initial amplitude of the velocity fluctuations in the ejecta and at high amplitudes ϵ can approach 100% [1]. These results are briefly summarized here and then a simple theory of outflows with Gaussian velocity fluctuations is developed. It is shown that the history of internal dissipation is governed by the spectrum of the fluctuations.

1 Γ-Fluctuations: Basic Parameters

Internal shocks are caused by an irregular radial profile of the Lorentz factor Γ in the launched outflow. The amplitude of Γ-fluctuations is $A = \Gamma_{\rm rms}/\Gamma_{\rm CM}$ where CM stands for the center-of-momentum of the ejecta and $\Gamma_{\rm rms}$ is the rms of Γ-fluctuations around $\Gamma_{\rm CM}$. Fluctuations are described by their spectrum that shows the contribution of different wavelengths to $\Gamma_{\rm rms}$. The spectrum extends from the largest scale corresponding to the outflow duration, $\Delta = t_b c \sim 10^{11} {\rm cm}$, to the smallest λ_m,

$$\frac{\Gamma_{\rm rms}^2}{\Gamma_{\rm CM}^2} = A^2 = \int_{k_\Delta}^{k_m} P_k \, dk. \tag{1}$$

Here $k = 2\pi/\lambda$ is the fluctuation wavenumber, $k_\Delta = 2\pi/\Delta$, and $k_m = 2\pi/\lambda_m$ is the maximum k which evolves as we discuss below. The initial $k_m = k_0$ can be associated with the size of the central engine ($\lambda_0 \sim 10^7 {\rm cm}$) and initial $A = A_0$.

2 Efficiency Depends on the Fluctuation Amplitude

The free energy of ejecta with a given amplitude of fluctuations $A < 1$ is [1]

$$U_{\rm free} = M c^2 \Gamma_{\rm CM} \frac{A^2}{2}, \tag{2}$$

where M is the mass of the ejecta. $U_{\rm free}$ is dissipated when the Γ-profile of the outflow is smoothed by internal collisions. Given an initial A_0 one immediately gets the dissipation efficiency $\epsilon = A_0^2/2$ without computing the details of the

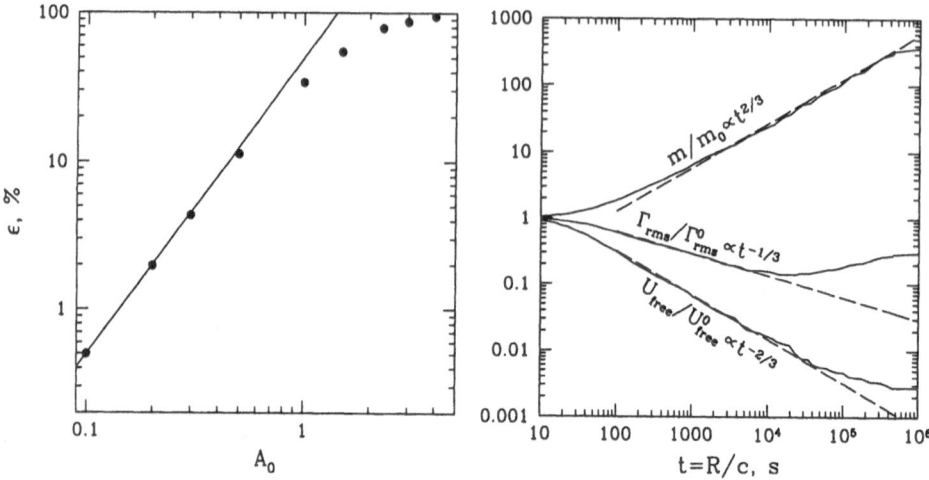

Fig. 1. Left: Dissipation efficiency versus the initial fluctuation amplitude for Poisson fluctuations with a log-normal distribution (see [1]). Solid line shows the law $\epsilon = A_0^2/2$. Right: Example of the dissipation history for fluctuations with $\alpha = 0$ and $A_0 = 0.2$.

stochastic internal shocks. This result is exact for almost any spectrum P_k provided $A_0 < 1$. At $A > 1$ the formula (2) does not apply (the fluctuations are relativistic in the CM frame and the system evolves in a non-linear regime, see [1]). At high A the free energy far exceeds the "regular" kinetic energy $Mc^2\Gamma_{CM}$. Full dissipation of U_{free} would then imply $\epsilon \to 1$. This behavior is confirmed by direct numerical simulations (Fig. 1).

The observed GRB energy may be smaller than the dissipated energy. Even if the dissipated heat is converted into radiation instantaneously (on a time-scale $\ll R/c$) a question remains whether the radiation escapes the ejecta freely. Indeed, the dissipation can start before the ejecta becomes transparent ($t < t_*$) and then the trapped photons loose energy in the process of adiabatic expansion. The lost energy depends on the history of dissipation before the transparency moment t_*.

3   Fluctuation Spectrum and the History of Dissipation

The history of dissipation depends on the initial fluctuations. Consider the simple case of Gaussian (random phase) fluctuations with a power-law spectrum $P_k \propto k^{-\alpha}$. Assume an initial amplitude $A_0 < 1$. Then the fluctuation velocity in the CM frame $\tilde{v}/c \approx (\Gamma - \Gamma_{CM})/\Gamma_{CM}$ is nonrelativistic and $\Gamma_{CM} = const$ (Γ_{CM} is not affected by dissipation). One can easily show that $P_k = \tilde{v}_k^* \tilde{v}_k / c^2 \Delta$ and $A = \tilde{v}_{rms}/c$, where $\tilde{v}_k$ is Fourier transform of $\tilde{v}$. Integration of equation (1) yields at $k_\Delta \ll k_m \leq k_0$

$$A^2 = A_0^2 \begin{cases} (k_m/k_0)^{1-\alpha}, & \alpha < 1, \\ \ln(k_m/k_\Delta)/\ln(k_0/k_\Delta), & \alpha = 1, \\ 1 - (k_\Delta/k_m)^{\alpha-1} + (k_\Delta/k_0)^{\alpha-1}, & \alpha > 1. \end{cases} \tag{3}$$

Internal dissipation tends to smooth the velocity profile starting from λ_0 and then moving to progressively larger scales. When the ejecta reaches a radius $R = ct$ fluctuations with $\lambda < \lambda_m$ have dissipated (on time-scale $< R/c$) and those with $\lambda > \lambda_m$ will be dissipated later on. The fluctuations with $\lambda \sim \lambda_m(R)$ are being dissipated now: their "collisional" time-scale $t_{coll} \sim R/c$. The t_{coll} can be estimated as

$$t_{coll} = \frac{\tilde{\lambda}_m}{\tilde{v}_{rms}} \, \Gamma_{CM} = \frac{2\pi \Gamma_{CM}^2}{\tilde{v}_{rms} k_m} = \frac{2\pi \Gamma_{CM}^2}{A c k_m}. \tag{4}$$

Equating $t_{coll} = R/c$ and using (3) we solve for $k_m(R)$. Then equation (3) gives $A(R)$. E.g. in the case of $\alpha < 1$ (white-type fluctuations) we get

$$\frac{k_m}{k_0} = \left(\frac{R}{R_0}\right)^{-2/(3-\alpha)}, \quad A = A_0 \left(\frac{R}{R_0}\right)^{-(1-\alpha)/(3-\alpha)}, \quad R_0 = \frac{\lambda_0 \Gamma_{CM}^2}{A_0}. \tag{5}$$

Here R_0 is the radius where the internal collisions begin. The mass scale of subshells with smoothed Γ-profile $m(R) \propto k_m^{-1} \propto R^{2/(3-\alpha)}$ and the remaining free energy $U_{free}(R) \propto A^2 \propto R^{-2(1-\alpha)/(3-\alpha)}$. The case of $\alpha = 0$ (white noise) was studied in [1] and the laws $m(R) \propto R^{2/3}$ and $A(R) \propto R^{-1/3}$ were confirmed by direct numerical simulations (Fig. 1).

Suppose the dissipated energy is immediately converted into radiation (the opposite adiabatic limit would imply a weak GRB) and let dissipation/emission start in the optically thick zone ($R_0 < R_* = ct_*$). Radiation produced at $R < R_*$ is trapped and cooled adiabatically as $(R/R_*)^{2/3}$ – the ejecta volume expands as $(R_*/R)^2$ before it becomes transparent. One then finds the net radiative efficiency

$$\epsilon = \frac{A_0^2}{2} \begin{cases} (2\alpha)^{-1} \left[(3-\alpha)(R_0/R_*)^{2(1-\alpha)/(3-\alpha)} - 3(1-\alpha)(R_0/R_*)^{2/3}\right], & \alpha \neq 0, \\ (R_0/R_*)^{2/3} \left[(2/3)\ln(R_*/R_0) + 1\right], & \alpha = 0. \end{cases} \tag{6}$$

The dissipation history crucially changes in the case of $\alpha > 1$ ("red" spectrum). Then A and U_{free} are dominated by small k and the dissipation is slow until k_m becomes comparable to k_Δ (see eq. 3). This normally happens far outside R_*. The case of $\alpha = 1$ (flicker noise) has a flat energy distribution over $\ln k$ and represents the boundary between the white-type and red-type fluctuations.

References

1. A.M. Beloborodov: ApJ, **539**, L25 (2000)
2. F. Daigne, R. Mochkovitch: MNRAS, **296**, 275 (1998)
3. S. Kobayashi, T. Piran, R. Sari: ApJ, **490**, 92 (1997)
4. P. Kumar: ApJ, **523**, L113 (1999)
5. M. Spada, A. Panaitescu, P. Mészáros: ApJ, **537**, 824 (2000)

Interaction between Internal Shocks and the Reverse Shock for a GRB in a Dense Stellar Wind

Frédéric Daigne[1] and Robert Mochkovitch[2]

[1] Max Planck Institut für Astrophysik, Garching, Germany
[2] Institut d'Astrophysique de Paris, Paris, France

1 Introduction

In the most popular scenario for GRBs a relativistic shell is produced by a compact central source (collapsar, BH + WD, BH + NS or NS + NS merger) with a highly non uniform distribution of the Lorentz factor. Internal shocks in the shell are then responsible for the gamma-ray emission. When the shell is decelerated by the surrounding medium a forward (external) shock is formed which accounts for the afterglow observed at lower energy from X-ray to radio frequencies. A reverse shock is also produced and propagates throughout the shell. In most studies, the internal and reverse shocks have been considered separately. However, in the case of a dense wind environment the internal and reverse shocks can be simultaneously present in the shell. We give examples of such situations and show that the burst profiles can be greatly affected. We obtain constraints on the burst parameters to maintain reasonably good profiles for GRBs in dense stellar winds.

2 The Deceleration Radius

At the deceleration radius the relativistic shell ejected from the central engine begins to "feel" the external medium. The deceleration radius r_d is given by the condition: *swept up mass* $\sim \frac{M_{\mathrm{shell}}}{\bar{\gamma}}$ where M_{shell} and $\bar{\gamma}$ are the initial shell mass and average Lorentz factor.

In a wind environment the density follows a r^{-2} law

$$\rho(r) = \frac{A}{r^2} \quad \text{with} \quad A = \frac{\dot{M}}{4\pi v_{\mathrm{w}}} \tag{1}$$

and the condition becomes

$$4\pi A\, r_d \sim \frac{E}{\bar{\gamma}^2 c^2} \quad \text{or} \quad r_d \sim 2\,10^{13} \frac{E_{52}}{A_* \gamma_{300}^2} \quad \text{cm} \tag{2}$$

where E_{52} is the total kinetic energy of the shell in units of 10^{52} ergs, $A_* = A/5\,10^{11}$ ($A = 5\,10^{11}$ being a typical value for the wind of a Wolf-Rayet star) and $\gamma_{300} = \bar{\gamma}/300$. For a wide range of burst and wind parameters, the deceleration radius is comparable and even smaller than the typical radius where internal shocks form. The reverse shock can therefore interact with internal shocks and affect the resulting burst profile.

3   Numerical Simulations

We have considered a burst with two pulses generated by an initial distribution of the Lorentz factor which oscillates between $\Gamma_{\min} = 100$ and $\Gamma_{\max} = 400$. Energy is injected during a period of 10 s at a constant rate $\dot{E} = 2\,10^{51}$ erg.s^{-1} so that $E_{52} = 2$ for this burst. The resulting profile is obtained following the method described in [4] where the relativistic shell is represented by a series of "solid" layers which interact only through direct collisions, while all pressure waves are neglected. We take into account the effect of the environment by progressively decreasing the Lorentz factor of the front layer as it moves into the stellar wind following Fenimore and Ramirez-Ruiz [5]. We computed profiles corresponding to wind environments with $A_* = 0.1$ and 1. The analytic results of the previous section indicate that one should then expect some interaction between the reverse and internal shocks. This is especially true in the case of a dense wind ($A_* = 1$) where the reverse shock travels all the way through the shell before any internal shock appears.

The early occurence of the reverse shock has important consequences on the burst profile. Instead of two pulses with a FRED shape one obtains a very different temporal evolution which is much less evocative of observed profiles (Fig 1). We believe that this represents a potential problem if a (large) fraction of bursts indeed comes from collapsars which have produced a dense stellar wind during the Wolf-Rayet phase.

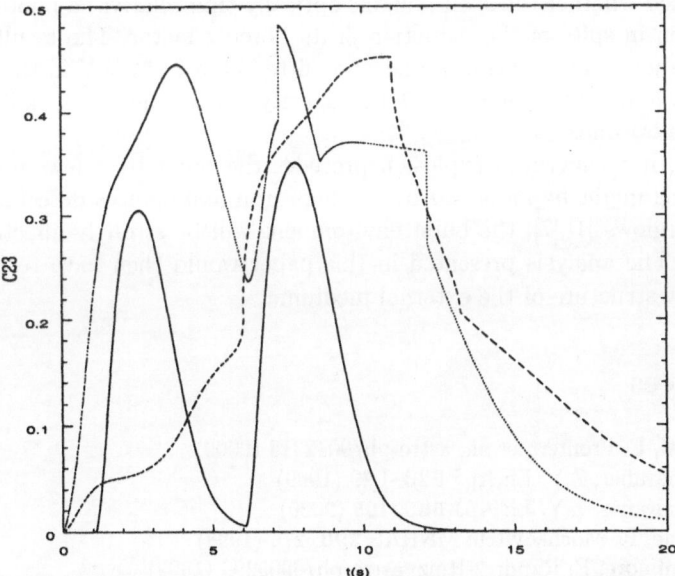

Fig. 1. Burst profiles in winds with $A_* = 0.1$ (dotted line), $A_* = 1$ (dashed line) and $E_{52} = 2$. The profile with no wind is shown for comparison (solid line).

4   Constraints on a GRB in a Dense Stellar Wind

The analytical estimates of Sect. 1 suggest two possibilities to recover burst profiles in better agreement with the observations.

- *An increase of the shell energy E*

In the two cases considered above we had E_{52}/A_* respectively equal to 2 and 20. If we increase E_{52}/A_* to 200 we again get two pulses which are not too different from the initial profile (with no external medium) represented in Fig. 1. To adopt a shell energy (in 4π sr) of 10^{53}–10^{54} erg is not a problem especially if beaming is present in most bursts. As shown by Kumar and Panaitescu [6] afterglows for jets propagating in a stellar wind do not exhibit obvious breaks in agreement with the light curves of GRB 970508, GRB 980326 and GRB 980519 for which Chevalier and Li [2],[3] find the best evidences for the presence of a wind. However, the fits obtained by Chevalier and Li for GRB 980519 and GRB 980508 respectively yield $E_{52}/A_* \sim 0.1$ and 1 which appears much too low to avoid reverse and internal shock interaction.

- *A reduction of the shell Lorentz factor*

Instead of increasing the energy another possibility is to decrease the average Lorentz factor so that the mass of the shell is larger and internal shocks take place at a smaller radius. Still with $E_{52} = 2$ and $A_* = 1$, but with an average Lorentz factor divided by a factor of 3 the effect of the reverse shock becomes hardly visible and the original shape of the profile is essentially recovered. We have checked whether the shell remains optically thin relative to photon-photon annihilation in spite of the reduction of its Lorentz factor. The results depend on the unknown fraction α of radiation emitted above 511 keV in the shell rest frame. We found that for $\alpha \lesssim 0.1$ the opacity $\tau_{\gamma\gamma}$ at the shock radius always remains below unity.

Finally, if a supernova explosion precedes the burst by a few months to a few years as might be indicated by the high iron abundances detected in some X-ray afterglows [1],[7], the burst environment will be strongly affected by the explosion. The analysis presented in this paper would then have to be redone for the new structure of the external medium.

References

1. L. Amati, F. Frontera et al.: astro-ph/0012318 (2000)
2. R.A. Chevalier, Z.Y. Li: ApJ **520**, L29 (1999)
3. R.A. Chevalier, Z.Y. Li: ApJ **536**, 195 (2000)
4. F. Daigne, R. Mochkovitch: MNRAS **296**, 275 (1998)
5. E.E. Fenimore, E. Ramirez-Ruiz: astro-ph/9909299 (1999)
6. P. Kumar, A. Panaitescu: ApJ **541**, L9 (2000)
7. L. Piro, G. Garmire et al.: astro-ph/0011337 (2000)

GRB Synchrotron-Self-Compton Emission Generated by Self-Consistent Electron Distribution

E.V. Derishev[1], V.V. Kocharovsky[1], Vl.V. Kocharovsky[1], and P. Mészáros[2]

[1] Institute of Applied Physics, Russian Academy of Science,
   46 Ulyanov Street, 603600 Nizhny Novgorod, Russia
[2] Department of Astronomy and Astrophysics,
   Pennsylvania State University, University Park, PA 16803

1 The Equation for Synchrotron-Self-Compton Cooling

The spectra of Gamma-Ray Burst (GRB) emission have been investigated theoretically by a number of authors (e.g., [1–3]) for different versions of a general fireball-shock scenario [4,5]. However, the existing analytical models give rather coarse predictions and suffer from many uncertainties.

Below we present a refined analytical model of synchrotron-self-Compton emission, in which we assume a monoenergetic injection of electrons (with the Lorentz factor γ_{max}) and calculate their distribution over the Lorentz factors taking into account both synchrotron and inverse Compton losses. The latter, in turn, can be calculated provided the synchrotron spectrum (defined by electron distribution function) is known. Then the spectrum of ultra-energetic (comptonized) photons is easy to find by means of perturbation theory. The second and higher cascades of Comptonization are negligible in the particular physical conditions, which we found in GRB emitting regions [6].

Let us introduce a function $\ae(\gamma)$, which is the ratio of synchrotron luminosity $\mathcal{L}_s$ to the total luminosity $\mathcal{L}_s + \mathcal{L}_{ic}$ (including Compton losses $\mathcal{L}_{ic}$) for an electron with the Lorentz factor γ, so that

$$1/\ae = 1 + \mathcal{L}_{ic}/\mathcal{L}_s. \tag{1}$$

By definition, $0 < \ae < 1$. The synchrotron and inverse Compton luminosity of a single electron are

$$\mathcal{L}_s = \frac{4}{3}\sigma_T \gamma^2 c \frac{B^2}{8\pi} \quad \text{and} \quad \mathcal{L}_{ic} = \frac{4}{3}\gamma^2 c \int_0^\infty \sigma(\omega) w_{sy,\omega}\, d\omega, \tag{2}$$

where ω is the photon frequency and $w_{sy,\omega}$ the energy density of the synchrotron radiation per unit frequency interval. Below we assume that synchrotron radiation of an electron is monochromatic with the frequency $\omega(\gamma) = 0.5\gamma^2(eB/m_e c)$ and $\sigma(\omega) = \sigma_T \Theta(\omega_* - \omega)$, where Θ is the step function.

The integral in Eq. (2) is then reduced to the product of σ_T and the energy density of synchrotron radiation emitted by all electrons with a Lorentz factor less than γ_*, where γ_* satisfies the equation $\omega(\gamma_*) = \omega_*$. If the electron distribution over their Lorentz factors is stationary, it can be found simply as

$$N_e(\gamma) = \frac{S(\gamma)}{\dot{\gamma}} = \frac{S(\gamma)\, m_e c^2}{\mathcal{L}_s + \mathcal{L}_{\rm ic}} \propto \frac{\text{æ}(\gamma)\, S(\gamma)}{\gamma^2}, \qquad (3)$$

where $S(\gamma)$ is the electron injection rate integrated over $\gamma' \geq \gamma$; $S(\gamma) = S_e \Theta(\gamma_{\rm max} - \gamma)$ in the specific case of monoenergetic injection. Now we can find the energy density of synchrotron radiation below the Klein-Nishina cut-off (for an electron with the Lorentz factor γ):

$$w_{\rm sy}(\omega < \omega_*) = \frac{\tau S_e m_e c^2}{V} \int_1^{\gamma_*} \text{æ}(\gamma')\, d\gamma'. \qquad (4)$$

Here τ is the variability timescale and V the volume of the emitting region.

For convenience, we introduce a new variable $x \equiv \gamma/\gamma_0$, where γ_0 satisfies the relation $\gamma_*(\gamma_0) = \gamma_0$, i.e.,

$$\gamma_0 = \left(\frac{2 m_e^2 c^3}{\hbar e B} \right)^{1/3}. \qquad (5)$$

For electrons with the Lorentz factor γ_0 the Klein-Nishina cut-off is at their own synchrotron frequency. In GRBs, one has $1 \ll \gamma_0 < \gamma_{\rm max}$. Finally, we obtain from relation (1) the following equation for $\text{æ}(x)$:

$$\frac{1}{\text{æ}(x)} = 1 + K \int_0^{1/\sqrt{x}} \text{æ}(x')\, dx' \qquad (6)$$

(strictly speaking, the lower limit in the integral is $1/\gamma_0$). The above equation is valid for $x_{\rm max}^{-2} \leq x \leq x_{\rm max}$; for $x < x_{\rm max}^{-2}$ one has $\text{æ} = (1 + \tau_{\rm ic})^{-1}$, where $\tau_{\rm ic} \simeq 4/3\gamma_{\rm max}^2 \sigma_T n_e L$ and L is the size of the emitting region. The parameter K is defined as

$$K \equiv \frac{E_r \gamma_0}{W_m \gamma_{\rm max}} = \frac{\tau_{\rm ic}}{\bar{\text{æ}}} \left(\frac{m_e c^2}{\varepsilon_p \gamma_{\rm max}} \right)^{1/3}. \qquad (7)$$

Here ε_p is photon energy at the peak of synchrotron spectrum (in the comoving frame), $W_m = (B^2/8\pi)V$ is the total energy of magnetic field in the emitting region and $\bar{\text{æ}}$ the average value of æ in the interval $(0, x_{\rm max})$. The value of $\bar{\text{æ}}$ shows what fraction of the total energy radiated by GRB is emitted in sub-MeV spectral domain (we assume $\bar{\text{æ}} > 1/2$). In GRBs, the parameter K is within $0.1 \lesssim K \lesssim 100$, but in a typical burst it is of the order of unity.

Let us introduce the reference point $x_{1/2}$ so that $\text{æ}(x_{1/2}) = 1/2$. In the absence of analytical solution, one can use approximate expressions

$$\text{æ}(x) > \frac{1}{1 + K/\sqrt{x}}, \quad \text{and} \quad \text{æ}(x) \simeq \left[1 + \frac{2}{3x^{3/4}} \right]^{-1}, \qquad (8)$$

which are valid for $x \ll 1/x_{1/2}^2$ and $x \gg x_{1/2}^4$ respectively. Both asymptotics join more or less smoothly at $x_{1/2}$ if $x_{1/2} < 1$, so that Eq. (8) gives nearly complete knowledge about the function $\text{æ}(x)$. Otherwise, there is an intermediate region where both approximations are unsatisfactory.

2 Synchrotron and Inverse Compton Spectra

The GRB sub-MeV spectrum is just blueshifted synchrotron spectrum, and – as a function of $x \propto \sqrt{\omega}$ – it is actually given by $\ae(x)$. Indeed, the luminosity (in the shock comoving frame) of electrons with the Lorentz factors between γ and $\gamma + d\gamma$ is $dI = \ae(x)S(x)m_e c^2 d\gamma$ and $d\gamma = (d\gamma/d\omega)d\omega$. As a result,

$$I_\omega \propto \frac{\ae(x)S(x)}{\sqrt{\omega}}, \qquad x = \left(\frac{2\hbar^2}{m_e c^3 eB}\right)^{1/6}\sqrt{\omega}. \qquad (9)$$

In particular, Eq. (9) gives the following power-law asymptotics: $\omega I_\omega \propto \omega^{3/4}$ for $1/x_{max}^2 < x \ll \min[x_{1/2}, 1/x_{1/2}^2]$, and $\omega I_\omega \propto \omega^{1/2}$ for $x < 1/x_{max}^2$ and $x \gg \max[x_{1/2}, x_{1/2}^4]$ if $\mathcal{K} \sim 1$.

The above spectrum extends down to $\omega_r = (\varepsilon_p/\hbar)(x_r/x_{max})^2$. Below this frequency I_ω reproduces the synchrotron spectrum of a single particle because the electron distribution is cut at $x = x_r$, as at smaller electron energies the cooling timescale becomes larger than the GRB variability timescale.

The spectrum of comptonized radiation is represented analytically as a convolution of syncrotron spectrum with the same spectrum but taken as a function of re-scaled frequency. As far as the resulting inverse Compton spectrum is softer than the spectrum of those synchrotron photons which undergo Comptonization, one may take an approximation assuming that the most energetic electrons produce monochromatic comptonized radiation with photon energy $\varepsilon \simeq \gamma m_e c^2/2$. For the hardest part of inverse Compton spectrum this gives:

$$\varepsilon I_\varepsilon \propto (1 - \ae)\, x \propto \varepsilon^{1/4}, \qquad (10)$$

where $x = x_{max}\varepsilon/\varepsilon_{ic}$, $\varepsilon_{ic} = \gamma_{max}m_e c^2$, and Eq. (8) is used to substitute $(1 - \ae)$. The inverse Compton spectrum has a very broad maximum at $\simeq \varepsilon_{ic}/2$.

Among the most significant results of our analysis is the conclusion that the cooling distribution of electrons depends on the Compton losses and can produce a broad range of spectral indices. The low energy portion of the synchrotron spectrum can be as steep as $I_\omega = const$. Still steeper spectra with $\alpha > 1$ may be obtained in alternative ways (e.g., [7]), but this leads to low radiative efficiency of GRBs and rises concerns about the visibility of emission from fireball photosphere against the non-thermal component.

References

1. R. Sari, T. Piran: MNRAS **287**, 110 (1997)
2. G. Ghisellini, A. Celotti: ApJ **511**, L93 (1999)
3. C.D. Dermer, J. Chiang, K.E. Mitman: ApJ **537**, 785 (2000)
4. P. Mészáros, M.J. Rees: MNRAS **257**, P29 (1992)
5. P. Mészáros, M.J. Rees: MNRAS **269**, L41 (1994)
6. E.V. Derishev, V.V. Kocharovsky, Vl.V. Kocharovsky: astro-ph 0006239 (2000)
7. A. Panaitescu, P. Mészáros: ApJ **544**, L17 (2000)

Effects of Free Neutrons on Gamma-Ray Bursts

E.V. Derishev[1], V.V. Kocharovsky[1], and Vl.V. Kocharovsky[1]

Institute of Applied Physics, Russian Academy of Science,
46 Ulyanov Street, 603600 Nizhny Novgorod, Russia

We consider a general case of nucleon (neutrons + protons) – loaded fireballs. In our analysis of the relative motion of neutrons and protons in the relativistic wind we pay particular attention to fireballs of cosmological Gamma-Ray Bursts (GRBs).

Specific effects of the neutron component depend on whether the final Lorentz factor of a plasma wind exceeds some critical value or not. If yes, decoupling of the neutron and proton flows takes place giving rise to an electromagnetic cascade induced by pion production in inelastic collisions of nucleons [2]. Otherwise, all nucleons in the wind behave as a single fluid. In both cases neutrons can strongly influence a GRB by changing dynamics of a shock initiated by protons in the surrounding medium [3].

The value of Lorentz factor corresponding to decoupling of the neutron flow is estimated to lie in the range expected for cosmological GRBs. Depending on whether neutron and proton flows decouple or not and whether neutrons decay before the shock of proton origin decelerates significantly or after that, four types of bursts are possible. Their lightcurves have different appearance and a number of lightcurve features are expected to be correlated. For example, the height, delay time and duration of the secondary pulse are related to each other. Multi-peaked lightcurves may be as common as single-peaked ones.

Effect of neutron decoupling is of particular interest in the case of cosmological GRBs because the decoupling threshold, Γ_*, falls in the expected range of fireball's Lorentz factors, 200–1500. Hence, a confirmation of the decoupling may give important clues to understanding physical conditions in the GRB source itself. Among proposed observational tests we find detection of energetic 100 GeV-photons by ground-based telescopes to be the most promising.

The photons at the maximum of cascade spectrum, which appear blueshifted to 1-2 GeV for an observer on the Earth, also carry approximately the same or even larger energy than the uprocessed quanta from pion decay. Effective area of GLAST telescope allows detection of the cascade photons from a GRB at redshift $z \sim 0.1$, provided the GRB energy was 10^{52} erg. Other fireball emission features, associated with neutron flow decoupling, such as ejection of $\sim 10^{52}$ positrons and burst of neutrinos of energy ~ 30 GeV, are too week to be detected by existing instruments. It should be noted that structured fireballs (i.e., those with internal shocks and/or sharp angular boundaries) may emit neutrinos [3] and produce pairs without decoupling.

Free neutrons may take most of the source's power, but they do not contribute to the energy budget of a shock wave in surrounding medium until they decay into charged particles – protons and electrons. Therefore, if the neutron-induced

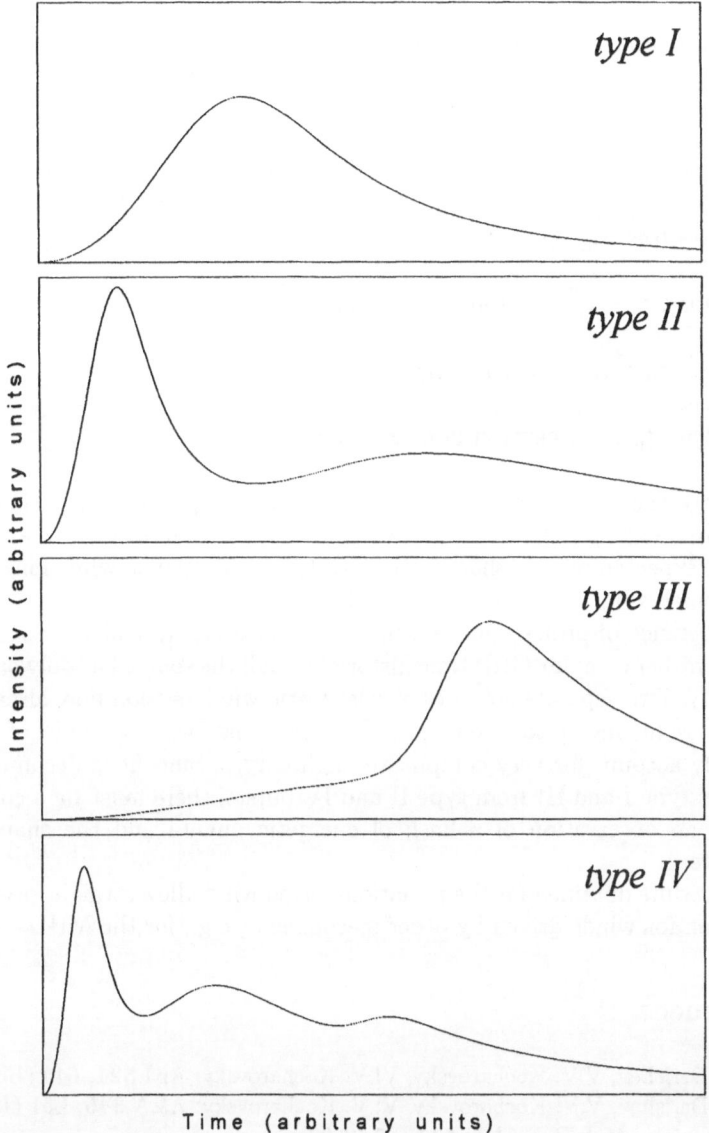

Fig. 1. Four types of GRB bolometric lightcurves. Proportions are not exactly maintained.

shock is visible in gamma-rays, duration of the main pulse of the burst should have a lower limit, $\sim 900/\Gamma_n$ s, contrary to the picture resulting from simple shock models [4–6].

The plasma component of a fireball pushes the surrounding medium and forms a shock at the interface, while neutrons propagate freely until they decay into charged particles. There are two independent alternatives: (i) the neutron

flow may decouple from the proton one or may not, and (ii) the lifetime of a free neutron, t_n, either exceeds or is smaller than the deceleration time of the proton shock, t_p. A source satisfying the condition of decoupling will be a typical GRB representative. The apparent lifetime of free neutrons, ~ 900 s $/\Gamma_n$, also appears to be comparable with duration of an ordinary burst. Therefore, the above two alternatives produce four combinations; each one gives rise to a distinct type of *First type:* $t_n < t_p$, no decoupling.
lightcurves (see Fig. 1):

Second type: $t_n < t_p$, neutron flow decouples.

Third type: $t_n > t_p$, no decoupling.

Fourth type: $t_n > t_p$, neutron flow decouples.

Clearly, there are no sharp boundaries between type I and type III bursts as well as between type II and type IV. However, lightcurves of first and second class are expected only in short GRBs with duration < 10 s, while longer bursts should possess lightcurves of third and forth class.

Interference of primary and secondary (from decayed neutrons) shocks may result in rather complex GRB time histories even if the source has only one period of activity. Two separate episodes of relativistic wind ejection may already yield a lightcurve having up to seven peaks. It is clear that two-flow model of a fireball can easily account for very complicated lightcurves. Since it is decoupling that separates type I and III from type II and IV bursts, there must be a correlation between an observation of a flash of energetic quanta and the shape of the lightcurve.

The results obtained for the radiation-driven wind allow straightforward generalization for winds driven by other mechanisms, e.g., for the MHD winds.

References

1. E.V. Derishev, V.V. Kocharovsky, Vl.V. Kocharovsky: ApJ **521**, 640 (1999)
2. E.V. Derishev, V.V. Kocharovsky, Vl.V. Kocharovsky: A&A **345**, L51 (1999)
3. P. Mészáros, M.J. Rees: ApJ **541**, L5 (2000)
4. M.J. Rees, P. Mészáros: ApJ **430**, L93 (1994)
5. R. Wijers, M.J. Rees, P. Mészáros: MNRAS **288**, L51 (1997)
6. E. Waxman, ApJ **489**, L33 (1997)

Bursts and Black Holes

Andreja Gomboc[1] and Andrej Čadež[1]

Faculty of Mathematics and Physics, University of Ljubljana, Slovenia

Abstract. We present numerical simulations of partial stellar capture by a black hole. The dominant phenomenon in the process is the rapid stripping of the stellar envelope, which occurs on a time-scale $\sim 10 R_{Schw}/c$, much shorter than the sound crossing time. Thus, the hot stellar interior is exposed just before the horizon crossing, which, combined with relativistic effects, produces a short, very intense and beamed burst, strongly intensity and Doppler modulated with respect to a far stationary observer. The time dependence of calculated intensity and color resemble those of observed gamma ray bursts.

1 The Model

In this work we study the accretion of a star by a black hole. It has been shown by Carter[1], Luminet, Marck[4] and Rees[5], among others, that tidal disruption rapidly gains momentum after the star crosses Roche radius $r_R = \left(\frac{M}{\rho_*}\right)^{1/3}$. At that moment tidal components of the stress energy tensor overwhelm the pressure and self-gravitation components. Therefore, we model the final stages of tidal disruption by neglecting self-gravity and pressure altogether and consider each gas particle of the star as freely falling. In this same spirit we also neglect exchange of energy between the gas and the photon reservoir. The photon gas temperature is assumed to remain constant with respect to the local rest frame of the gas until photons random walk out of the star through scattering on plasma electrons. This approximation would be perfectly justified if the proper time between the Roche radius crossing and total disruption was short compared to the dynamic time scale of the star. However, it can be shown that in our case the ratio of the two time scales is of order 1, so a nonnegligible amount of energy may be exchanged between the gas and photon energy reservoir. Nevertheless, we neglect this exchange for the sake of keeping the model simple. In fact, the models predict interesting gravitational focusing phenomena on characteristic time scales that are two orders of magnitude shorter than the dynamical time scale, so that they are independent of the mentioned energy exchange, after all. The only remaining uncertainty in our model is thus the amount of available energy.

The actual numerical model starts with a spherically symmetric star of radius r_* and mass M_* consisting of N equally massive parts ($m_i = M_*/N$, $N \sim 10^6$) distributed randomly but in such a way that in the average their mass distribution corresponds to the model mass distribution. Each part has a scattering cross section σ and is assigned a temperature according to its initial position inside a star. For the star we use the $n = 5$ polytrope model, which, because of its algebraic simplicity, allows analytic approximations that can be checked against numerical calculations. Thus:

$$T(r) = \frac{T_c}{\sqrt{1 + \frac{1}{3}(\frac{r}{r_0})^2}} \tag{1}$$

$$\rho(r) = \frac{\rho_c}{\left[1 + \frac{1}{3}(\frac{r}{r_0})^2\right]^{\frac{5}{2}}} \tag{2}$$

All parts of the star start with the same velocity, that corresponds to parabolic velocity of the center of mass, which is placed at a distance r_R from the black hole. Subsequently the position of the free falling stellar parts is calculated at later discrete times (t_i) and images of the star with respect to the far observer are formed as follows: photon trajectories and the time of flight between each stellar part and the observer are calculated. Only two trajectories connecting two space points are considered – the shortest one and the one passing the black hole on the other side, while those winding around the black hole by more than 2π are neglected. It has been shown before[2], that light following the winding trajectories contributes little. The beam contributions are sorted into image frames according to the arrival time. This also takes care of the obscuration of deep layers by the surface. The intensity of each beam is calculated assuming that each stellar part emitts in its own rest frame as a black body at its temperature. Successive stellar images so formed are pasted in a movie and the apparent luminosity and effective temperature (Wien's law) of the star as a function of time (with respect to the chosen observer) is calculated.

We point out that at present our numerical simulation can not quite model the idealized capture of a star presented above, since the average opacity coefficient $\kappa = \sigma N/M_*$ can not be made high enough to match the opacity coefficient of gas in an actual star ($\kappa \sim 1 \frac{cm^2}{g}$). Namely, if $\sigma = \pi r_*^2/N'$, one can write $\kappa = \frac{\pi r_*^2}{M_*} \frac{N}{N'} \sim 8 \times 10^{-12} \frac{cm^2}{g} \frac{N}{N'}$ and it is clear that it is impossible to attain the necessary ratio $\frac{N}{N'} \sim 10^{11}$ in a numerical calculation (for the Sun $\frac{\pi r_*^2}{M_*} \approx 8 \times 10^{-12} \frac{cm^2}{g}$). Therefore, all our model stars are too cool in their centers. We remedy this deficiency to some extent by calculating similar models with different N. It turns out that many model results scale with powers of N, so that it is possible to extrapolate them to the regime of the correct opacity.

Simulations show that the tenuous envelope of the star is sheared away from the hot central core, which becomes increasingly exposed to the far observer. As a result the stellar luminosity increases. The amount of the increase depends on the angular momentum of the star with respect to the black hole. It is quite dramatic, if the dimensionless stellar angular momentum is close to 4, i.e. if the star approaches the black hole so much that it is partially swallowed.

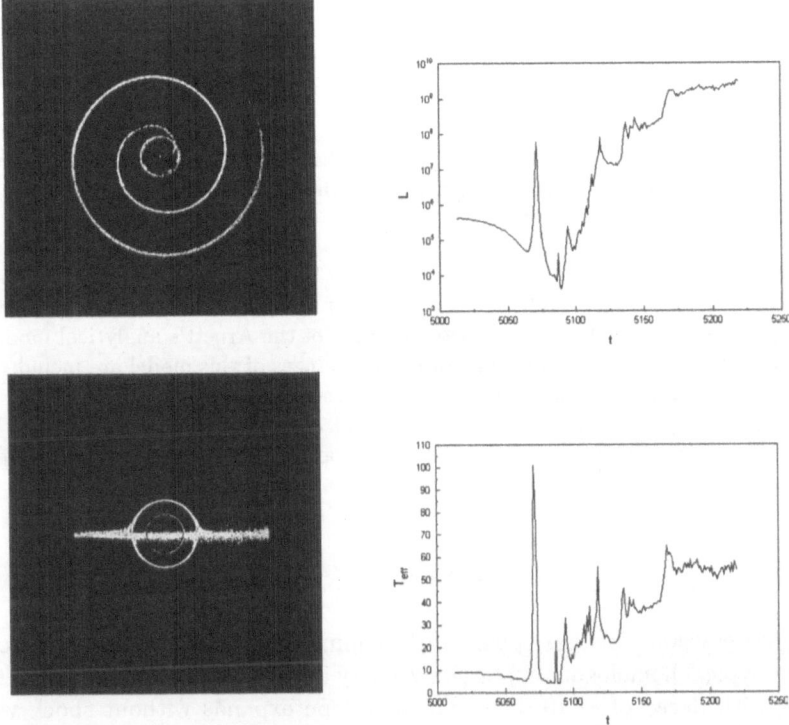

Fig. 1. A disrupted star seen perpendicular to the orbital plane and from the orbital plane (left), and the time dependence of the luminosity and of the effective temperature (right). More data can be obtained at *http://weber.fiz.uni-lj.si/ãndreja/*.

2   Concluding Remarks

Our model, although still crude, predicts that a medium size galactic black hole can completely disrupt a normal star by critical capture. In the process a large fraction of stellar thermal energy is released in a short burst, which is triggered just before half of the star crosses the horizon of the black hole. The appearance of such an event is calculated using analytic ray tracing in the gravitational field of the black hole. Some related ray-tracing results are in [3], [2] .

References

1. B. Carter, J.P. Luminet: *MNRAS* 212, 23 (1985).
2. A. Čadež, & A. Gomboc: *A&A Suppl. Ser.*119, 293 (1996).
3. A. Gomboc, & A. Čadež: *Variable Stars and the Astrophysical Returns of Microlensing Surveys*, R. Ferlet, et al. (eds) 413 (1996).
4. J.P. Luminet, J.A. Marck: *MNRAS* 212, 57 (1985).
5. M.J. Rees: *Nature* 333, 523 (1988).

Light Curves of Optical Afterglows

Filip Hroch[1,2]

[1] Faculty of Science, Masaryk University, Kotlarska 2, 611 37 Brno, Czech Republic
[2] Astronomical Institute, Academy of Science of the Czech Republic,
   251 65 Ondrejov, Czech Republic

Abstract. An alternative model for description of the light curve of the optical after-glow is presented. It has been constructed on base of the Arnett's analytical model of the light curve of the supernovae. The principal changes of this model are included by omitting of the radioactive heating and the relativistic expansion of the envelope. The shape of the computed light curve is more complicated than usual simple power law. The predicted light curves are compared to the optical light curves of the afterglow GRB 980508.

1 The Supernova Envelope

Supernova envelope is a gas sphere with domination of radiation pressure over gas. The typical homologous expansion velocity is $\sim 10^3 \mathrm{km/s}$ and the radioactive heating with decay of $\sim 10^2$ days. This envelope expands without shock wave after first few hours. Luminosity without fluency of recombination wave is:

$$L(t) = \frac{4\pi}{3} c \frac{R_0 E_0}{\kappa_0 M} (\alpha^2 I_M) \Phi(t)$$

where

$$\Phi(t) = e_1 e^{-\alpha^2 (t/\tau_0 + t^2/2\tau_h^2)} \int_0^t e^{-t/\tau_1} \left(1 + \frac{t}{\tau_h}\right) e^{\alpha^2 (t/\tau_0 + t^2/2\tau_h^2)} dt$$

and photospheric velocity is

$$v_f = x_e v_0 \approx \frac{T_0}{T_i} \frac{v_0}{1 + t/\tau_h}$$

This solution contains both parts: the cooling by the envelope expansion and the heating by the radioactive decay.

2 The Afterglow Envelope

Afterglow envelope also is a gas sphere with domination of radiation pressure over gas. The homologous expansion velocity is more greater $v \leq c$ and no radioactive heating opposite to supernova envelope. This envelope expands without shock

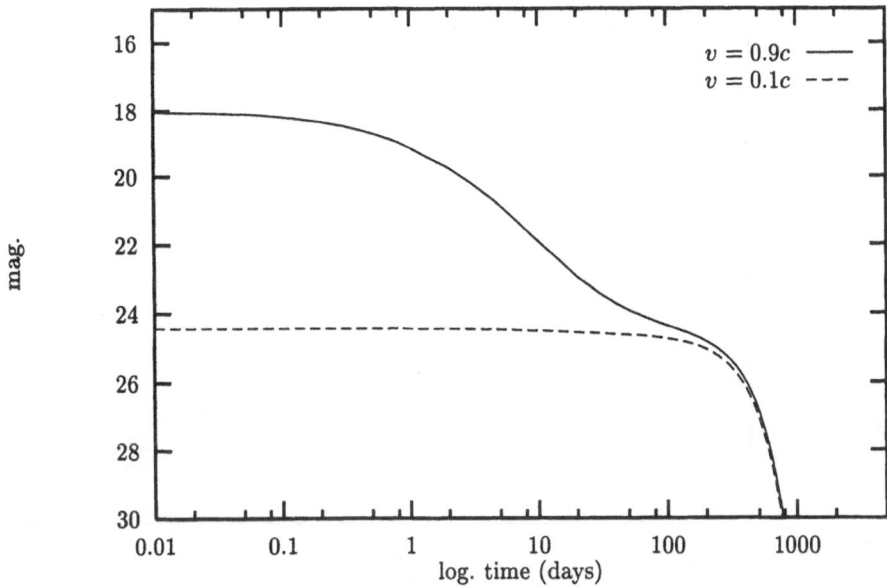

Fig. 1. This graph shows the calculated light curve of optical afterglow with parameters like GRB 970508. The three main parts are presented in logarithmic time scale: the constant part, the "power law" and the rapid fall of. The full line is for velocity 0.9c, the dashed line for 0.1c. Note, that the SN light curves are usually not plotted in logarithmic scale.

wave after first few hours and its luminosity (without fluency of recombination wave) is:

$$L(t) = \frac{2\pi c}{3} \frac{E_0 \mathcal{R}_0}{a \kappa_0 M} (\alpha^2 I_M) F(t)$$

where

$$F(t) \equiv \int_0^{\pi/2} \Phi(\tau) \frac{(1 - v^2)^{(\kappa-1)/2}}{(1 - v^2 \cos \vartheta)^{(3+\kappa)/2}} \sin \vartheta \cos \vartheta d\vartheta$$

photospheric velocity is $v_0/v_f = 1 + (t - t_0)(1 - v_f \cos \vartheta)/(\tau_h \sqrt{1 - v_f^2})$. Solution describes only the cooling by the expansion and the diffusive radiation.

3 Discussion

The present model was developed as generalisation of the supernova light curve model by W.D. Arnett and can reveal a few secrets of the optical afterglow of the GRB's. The important characteristics of this model is unified view to the main power law decay and the plateau (flattennig) at the end of the light curve. This model represents an alternative explanation of the flattenings observed at late stages of the light curves of OAs of some GRBs. This poster is a preliminary version of a paper which will describe a full way of the computing details.

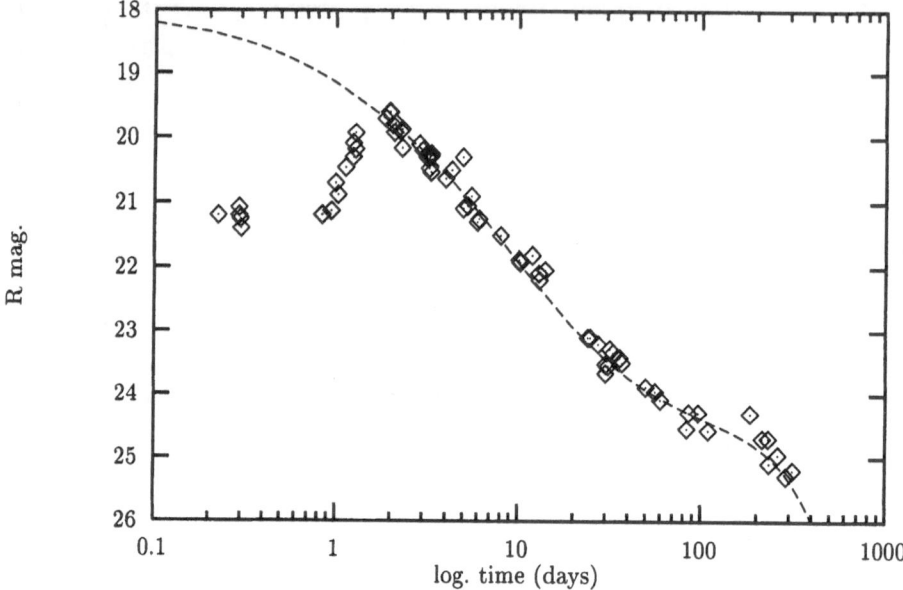

Fig. 2. The light curve of afterglow of GRB 970508. The expansion velocity is 0.9c and diffuse time in observer's frame is 3×10^4 days. The diamonds represent the data from [2]. The dashed line is plotted on base of this model.

References

1. Arnett, W.D.: Supernovae and Nucleosynthesis, Princeton University Press, 1996
2. Garcia et al.: Power-Law Decays in the Optical Counterparts of GRB 970228 and GRB 970508, Astrophys. J. **500**, L105–L108, (1998)
3. Hroch, F.: Light Curves of Optical Afterglow, PhD Thesis, MU Brno, 2000

Ultra Efficient Internal Shocks

Shiho Kobayashi[1] and Re'em Sari[2]

[1] Dept. Earth and Space Science, Osaka University, Toyonaka 560, Japan
[2] Theoretical Astrophysics, Caltech, Pasadena, CA91125, USA

Abstract. We define ultra efficient internal shocks as those in which the fraction of emitted energy is larger than the fraction of energy given to the radiating electrons at each collision. In our model, colliding shells which do not emit all their internal energy are reflected from each other and it causes subsequent collisions, allowing more energy to be emitted. As an example, we obtain about 60% overall efficiency even if the fraction of energy going to electrons is $\epsilon_e = 0.1$ provided that the shells' Lorentz factor varies between 10 and 10^4.

1 Introduction

It is argued that the conversion efficiency by the internal shocks from the bulk motion to gamma-ray is only 1% [4]. The argument is based on three points: (I) The hydrodynamic efficiency is typically 10% as we had estimated [3]. (II) It is only the electrons that are radiating. Even in equipartition among protons, magnetic field and electrons, the latter electrons only have a third of the internal energy. (III) The amount of the radiated energy within the gamma-ray band is about a third of the total radiated one. Combining these three factors give the low efficiency.

Such a low efficiency results in sever energy demands on the source. Moreover, it has problems explaining the energy ratio between the GRB and its afterglow. According to the internal-external shock model, the remaining kinetic energy, which was not converted to radiation by internal shocks, is radiated during the afterglow stage. External shock does not suffer from problem (I), and the energy released in the afterglow should be considerably higher than that in the GRBs. However, it seems to be that the energy during the afterglow is only a tenth of that during the GRB, rather than ten times larger.

In this paper, we suggest a simple solution, which overcomes the problems. We show that if the distribution of the Lorentz factor is not uniform, but instead its logarithm is distributed uniformly, then the typical ratio of Lorentz factors between neighboring shells is considerably larger. Then, the hydrodynamic efficiency can be close to hundred percents, even for a reasonable spread of Lorentz factors (Similar calculation was recently done by Beloborodov 2000). The main point of this paper is the possibility of "ultra efficient" internal shocks. We define ultra efficient internal shocks as a scenario in which the emitted fraction of kinetic energy is larger than ϵ_e, the fraction of internal energy that is going into electrons (and then radiated) at each collision.

2 "Ultra Efficient" Internal Shocks

Internal shocks could occur within a variable relativistic wind produced by a highly variable source. We represent the irregular wind by a succession of relativistic shells with a random distribution of Lorentz factors in a similar manner as in [3]. Beloborodov (2000) has shown that the internal shocks can convert most of the kinetic energy to internal energy if the fluctuation of the initial Lorentz factors $A^2 = (\langle \gamma^2 \rangle - \langle \gamma \rangle^2)/\langle \gamma \rangle^2$ is large. Though it is maximally $1/3$ if the initial Lorentz factors take random values between γ_{min} and γ_{max}, it is not limited if the distribution is uniform in logarithmic space between $\log \gamma_{min}$ and $\log \gamma_{max}$.

The efficiency can be estimated by an equation similar to equation 19 in [3]. The most efficient case is that the masses of the shells are taken equal, and that the efficiency is given by a simple form $\langle \epsilon \rangle \sim 1 - a^{1/2} \log a/(a - 1)$ where $a = \gamma_{max}/\gamma_{min}$. This analytic estimate fits the result of our numerical simulation, and has the asymptotic form $\langle \epsilon \rangle = A^2/2$ for a small fluctuation $A^2 = -1 + (a + 1) \log a/2(a - 1)$.

Though we have seen that a large hydrodynamic efficiency is possible if the fluctuation of the initial Lorentz factors is large, it is not reasonable that all the internal energy is emitted after each collision, since electrons do not have most of the internal energy. Even at equipartition with protons $\epsilon_e = 1/2$. Under these circumstances the total emitted energy is still limited by ϵ_e. However, if $\epsilon_e < 1$ the merger produced by a collision is expected to stay hot after the emission. As a result, the merger will spread to transform the remaining internal energy back to the kinetic energy. A simplified description of this process is to assume that the two shells reflect with a smaller relative velocity. The difference of the kinetic energy before and after the collision is the emitted internal energy (see [2] for the formula).

Since the reflecting shell collide into the outer neighbor shell, the index of the shell which has a high Lorentz factor propagates outward until the high value decays by the radiation loss. The shell itself which has initially a high Lorentz factor might not go through many collisions, but its high "kinetic energy" does. Therefore, the internal shock process is very efficient even if the fraction of internal energy emitted at each collision is small. Previously, the efficiency in the case of $\epsilon_e < 1$ had been estimated as smaller by a factor of ϵ_e than that in the corresponding fully radiative case. Our ultra efficient internal shocks scenario shows this to be a significant underestimate. Though the efficiency that we have estimated is bolometric efficiency, the efficiency from the kinetic energy of the shells to gamma-ray radiation is also high if the faction of the energy radiated in the BATSE band is not very small.

Numerous collisions happened in our ultra-efficient internal shocks model, it made the peak width wider than in the previous internal shocks model. However, the number of main peaks is still almost the same as the number of shells that the source emitted. There is a strong correlation between the time at which we observe a pulse and the emission time of the corresponding shell from the source.

This correlation persists even for a small ϵ_e case where larger number of collision happen. The temporal structure reproduces the activity of the source.

We have shown that the efficiency of the internal shock process is not limited by ϵ_e, while that of the external shock is so. If a fraction ζ of kinetic energy of an explosion is converted to the radiation by the internal shocks, all the remaining one is converted to the thermal by external shock in the afterglow stage, and a fraction ϵ_e of the thermal is emitted. Then, we can roughly estimate the ratio between the energy released in afterglow and that in GRB as $\sim \epsilon_e(1 - \zeta)/\zeta$. Assuming $\epsilon_e = 0.1$, the ratio is $1/10$ for $\zeta = 0.5$ and decrease as ζ increase. If the efficiency of internal shocks is indeed very large, the luminosity of GRB and the afterglow are expected to be anticorrelated.

References

1. A. Beloborodov: Astrophysical Journal Letter **539**, L25 (2000).
2. S. Kobayashi and R. Sari, Astrophysical Journal (2001) in press.
3. S. Kobayashi, T. Piran and R. Sari, Astrophysical Journal **490** 92 (1997).
4. P. Kumar, Astrophysical Journal Letter **523**, L113 (1999).

Light Curves of GRB Optical Flashes

Shiho Kobayashi

Department of Earth & Space Science, Osaka University, Toyonaka 560, Japan

1 Introduction

The emission from the reverse shock is sensitive to the initial properties of the fireball. The observations can provide some important clues on the nature of the GRB source. Previous studies focused on the emission at the peak time. In this paper we calculate the full light curves for several frequency regimes. We also make some comments on the lack of the prompt detections by ROTSE for GRB 981121 and GRB 981223.

2 Hydrodynamics of a Relativistic Shell

We consider a relativistic shell with an energy E, a Lorentz factor η and a width in laboratory frame Δ_0 expanding into a surrounding medium (ISM) with a particle number density n_1. The evolutions with different parameters are classified into two cases, the thick shell case and the thin shell case [5]. In this paper we discuss only the thick shell case $\Delta_0 > l/2\eta^{8/3}$ where l is the Sedov length $l = (3E/4\pi n_1 m_p c^2)^{1/3}$ (see [3] for the thin shell case).

In this case, the reverse shock becomes relativistic before it crosses the shell. The scalings of the hydrodynamic variables in terms of the observer time $t = R/2c\gamma^2$ are given by $\gamma \propto t^{-1/4}, n \propto t^{-3/4}, p \propto t^{-1/2}$ and the number of the shocked electrons $N_e \propto t$. After the reverse shock crosses the shell, the profile of the forward shocked ISM begins to approach the Blandford-McKee (BM) solution [2,5,4], $\gamma \propto t^{-7/16}$, $n \propto t^{-13/16}$, $p \propto t^{-13/12}$ and $N_e = \text{constant}$.

3 Light Curves of the Reverse Shock Emission

We consider now the synchrotron emission from the reverse shocked shell. The shock accelerates electrons in the shell material into a power law distribution. We assume that constant fractions ϵ_e and ϵ_B of the internal energy go into the electrons and the magnetic field respectively. The spectrum is given by the broken power laws discussed in [9], and has two breaks at the typical synchrotron frequency ν_m and at the cooling frequency ν_c. The peak flux $F_{\nu,max}$ is obtained at the lower of the two frequencies. The scalings before and after the shock crossing are given by $t < T$: $\nu_m = \text{constant}$, $\nu_c \propto t^{-1}$, $F_{\nu,max} \propto t^{1/2}$ and $t > T$: $\nu_m \propto t^{-73/48}$, $\nu_c \propto t^{1/16}$, $F_{\nu,max} \propto t^{-47/48}$. It is interesting that ν_m is constant during the shock crossing.

With the typical values of the parameters, the spectrum is the slow cooling one throughout the evolution. The light curves are different among three frequency regimes separated by two frequency, ν_m and ν_c at the shock crossing time. The typical light curves are shown in [3]. The flux at a frequency above $\nu_c(T)$ disappears at the crossing time because no electron is shocked anymore, while the flux at $\nu < \nu_c(T)$ vanishes when the cut off frequency decreasing as $t^{-73/48}$ due to the adiabatic expansion crosses the frequency.

4 GRB 981121 and GRB 981223

The theory succeeded for GRB 990123 [8,4], but the optical flash was detected only for it so far. Akerlof et al. (2000) reported no detections of the optical flashes to six GRBs with small localization errors. Especially, GRB 981121 and GRB 981223 are the most sensitive bursts in the sample. If the optical flashes are correlated with the GRB fluences, the optical emission should be more than 2 mag over the ROTSE detection thresholds. The thick line in figure 1a shows the expected light curve for GRB 981121.

GRB 990123 was an exceptionally energetic burst. The energies of GRB 981121 and GRB 981223 should be considerably lower to explain the lower fluences. Hereafter we assume $E = 10^{52}$ ergs for the two bursts. Since GRB 990123 is a marginal case, a burst with a lower E and a comparable n_1, η and T is classified into the thick shell case in which $F_{\nu,max}$ is proportional to $E^{5/4}$ instead of E. The thin line in figure 1a depicts the corrected light curve which is still above the thresholds.

A possible solution to the problem is to assume that the reverse shock energy is radiated at a non-optical frequency, $\nu_m \ll \nu_R$ or $\gg \nu_R$. The typical frequency ν_m is proportional to $\epsilon_e^2 \epsilon_B^{1/2} n_1^{1/2} \eta^2$, but the values of ϵ_e and ϵ_B are determined by the micro-physics and are likely to be universal. Then, n_1 or η for the bursts should be different from the correspondence for GRB 990123, $n_1 = 0.2$ protons/cm^3 and $\eta = 270$ [3].

If $\nu_m < \nu_R$, the no detections give upper limits on n_1 and η. Assuming $n_1 = 0.2$ protons/cm^3 (or $\eta = 270$), we get $\eta < 135$ (or $n_1 < 0.07$ protons/cm^3) for GRB 981121. The light curves with a low η or n_1 are shown as the thin lines in figure 1b and 1c. If $\nu_m > \nu_R$, assuming $n_1 = 0.2$ protons/cm^3 (or $\eta = 270$), we get $\eta > 400$ (or $n_1 = 2 \times 10^5$ protons/cm^3). The light curve for a large η or n_1 is shown as the thick lines in figure 1b and 1c.

A large ISM density is an unlikely reason to explain the no detections, because we need to require several order larger density, and with the large density the peak power $F_{\nu,max}$ itself is very large. However, there is a possibility of the extinction. Though we have normalized the optical flux according to the GRB fluence, gamma-rays do not suffer any kind of extinction. Within the six bursts reported by the ROTSE group, only the location of GRB 980329 was determined precisely by BeppoSAX, and the optical afterglows were observed hours later. However, the ROTSE observation on this burst was not so sensitive. Future observations will provide additional information to this effect.

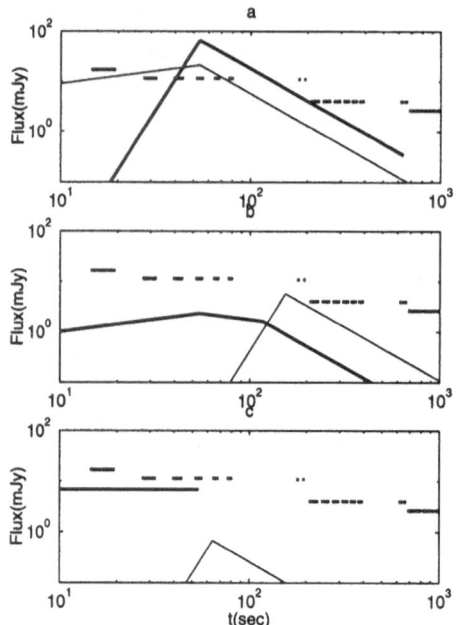

Fig. 1. GRB 981121: ROTSE detection thresholds (segments) and the theoretical light curves

5   Conclusions

We have constructed the full light curves of the reverse shock emission. The typical synchrotron frequency is constant in the thick shell case while it increases rapidly as t^6 in the thin shell case [3].

The lack of the prompt optical detection by ROTSE for GRB 981121 and GRB 981223 can be explained if the initial Lorentz factors or the surrounding medium densities are slightly different from that in GRB 990123 case.

References

1. Akerlof,C.W. et al.: ApJ, **532**, L25 (2000)
2. Blandford,R.D. & McKee,C.F.: Phys of Fluids, **19**, 1130 (1976)
3. Kobayashi,S. 2000, ApJ, **545**, 907 (2000)
4. Kobayashi,S. & Sari,R. : ApJ, **542**, 819 2000
5. Kobayashi,S., Piran,T. & Sari,R. : ApJ, **513**, 669 (1999)
6. Mészaros,P. & Rees,M.J. : ApJ, **476**, 231 (1997)
7. Sari,R. & Piran,T. : ApJ, **520**, 641 (1999a)
8. Sari,R. & Piran,T. : ApJ, **517**, L109 (1999b)
9. Sari,R. & Piran,T. & Narayan,R. : ApJ, **497**, L17 (1998)

100 GeV Photons
from Gamma-Ray Burst Fireballs

Vl.V. Kocharovsky[1], E.V. Derishev[1], and V.V. Kocharovsky[1]

Institute of Applied Physics, Russian Academy of Science,
46 Ulyanov Street, 603600 Nizhny Novgorod, Russia

Abstract. We find that there is an efficient way of generating high energy photons (from sub-GeV to 100 GeV) in fireballs of Gamma-Ray Bursts (GRBs). We present analytic theory of the phenomenon, obtain the expected spectra and derive their parameters. The radiation of this kind carries an essential information about GRB sources and can be detected by means of existing (or planned) instruments.

1   Introduction

In an electromagnetic cascade, initial high-energy photons are absorbed by the bulk GRB radiation producing primary electron-positron pairs. Bulk photons gain energy due to inverse Compton scattering off these energetic particles and may give rise to another generation of pairs, etc. In general, this complicated problem can be solved by numerical methods only.

However, we find its analytic solution for a one-step regime [1], when the photons upscattered by the primary pairs are not enough energetic to create secondary pairs. It is the case in fireballs of cosmological GRBs and microquasars (Galactic sources with relativistic jets). In both cases the cascade is started by gamma-quanta from the decay of neutral pions, which in turn result from inelastic proton-neutron collisions in a relativistic wind [2]. The energy spent for production of sub-relativistic pions may be as high as 10% of the total energy.

The post-cascade spectrum is composed of two power-laws with a break at the cascade threshold energy, and most of the emission is concentrated near this break. The spectrum, which in the wind comoving frame peaks at several MeV, is blueshifted by a factor $\Gamma \sim 10^3$ (the typical Lorentz factor of GRB fireballs) to GeV spectral domain.

The photosphere for unprocessed quanta from the pion decay differs from the Thomson photosphere for low-energy quanta and is defined by two-photon absorption on the bulk radiation. The self-consistent solution of the problem allows us to determine both the location of the absorption photosphere and the fraction of the total energy transported by the unprocessed quanta.

2   The Fireball Structure

In a global view, a fireball has an onion-like structure [2]. As the plasma wind streams outwards it first encounters a radius R_d where the neutron flow decouples, ceases accelerating, and the pion-induced electromagnetic cascade begins.

At a larger distance R_s the Lorentz factor of the proton flow also saturates and the energy of proton-neutron collision approaches its limiting value. Beyond the saturation radius a last annihilation surface is located; starting from this point, the total positron flux practically does not change with radius.

The photospheric radius R_{ph} (typically in the range $3 \cdot 10^{10} - 3 \cdot 10^{11}$ cm) is almost 16/3 times larger than that of the last annihilation surface. The effective photosphere for energetic quanta from π^0 decay is located at an even greater distance, $R_a \sim 3R_{ph}$. It is the neutron component present in a fireball that is responsible for the development of electromagnetic cascade and for the appearance of new spatial scales – the effective photosphere for energetic quanta, the last annihilation surface, and the characteristic radius of neutron flow decoupling. The effects of neutrons in the fireball dynamics are analyzed in detail in the papers [2,3].

3   Pion-Induced Electromagnetic Cascade

Decoupling of proton and neutron flows in ultrarelativistic wind causes pion production in inelastic collisions between protons and neutrons[2]. Decay of both charged and neutral pions leads to the injection of electrons and positrons with energies $\gtrsim 35$ MeV, since the bulk radiation is very opaque for 70 MeV photons from π^0 decay. Charged pions are also responsible for the emission of ~ 30 GeV neutrinos.

Ultrarelativistic electrons and positrons scatter off bulk photons, thus initiating electromagnetic cascade – multiplication of energetic quanta and e^+e^--pairs. Description of the process may be considerably simplified when it proceeds in one step, i.e., an injected electron has a Lorentz factor high enough to produce photons with the energy above the threshold value ε_t, but secondary electrons have not. The corresponding conditions are met in the GRB fireballs. Also, under these conditions the scattering proceeds in the classical (Thomson) regime.

Transferring energy to bulk photons, an electron (positron) gradually decelerates and the characteristic deceleration time is smaller than the annihilation lifetime. So, a positron (electron) in a fireball becomes non-relativistic long before it annihilates. Due to large number of cold leptons present in a fireball the value of ε_t, that marks the lowest energy at which two-photon absorption losses still prevail over Comptonization, appears to be only several times greater than $m_e c^2$.

4   The Cascade Output

Location of the effective photosphere for energetic quanta from pion decay having $\varepsilon_i \gg \varepsilon_t$ is defined by absorption on the soft cascade photons. Setting absorption depth for energetic photons to unity, we obtain a fraction of the total fireball energy carried away by these photons:

$$\delta E_{\gamma\pi} \sim 10^{-3} \left(\frac{\Gamma_p}{\Gamma_*}\right)^{7/30}, \tag{1}$$

where Γ_p is the terminal Lorentz factor of proton component and Γ_* the neutron decoupling threshold.

Most of the cascade output is concentrated near ε_t since for photons of that energy the photosphere location is closest to the central source. When absorbed, each quantum from pion decay produces ~ 10 photons with energy around ε_t. After correction for different photospheris radii at ε_t and ε_i, we obtain that photon fluence at the maximum of cascade spectrum is related to the fluence of unprocessed quanta in the following way:

$$F_{\text{cas}} \sim 15 \left(\frac{\Gamma_p}{\Gamma_*}\right)^{4/3} F_{\gamma\pi}. \tag{2}$$

Thus, photons at the maximum of cascade spectrum, which appear blueshifted to 1-2 GeV for an observer on the Earth, carry approximately the same or even larger energy than the uprocessed quanta from pion decay. Effective area of GLAST telescope allows detection of the cascade photons from a GRB at redshift $z \sim 0.1$, provided the GRB energy was 10^{52} erg.

References

1. Derishev E.V., Kocharovsky V.V., Kocharovsky Vl.V. Cosmological GRBs from collapse of a neutron star induced by a primordial black hole // Proceedings of the 19th Texas Symposium on Relativistic Astrophysics, Paris, France, December 14-18, (1998).
2. E.V. Derishev, V.V. Kocharovsky, and Vl.V. Kocharovsky, ApJ 521, 640, (1999).
3. Derishev E.V., Kocharovsky V.V., Kocharovsky Vl.V., A&A 345, L51, (1999).

New Results on the Temporal Structure of GRBs

E. Nakar and T. Piran

Racah Institute, Hebrew University, Jerusalem 91904, Israel

Abstract. We analyze the temporal structure of long ($T_{90} > 2sec$) and short ($T_{90} < 2sec$) BATSE bursts. We find that: (i) In many short bursts $\delta t_{min}/T \ll 1$ (where δt_{min} is the shortest pulse). This indicates that short bursts arise, like long ones, in internal shocks. (ii) In long bursts there is an excess of long intervals between pulses (relative to a lognormal distribution). This excess can be explained by the existence of *quiescent times*, long periods with no signal above the background that arise, most likely, from periods with no source activity. The lognormal distribution of the intervals (excluding the *quiescent times*) is similar and correlated with the distribution of the pulses width, in agreement with the predictions of the internal shock model.

Introduction

The variability of GRBs provided the main evidence for the internal-external shocks scenario. External shocks cannot produce efficiently such variability [1]. Internal shocks can produce such temporal structure provided that there are two time scale within the "inner engine" – a short time scale that produces the variability and a long time scale that determines the duration of the burst. So far variability was shown only for long bursts. It is an open question whether short bursts arise in internal shocks as well. Using a new algorithm [2] we study their variability. We also present some new results on the temporal structure of long bursts. Our results provide further support for the internal shocks scenario and show that three different time scales operate within the "inner engine".

Variability of Short Bursts

We analyze the distribution of $\delta t_{min}/T$ (where δt_{min} is the duration of the shortest pulse in a burst, and T is the total duration of the burst) in a sample of the brightest 33 short bursts (peak flux in 64ms>$4.37ph/(sec \cdot cm^2)$) with a good TTE data coverage (for BATSE data types review [3]). The TTE data is binned into 2msec time bins. We compare the results to a sample of 34 long bursts with the same peak flux, using the 64ms concatenate data to which we have added a Poisson noise so that the signal to noise ratio of both samples would be similar. We call this later sample the *'noisy long'* set.

Fig 1 depicts $\delta t_{min}/T$ in both data sets: *'short'* and *'noisy long'*. In the *'short'* set the median of $\delta t_{min}/T$ is 0.25. 35% of bursts have $\delta t_{min}/T < 0.1$ and 35% of the bursts show a smooth structure ($\delta t_{min}/T = 1$). This result could

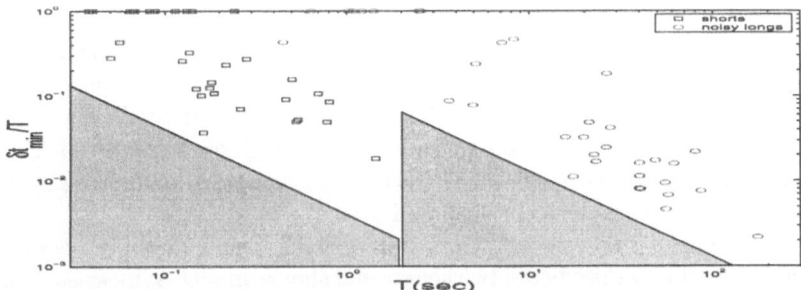

Fig. 1. The shortest pulse-δt_{min} represented as a function of the total duration of the burst. The shaded areas are excluded because of the data resolution (4ms for shorts and 128ms for noisy longs) or the data definition ($\delta t_{min} \leq T$). Notice that: (i) δt_{min} and T are not correlated. (ii) Some of the results are approaching the resolution limit. (iii) In some of the bursts $\delta t_{min} = T$.

mislead us to the conclusion that a significant fraction of the short bursts have a smooth time profile. But a comparison with the *'noisy long'* results show that also in this group more than 20% of the bursts are single pulsed, while there were no such bursts in the original (without the added noise) *'long'* set.

We conclude that short bursts are variable and hence are most likely produced in internal shocks. While the observed variability is not as large as seen in long bursts one has to remember that when studying variability of short bursts we are approaching the instrumental limitations both in terms of the time scales and of the signal to noise ratio. It is possible that 10%-20% of the short bursts are produced by external shocks.

The Pulses' Width and the Intervals between Pulses

According to the internal shocks model [4] the source ejects relativistic shells with different velocities and shocks arise when faster shells catch slower ones. We show in [5] that both the pulses' width δt and the intervals between pulses Δt are proportional to the same parameter – the separation between two following shells, namely the variability time scale of the "inner engine". Therefore both distributions should be similar. Moreover, any interval should be correlated to the width of its neighboring pulses.

We have applied our algorithm to a sample of the 68 brightest long bursts in BATSE 4B catalog (peak flux in 64ms>$10.19ph/(sec \cdot cm^2)$). This resulted in 1330 pulses (1262 intervals). Our null hypothesis was that both δt and Δt, have lognormal distributions. The χ^2 test gives a probability of 0.52 that the pulses width were taken from a lognormal distribution with $\mu = 0.065 \pm 0.04$ ($\overline{\delta t} \approx 1sec$) and $\sigma = 0.77 \pm 0.03$ (1σ corresponds to δt between 0.5 and 2.3sec).

The Δt distribution shows, however, an excess of long intervals relative to a lognormal distribution. The χ^2 probability for a lognormal distribution is $1.2 \cdot 10^{-10}$. McBreen[6,7] and Li & Fenimore[8] suggest that this deviation is due to the limited resolution (64ms). However, fitting the intervals above the median with a half Gaussian fails. The inconsistency is not due to the resolution.

Many of the long intervals are dominated by a quiescent time: periods within the burst with no observable counts above the background noise. When excluding *all* the intervals that contained a quiescent time the χ^2 probability that the data is lognormal is 0.27, with $\mu = 0.257 \pm 0.051$ ($\overline{\Delta t} \approx 1.3 sec$) and $\sigma = 0.90 \pm 0.04$ (1σ corresponds to Δt between 0.53 and 3.1sec).

The similarity between the parameters of both distributions is remarkable. Moreover, we find, as predicted by the internal shocks model, a linear correlation, r, between intervals and the following pulses. The average r is 0.48, showing a strong correlation.

Conclusions

For most short bursts $\delta t_{min}/T \ll 1$. This suggests that these bursts are produced by internal shocks. If, later, the ejecta encounters a surrounding ISM then we expect it to produce an external shock and emit an afterglow. For some (30% of our sample) short bursts $\delta t_{min} \approx T$. However, a comparison with the *'noisy long'* set, shows that this feature could very well be due to the noise. We cannot rule out the possibility that 10%-20% of the short bursts are produced by external shocks or by a single internal collision.

The distribution of interval between pulses shows an excess of long intervals relative to a lognormal distribution. After removing intervals that include quiescent times the distribution is consistent with a lognormal distribution with comparable parameters to the pulse width distribution. This result suggests that the Δt distribution is made from the sum of two different distributions: A lognormal distribution that is also compatible with the δt distribution and the quiescent times distribution. As Δt reflects the central engine behavior, this suggests that there are two different mechanisms operating within the source. A short time scale mechanism, with a lognormal distribution and a longer time scale mechanism that turns the central engine on and off and is responsible for the quiescent times. The correlation between the intervals width and the duration of the neighboring pulses width, confirms this suggestion and is in an excellent agreement with the internal shocks model.

References

1. Sari, R. & Piran, T., 1997, ApJ 485, 270
2. Nakar, E. & Piran, T., 2001A in preparation
3. Scargle, J. D. 1998, ApJ v.504, p.405
4. Rees. M. J. & Meszaros, P.,1994, ApJ, 430, L93
5. Nakar, E. & Piran, T., 2001B in preparation
6. McBreen B. personal communication
7. McBreen, B., et. al. 1994, MNRAS, 271, 662
8. Li, Hui & Fenimore, E., 1996, ApJ 469, L115

Signature of a Highly Magnetized Millisecond Pulsar in GRB Afterglows

Bing Zhang and Peter Mészáros

Pennsylvania State University, PA 16802, USA

Abstract. We discuss the consequences of a continuously injecting central engine on the Gamma-Ray Burst afterglow emission, and present a possible signature of a magnetar millisecond pulsar in gamma-ray bursts afterglow lightcurves.

Present fireball models usually assume prompt energy injection into the fireball, while in some types of central engines, such as a fast-rotating high-field pulsar (magnetar) or a black hole plus a long-lived debris torus system, energy inputs into the fireball may continue for a longer time scale. Therefore, there is a need to investigate a continuously-fed fireball in more detail. A more detailed analysis on this topic is presented in [8], and previously, also in [3].

1 Continuous-Injection Dynamics

We consider a central engine which emits both an initial impulsive energy input $E_{\rm imp}$ as well as a continuous luminosity, the latter varying as a power law in the emission time. In terms of time measured by observers, the energy conservation equation can be expressed as [2,8]

$$E = \left(\frac{L_0}{\kappa+q+1}\right)\left(\frac{T}{T_0}\right)^q T + E_{\rm imp}\left(\frac{T}{T_0}\right)^{-\kappa}, \quad T > T_0. \tag{1}$$

Here $L = L_0(T/T_0)^q$ is the injection luminosity from the central engine, T_0 is the characteristic timescale for the formation of a self-similar solution, which is roughly equal to the time for the external shock to start to decelerate due to the back-reaction of the external medium, q and κ are dimensionless constants, and $q \neq -1 - \kappa$. Setting $T = T_0$, the total energy at the beginning of the self-similar expansion is the sum of two terms, $E_0 = L_0 T_0/(\kappa+q+1)+E_{\rm imp}$. The first term, for $q > -1 - \kappa$, is the accumulated energy from the continuous injection (with radiative corrections) before the self-similar solution starts, while the second term, $E_{\rm imp}$, is the energy injected impulsively by the initial cataclysmic event. When $q > -1 - \kappa$, the first term in (1) will eventually dominate over the second term after a critical time

$$T_c = \text{Max}\left\{1, \left[(\kappa+q+1)\left(\frac{E_{\rm imp}}{L_0 T_0}\right)\right]^{1/(\kappa+q+1)}\right\} T_0. \tag{2}$$

In this regime, the blastwave dynamics is dominated by injection, and one has $m = (2 - q)/(2 + q)$ (cf. [1]), if $\Gamma^2 \propto t^{-m}$ is assumed, where t is the time measured in the fixed frame. The blastwave dynamics is then changed to $\Gamma \propto r^{-(2-q)/2(2+q)} \propto T^{-(2-q)/8}$ and $r \propto T^{(2+q)/4}$, and the temporal indices of the afterglow lightcurves are altered during the injection dominated phase. Central engines endowed with a continuous injection term may, in addition, have its own characteristic timescale $\mathcal{T}$, e.g. at which the continuous injection power law index (say $q_1 > -1 - \kappa$) switches to a lower value $q_2 < -1 - \kappa$. The interval of $\mathcal{T}_c < \mathcal{T} < \mathcal{T}$ is the regime when the continuous injection law has a noticeable effect on the afterglow light curve.

2   Magnetar Millisecond Pulsar as the Central Engine

Although a wide range of GRB progenitors lead to a black hole–debris torus system, some progenitors may lead to the formation of a highly-magnetized rapidly rotating pulsar. During their formation, an impulsive energy component due to either $\nu\bar{\nu}$ annihilation, dissipation of initial differential rotation, or phase conversion, is released accompanied with a continuous injection energy component due to the spinning-down of the star via (mainly) electromagnetic dipolar radiation. The luminosity law of the continuous injection is

$$L(T) = L_{em,0} \frac{1}{(1+T/\mathcal{T}_{em})^2}$$
$$\simeq \begin{cases} L_{em,0}, & T \ll \mathcal{T}_{em} \\ L_{em,0} \left(\frac{T}{\mathcal{T}_{em}}\right)^{-2}, & T \gg \mathcal{T}_{em}, \end{cases} \tag{3}$$

with

$$\mathcal{T}_{em} = \frac{3c^3 I}{B_p^2 R^6 \Omega_0^2} = 2.05 \times 10^3 \text{ s } I_{45} B_{p,15}^{-2} P_{0,-3}^2 R_6^{-6}, \tag{4}$$

The injection dominated phase is then $\mathcal{T}_c < \mathcal{T} < \mathcal{T}_{em}$, which constraints the phase space of the initial parameters of the pulsar at birth. For $\kappa = 0$ and $q = 0$ (the initial pulsar case), Eq.(2) is simplified to $\mathcal{T}_c = \text{Max}(1, E_{imp}/L_0 T_0)T_0$. When $E_{imp} \lesssim L_0 T_0$ (regime I), we have $\mathcal{T}_c = T_0 \simeq 0.33\text{s } B_{p,15} P_{0,-3}^{-2} R_6^3 (\Gamma_0/300)^{-4} n^{-1/2}$. The condition $\mathcal{T}_{em} > T_0$ then implies

$$B_{p,15} < 18.4 P_{0,-3}^{4/3} I_{45}^{1/3} R_6^{-3} (\Gamma_0/300)^{4/3} n^{1/6}. \tag{5}$$

When $E_{imp} > L_0 T_0$ (regime II), we have $\mathcal{T}_c = E_{imp}/L_{0,em} = (2E_{imp}/I\Omega_0^2)\mathcal{T}_{em}$. The condition $\mathcal{T}_{em} > \mathcal{T}_c$ is simply $E_{imp} < (1/2)I\Omega_0^2$, or

$$P_{0,-3} < 4.4 I_{45}^{1/2} E_{imp,51}^{-1/2}. \tag{6}$$

Since both $\mathcal{T}_{em}$ and $\mathcal{T}_c$ are large in this case, the continuous injection term may dominate at a later time. The lines (5), (6) together with the dynamically stable condition of the pulsar, i.e., $P_0 > P_0(\text{min})$, define a region of the $P_0, B_{p,0}$ initial parameter space of the pulsar, inside which the continuous injection has the pulsar signature. (see Fig.1 of [8]). The separation between the two regimes (I,II) is defined by the condition $L_{em,0}T_0 \simeq 10^{51}E_{imp,51}$ erg, or $B_{p,15} \simeq 6.7 P_{0,-3}^2 R_6^{-3}(\Gamma_0/300)^{4/3} n^{1/6} E_{imp,51}^{1/3}$.

3 Signature

The pulsar signature shows up when $T_c < T < T_{em}$. Starting from T_c, the lightcurve starts to flatten until reaching a slope defined by the $q = 0$ dynamics. After T_{em}, the light curve will steepen and resume the original dynamics. The temporal slope may be complicated by whether the forward or the reverse shock is dominating the emission in a given band. As an example, we discuss a spherical adiabatic blastwave propagating into a constant density medium in the slow cooling regime. Assuming either the forward shock or the reverse shock dominates the emission throughout the signature, the temporal decay index will gradually change from α_1 before T_c to $\alpha_2 = (2/3)\alpha_1 + 1$ after T_c, and changes back to α_1 after T_{em}. The temporal index in this approximation is related to the spectral slope β through $\alpha_1 = (3/2)\beta$ for the forward-shock-dominated case, and $\alpha_1 = (6\beta - 3)/8$ for the reverse-shock-dominated case. The signature bump is *achromatic* due to the change of the blastwave dynamics.

4 Discussion

At present, the early afterglow data is insufficient to provide good tests for this feature. An interesting possibility is the achromatic bump observed in the afterglow of GRB 000301C [5], which was successfully modeled by the microlensing model [4]. Alternatively, this bump may be also caused by the pulsar signature discussed here, with pulsar parameters $P_0 \sim 3.4$ ms and $B_{p,15} \sim 0.27$. Taking $\alpha_1 \sim -1.28$ for the principal temporal index (before the bump and before the decay ascribed to a jet transition), the temporal index during the pulsar signature is expected to be $\alpha_2 = (2/3)\alpha_1 + 1 \sim 0.15$, which seems reasonable to fit the achromatic bump. Detailed χ^2 fit to the data would be necessary to validate this proposal. Another relevant observation may be the recent Fe line detection in the X-ray afterglow of GRB 991216 [6], which in one interpretation would require a continuously-injecting central engine [7]. To be consistent with the afterglow observations, a luminosity of $L_{em} \leq 3 \times 10^{46}$erg s^{-1} from a magnetar pulsar is allowed to explain the Fe features detected 37 hours after the burst trigger. The direct detection of or upper limits on the above-mentioned characteristic afterglow bumps by missions such as HETE2 or Swift may be able to provide interesting constraints on magnetar GRB models and their progenitors.

References

1. R. Blandford, C. McKee: Phys. Fluids, 19, 1130 (1976)
2. E. Cohen, T. Piran: ApJ, 518, 346 (1999)
3. Z. G. Dai, T. Lu: A&A, 333, L87 (1998)
4. P. Garnavich, A. Loeb, K. Stanek: ApJ, 544, L11 (2000)
5. N. Masetti, et al.: A&A, 359, L23 (2000)
6. L. Piro, et al.: Science, 290, 955 (2000)
7. M. J. Rees, P. Mészáros: ApJ, 545, L73 (2000)
8. B. Zhang, P. Mészáros: ApJ Letters, submitted (2001) (astro-ph/0011133)

Part V

Experiments: Present and Future

The Swift Panchromatic GRB Mission

Neil Gehrels

NASA/Goddard Space Flight Center, Mail Code 661, Greenbelt, MD 20771, USA

on behalf of the Swift Team

1 Introduction

The discovery by BeppoSAX and ground observers [1–3] of afterglow in gamma ray bursts (GRBs) has shown that they are cosmological, involving the most powerful explosions known. These explosions are thought to create super-relativistic blast-waves resulting in afterglow that fades from gamma-rays to radio. However, important information on the afterglow is lost between the initial burst and follow-up observations, which typically take ∼ 8 hours.

Swift, scheduled for launch in Ocotober 2003, is a multiwavelength observatory that exploits the afterglow characteristics of GRBs to make a comprehensive study of ∼ 1000 bursts. It will determine the origin of GRBs, tell us how the blast wave interacts with its suroundings, and identify classes of bursts. Swift will also investigate how GRBs can be used to study the early Universe.

2 Swift Instruments

The Swift instrumetation was carefully chosen for GRB discovery. It incorporates a wide-field GRB detector, plus two sensitive narrow-field telescopes for identifying and observing the X-ray, UV, and optical afterglow.

The *Burst Alert Telescope (BAT)* covers the 10-150 keV energy band and will detect ∼ 300 bursts per year. The GRB detector has a CdZnTe (CZT) detector array with an area of 5200 cm^2 and a coded aperture mask covering 2 sr of the sky. The mask is positioned one meter away from the detectors and will provide positions of 1-4 arcmin accuracy depending on burst brightness. The large detector area and sophisticated triggering system will allow BAT to detect bursts of all durations to a sensitivity 5 times better than BATSE (BAT threshold ∼ 10^{-8} erg cm^{-2} for a 1 sec GRB). The instrument is being developed at NASA's Goddard Sapce Flight Center.

The *X-Ray Telscope (XRT)* will locate bursts to 5.0 arcsec accuracy using flight-spare optics from the JET-X instrument on the Spectrum X-Γ mission. This mirror has a 15 arcsec half-power diameter at 1.5 keV. The detector is a 600 square pixel CCD from the XMM program, giving a FOV of 24 arcmin square in an energy range of 0.2-10.0 keV. Compared to the BeppoSAX X-ray telescope, the XRT has twice the effective area (∼ 110 cm^2 @ 1.5 keV) and four times better angular resolution. The instrument is being developed at University of Leicester, Osservatorio Astronomico di Brera, and Penn State University.

The *UV/Optical Telescope (UVOT)* is a 30 cm diameter modified Ritchey-Crétien equipped with an image intensified CCD covering 170 to 650 nm. It has a FOV 17 arcmin square and is based closely on the design of the XMM Optical Monitor (OM). The UVOT is capable of reaching $m_B = 24$ in 100 s (open filter). A filter wheel provides 6 colors plus two grisms and a 4× magnifier. The grisms will obtain spectra with resolving power of $\lambda/\Delta\lambda = 200-400$ for sources brighter than $m_B = 17$. The optical point spread function of the telescope is 0.3 arcsec allowing for excellent astrometry. By registering the field against foreground stars, the UVOT will provide < 0.3 arcsec positions. The instrument is being developed at Mullard Space Science Laboratory and Penn State University.

3   Swift Mission

The strategy of the Swift mission is to slew to each new GRB position as soon as possible and to follow the GRB afterglows as long as they are visible. To observe the earliest possible phase of the afterglow, new BAT positions will trigger an autonomous slew of the spacecraft followed by a programmed sequence of observations with the XRT and UVOT. The slew time will be ~ 1 minute.

Each of the three Swift instruments rapidly produces alert messages after a GRB is detected. To ensure prompt delivery, these messages are sent through a real-time TDRSS downlink to the ground, and routed immediately to the GRB Coordinates Network (GCN) [4] for delivery to the community.

While not performing observations of the most recent bursts, Swift will follow a one week schedule uploaded from the ground each working day and as needed. This schedule will provide for long term follow-up of afterglows and other science.

4   Swift Science

Recent GRB discoveries have shown that X-ray, optical, and radio afterglows exist, continuing for days after the bursts, but fading quickly (t^{-1} to t^{-2} is typical). Better data on faster time scales for many more bursts is needed. (See [5] for discussion of requirements for future GRB missions.) The Swift mission provides the needed capability to answer the following four key science questions:

What are the progenitors of GRBs? To determine the origin of GRBs, three parameters are needed: the total energy released, the nature of the host galaxy (if one exists), and the location within the host galaxy. The Swift mission is optimized to measure all three of these for many hundreds of bursts.

How does the blast-wave evolve and interact with its surroundings? Afterglow is thought to be produced by the interaction of an ultra-relativistic blast-wave with the interstellar or intergalactic medium. The blast-wave model [6] predicts a series of stages as the wave slows. A key prediciton is a break in the spectrum that moves from the gamma to optical band, and is responsible for the power law decay of the source flux. This break moves through the X-ray band in a few seconds, but takes up to 1000 s to reach the optical. Thus observations within the first 1000 s in the optical and UV are crucial to see this early phase.

Are there different classes of bursts with unique physical processes at work? While some evidence of sub-classes has been obtained (e.g. bimodal duration distribution, possible correlation of hardness and logN-logP shape, short bursts having V/V_{max} consistent with a Euclidean distribution), it is not clear if these are real differences or, rather, the result of the distribution fuction of GRB properties such as beaming angle, density of the local medium, or initial energy injection. Swift data will determine locations, redshifts, and afterglow properties of the different classes and thus allow physical understanding of their nature.

What can GRBs tell us about the early Universe? Since GRBs are the most luminous objects in the Universe, they provide a unique opportunity to probe the intergalactic medium (IGM) and the ISM of the host galaxies via measurment of absorption along the line of sight [7]. Depending on evolution, GRBs might originate from redshifts up to ~ 15 and have a median redshift > 2, larger than that of any other observable population. By rapidly providing both accurate positions and optical brightness, Swift will enable the immediate follow-up of those GRBs bright enough for high resolution optical absorption line spectroscopy at redshifts large enough to study the reionization of the IGM [8]. This information on the high-z Ly-α forest will be unique because there are currently no known bright ($m < 17$) galaxies or quasars at $z > 4.8$ [7].

5   Ground System and Data Analysis

A layered data analysis approach will be used to achieve rapid dissemination of Swift results and data to the community. The most urgently needed results, namely GRB positons, are produced on the spacecraft. Quicklook results, including optical finding charts and multiwavelength light curves, are produced in the Penn State Mission Operations Center (MOC) in near real-time and distributed using the GCN. Definitive standard products, including spectra, multi-band light curves, and images, will be made into production FITS files.

All this Swift data will be processed at the Swift data center at Goddard and will be made available to the general public through the HEASARC in the US and data centers in the UK and Italy. The end result will be easy access for the entire community to a broad range of timely information on GRBs.

References

1. Costa, E., et al.: Nature 387, 783 (1997)
2. Van Paradijs, J., et al.: Nature 386, 686 (1997)
3. Frail, D.A., et al.: Nature 389, 261 (1997)
4. Barthelmy, S. et al.: *Proc. Fourth Huntsville GRB Workshop*, ed. C. Meegan, R. Preece, and T. Koshut (AIP, New York, 1998) p 139.
5. Gehrels, N. and Macomb, D.: *Cosmic Explosions*, ed. S. Holt and W. Zhang (AIP, New York, 2000).
6. Mészarós, P. and Rees, M.: ApJL 418, L59 (1993)
7. Lamb, D.Q. and Reichart, D.E.: ApJ 536, 1 (2000)
8. Miralda-Escudé, J.: ApJ 501, 15 (1998)

The Italian Contribution to the Swift Mission

Guido Chincarini[1,2], Gianpiero Tagliaferri[2], and Filippo Maria Zerbi[2]

[1] Università degli Studi Milano Bicocca,
    P.za dell'Ateneo Nuovo 1, 20126, Milano, Italy
[2] Osservatorio Astronomico di Brera, Via Bianchi 46, 23807 Merate Italy

Abstract. Italy participates to the Swift Mission contributing the Mirror Modules of the X Ray Telescope, the XRT data analysis software and the Malindi Ground Station. We will participate to the operation of the satellite and coordinate a follow-up team at the European level. Italy proposed also to build a 60 cm NIR Robotic Telescope (REM) in order to detect and monitor in the NIR the GRBs soon after (20 to 60 s) the detection by the Burst Alert Telescope on board of Swift. ASI supports the Italian participation to the Mission. Murst and CNAA support the construction of REM.

1 The Mirror Module and the X Ray Telescope (XRT)

As stated by Gehrel at this meeting the XRT is provided by the Penn State University, the Leicester University and the Osservatorio Astronomico di Brera. In particular Brera will supply the Mirror Module (MM),which is a spare copy built for the JET-X experiment on board the SXG Mission. This MM is perfectly suited for the Science requirements of Swift. It provides excellent image quality on the full field of view (FOV). At 1.5 and 8 keV the Half Power Diameter (HPW) is of ~ 20 and ~ 24 arcsec, respectively, at 10 arcmin off axis angle (to be compare with the on axis values given in Table 1). The full XRT FOV is of 24x24 arcmin. XRT will measure the position of the event with an accuracy of 2-3 arcsec. The sensitivity of XRT (in 10^5 s it will detect a source flux of $\sim 4 \times 10^{-15}$) $erg\ m^{-2}\ s^{-1}$) is well suited for the observations of the GRB X-ray afterglow, allowing to monitoring their decay from few seconds soon after the burst up to weeks. However, since this MM has been in a long storage, about 4 years, with another 3 years to go before the Swift launch, we tested it again at the Panter X-ray facility (MPE) in Münich. The results of the new tests and of those carried out 4 years ago are listed in Table 1. We have no sign at all of deterioration, both in effective area (A_{eff}), or image quality (HPW).

2 The Malindi Ground Station

The Base Camp of Malindi is being operated by CRPSM of the University La Sapienza (Roma). It is currently used by ASI as the ground station for the Beppo-SAX satellite. Malindi will be the primary Station for Swift with another one in Hawaii as a back up. Due to the very favorable geographic location of Malindi and the future activity of the Space Research in Italy the Station might become very busy with other spacecrafts, rising the problem of operating various satellites at

Table 1. XRT Mirror Module characteristics

Energy	A_{eff} 1996	A_{eff} 2000	Theory	Required by Swift	HPW 1996	HPW 2000	Required by Swift
keV	cm^2	cm^2	cm^2	cm^2	arcsec	arcsec	arcsec
1.49	162.5	161.4	164.6	> 140	15.6	14.1	< 20
8.05	69.6	68.3	72.9	> 45	18.8	20.4	< 30

the same time. Detailed simulations will specify the eventual additional needs and show how the second dish, property of the University of Rome, could play a role when needed. A good aand convenient development for ASI is probably to build a second large antenna, about 11 meters diameter, as to handle properly an increasing activity for Malindi. This would be excellent also for Swift and solve any possible conflict with other satellites. The Swift data will be transmitted from Malindi to the Penn State Mission Operation Center (MOC) via Intelsat. From here the science data will be sent to the Swift Data Center at the GSFC and then to the Data Archives and Analysis Centers in the US (HEASARC), Italy (ISAC) and UK (UKDC).

3   A NIR Robotic Telescope: Rapid Eye Mount (REM)

Near infrared (NIR) observations will have a fundamental role in understanding the GRBs relation to the host galaxies and their distribution as a function of the cosmic time. NIR observations are an obvious addition to the Swift Instrumentation that covers the electromagnetic spectrum from 6000 Å to 150 keV.

¿From the science point of view the need of the NIR detection is a must. We know that optically we detect only about 50% of the bursts detected in the Gamma and X rays. We also know that looking toward the Center of our Galaxy while we detect sources emitting high energy photons we suffer about 30 magnitudes absorption in the optical region of the electromagnetic spectrum. We are better off, even if not completely transparent, in the NIR. Finally due to the cosmological redshift the only chance we have to detect high z objects is to observe in the Infrared region of the spectrum. These are key points.

It is reasonable to assume that the GRBs are events related to the evolution of stars and with the shape of the stellar Luminosity Function. The shape is obviously time dependent and if very massive objects are involved, this would call for a very strong dependence on cosmic time as well. So that the obvious question is how far in the Universe can we detect them? Could they give information on the Universe beyond $z \sim 6$ and up to $z \sim 15$? We know that while the starburst activity is very high till about $z \sim 1.2$, star formation might have started at $z > 15$ following a period of re-ionization at $10 < z < 15$ and galaxy formation. We can not exclude, therefore, the possibility of GRBs being related to massive objects likely to form also at very large z. A NIR telescope would have the capability to detect them.

Therefore, to complement UVOT and observe the GRBs light curve also in the NIR, to detect bursts occurring in highly absorbed regions or at extremely high redshift, we decided to build a Robotic Infrared Telescope and to locate it in the Southern Hemisphere. To this end the Observatory of Brera participates to a joint venture with the Observatories of Catania and Roma, with the High Energy Astrophysics group of Saclay and with many other Italian and European scientists involved in the European Swift follow up team. The telescope will be on ESO ground following agreements and collaboration with ESO.

Table 2. REM expected limit magnitudes, in the various infrared band

T int.	J S/N=10	J S/N=5	H S/N=10	H S/N=5	K S/N=10	K S/N=5
5 s	15.7	16.5	14.7	15.4	13.5	14.3
30 s	16.7	17.5	15.6	16.4	14.5	15.3
600 s	18.3	19.1	17.3	18.0	16.2	16.9

The main specification of the telescope is the capability of being on the target in about 20-60 sec after the Swift alert. It should be capable to observe simultaneously at various wavelengths and have a fast readout system in order to allow high time resolution photometry. The telescope, 60 cm aperture with alt-azimuth mounting, will be fully automatic. In Table I we list the REM sensitivity in the J, H and K bands. The data will be processed and made public as soon as possible.

Acknowledgements: As any large project this is also the output of a large collaboration of scientists with different expertise and located in many Research Centers. We would like in particular to thank all our colleagues who are part of the collaboration, participated to the preliminary discussion, wrote with us the proposal and encouraged in various ways at the beginning of this effort. Among this Ghisellini, Stella, Jacque Paul etc. A particular thank goes to those referees that prized and approved our proposal with enthusiasm. We appreciated the interest expressed by the previous Director Generals of ESO, Riccardo Giacconi, who encouraged our effort of organizing a follow up team for the GRBs alerted by Swift and suggested an ESO ToO proposal. One of us is grateful to Catherine Cesarsky who was very patient in listening the enthusiastic defense about the possibility of locating at ESO a small NIR telescope for GRBs observations. Alvio Renzini, who became our ESO reference point for the telescope, should be thanked for the capability of understanding the temper of one of us (GC) and being unusually patient. ASI for supporting with generosity the program.

The INTEGRAL Burst Alert System

Sandro Mereghetti[1], Davide I. Cremonesi[1], and Jurek Borkowski[2]

[1] Istituto di Fisica Cosmica G.Occhialini, CNR, Milano, Italy
[2] INTEGRAL Science Data Center, Versoix, Switzerland

Abstract. We describe the INTEGRAL Burst Alert System (IBAS): the automatic software developed at the INTEGRAL Science Data Center to allow the rapid distribution of the coordinates of the Gamma-Ray Bursts detected by INTEGRAL.

More than one GRB per month is expected in the field of view of the main INTEGRAL instruments. Positions with an accuracy of a few arcminutes will be distributed to the community for follow-up observations within a few tens of seconds of the event.

1 Introduction

The INTEGRAL gamma-ray astronomy satellite (Winkler 1999) carries two main imaging instruments operating in the $\sim$20 keV–10 MeV energy range (SPI and IBIS) complemented by an X-ray (JEM-X) and an optical (OMC) monitor. The launch is scheduled in April 2002. The coded mask instruments on board INTEGRAL have large fields of view and offer the possibility of rapidly obtaining accurate positions of GRB's.

No on-board GRB triggering system is present on the INTEGRAL satellite. Since the data are continuously transmitted to the ground, the search for GRB's will be performed by means of a near real time analysis software running at the INTEGRAL Science Data Center (ISDC, Courvoisier et al. 1999). Here the relevant data packets will be extracted from the telemetry stream and fed into a dedicated software system (Mereghetti et al. 1999). The first step of the GRB search will be based on a simple monitoring of the IBIS and SPI count rates, without resorting to more complex image analysis. This will be done by looking for significant excesses with respect to a running average, in a way similar to traditional on-board triggering algorithms. The search will be simultaneously performed in different time scales and energy ranges, to optimize the sensitivity to GRB's with different characteristics. When a candidate event is detected, a process of image analysis shall start to verify the origin of the count rate variation and to ensure that the event was not caused by an instrumental malfunctioning or by a background variation. Images shall be accumulated for different time intervals, deconvolved with very fast algorithms, and compared to the pre-burst reference images in order to detect the appearance of the GRB as a new source. If the event is genuine, the satellite attitude information will be applied to derive a sky position that is then automatically transmitted to all the subscribed users. In addition, if the GRB is located in the sky region covered by the OMC, an appropriate telecommand will be sent to the satellite to reconfigure its observing parameters in order to optimize the possible detection of optical emission.

Because full event validation and localization might require a longer time, we foresee different levels of alert messages providing increasingly accurate and reliable information. These messages will be configured in such a way to allow an easy filtering by the users in order to react only to the situations that best fit their needs.

2   Number of Expected GRB's

A simple estimate of the number of GRB's expected within the field of view of the INTEGRAL instruments can be done by scaling the total rate of events measured by BATSE ($\sim$666 GRB year^{-1}, Paciesas et al. 1999). For the field of view of IBIS ($\sim$30$°\times$ 30$°$), this yields 666 $\times$ (0.23 sterad / 4π) $\sim$ 12 GRB year^{-1}.

A more accurate estimate must take into account the varying sensitivity within the IBIS field of view and the different energy range of the BATSE detectors. This can be done by convolving the IBIS sensitivity and solid angle as a function of off-axis angle with the BATSE LogN-LogP relations, converted to the appropriate energy range assuming an average GRB spectral shape. Such computations do not significantly change the above rough result. In fact we obtain expected rates of $\sim$13, 10 and 8 GRB year^{-1}, within the IBIS field of view, adopting respectively the LogN-LogP corresponding to BATSE trigger times ΔT of 1 s, 256 ms and 64 ms. The major uncertainty on the derived GRB rates is related to the extrapolation of the BATSE LogN-LogP curves down to the IBIS sensitivity ($\sim$0.1 ph cm^{-2} s^{-1}, 50-300 keV peak flux for ΔT=1 s). Such an extrapolation depends on the poorly known spectral shape of the faintest GRB's.

These estimates do not take into account the possible existence of GRB's with different characteristics than those observed by BATSE. BATSE had a limited sensitivity for events shorter than 64 ms, as well as a strong bias against the detection of long, slowly rising GRB's. It is therefore very likely that the figures reported above are underestimated, and that a few events per month will be available through IBAS for rapid multi-wavelength follow-up observations.

3   IBAS Performances

3.1   Location Accuracy

The source location accuracy (SLA) of coded mask imaging systems depends on the signal to noise ratio of the source. For sources detected with a high statistical significance the SLA can be a small fraction of the angular resolution. Theoretical evaluations, confirmed by several independent simulations, have shown that for IBIS a SLA smaller than 30$''$ (90% confidence level) can be obtained for a signal to noise ratio of 30. For most of the time (i.e. except during satellite slews) the INTEGRAL attitude accuracy will be $\lesssim$ 30$''$. Actually, most of the detected GRB's will have relatively small signal to noise ratios, resulting in typical uncertainties, dominated by the photon statistics, of the order of $\sim$2-3$'$.

3.2   Time Delay

Using simulated telemetry data, we have obtained the following performance figures for the current prototype version of IBAS running on a SUN ULTRA 10 workstation (440 MHz clock, 256 Mbytes RAM – the final version will run on a faster workstation exclusively dedicated to the IBAS system).

The first IBAS steps (telemetry receipt, extraction and sorting of the relevant data packets, photon binning for the trigger search) require less than 0.1 s. The speed of the triggering algorithm depends on the duration of the smallest time interval considered and on the number of timescales. For example, sampling on bins of 1, 4, 16, 64 and 256 ms, with a resolution of 1 ms it is possible to process the telemetry at a speed twice faster than that of the real-time incoming data. The most time consuming tasks are those related to the image analysis. For this reason, the first part of imaging after a trigger is based on a detection algorithm able to discover new sources by only considering (ghost) peaks in the fully coded field of view. This can be done within ~200 ms, allowing to discriminate between false triggers and true events. Only when a likely point source is detected, a more thorough analysis will be done to determine its position. The deconvolution of the whole (totally plus partially coded) field of view and localization of the true source peak currently requires ~5 s. Thus it should be possible to send the first GRB alert within a few seconds after the trigger time. The delay between the trigger time and the GRB onset is of course dependent on the intensity and time profile of the event, but the IBAS simultaneous sampling in different timescales should ensure a minimum delay in most cases. To this time budget one has to add the time required for the telemetry transmission from the satellite to the ISDC, that under normal circumstances should be smaller than ~30 s. Thus, in many cases, we foresee to be able to generate *first level* alerts while the GRB is still ongoing.

Acknowledgments

The IBAS development is supported by the Italian Space Agency. JB was supported by the Polish Committee for Scientific Research (KBN) under grant No. 2P03C00619p02.

References

1. Courvoisier T., et al.: *Proc. 3nd INTEGRAL Workshop – Astro. Lett. and Communications* **39**, 355 (1999).
2. Mereghetti S., et al.: *Proc. 3nd INTEGRAL Workshop – Astro. Lett. and Communications* **39**, 301 (1999).
3. Paciesas W.S., et al.: ApJSS **122**, 465 (1999).
4. Winkler C.: *Proc. 3nd INTEGRAL Workshop – Astro. Lett. and Communications* **39**, 309 (1999).

AGILE and Gamma-Ray Bursts

M. Tavani[1], G. Barbiellini[2], A. Argan[1,3], N. Auricchio[4], P. Caraveo[1],
A. Chen[1,3], V. Cocco[3,5], E. Costa[6], G. Di Cocco[4], G. Fedel[2], M. Feroci[6],
M. Fiorini[1], M. Galli[7], A. Giuliani[1], C. Labanti[4], I. Lapshov[6], P. Lipari[8],
F. Longo[3,9], S. Mereghetti[1], E. Morelli[4], A. Morselli[5], A. Pellizzoni[1],
F. Perotti[1], P. Picozza[5], C. Pittori[3,5], C. Pontoni[2,3], M. Prest[2,3],
M. Rapisarda[10], E. Rossi[4], A. Rubini[6], P. Soffitta[6], M. Trifoglio[4], E. Vallazza[2],
S. Vercellone[1], and D. Zanello[8]

[1] Istituto Fisica Cosmica, CNR, Milan (Italy)
[2] Università di Trieste and INFN, Trieste (Italy)
[3] Consorzio Interuniversitario Fisica Spaziale, Turin (Italy)
[4] Istituto TESRE, CNR, Bologna (Italy)
[5] Università di Roma "Tor Vergata" and INFN, Roma (Italy)
[6] Istituto di Astrofisica Spaziale, CNR, Roma (Italy)
[7] ENEA, Sez. Bologna (Italy)
[8] Università di Roma "La Sapienza" and INFN, Roma (Italy)
[9] Università di Ferrara and INFN, Ferrara (Italy)
[10] ENEA, Sez. Roma (Italy)

Abstract. The AGILE astrophysics mission planned to be operational in 2003 will significantly contribute to the study of Gamma-Ray Bursts. AGILE is equipped with state-of-the-art imaging Silicon detectors in the 10-40 keV and 30 MeV–50 GeV ranges. In addition, also a CsI Mini-Calorimeter can independently detect transient events. Among the instrument characteristics most useful to GRB science we mention: (1) the excellent positioning and sensitivity at large off-axis angles; (2) the very small instrumental deadtimes ($\sim 100\,\mu$s above 30 MeV, and $\sim 5\,\mu$s below 10 MeV); (3) the on-board capability to trigger sky map acquisition in the 10-40 keV range and to transmit burst coordinates within a few seconds from the event. A special GRB search procedure will be implemented on-board for a broad range of trigger timescales from 1 millisecond to 100 seconds. AGILE will be the only Mission dedicated to astrophysics above 30 MeV and operating during the years 2003–2006.

1 The AGILE Mission

The AGILE Mission (the first of the Italian Space Agency Small Scientific Missions) is currently in the final engineering design phase, and is planned to be operational during the first half of the year 2003. The spacecraft will be of the MITA class, for a total satellite weight of ~ 220 kg. Three-axis satellite stabilization and the use of a pair of star sensors will provide a positional reconstruction of ~ 1 arcmin. The AGILE scientific instrument [5] consists of two imaging detectors: the Gamma-Ray Imaging Detector (GRID) operating in the energy band ~ 30 MeV–50 GeV (made of a Silicon Tracker [1] plus a Mini-Calorimeter [2]), and Super-AGILE operating in the 10–40 keV band [3].

2 AGILE and Gamma-Ray Bursts

We can summarize the main expected capabilities as follows:

1. The imaging AGILE-GRID with a large field-of-view (~ 3 sr) and good effective area ($\sim 350\,cm^2$ at 400 MeV and 50° off-axis) is expected to detect $\gtrsim 10$ GRBs per year in the range above 30 MeV.

2. The imaging Super-AGILE will provide GRB position uncertainties of 2-3 arcminutes for intense events, and ~ 6 arcmin for medium-weak intensity events. For a field-of-view of ~ 0.9 sr, about 20–30 bursts per year are expected to be detectable and fully imaged by Super-AGILE in the energy range 10–40 keV [4].

3. The Mini-Calorimeter can detect GRBs in the energy range $\lesssim 1$ MeV–200 MeV.

4. AGILE will provide microsecond photon time tagging and unprecedently small deadtimes ($\sim 100\,\mu s$ for the GRID, and $\lesssim 5\,\mu s$ for the Mini-Calorimeter and Super-AGILE). Fast timing is crucial for GRB studies [6].

5. The instrument is capable of processing on-board Super-AGILE pre- and post-burst sky images, and to provide GRB coordinates within 5-10 seconds from trigger. A (currently under study) fast communication channel can allow transmission of these data to the ground for rapid dissemination.

6. The independent or joint GRB detection by the GRID (30 MeV–50 GeV) and Super-AGILE (10–40 keV) will be of great importance for ground-based radio/optical/TeV observations. AGILE has a great potential for joint work with INTEGRAL and SWIFT.

References

1. Barbiellini, G., Prest, M., et al.: Nucl. Instr. & Methods, to be submitted (2001)
2. Di Cocco, G., et al.: these Proceedings (2001)
3. Feroci, M., et al.: these Proceedings (2001)
4. Preger, B., Fenimore, E., and Costa, E.: these Proceedings (2001)
5. Tavani M., et.al.: in *Proceedings of the 5th Compton Symposium*, (AIP Conf. Proceedings no. 510, 2000), ed. M. McConnell, p. 746
6. Tavani, M.: in preparation (2001)

Gamma-Ray Bursts with SuperAGILE

M. Feroci[1], E. Costa[1], E. Del Monte[1], I. Lapshov[1], M. Rapisarda[2], P. Soffitta[1],
G. Barbiellini[3], M. Prest[3], E. Vallazza[3], S. Mereghetti[4], M. Tavani[4],
S. Vercellone[4], A. Morselli[5,6], and F. Longo[7]

[1] Istituto di Astrofisica Spaziale, CNR, Rome, Italy
[2] ENEA, Frascati, Italy
[3] INFN, Sezione di Trieste, Italy
[4] Istituto di Fisica Cosmica, CNR, Milan, Italy
[5] Dipartimento di Fisica, Universita' di Roma *Tor Vergata*, Italy
[6] INFN, Sezione di Roma 2, Italy
[7] Dipartimento di Fisica, Universita' di Ferrara and INFN, Sezione di Ferrara, Italy

Abstract. We present the capabilities for Gamma Ray Bursts of SuperAGILE, the
hard X-ray (10-40 keV) monitor of the gamma-ray mission AGILE. SuperAGILE has a
wide field of view (~1.8 sr), coded by one-dimensional coded masks in two orthogonal
directions, read-out by single-sided silicon microstrip detectors. Based on our simulations, we expect to detect and localize ~1-2 GRBs/month to better than 6 arcmin.

1 Introduction to SuperAGILE

SuperAGILE is composed by a Detection Plane (DP), a Collimator equipped
with four Coded Masks, Front-End Electronics and an Interface Electronics
(SAIE). Table 1 summarizes the main instrument characteristics (see also [3] for
a description of SuperAGILE and [4] for a description of the AGILE mission).
The DP is made of 4 detection units (DUs), placed on the same Al honeycomb
plane support, two for the X-direction and the other two for the Y-direction.
Each DU is composed by 4 Si microstrip tiles, bonded in pairs so that the effective length of each strip is approximately 19 cm. They are read-out through
a set of IDEAS-XAA1 ASIC chips, 12 for each of the DUs. The collimator is
mounted on the same tray supporting the DP, and in turn supports the 4 orthogonal, one-dimensional coded masks. The coded masks have a 50% covering
factor. They will be manufactured either of Gold or Tungsten. The SAIE is in
charge of interfacing SuperAGILE with the AGILE Data Handling unit (DH),
allowing an event-by-event transmission with timing resolution better than 5 μs
and 1 keV energy bins. The combined capabilities of the SAIE and the DH (see
also [2]) allow the transmission to the ground of the scientific housekeeping data,
including ratemeters and detector images.

2 On-board Transient Events Triggering and Localization

The DH will be able to perform a continuous automatic search for transient
events (e.g. gamma-ray bursts, AGN flares or others) simultaneously on several

Table 1. The Characteristics of SuperAGILE

Detector Type	400 μm thick Silicon microstrip
Basic Detection Unit	4 Si Tiles, 19cm x 19cm
Total Geometric Area	1444 cm^2
On-Axis Effective Area	320 cm^2 (13keV)
Detector Strip Size	121 μm
Nominal Energy Range	10-40 keV
Energy Resolution (FWHM)	$\sim$ 3-4 keV
Timing Accuracy	$\sim$ 5 μs
Collimator Materials	75 μm Tungsten-Coated Carbon Fiber
Mask Size	1444 cm^2
Mask-Detector Distance	14 cm
Mask Transparency	50%
Mask Material and Thickness	Tungsten, 100 μm
Mask Element Size	242 μm
Field of View (FWZR, each direction)	107° x 68°
On-Axis Angular Resolution	5.9 arcmin
Source Location Accuracy	$\sim$2 arcmin for bright sources
Point Source Sensitivity	5 mCrab on axis (5-σ, 50 ks)

timescales from 1 ms to 100 s. Once a transient event is triggered onboard, the DH will be able to provide its attitude-corrected sky images, determining the location of the transient source on the sky. In Figure 1 we show the simulation of the detection of a 10-s long GRB, with the same photon fluence (over 10 s) as the Crab Nebula. The left panel shows the X and Y sky images accumulated over the 10-s of the event, showing both the GRB and the Crab Nebula. The right panel shows the same images, once the "background sky images" have been subtracted, thus removing the persistent sources (in this case the Crab only). From the latter type of image, the DH will be able to derive the X and Y coordinates of the transient event. The possibility to distribute in almost real time to the worldwide scientific community the coordinates through a fast link (e.g., TDRSS or similar) is currently under study.

3   Expected Number of Dectected GRBs

In Figure 2 (left panel) we show a map of the number of GRBs detected by SuperAGILE as a function of their position within the field of view (FOV). We have taken the first 23 GRBs detected and localized by the BeppoSAX Wide Field Cameras (WFCs), put them in every location of the part of the SuperAGILE FOV for which we have the bi-dimensional capability, and studied their detectability at a 5-σ level. We see from the plot that in a field of 40° by 40°, corresponding to the WFC FOV, SuperAGILE can detect 22 out of 23 events.

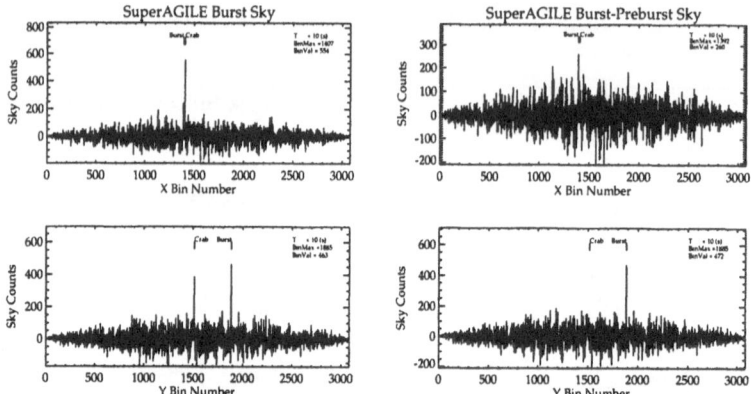

Fig. 1. *Left*: Simulated X and Y sky images of a GRB detected in the same field of the Crab Nebula. *Right*: The same GRB as above, when the "background sky image" is subtracted, and only the transient event is left.

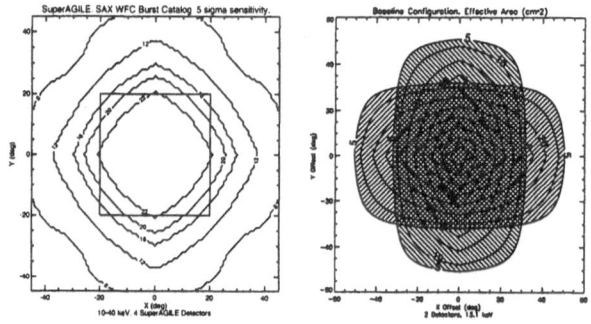

Fig. 2. *Left*: Simulation of the number of GRBs (from a BeppoSAX WFC catalog of 23 events) detected by SuperAGILE, as a function of their position in the field of view. The central square shows the WFC field of view. *Right*: the map of the effective area (at 13 keV) of 2 orthogonal SuperAGILE units.

The outer areas of the FOV will be able to detect a smaller but not negligible number of events, bringing the total number of GRBs detected (and localized in 2 dimensions) in one year to approximately twice as much as those detected by the BeppoSAX WFCs. In addition to that, as can be seen from the map of the effective area of two orthogonal detectors shown in Figure 2 (right panel), more events can be detected in the outer regions of the FOV, where SuperAGILE has only one-dimensional imaging capabilites.

References

1. Barbiellini, G., et al. 2000, *Proc. 5th Compton Symp., AIP 510, p. 750*
2. Morselli, A., et al. 2000, *Proc. SPIE conference 4140, p.493*
3. Soffitta, P., et al. 2000, *Proc. SPIE conference 4140, p.283*
4. Tavani, M., et al. 2001, these proceedings

The GLAST Burst Monitor (GBM)

G.G. Lichti[1], M.S. Briggs[3], R. Diehl[1], G. Fishman[2], R. Georgii[1],
R.M. Kippen[3], C. Kouveliotou[2], C. Meegan[2], W. Paciesas[3], R. Preece[3],
V. Schönfelder[1], and A. von Kienlin[1]

[1] Max-Planck-Institut für extraterrestrische Physik, 85748 Garching
[2] NASA/Marshall Space-Flight Center, Mail Code SD50, Huntsville, AL 35812
[3] University of Alabama in Huntsville, AL 35899

Abstract. The selection of the GLAST burst monitor (GBM) by NASA will allow the investigation of the relation between the keV and the MeV-GeV emission from γ-ray bursts. The GBM consists of 12 NaI and 2 BGO crystals allowing a continuous measurement of the energy spectra of γ-ray bursts from ~ 5 keV to ~ 30 MeV. One feature of the GBM is its high time resolution for time-resolved γ-ray spectroscopy. Moreover the arrangement of the NaI-crystals allows a rapid on-board location ($< 15°$) of a γ-ray burst within a FoV of ~ 8.6 sr. This position will be communicated to the main instrument of GLAST making follow-up observations at high energies possible.

1 Introduction

It was in 1994 that EGRET observed a γ-ray burst which showed a γ-ray emission above 50 MeV up to ~ 1.5 hours after the start of the burst. The γ-quantum with the highest energy was observed after ~ 1.3 hours with an energy of 18 GeV [4]. This was an unexpected and very surprising result and as of yet the relation between this high-energy and the low-energy emission is not understood. It is a goal of the GLAST mission to investigate this relation and to unravel the underlying physical processes.

2 Characteristics of Energy Spectra of γ-Ray Bursts

A typical spectrum of a γ-ray burst is characterized by a broken power law (see Figure 1). Below a certain break energy E_p the spectrum can be described by a power law E^α with an exponential decline, whereas above this energy it is a pure power law E^β indicating a non-thermal origin of this part of the spectrum. The break-energy E_p at which the luminosity reaches a maximum has a log-normal distribution around an energy of 250 keV [5]. In order to determine E_p well enough one needs a long lever arm on both sides of this energy, accentuating the importance of low- and high-energy measurements.

The spectral index α of the low-energy emission is distributed around a value of -1 spanning the range from -2 to 0. This distribution strongly constrains popular burst-emission models like the synchrotron emission from shocked electrons ([7] and [6]) or the blast-wave model [2] strongly. The distribution of the spectral index β of the high-energy emission reaches a maximum around -2.3. It extends

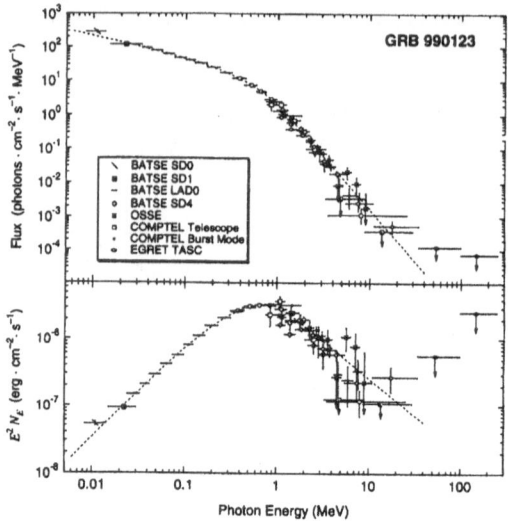

Fig. 1. The energy spectrum of GRB990123

from –3.5 to -1.4 [5]. The spectra with β-values >-2 are of special interest because in this case the spectrum would diverge for E → ∞. Therefore a cut off at high energies must exist. An interersting question which will be answered by GLAST is, at which energy this cutoff occurs.

3   The GLAST Mission

GLAST will continue the successful measurements of EGRET in a wider energy range, with a higher sensitivity and with a better location accuracy. Its main instrument, the Large-Area Telescope (LAT), will use the same physical processes as EGRET for the detection of γ-rays, but using an advanced technical concept. It will consist of an array of towers of pair-conversion chamber stacks made from Silicon-strip detectors.

The LAT measures γ-rays in the energy range ~15 MeV to ~500 GeV, reaching a point-source sensitivity of better than $4 \cdot 10^{-9}$ photons/(cm^2 s) above 100 MeV for an observation time of one year. It will therefore be more than 30 times more sensitive than EGRET. Within its large FoV of <3 sr it will locate point sources from 5' down to 30". With its fairly good energy resolution of ~10% it will measure the energy spectra of sources with a high accuracy. The LAT is devoted to the study of particle acceleration in the universe as it takes place in the nuclei of active galaxies, at or near pulsars, in supernova remnants and in interactions of the cosmic rays with the interstellar matter. In addition, it will detect ~50–150 γ-ray bursts per year. For a description of GLAST see [3].

From the latter measurements it is hoped to solve the afore-mentioned problem of the high-energy burst emission. However, the LAT suffers from some deficiencies because high-energy measurements alone do not allow a unique classification of γ-ray bursts, because the break energy $E_{\rm p}$ of a burst lies below

GLAST's energy threshold of ~15 MeV. Therefore without low-energy measurements a classification of the bursts is difficult, since most bursts were measured by BATSE at these energies and the connection to the BATSE data archive cannot be established. Another deficiency is that the trigger conditions of the LAT for weak bursts are unfavourable because of the rather high background rate. In order to overcome these deficiencies a secondary instrument was proposed, the GBM. It will extend the energy range of GLAST towards lower energies and it will have a much larger FoV than the LAT. Therefore it will detect more bursts than this instrument. The GBM will communicate the positions of these bursts to the LAT which then will, after reorientation if needed, search for high-energy γ-rays. Moreover, for weak bursts the LAT will use GBM-provided information to reduce background by eliminating events with directions far frm the GBM burst location.

4   Description of the GBM and Its Performance

The goals of the GBM described above can be achieved by an arrangement of 12 thin NaI discs which are oriented such that from the relative count rates the direction to the burst can be derived (KONUS/BATSE principle). They will in addition measure the burst spectra in the energy range 5 keV–1 MeV.

In order to obtain spectral overlap with the LAT, two cylindrical BGO crystals will be mounted to the GLAST spacecraft which are sensitive to γ-rays in the energy range from 150 keV–30 MeV. A more detailed description of the GBM can be found in [8]. Within the large FoV of ~8.6 sr ~215 bursts/year will be detected by the GBM. Most of them will be located on board in real time with an accuracy < 15°. On ground much better locations (~3°) can be derived. The 50–300 keV sensitivity for nominal on-board triggers will be ~0.6 photons/(cm^2 s), whereas an ultimate 5σ sensitivity of ~0.35 photons/(cm^2 s) can be achieved on ground.

5   Scientific Goals and Expected Results

With the GBM continuous measurements of energy spectra from ~5 keV to ~30 MeV will be performed. Apart from spectra also the light curves of bursts will be recorded with a time resolution in the μs-range. The GBM will serve as a sensitive burst trigger for the LAT and will communicate very rapidly (<5 s) the burst location to it. This trigger will initialize data-reduction modes in the LAT which will then observe the burst and localize it with a much better accuracy (<3'). This precise location will be communicated within less than 10 s to ground in order to allow a search for objects at other wavelengths. The burst data collected by the GBM will preserve the continuity to the BATSE data and enlarge this important archive. Moreover the GBM will be part of the IPN as a near-earth burst detector.

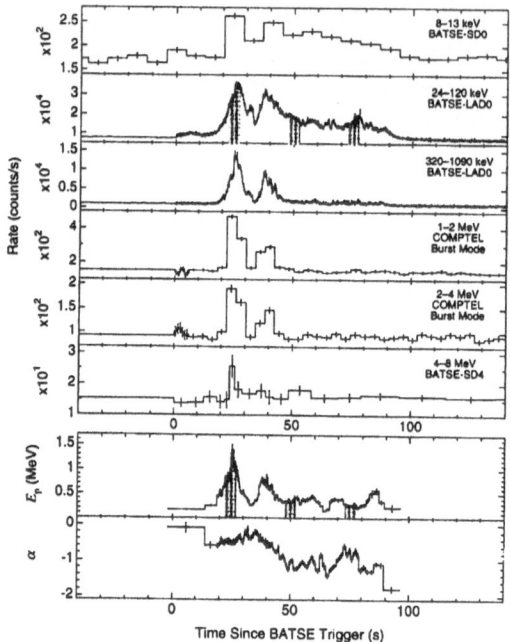

Fig. 2. The energy-resolved light curves of GRB990123 and the evolution of E_p and α

With the GBM the relation between the keV-MeV-GeV emission can be investigated in great detail. Time-resolved energy spectra will be measured allowing time-resolved spectroscopy (see Figure 2 of [1]). This permits the investigation of the hard-to-soft evolution of the power-law index α and the hardness-intensity correlation and tackles the problem of the narrowing of the peaks with energy. It may also give an answer to the question why the low-energy emission lasts longer than the high-energy one. Together with the LAT it will be possible to investigate these correlations to high energies with the aim to measure the suspected cutoff and to find a possible evolution of the spectral index β.

References

1. Briggs, M. S. et al., Ap. J. **524**, 82, 1999
2. Cen, R., Ap. J. L. **517**, L113, 1999
3. Gehrels, N. et al., Astroparticle Physics **11**, 277, 1999
4. Hurley, K. et al., Nature **372**, 652 (1994)
5. Preece, R. D. et al., Ap. J. Suppl. Ser. **126**, 19-36, 2000
6. Rees, M. J. and P. Meszaros, MNRAS **258**, 41, 1992
7. Tavani, M., Ap. J. **466**, 768, 1996
8. von Kienlin, A. et al., Proc. 4th INTEGRAL Workshop, Alicante, 2000

The IPN I: From the Past to the Future

T.L. Cline[1], K.C. Hurley[2], S. Barthelmy[1], P. Butterworth[1], M. Feroci[3],
F. Frontera[4], S. Golenetskii[5], E. Mazets[5], and J. Trombka[1]

[1] NASA's Goddard Space Flight Center, Greenbelt MD 20771, USA
[2] SSL, University of California, Berkeley CA 94720, USA
[3] Istituto di Astrofisica Spaziale, CNR, Rome, I-00133 Italy
[4] Istituto di Tecnologie e Studio delle Radiazioni Extraterrestri, CNR,
   40129 Bologna, Italy
[5] Ioffe Physico-Technical Institute, St. Petersburg 194021, Russia

Abstract. Interplanetary spacecraft have been used with orbiting satellites for over 25 years to precisely localize gamma ray transients by the measurement of their time-delay geometry. The first interplanetary network (IPN) made both discoveries and controversies, and the latest is making possible a significant number of GRB counterpart observations. The IPN technique was pursued with dedicated payloads, with piggy-back experiments, and by the creative modifications of other experiments. The achievement of the NEAR in-flight software revision added a distant vertex to the array of Ulysses and the near-Earth group of GGS-Wind Konus, Beppo-Sax and Rossi-XTE. This 3-way long-baseline network culminated IPN history by, in the year 2000 alone, enabling over one-third of the afterglow searches and 5 redshift measurements. Future IPN possibilities are also outlined.

1   Introduction

Gamma-ray burst (GRB) study has been a rapidly evolving field in the last few years, featuring the Compton-GRO BATSE phenomenology and the BeppoSax discovery of the long-lasting soft x-ray component that made possible more precise alerts, enabling the discovery of GRB counterparts. A dramatic advance was the robotic, wide-angle telescope detection of a simultaneous optical event on January 23, 1999, triggered by an automatic, electronic GCN rapid alert. Yet, in fact, the year 2000 has seen four additional significant milestones in GRB efforts. In addition to the end of nine fruitful years of Compton-GRO, a second milestone is the initiation of the Swift mission, a next-generation GRB spacecraft that will be able to make many of its own counterpart studies. Another is the successful launch of HETE-2, a mission dedicated to localizing GRBs precisely with rapid data distribution. Finally, the fourth milestone is a full year of precise localizations from the reconfigured IPN, averaging one GRB per week. These milestones could be viewed as, respectively, the death, the conception, the birth, and the maturity of four GRB projects. The IPN results, enabling several dozens of counterpart searches and producing at least five red-shift observations, are the present culmination of the program briefly reviewed here, as outlined in the companion paper [1].

2   History

The first interplanetary network (IPN) began just over 25 years ago with the launch of Helios-2 in January 1976, carrying the first experiment purposefully built for GRB studies. As the only GRB detector at a great distance, its initial source fields were long, thin annuli and one precise, small location determined from the intercept of a one-dimensional Ariel x-ray source region. It was found that none of these regions included an interesting candidate source object. By late 1978, the NASA launch of Pioneer Venus Orbiter and the Soviet launches of Venera-11 and -12 completed an IPN with up to 5 mutually distant vertices. The precise localizations made for several years continued to lack source-candidate consistency – effectively launching the major mystery of gamma ray bursts. Given the awkwardness of long-range collaborations with the technology of that era, the analyses were not made in today's rapid-alert fashion, so the delayed optical searches then performed never produced counterparts.

The second advance of the first IPN was to establish the two-component gamma-ray transient controversy, lasting over thirteen years, with the observation of N49 as candidate source object of a clearly anomalous gamma-ray transient. This was not at the time widely accepted as an identification, due to the 'violation of the economy of hypotheses' with two apparent populations of gamma-ray transient sources rather than one. The 1979 March 5 event could be interpreted either as evidence that GRBs are galactic neutron-star hard x-ray events with one accidental coincidence on a distant snr, or as evidence that the repeaters, found in 1979 with other, more sensitive Soviet instruments, along with the March 5 event, may come not from nearby interstellar regions but from greater distances, even from the LMC, while 'classical' GRBs remain of unknown origin. These views could not be resolved until the third major contribution of the early IPNs enabled its solution – by providing a precise SGR location later found to fit another very distant radio snr, in turn enabling the real-time detection of an outburst from that same SGR with the absorption characteristic of a source distance across the galaxy. These discoveries could have been possible only due to the precision of the IPN.

However, the array dwindled in the early 1980s when the Helios-2 and Venera-11 and -12 missions expired and the Venera-13 and -14 missions were also terminated. The vagaries of Fate prevented a return of the IPN to full strength for many years. The reduction in data rate and eventual termination of ICE limited its IPN utility. PVO was left orbiting Venus without another distant IPN vertex for over a decade. In the late 1980s, two Soviet Phobos missions failed unexpectedly upon their arrivals at Mars. Even the number of near-Earth detectors with GRB capability became limited, with only SMM and some surviving Velas until C-GRO was launched. The major historic setback to the IPN was the cancellation of the NASA half of the International Solar Polar Mission, leaving the European spacecraft Ulysses as survivor. Of course, if this had not been necessary, the IPN would now have enjoyed ten years of precise GRB event localization, instead of one. Mars Orbiter mission also failed on its arrival at Mars in 1993, after a one-year trajectory during which GRB detection, with a single exception, could not be performed. Finally, when the operating lives of Ulysses and PVO did overlap in the early 1990s, Ulysses, on its way to an assisting bounce off Jupiter's gravitational field, happened to be along an almost direct extension of the line of sight to Venus. The lovely post-sunset view of the two planets was the visual demonstration of a collapsed IPN 'tripod' – permitting only annular segments as source loci. After PVO entered the Venusian atmosphere, Ulysses was left without a mate for several years, until 1999.

3   Present IPN Status

The Ulysses GRB source annuli, available throughout the 1990s, were used to limit the BATSE source fields, and also precisely confirmed a number of the BeppoSax localization measurements in the late-1990s era of the initial counterpart detections. The failure of the Compton-GRO tape recorder had made possible the construction of a system to extract BATSE event data, determine an approximate GRB source direction, and electronically distribute it, instantly, to the global community. An extended version, called GCN (for GRB Coordinates distribution Network), automatically distributes all GRB and other available transient event data upon receipt. The Ulysses data were incorporated into the GCN, with Ulysses-BATSE source annuli automatically calculated and distributed essentially upon receipt. Soon afterwards, BeppoSax data were incorporated into the GCN as well, and the Ulysses-BeppoSax annuli continued to confirm and limit GRB source fields.

Konus, a continuously sensitive and nearly omnidirectional GRB experiment, was launched on the GGS-Wind spacecraft in late 1994 to regions within a few light-seconds from the Earth. It provided a near-Earth augmentation to the IPN and, with the demise of the Compton-GRO mission in mid-2000, became the primary near-Earth IPN vertex. Konus event data also became incorporated into the GCN for automatic distribution. The giant flare of August 27, 1998, the first March-5-like event since 1979, provided a precise, 2-msec calibration with its detection by Konus, Rossi-XTE, and Ulysses.

The Near Earth Asteroid Rendezvous (NEAR) mission was launched without a GRB capability, but a courageous and unfunded software modification later enabled NEAR to be converted to full IPN partnership. After an Eros encounter was missed on first attempt, data were limited until after a solar orbit and successful orbit insertion were achieved. Thus, a 3-way long-baseline IPN was finally rebuilt, that has, throughout the year 2000, localized GRBs at an average weekly rate, contributing dozens of counterpart searches and at least five redshift measurements [1]. HETE-2 should soon be incorporated into the GCN and the IPN as well, and Mars-2001 should replace NEAR as the second distant vertex after a successful launch. Thus, HETE-2, the European INTEGRAL mission and the ongoing IPN should carry GRB studies until the era of Swift.

References

1. K. Hurley et al.: These proceedings

One Year of Rapid, Precise Gamma-Ray Burst Localizations by the Interplanetary Network

Kevin Hurley[1], T.L. Cline[2], S. Barthelmy[2], P. Butterworth[2], M. Feroci[3], F. Frontera[4,5], E. Montanari[5], C. Guidorzi[5], S. Golenetskii[6], E. Mazets[6], and J. Trombka[2]

[1] SSL, University of California, Berkeley CA 94720, USA
[2] NASA's Goddard Space Flight Center, Greenbelt MD 20771, USA
[3] Istituto di Astrofisica Spaziale, CNR, Rome, I-00133 Italy
[4] Istituto di Tecnologie e Studio delle Radiazioni Extraterrestri, CNR, 40129 Bologna, Italy
[5] Dipartimento di Fisica, Universita di Ferrara, 44100 Ferrara, Italy
[6] Ioffe Physico-Technical Institute, St. Petersburg 194021, Russia

Abstract. We review the performance of the interplanetary network over the past year, emphasizing the GRB detection rate, and the speed and accuracy of the localizations. Two scientific highlights, the burst with the highest redshift to date, and the first rapid, precise localizations of short bursts, are outlined.

1 Speed, Accuracy, and GRB Detection Rates

Since December 1999, the 3rd Interplanetary Network (IPN) has consisted primarily of *Ulysses* , in heliocentric orbit, the Near Earth Asteroid Rendezvous (NEAR) mission, in orbit about the asteroid Eros, Konus-Wind, at the L_1 Lagrange point, and BeppoSAX (the GRBM) and the Rossi X-Ray Timing Explorer, in low-Earth orbit [1]. Using a combination of triangulation and the directional response of one or more of the near-Earth spacecraft, about one burst per week can be localized to a single error box whose typical size is 10 square arcminutes or greater. Because interplanetary spacecraft downlink their data typically only once a day during an 8 hour tracking pass, the inherent delays in producing error boxes are 10 hours and more.

Figure 1 shows the sizes of 56 IPN error boxes as a function of the delay to obtain them. For each of these bursts a Global Coordinates Network (GCN) message was sent out. The delay is the length of time between the arrival time of the burst at Earth and the emission of the GCN notice. These bursts were detected by the IPN in an $\approx$ 15 month long period from December 1999 to February 2001.

2 Some Statistics

Approximately 60 % of the IPN bursts were followed up in the optical, radio, and/or X-ray ranges. 9 counterparts were identified, bringing the total number

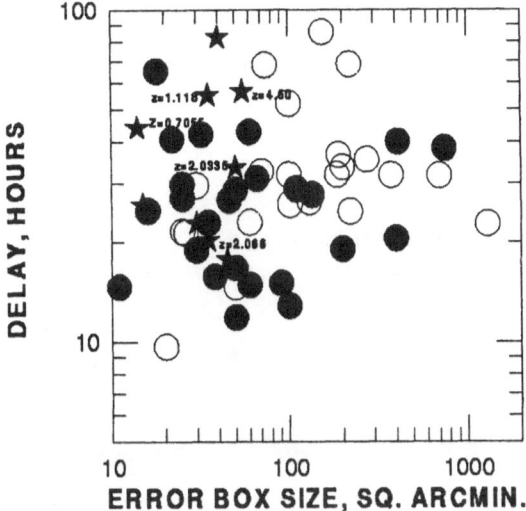

Fig. 1. Fifty-six IPN error boxes. The black dots are bursts for which follow-up observations were performed, but no counterpart was identified. The circles are bursts for which no follow-up observations were performed. The stars are bursts for which follow-up searches identified a counterpart. For those bursts whose redshifts were measured, z is given

of optical and/or radio counterparts to 28. Five redshifts were measured for these counterparts, bringing the total number of redshifts to 14. Both totals are up by ≈ 50 % due to the IPN results alone. The success rate for counterpart searches is ≈ 30 %, which is comparable to that for BeppoSAX, considering the small numbers involved.

3 Two Scientific Highlights

GRB000131 was localized by the IPN to a 110 square arcminute error box which was circulated 57 hours after the burst. Andersen et al. [2] have identified the counterpart with the ESO VLT; its redshift is 4.5, the highest of any burst to date.

To date, the only optical and radio counterparts found have been for the long bursts. The IPN is the first mission to provide rapid, precise localizations of short bursts, which are believed to form a separate class. GRB000607, 001025, 001204, and 010119 all had durations less than 1 second, placing them securely in the short burst category. In addition, observations by NEAR in the >100 keV range demonstrated that they had hard energy spectra. They were localized to error boxes in the 11-110 square arcminute range, with delays between 15 and 29 hours. Radio and optical searches were carried out which should have revealed counterparts, but failed to do so. If these four short bursts were like the long ones, of which ≈ 60 % are "dark" (i.e., they do not display optical or radio counterparts) we would have expected 2.4 not to display counterparts. The fact that counterparts were not detected for all four is not statistically significant, but it does hint that, in this one respect (lack of counterparts), the short bursts may indeed be like the long ones.

References

1. T. Cline et al.: These Proceedings (2001)
2. M. Andersen et al.: Astron. Astrophys., in press (2001)

GRBs of Energy E >10 GeV with ARGO-YBJ

Silvia Vernetto, for the ARGO-YBJ Collaboration

Istituto di Cosmogeofisica del CNR, Torino, Italy

Abstract. ARGO-YBJ is an air shower experiment optimized to detect small showers, under construction at the Yangbajing Laboratory (Tibet) at an altitude of 4300 m. One of the aims of the experiment is the study of high energy gamma-ray bursts, with a sensitivity to energy fluences as low as $F \sim 10^{-6} \div 10^{-5}$ erg cm^{-2}, in the 1-100 GeV range. This will be achieved using the "single particle" tecnique, more profitable in the energy region $E < 50$ GeV, and the "low multiplicity" technique, suitable to observe GRBs with spectra extending at higher energies.

1 High Energy Gamma Rays from GRBs

So far the Gamma Ray Burst radiation above 1 GeV has been very poorly studied. EGRET, the high energy detector aboard the Compton Gamma Ray Observatory, covering the $\sim$ 30 MeV–30 GeV range, detected only 3 bursts containing GeV photons, with a maximum energy of 18 GeV in GRB940217 [1]. In spite of this small number of observations, all GRBs could contain a GeV (or even a TeV) energy component, so far not measured because of the limited sensitivity of the existing instruments. The region above 1 TeV could be hardly accessible to our observations due to the absorption of gamma rays in the intergalactic space caused by pair production of high energy gamma rays off infrared photons emitted by stars and dust. According to calculations [2] the optical depth becomes equal to 1 for energies as low as 40-70 (200-400) GeV for a distance z=1(0.2). The redshifts of GRBs hosts measured so far (about 15 events) range between 0.4 and 4.5, clustering around $z \sim 1$. This implies that even if GRBs emit TeV-PeV gamma-rays we could not detect them, unless we are so lucky to observe a rare event occurring in our neighbourhood. As a consequence the study of the high energy component of GRBs requires detectors with a high sensitivity at energies less than 1 TeV.

Ground based experiments can observe high energy gamma rays detecting the secondary particles of air showers generated in the atmosphere. Cerenkov Telescopes detect the Cerenkov photons emitted by the shower particles in the high atmosphere, while Air Shower Arrays detect the charged particles (mainly electrons and positrons) reaching the ground level. Cerenkov Telescopes are not suitable to detect transient events as GRBs, because of their small field of view (few squared degrees) and their limited duty cycle ($\leq 10\%$). The probability for a GRB to happen by chance in their field of view is only $\sim 10^{-4}$. On the contrary, Air Shower Arrays have a much larger field of view ($\sim \pi$ sr) and their duty cycle, not limited by darkness or weather conditions, can be $\sim 100\%$. Traditionally Air Shower arrays consists of several detectors scattered over a large area, catching a small fraction (~ 0.1-1%) of the shower particles reaching the ground. Due to this sampling technique they can detect large showers with at least $\sim 10^4$-10^5 particles, that implies a primary energy threshold of ~ 10-100 TeV, too high to observe cosmological GRBs. In the last years, the importance of studing gamma-ray sources at energies < 10 TeV grew up, and a new type of air shower array has been conceived, able to detect small air showers: the *full coverage* arrays, characterized by a large and continuos sensitive area, in order to detect a large fraction of the particles of the shower front. These detectors can lower the energy threshold by a factor up to ~ 100 with respect to sampling arrays and can reach energies as low as ~ 100 GeV by operating at high altitude, where the number of detectable particles is strongly increased (see Fig.1). Hence a *full coverage* array located at mountain altitude is the best ground based detector suitable to search for high energy GRBs. An interesting TeV GRB candidate has been observed by the prototype Milagrito, in coincidence with GRB970417 [3].

2   The ARGO-YBJ Detector and Its Sensitivity to GRBs

The ARGO-YBJ experiment is a *full coverage* air shower array under construction at the Yangbajing High Altitude Cosmic Ray Laboratory (Tibet, China) at an altitude of 4300 m a.s.l. It consists of a central core made by a single layer of Resistive Plate Chambers (RPC's) covering an area of 71×74 m^2, sorrounded by an outer ring of 28 cluster of RPC's of 42 m^2 each, for a total sensitive area $A_d \sim 6100$ m^2 [4]. The detector is uniformely covered by a layer of lead 0.5 cm thick, in order to increase the number of charged particles by converting a fraction of the secondary photons, and to reduce the time spread of the shower front. A subset of the detector will start operating in 2001.

The main fields of research of ARGO-YBJ are gamma-ray astronomy in the energy range $E > 100$ GeV and gamma-ray bursts physics above 10 GeV. The detection of GRBs will be performed with two different techniques [5].

1) The *low multiplicity technique* (LMT) consists in the detection of very small air showers, by requiring at least ~ 10 detected particles per shower. The angular resolution for such small showers according to simulations is $\sigma \sim 3.2°$.

2) The *single particle technique* (SP), suitable to observe transient "low energy" events as GRBs [6-8], consists in recording *all* the particles hitting the detector, not requiring any coincidence among the particles arrival times, as it is usually done to detect air showers. In this way one can detect the lonely survivals of very small air showers otherwise under the detection threshold. Obviously with only one particle per shower it is not possible to reconstruct the arrival direction nor the energy of the primaries and a GRB is observable only as a short duration counting excess over the cosmic ray all sky background, possibly in coincidence with a GRB satellite detection.

To evaluate the ARGO-YBJ sensitivity to GRBs, we assumed for simplicity a burst of duration $\Delta t = 1$ s, with a zenith angle $\vartheta = 20°$ and a power law energy

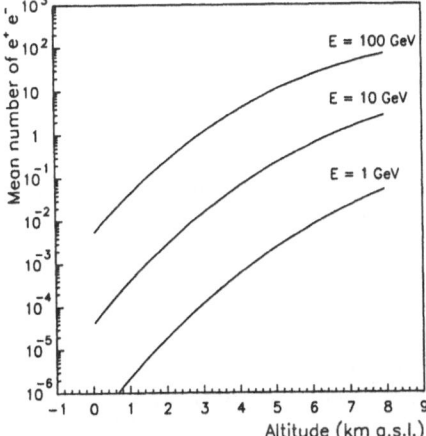

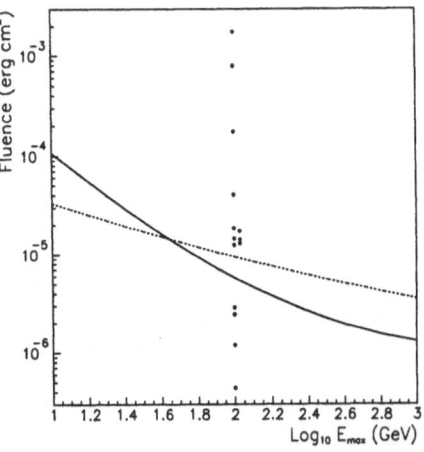

Fig. 1. Mean number of secondary $e^{\pm}$ reaching the ground from gamma rays of different energies, as a function of the altitude above the sea level of the observer (assuming a gamma ray zenith angle $\vartheta = 30°$).

Fig. 2. Minimum fluence in the 1 GeV-E_{max} range observable by ARGO-YBJ, as a function of the cutoff energy E_{max}, for the LMT (continuous line) and the SP (dashed line) tecniques. The points are from EGRET data (see text).

spectrum of slope $\alpha = 2.0$, extending up to a variable cutoff energy E_{max}. Fig.2 shows the minimum energy fluence between 1 GeV and E_{max} necessary to make a GRB observable (with a significance of 4 σ) as a function of E_{max}, using the SP and the LMT techniques. The minimum fluence for a different duration Δt scales as $\sqrt{\Delta t}$. One can see that ARGO-YBJ could detect GRBs with energy fluence between 1 and 100 GeV as low as $\sim 5 \ 10^{-6}$ erg cm^{-2}, and that the SP technique is more profitable for $E_{max} < 50$ GeV.

To compare the ARGO-YBJ sensitivity with real fluxes, in the same figure we report the fluences in the 1-100 GeV energy range obtained by extrapolating up to 100 GeV (with their observed slope) 14 spectra measured by EGRET [1]. Most of the events have an energy fluence larger than the ARGO limits, suggesting that ARGO-YBJ could succesfully detect a few high energy GRBs per year.

References

1. J.R.Catelli, B.L.Dingus, E.J.Schneid: AIP Conf.Proc., **428**, 309 (1997)
2. M.H.Salamon and F.W.Stecker: Ap.J., **493**, 547 (1998)
3. R.Atkins et al.: Ap.J.Lett., **533**, L119 (2000)
4. Abbrescia M. et al.: Proposal of the ARGO experiment (1996), and Bacci C. et al.: Addendum to the ARGO Proposal (1998), unpublished, can be downloaded at the URL http://www1.na.infn.it/wsubnucl/cosm/argo/argo.html
5. S.Vernetto et al: Proc.26th ICRC, **4**, 28 (1999)
6. S.Vernetto: Astrop.Phys, **13**, 75 (2000)
7. M.Aglietta et al.: Ap.J., **469**, 305 (1996)
8. R.Cabrera et al.: Astr. and Astroph. Suppl.Series, **138**, 599 (1999)

Progress on MARGIE, a Gamma-Ray Burst Ultra-long Duration Balloon Mission

D. Band[1], M. Cherry[2], J. Stacy[2], T. Guzik[2,3], S. Kappadath[2], J. Buckley[4], P. Hink[4], J. Macri[5], M. McConnell[5], J. Ryan[5], and J. Matteson[6]

[1] X-2, Los Alamos National Laboratory, Los Alamos, NM, 87545, USA
[2] Louisiana State University, Baton Rouge, LA, 70803, USA
[3] Southern University, Baton Rouge, LA, 70813, USA
[4] Washington University, St. Louis, MO, 63130, USA
[5] University of New Hampshire, Durham, NH, 03824, USA
[6] CASS, University of California at San Diego, La Jolla, CA, 92093, USA

Abstract. We are designing the Minute of Arc Resolution Gamma-ray Imaging Experiment (MARGIE) as a 100 day Ultra Long Duration Balloon (ULDB) mission to: a) detect and localize gamma-ray bursts; and b) survey the hard X-ray sky. Major advances in designing the CZT detectors increase the senstitivity to higher energy. Design of the gondola has also progressed.

1   Introduction

The MARGIE project is a synthesis of advanced mask construction, detector material and gondola control technologies under development at the various institutions in this collaboration. In addition to performing astrophysically relevant observations, MARGIE will be a test bed for these new technologies. Since the burst-detection capabilities were reported at a previous gamma-ray burst meeting[1], here we provide a progress report on technical advances.

The MARGIE mission will consist of a central imager with a small (8.3′ half angle) field-of-view (FOV) and 4 large (26.1′ half angle) FOV detectors whose orientations will be offset from that of the central detector. With areas of $\sim 1900\,\mathrm{cm}^2$, the detector plane will most likely be Cadmium-Zinc-Telluride (CZT), with segmented CsI scintillators as a backup. With 0.83 mm (0.5 mm) pixels for the central (side) detector and a detector-mask distance of 150 cm (45 cm), the geometric resolution will be 1.9′ (3.8′). Imaging will be achieved using a tungsten coded mask consisting of a 2 dimensional URA pattern with 2 cycles in each direction. The burst position will be calculated onboard and disseminated in near-real time, while information about every count will be stored for further analysis.

2   The Detector Plane

We plan to use the CZT detector design developed at the Universities of New Hampshire and Montreal[2,3]. In this design, the cathode and anode are on the source and back sides of the detector, respectively. The anode consists of orthogonal coplanar electrodes. Each of the N rows of anode electrodes consists of N interconnected pixels, while each of the N columns is a single conducting strip. The pixels are circles punched out of the conducting strips which make up the columns, with a contact bonded to the surface of the CZT in the middle of the pixel. This geometry permits an N×N pixel array to be read out by 2N electronic channels, reducing the complexity of the electronics. A gamma ray absorbed in the CZT produces electron-hole pairs. The anode geometry relies on the relatively mobile electrons for the energy and row/column location measurements. Processing of the shape of the strip (column) signal exploits hole trapping in the CZT to measure the depth of interaction. Since the strips surrounding the pixels are biased at a voltage intermediate between the cathode on the source side and the pixels, the pixels collect the electrons but the strips surrounding the pixels register a transient signal as the electron cloud approaches the anode plane. Since the electrons produce a signal in more than one strip or row of pixels, the position of the absorbed photon can be resolved to 0.1 mm along a row and 0.3 mm along a strip, even though the pixels are 1 mm across. The depth can be resolved to less than 1 mm. Because the horizontal position is determined by the electrons at the anode, and detection of the holes is secondary, the CZT can be thicker (up to 10 mm), increasing the detector's high energy efficiency (e.g., to $E > 1\,\mathrm{MeV}$). The gamma-ray energy deposited is measured by the strength of the electron signal. The measured spectral resolution is currently 5.7% at 60 keV, 2.6% at 122 keV and 0.94% at 662 keV, within a factor of 3 of the theoretical limit.

The technology of packaging the CZT detectors and bonding the leads connecting the pixels has been developed and demonstrated. The major impediment is the cost of the CZT ($3000 cm^{-2}); this cost will undoubtedly come down with time, particularly for a large quantity. However, as an alternative we have developed and will test segmented BGO scintillator detectors.

3   The Gondola

A ULDB is essentially a spacecraft mission without the paperwork. We will use the experience of the team members in other ULDB projects, specifically the TIGER program at Washington University which will be the first ULDB flight. The MARGIE detectors will be oriented vertically within a spherical KEVLAR pressure shell. Such a pressure shell was successfully flown on the recent 17-day long duration Antarctic flight of the ATIC cosmic ray experiment[3]. This pressure shell will sit within a frame of I beams. Thus the central detector will be constrained to point vertically. The rotor connecting the gondola to the balloon will provide the azimuthal orientation. Sub-arcminute resolution aspect determination will result from fiber optic gyroscopes, CCD cameras and a GPS receiver. The data will be stored primarily on board with the ability to upload commands and download data when the mission is within line-of-sight of a groundstation or a chase aircraft. However, MARGIE will also have a TDRSS link which will be used for time-critical communication, such as downloading burst positions (which can then disseminated through GCN).

The science package (detectors, pressure vessel, frame but not NSBF components) will weigh ∼950 kg. The weight and power, as well as the payload durability, all meet ULDB requirements.

4   Burst Sensitivity

We discussed MARGIE's burst sensitivity previously [1], and thus here we only present a summary. Bursts will be detected by a statistically significant increase (e.g., by 5σ) in the count rate. We find a BATSE-equivalent (i.e., 50–300 keV) threshold peak flux of $\psi_{\mathrm{cen}} = 0.209t^{-1/2}(\sigma_b/5)\,\mathrm{ph\,s^{-1}\,cm^{-2}}$ and $\psi_{\mathrm{side}} = 0.376t^{-1/2}(\sigma_b/5)\,\mathrm{ph\,s^{-1}\,cm^{-2}}$ for the central and each side detector, respectively. Based on the burst rate, the FOV, and a mission duration of $M \sim 100$ days, we find that MARGIE should detect $N_{\mathrm{cen}}(> \psi_{\mathrm{cen}}) = 1.1t^{0.4}(\sigma_b/5)^{-0.8}(M/100\,\mathrm{d})$ and $N_{\mathrm{side}}(> \psi_{\mathrm{side}}) = 7.1t^{0.4}(\sigma_b/5)^{-0.8}(M/100\,\mathrm{d})$ bursts in the central and each of the side detectors, respectively. Accounting for overlapping FOVs, ~ 22 bursts should occur in some detector's fully coded FOV. Of course, the detectors will also detect bursts in the partially coded FOVs. Therefore, we should detect at least 40 bursts. We assume that each detector triggers independently.

For the fully-coded FOV, the brightest burst expected in 100 days (i.e., ψ_b such that $N(> \psi_b) = 1$) will have a peak flux of $\psi_{b,\mathrm{cen}} = 0.25\,\mathrm{ph\,cm^{-2}\,s^{-1}}$ and $\psi_{b,\mathrm{side}} = 4.36\,\mathrm{ph\,cm^{-2}\,s^{-1}}$. Of course, with the much larger area covered by all 4 side detectors, including the partially coded regions, the "burst of the mission" will probably be somewhat larger.

With the small number of counts accumulated during a burst and distributed over a much larger number of pixels, backprojection will be the most efficient image reconstruction technique to localize the bursts. At the burst threshold there is a $\sim 10\%$ chance of a spurious localization in a side detector but this probability rapidly becomes negligible for brighter bursts. Similarly, persistent sources in the FOV should not be bright enough to compromise the localizaton, although images before and after the burst should identify such background sources. Sources can be localized to less than the geometric resolution; we anticipate localizations of $\sim 1.7'$ in the side detectors.

References

1. D. Band, et al.: 'MARGIE, a Gamma-Ray Burst Ultra-Long Duration Balloon Mission'. In: *Gamma-Ray Bursts, Proc. of the 5th Huntsville Symposium*, ed. by R. M. Kippen, R. Mallozzi, and G. Fishman (AIP, New York, 2000), pp. 696-697
2. McConnell, M. L., et al.: 'Three dimensional imaging and detection efficiency performance of orthogonal coplanar CZT strip detectors'. In: *Proc. SPIE* **4141**, 157 (2000)
3. P. Altice, et al.: 'MARGIE'. In: *Proc. 33rd COSPAR Scientific Assembly, Warsaw, Poland, 16-23 July 2000* (2001), in press
4. G. Guzik, et al.: Advances in Space Research **19**, 711 (1997)

Simultaneous Detection of the High Energy and Optical Transients by Čerenkov Telescopes

Gregory Beskin[1,4], Corrado Bartolini[2], Adriano Guarnieri[2],
Adalberto Piccioni[2], Sergej Biryukov[3], David Eichler[3], David Faiman[3]

[1] Special Astrophysical Observatory, Nizniji Arkhyz, Russia
[2] Dipartimento di Astronomia, Università di Bologna, Italy
[3] Ben-Gurion University, Beer-Sheva, Israel
[4] Isaac Newton Institute of Chile, SAO Branch

Abstract. In order to detect and investigate short optical flares connected with GRBs, the use of very large, wide field "bad" mirrors is proposed and justified. The possibility of simultaneous observation in the optical and in the VHE gamma-ray bands with the same mirror is analyzed.

To understand the physics that underlies the GRB events, to do a selection among their models and to validate by proper tests a hypothesis, the observational opportunities must be extended in time and frequency to obtain a full description of the development of the GRB light curves:

1. before and during the GRB emission phase;
2. in multifrequency, performing simultaneous observations in all possible spectral bands;
3. with high time resolution.

The first condition was never fulfilled; only in the case of GRB990123 a very close observation was performed by ROTSE (Akerlof & McKay, 1999) that confirmed forecasts of strong OT's during the gamma events; however, also in this case, the first exposition has been started 22 s after.

The second condition, highlighted by the observed radio (Frail et al., 1997) and high energy tails (Hurley et al., 1994) is to be considered a priority owing to the non-thermal characteristic of the energy distribution of the flux associated with the GRB events.

The third condition allows to study fine time structure of GRBs. We discussed the possibilities of solving the above-mentioned problems by big telescopes with "bad" mirrors and wide fields (Beskin et al., 1999). We develop this approach here.

The situation resulting from these opportunities can be so summarized:

- the main need is to perform on-time observations with high temporal resolution.
- the high fluence of GRB990123, (the highest ever measured in any spectral band from the start of BATSE operation) seems to be a very rare situation.
- the high spatial resolution is not a priority request because the exact position of sources can be obtained by the slow CCD observations of the afterglows.

- to obtain good observations with high time resolution, a high photon flux is needed because the phenomenon under observation is neither periodic nor repetitive.

On this basis we have a new approach to the problem requiring:

1. large mirrors in order to have high photon counts for faint OT's
2. photoelectric, position sensitive cameras, to obtain the maximum possible time resolution
3. combination of different techniques of measurement to have simultaneous observations of bursts ground based and by satellites in different energy bands
4. use of the existing instrumentation to reduce costs and times
5. continuous tracking of the field observed by a gamma satellite (the sky area corresponding to the best sensitivity beam) with a few wide field instruments to enhance the probability of covering the OT.

The most suitable existing instrumentation for these kind of observations can be found on paraboloidal mirror arrays used for Čerenkov light pulse measurements (Weeks, 1988). The VERITAS project (Lessard, 1999a) is an example of a highly developed system of this kind, consisting of 7 large (10 m) Čerenkov paraboloids. The HEGRA IACT system with 5 smaller Čerenkov telescopes, each of them featuring a collecting surface of $8.5m^2$ and a field of view of $4.3°$, could also be considered. These huge light collectors are normally used to perform measurements of atmospheric showers generated by high energy (from $10GeV$ up to $100TeV$) gamma rays and cosmic ray particles; their spatial resolution in the optical band is not as good as required for direct imaging of astronomical objects. Resolutions of the order of 5–10 arcmin are common to the field central region of VERITAS telescopes ($0.15°$, Lessard 1999b) , MAGIC ($0.1°$, Martinez, 1999) and other instruments such as PETAL 25m solar energy concentrator (Biryukov et al., 1999), recently reconverted to detect Čerenkov showers.

It is clear from Fig. 1 that detection limit of these telescopes is similar with one of small "real" optical telescopes. However there are several advantages of Čerenkov telescope. By technological reasons optical telescope have field of view of about $1°$ – there are only four with field of view of several degrees. Because of their short focal lengths the focal planes can be equipped only by CCD (not photomultiplier tubes) therefore time resolution is worse than 0.1 s. At last only Čerenkov telescopes are able to detect optical and VHE flares simultaneously.

The possibility of doing simultaneous Čerenkov and optical measurements can be realized by using only Čerenkov electronic camera, splitting the signals by a twin trigger system. The existing Čerenkov electronic camera will work simultaneously as optical and as Čerenkov imager. The arrival times and the position of the single photon events, discarded by the Čerenkov instrumentation threshold system, should be stored to allow the following temporal analysis. Anticoincidence gates should be used to divide optical photons from Čerenkov ones produced by primary gamma rays in the high atmosphere. Very fast, low gain PMT and high speed electronics are required with a special acquisition system to

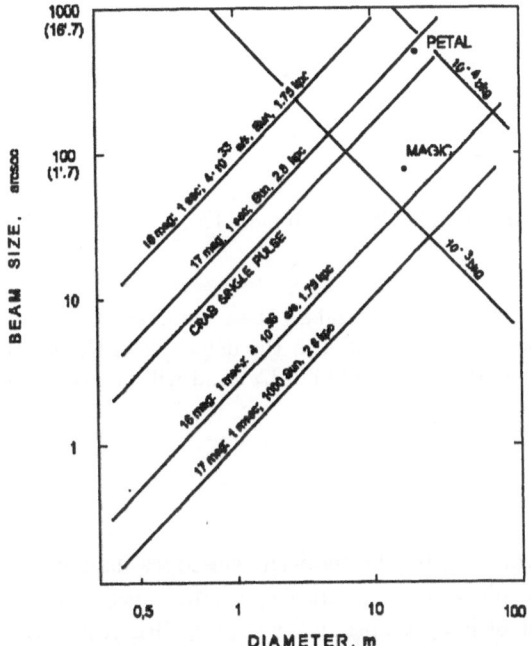

Fig. 1. Flare detection level. The two line families, computed for $m_{bg} = 20.5$, represent the two limiting conditions coming from background photons Poissonian noise (continuous lines) and background variability (dashed lines).

store the optical photon flux; for this purpose the structure of QUANTOCHRON (Beskin et al., 1997), modified to attain a better performance, could be the right solution.

This investigation was supported by the Italian Ministry of Foreign Affairs, by the University of Bologna (Funds for Selected Research Topics) by the Russian Foundation of Fundamental Researches (grant 98-02-17570).

References

1. Akerlof C.W. and McKey T.A.: I.A.U.Circ. **7100**, (1999)
2. Beskin et al.: A&A, **138**, 589, (1999)
3. Beskin et al.: Exp. Astron., **7**, 413, (1997)
4. Biryukov S., Eichler D., Faiman D., 1999, private communication.
5. Frail D.A. et al.: Nature **389**, 261, (1997)
6. Hurley et al.: Nature, **372**, 652, (1994)
7. Lessard R.: The VERITAS Collaboration WEB page, (1999a)
8. Lessard R.: VERITAS Telescope Design, 3, 18, WEB page, (1999b)
9. Martinez M.: XXVI Int. Cosmic Ray Conf., in press, (1999)
10. Weeks T.C.: Physics Reports, /textbf160, 17, (1988)

PREPROCESS:
A Fast Image Processing Software Tool

Andrea Di Paola

Rome Astronomical Observatory, Monte Porzio Catone, Rome, ITALY

Abstract. PREPROCESS is a tool to perform standard data reduction on astronomical images in a very short time. It has been used in the detection of the GRB afterglows at the Campo Imperatore Observatory (AQ-ITALY) and will be also used within R.E.M. project.

1 Introduction

The large amount of raw data produced by the modern telescopes usually requires a long time to be processed but a lot of interesting studies need a faster processing. The reduction of a set of images obtained with the dithering technique, or even the reduction a single frame, requires a certain number of standard operations that, if performed in a traditional way by an astronomer using an interactive software, are usually time consuming.

PREPROCESS has been realized to automate many standard operations. It produces in few seconds images ready for photometrical measurements or detection of new sources and moving objects. It has been very useful for the GRB optical couterparts observation as well as for normal observations.

2 Features

PREPROCESS is optimized to perform the following operations on a set of raw images:

- Flat-fielding;
- Dark, Bias, Sky subtraction;
- Bad Pixels removal;
- Gaussian smoothing and/or FWHM adaptation (through convolution);
- Recentering of the sources with the shift of the images to a subpixel accuracy;
- Forced shift on the basis of the acquisition time or processing order for moving or variable objects study;
- Fluxes adaptation by scaling or offsetting each image depending on sky level or sources' photometry in the reference image;
- Composition by mean MEDIAN, AVERAGE, SUM or DIFFERENCE;
- New sources detection using FWHM adaptation and DIFFERENCE on two images obtained at different times;
- Final image resize to the INTERSECTION or the UNION of the originals.

All the available operations may be easily selected using an ASCII file. This file contains, for each allowed command, a short explanation and both allowed and default values.

PREPROCESS can be used on images having the same pixel scale, the same FITS numerical type and which are not field rotated.

It means that PREPROCESS only works with images obtained from one selected instrument at time. Definitely PREPROCESS is tuned to work with the acquisition software to get out from the acquired images their scientific content in an extremely short time.

3 A Real Case

In the following I present an example of a typical data reduction session with PREPROCESS.

Suppose to have 5 dithered J band images obtained from a scientifical target, 5 images from a nearby as empty as possible field to be used as sky, 1 library flat-field image and the list of all the bad pixels of the used detector (Fig. 1).

Fig. 1. A schematic representation of the data to be furnished to PREPROCESS

Two steps are required to obtain the final image. The first step creates the sky image. Sky image is obtained by composing the raw sky images with the MEDIAN operator. MEDIAN operator is applied without shifting the dithered images in order to make stars disappearing. This can be obtained using the following configuration file:

```
;
; SKY.PPR configuration file for PREPROCESS version >= 1.10
;
; Explanation:
;   Group.Keyword = value              // Mm * [type] (default) comment
;                                          || | |       |
;   Mandatory keyword ----------------------+| | |       |
;   Mandatory if its Group is selected -----+ | |       |
;   May appear more than once ----------------+ |       |
;   Value type       ---------------------------+       |
;   Default value    -----------------------------------+
;
; The Group keyword not starting at the first column are not considered.
; Parameter names and selection constants are all case insensitive.
; The parameter <NameA> may only be used in the value fields with the
;   following syntax: $(NameA).
; Parameter values can only contain parameters declared in a previous line.

Param.Filter     = j           // * [string]        everywhere $(Name) will be substituted by value
Param.File       = sky_$(Filter) // * [string]      everywhere $(Name) will be substituted by value
```

```
Param.Camera        = swircam        //  * [string]        everywhere $(Name) will be substituded by value
Source              = $(File).*.fits // M * [filename]     images file name to be used
Compose             = TRUE           //    [boolean]  (FALSE) FALSE or TRUE
Compose.Technique   = MEDIAN         //    [selection] (MEDIAN) MEDIAN, AVERAGE, ADVAVERAGE, SUM or DIFF
Compose.OffsetBy    = BACKGROUND     //    [selection] (NONE) NONE or BACKGROUND
Compose.FinalImage  = $(File).fits   // m  [filename]     processed image
```

The second step creates the final image. Final image is obtained by performing sky subtraction, flat-fielding, recentering (to match sources) and composition with the AVERAGE operator. The bad pixels are automatically skipped during composition. This can be obtained using the following configuration file:

```
;
; TARGET.PPR configuration file for PREPROCESS version >= 1.10
;
Param.Filter              = j                   //  * [string]        everywhere $(Name) will be substituded by value
Param.File                = target_$(Filter)    //  * [string]        everywhere $(Name) will be substituded by value
Param.Camera              = swircam             //  * [string]        everywhere $(Name) will be substituded by value
Source                    = $(File).*.fits      // M * [filename]     images file name to be used
Compose                   = TRUE                //    [boolean]  (FALSE) FALSE or TRUE
Compose.Technique         = AVERAGE             //    [selection] (MEDIAN) MEDIAN, AVERAGE, ADVAVERAGE, SUM or DIFF
Compose.NormalizeOn       = NONE                //    [selection] (NONE) NONE, SOURCES or BACKGROUND
Compose.OffsetBy          = BACKGROUND          //    [selection] (NONE) NONE or BACKGROUND
Compose.ResultType        = INTERSECTION        //    [selection] (UNION) UNION or INTERSECTION
Compose.FinalImage        = $(File).fits        // m  [filename]     processed image
Recenter                  = TRUE                //    [boolean]  (FALSE) TRUE or FALSE
Recenter.Accuracy         = FULL                //    [selection] (FULL) PIXEL or FULL
Recenter.Tolerance        = 2                   // m  [pixel]        the maximum allowed distance of two images
Recenter.MaxShift         = 100                 //    [pixel]    (inf) the maximum allowed distance of two images
Recenter.MinObjectsNumber = 3                   //    [integer]  (3) the minimum number of objects to match (>=1)
Recenter.ResamplingTecnique = LINEAR            //    [selection] (LINEAR) LINEAR
Recenter.SextractorFile   = $(Camera).sex       // m  [filename]     sextractor configuration file
Process                   = TRUE                //    [boolean]  (FALSE) TRUE or FALSE
Process.Flat              = flat_$(Filter).fits //    [filename]     flat filename
Process.Sky               = sky_$(Filter).fits  //    [filename/DN]  sky file name or DN #number
Process.BadPixels         = $(Camera).bad       //    [filename]     ascii list (one x,y point per line)
```

The results, coming from a set of images acquired with the 256×256 pixels IR camera (SWIRCAM) installed at the AZT-24 telescope of Campo Imperatore, have been obtained in about 5 seconds on a Pentium III 500 MHz PC (Fig. 2).

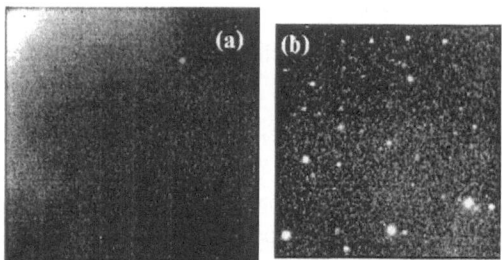

Fig. 2. The sky image (a) and the final image (b) produced by PREPROCESS.

4   Availability

PREPROCESS is a multiplatform software available in binary format for OS/2, Win32, Linux and AIX. It is available for free download at:

http://www.mporzio.astro.it/cimperatore/WWW/

The HETE Triggering Algorithm

E.E. Fenimore and M. Galassi

Los Alamos National Laboratory, Los Alamos, NM, USA

Abstract. The High Energy Transient Explorer uses a triggering algorithm for gamma-ray bursts that can achieve near the statistical limit by fitting to several background regions to remove trends. Dozens of trigger criteria run simultaneously covering time scales from 80 msec to 10.5 sec or longer. Each criteria is controlled by about 25 constants which gives the flexibility to search wide parameter spaces. On orbit, we have been able to operate at 6σ, a factor of two more sensitive than previous experiments.

Gamma-ray bursts (GRBs) occur at unpredictable times and satellite telemetry bandpasses are too limited to send every photon to the ground in real time. Thus, GRB experiments usually have on-board triggering systems to detect when a GRB is occurring and to switch operation modes to capture the event with maximum time and energy resolution. Most previous experiments (Vela, PVO, ISEE-3, Ginga, BATSE) employed a system that looked for "significant" increases in the photon count rate over a background count rate. Such increases are usually searched for over a few (e.g., 3) time scales that have ranged from 0.064 sec to 4 sec. For the background count rate, these systems took an average of the count rate from a period assumed to be well before the bursts (e.g., from 16 sec to 30 sec before the trigger sample).

The definition of a "significant" trigger has usually been how many standard deviations ("σ") the candidate time sample exceeds the expected count rate *assuming Poisson statistics*. Because of the use of Poisson statistics in the definition of the trigger, it is a common misconception that trigger algorithms are guarding against statistical fluctuations. The σ level is usually never set below ~ 11 and, yet, there are still many false triggers. Obviously, the cause of false triggers is not statistical fluctuations. In most experiments (e.g., PVO, Ginga, ISEE-3), the σ level was 11 and 90% of the triggers were, in fact, not GRBs. BATSE also had a threshold equivalent to $\sim 11\sigma$; it used 5.5 σ in the second brightest illuminated detector which translates to $\sim 11\sigma$ in the brightest illuminated detector. BATSE achieved $\sim 50\%$ false trigger rate because many triggers could be rejected on-board by crude locating which was able to nullify many false triggers when the source appeared to be inside the satellite (i.e., particle events) or coming from the sun.

Rather than guarding against statistical fluctuation, on-board triggers need to be designed to avoid false triggers, often caused by trends in the data. Consider the scenario in the figure. If one only has a single background region before the burst (labeled "Back$_1$"), a slight trend can make the count rate in a candidate trigger period to appear to be statistically significant. This situation becomes worst for larger experiments. For example, the 5000 cm^2 Burst Alert Telescope on the Swift satellite will have a background rate of $\sim$ 17 kHz. A 4% trend in the background is enough to have the appearance of a 5.5 σ statistical fluctuation. Variations in the particle flux in low earth orbit can easily make a factor of two variation in tens of seconds. Eleven σ was selected in the past because such a threshold would eliminate triggers from most trends over the time scales used in the triggering, a reason that does not involve statistical fluctuations.

In contrast to BATSE which required a uniform and easily understood trigger to have bursts with well defined properties, our goal is to capture as many and as varied GRBs as possible. Thus, on the High Energy Transient Explorer (HETE) we have implemented an extremely flexible triggering algorithm designed to remove trends and achieve triggering close to the Poissonian limit. A large number of triggers run simultaneously, each defined by $\sim$ 25 constants. HETE was launched with 31 such triggers, controlled by $>$ 700 constants. As we learn more about the actual background, we can upload new triggers or new constants for existing triggers.

HETE consists of three instruments. The FREGATE gamma-ray scintillators were built by CESR of France and covers the energy range 6 to 400 keV. The Wide Field X-ray Monitor (WXM) is a proportional counter-based coded aperture built by the RIKEN Institute of Japan and Los Alamos. It covers 2 to 25 keV and can provide locations with a point spread function of 34 arc minutes. The Soft X-ray Cameras (SXC) are CCD-based coded apertures built by MIT that cover 0.5 to 10 keV and can provide locations with a point spread function of 30 arc seconds. SCX and FREGATE have their own triggering systems. In

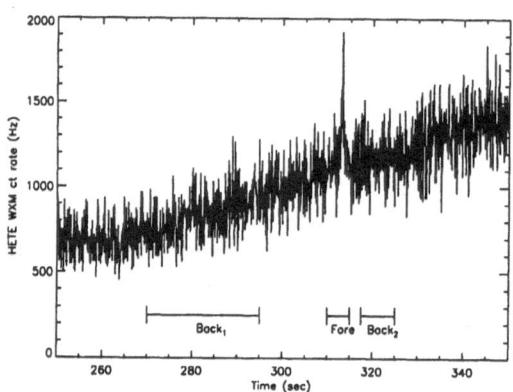

Fig. 1. A triggering scenario where a trend causes a false trigger. Previous on-board triggering systems have only used a single background period (e.g., "Back$_1$") before the candidate trigger sample (i.e., "Fore"). The trend can give a false trigger. In HETE, we use two background regions which can either be both before the trigger sample or bracket the trigger sample. Many triggers run simultaneously, covering a wide temporal and energy parameter space.

this paper, we describe the WXM triggering software which is also applied to FREGATE. Each of the three instruments can trigger the other instruments.

The HETE trigger algorithm works in the past. It can use either one background region or two and they can either be both before the candidate trigger sample (and therefore extrapolate to remove a trend) or they can bracket the time sample (and therefore interpolate to remove a trend, the situation depicted in the figure). The background regions are used to predict the count rate during the candidate foreground sample and the net counts are characterized by the number of σ the deviation appears to be.

The triggering runs asynchronously with the data collection (which is based on 80 ms samples). When invoked (typically once per sec but perhaps longer), the triggering algorithm tests all the samples which have occurred in each trigger since the last time it was invoked (perhaps 20 to 40 new samples each second) and reports the best one. The trigger algorithm can sense increases in either the FREGATE scintillators, or the WXM proportional counters, or a combination of both. We can control which of the four FREGATE detectors or four WXM detectors are used in each trigger.

The following is a summary of the major parameters that define each trigger. A set of flags indicate whether the trigger applies to WXM, FREGATE, both WXM and FREGATE, or neither. For both FREGATE and WXM we specify the range of detectors to be used and the range of energies to be used. We usually use all detectors for both systems. For energy ranges, we currently use 2 to 20 keV for all of the WXM triggers and 20 to 300 for all of the FREGATE triggers. One parameter tells if there will be one background region or two.

We can specify the start times and durations of each of the backgrounds relative to the current time. Typically, the background before the candidate sample is about 16 sec and the background afterwards is 1 or 2 sec. We also specify how frequently the backgrounds must be recalculated. (For triggers looking for short bursts it is unnecessary to recalculate the background for each new candidate sample.) Other parameters define the start, duration, and the spacing between the foreground samples. Typically, we use foreground candidate durations of 80 msec to 10.5 sec. The spacing between evaluations is adjustable which allows us to have samples of length ΔT and sample them more frequently than ΔT to check more phases. The threshold for declaring a trigger is set in units of σ^2. The trigger algorithm provides to the WXM imaging algorithm the start and stop times of the first background region and the candidate trigger region. These are used to determine, on-board, the location of the GRB.

The bracketing background works best. The single background cannot tolerate a slope and the two backgrounds before the candidate time can produce false triggers whenever there is a change in slope. One disadvantage of the bracketing system is the determination of the on-board location (and subsequent report to the ground) is delayed until after the second background period.

This triggering algorithm is capable of using all available on-board computing power and, as such, we often will only operate a subset of the available triggers.

The Status of the Ondřejov BART Experiment

René Hudec[1], Martin Nekola[1], Petr Kubánek[2], Cyril Polášek[1], and Alberto J. Castro-Tirado[3,4]

[1] Astronomical Institute, Academy of Sciences of the Czech Republic, CZ-251 65 Ondřejov, Czech Republic
[2] Department of Informatics, Faculty of Mathematics, Charles University, Prague, Czech Republic
[3] Instituto de Astrofisica de Andalucia (IAA-CSIC), Granada, Spain
[4] Laboratorio de Astrofisica Espacial y Fisica Fundamental (LAEFF-INTA), Madrid, Spain

1 Introduction

The BART (Burst Alert Robotic Telescope) is an dedicated automated small aperture telescope in the test operation at the Ondrejov Observatory. The device represents remotely controlled small aperture telescope with attached wide field camera. The BART serves as a dedicated optical telescope for gamma-ray bursts and high energy astrophysics. At the same time, BART is a low-cost device based on commercially available parts. Due to the low cost of the hardware parts, the system is suitable for networking and duplications.

2 Science with BART

The BART system has been designed to focus on the following scientific tasks.

- optical follow-up observations of Gamma Ray Bursts (GRBs) detected by satellites (HETE2, INTEGRAL, ...) with automated rapid response (~ 1 min depending on position)
- dense optical follow-up observations of GRBs, astrometry and photometry
- simultaneous and quasisimultaneous optical data for satellite observations and campaigns
- secondary science: photometry and optical monitoring of selected triggers (X-ray stars, AGNs, blazars, QSOs, SNe, CVs, targets of opportunity triggers, ...)

3 Technical Solution

- optical tube: Meade LX200 Schmidt-Cassegrain telescope, aperture 25 cm LX200 Meade mount, 8 degrees/sec
- attached wide-field camera: Meopta lens, FOV 5 degrees diameter (identical with the INTEGRAL OMC Test Device)
- SBIG ST7 and ST8 CCD cameras
- resulting limiting magnitudes 18.5 for optical tube and 15.5 for the wide-field camera

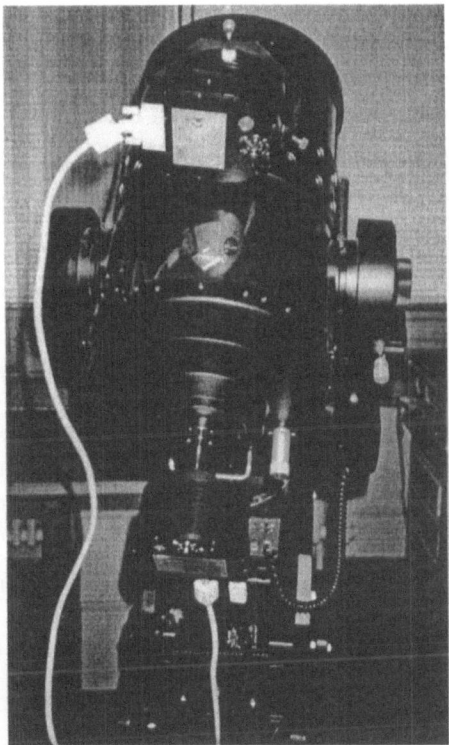

Fig. 1. The BART system located at the Ondřejov Observatory

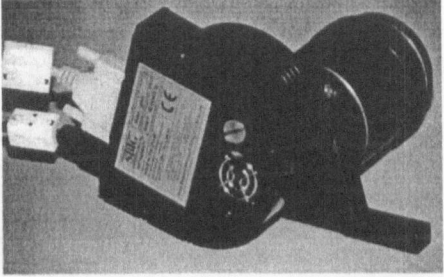

Fig. 2. The wide-field camera equipped with a Meopta lens is identical with the OMC INTEGRAL experiment Test Device

- controler software written in Python (http://www.python.org), a high level scripting language with objects features
- cameras driver written in C, both cameras can shoot and download the images simultaneously
- control software runs on a Pentium PC under Linux RedHat 6.0 operating system

- user access is provided by simple web-based interface, obtained images can be downloaded by FTP
- users can create their own scripts, which controls the telescope movement and cameras exposition during automated observation

The system (including the newly designed software packages) is in test operation at the Ondřejov observatory. Attempts are done to improve the guidance and tracking capabilities of the LX200 mount.

4   BART and BOOTES

The BART telescope is closely related to the BOOTES experiment (Castro-Tirado et al., 2001). Both systems are based on commercially available hardware parts and are operated in an analogous way. The Ondřejov site (easily accessible) has been used primarily for testing of various hardware parts and software packages, while the El Arenosilo site (with essentially more preferable weather conditions) is primarily used for routine scientific observations.

- BOOTES: Burst Observer and Optical Transient System
- modified and advanced BART system
- 30 cm aperture Meade LX200 optical tube with SBIG ST8 CCD camera and various WF cameras
- in routine operation in El Arenosilo, Spain
- 2nd station in preparation at the 200 km distance
- for more details see Castro-Ceron et al., 2001.

5   Conclusion

There are 2 (and soon 3) new small aperture automated telescopes with attached wide-field cameras for rapid response to GRB satellite triggers with limiting magnitudes 18–19. The systems are also suitable to provide dense observations (photometry) of fading optical counterparts of GRBs, as well as of secondary science triggers of various physical nature, as well as to provide simultaneous and quasisimultaneous optical observations for satellite observations. The experience gained is expected to be applied also for larger aperture telescopes such as the high-speed 62 cm CCD telescope recently under development at the Ondřejov Observatory (supported by the Academy of Sciences of the Czech Republic).

Acknowledgements. We acknowledge the support provided by the Grant Academy of the Czech Republic (grant 205/99/0145), by the Ministry of Education of the Czech Republic (Projects ES02 and ME137) and by the ESA PRODEX (Contract 14527/00/NL).

References

1. Castro-Ceron, J. M. et al., this volume, 2001.

Progress in Lobster Eye X-Ray Telescope Development

René Hudec[1], Adolf Inneman[2], and Ladislav Pina[3]

[1] Astronomical Institute, Academy of Sciences, 251 65 Ondřejov, Czech Republic
[2] Centre for Advanced X-Ray Technologies, Reflex, Prague, Czech Republic
[3] Czech Technical University, Faculty of Nuclear Science, Prague, Czech Republic

1 Introduction

The lobster-eye (LE) geometry X-ray optics offer an excellent opportunity to achieve very wide fields of view (1 000 square degrees and more) while the widely used classical Wolter grazing incidence mirrors are limited to roughly 1 deg FOV. Wide field X-ray telescopes with imaging optics are expected to represent an important tool in future space astronomy projects, especially those for deep monitoring and surveys in X-rays over a wide energy range Wide field X-ray optics has been suggested in 70ies by Schmidt [2] (orthogonal stacks of reflectors) and by Angel [1] (array of square cells) but has not been constructed yet. Up to 180 deg FOV may be achieved.

2 Observing GRB X-Ray Afterglows with LE Telescopes

The LE X-ray telescopes are extremely important since the discovery of X-ray afterglows of Gamma Ray Burst (GRBs) sources in 1997.

Almost every GRB is accompanied by a X-ray afterglow. The expected rate of GRBs is 1 per day, however the theoretical prediction assumes larger beaming angle in X-rays if compared with gamma rays, hence the actual rate of X-ray afterglows is expected to be larger than the rate of GRBs.

The calculated sensitivity of LE telescopes is sufficient enough to detect the recently discovered X-ray GRB afterglows. For pointed observations, limits better than 10^{-14} erg s^{-1} cm^{-2} (0.5 ... 3 keV) can be obtained (10^{-12} erg s^{-1} cm^{-2} for daily observations). The localization accuracy of the LE telescopes is of order of 1 arcmin, substantially exceeding the recent localization accuracy of most gamma ray instruments (2 deg and more). This is sufficient to localize the fading X-ray GRB counterpart to 1 arcmin or better, as well as obtain the light curve. The LE telescopes are expected to provide a substantial contribution to the science and statistics of GRBs.

The additional science of LE X-ray telescopes includes supernova explosions, high energy binary sources, AGNs, blazars, X-ray novae, X-ray flares on stars, X-ray transients, cataclysmic variables etc. The use of LE telescopes will allow these objects to be detected and studied by sky patrol monitoring.

Fig. 1. The advanced Mini-Schmidt Lobster-Eye telescope prototypes based on 23x23 mm plates 100 microns thick

3   LE Telescope Prototypes

First Lobster-eye X-ray telescope prototypes have been finished. The 1st prototype of the Schmidt geometry represents one module and consists of two perpendicular arrays of double-sided X-ray reflecting flats (36 and 42 double-sided flats 100 x 80 mm each). The flats are 0.3 mm thick and gold-coated. The microroughness is below 1 nm. The focal distance is 400 mm from the midplane. The FOV of one module is about 6.5 degrees. More such modules may create an array with substantially larger FOV. The optical and X-ray tests indicate performance close to those calculated and expected (e.g. by ray tracing).

The 2nd and 3rd Schmidt prototypes are based on 0.1 mm thick glass plates 23 x 23 mm, gold coated, spaced at 0.3 mm. 60 such plates are used for one module, the double focusing device is created by two such modules. The aperture/length ratio is 80, the reflecting surface microroughness amounts to 0.2 ... 0.5 nm. The FOV of the module is 2.5 deg.

For the Angel geometry, numerous square cells of very small size (about 1x1 mm or less at lengths of order of tens of mm, i.e. with the size/length ratio of 30 and more) are to be produced. This demand can be also solved by modified innovative replication technology. First test modules with LE Angel cells have been succesfully produced. First linear test module has 47 cells 2.5 x 2.5 mm, 120 mm long (i.e. size/length ratio of almost 50), surface microroughness 0.8 nm, f = 1.3 m. Second test module is represented by a L-shaped array of 2 x 18 = 36 cells of analogous dimension. The third test module with 6 x 6 = 36 cells is finished recently. The fourth test module with 96 x 96 i.e. 9216 cells is in development. The surface microroughness of the replicated reflecting surfaces is better than 1 nm. An innovative technique for production of 120 x 120 mm sized modules with large number of 3 x 3 mm cells, 120 mm long, is under development.

The first prototypes of lobster eye X-ray lenses of both the Schmidt as well as Angel geometries demonstrate the feasibility that the wide field telescopes may be constructed in the future based on this type of reflective X-ray optics.

4   Discussion

The use of very wide field X-ray imaging system could be without doubts very valuable for many areas of X-ray and gamma-ray astrophysics.

Results of analyses and simulations of lobster-eye X-ray telescopes have indicated that they will be able to monitor the X-ray sky at an unprecedented level of sensitivity, an order of magnitude better than any previous X-ray all-sky monitor. Limits as faint as 10^{-12} erg s^{-1} cm^{-2} for daily observation in soft X-ray range are expected to be achieved, allowing monitoring of all classes of X-ray sources, not only X-ray binaries, but also fainter classes such as AGNs, coronal sources, cataclysmic variables, as well as fast X-ray transients including gamma-ray bursts and the nearby type II supernovae. For pointed observations, limits better than 10^{-14} erg s^{-1} cm^{-2} (0.5 to 3 keV) could be obtained, sufficient enough to detect X-ray afterglows of GRBs.

The various prototypes of both Schmidt as well as Angel arrangements have been produced and tested successfully for the first time, demonstrating the possibility to construct these lenses by innovative but feasible technologies. This makes the proposals for space projects with very wide field lobster eye optics possible.

The future steps toward the real lobster eye telescope in both Schmidt and Angel arrangements should be focussed to application of additional layers, to extend the energy range to higher energies, to further improve the surface quality (microroughness, slope errors), as well as the reflectivity and the angular resolution. The construction of larger or multiple modules to achieve a larger FOV of order of at least 1 000 square deg or more will be necessary as well as the further reduction of the cell apertures (Angel) and/or spacing and plate thickness (Schmidt), to enhance the length/aperture ratios necessary for a better angular resolution.

Acknowledgements

The work has been supported by grant of the Grant Agency of the Czech Republic No. 102/99/1546. Reflex Prague acknowledges the support provided by the Ministry of Industry and Trade of the Czech republic withi the project FB-C3/29 Centre of Advanced X-ray Technologies.

References

1. Angel J. R. P., ApJ, 233, 364-373, 1979.
2. Schmidt W. K. H., Nucl. Isntr. Methods, 127, 285-292, 1975.

On the Feasibility of Independent Detections of Optical Afterglows of GRBs

René Hudec

Astronomical Institute, Academy of Sciences of the Czech Republic,
CZ-251 65 Ondřejov, Czech Republic

1 Introduction

The recent detection of optical afterglows (OAs, optical delayed emission) and optical transients (OTs, optical prompt emission) of gamma ray bursts (GRBs) allows to consider optical ground-based independent detection of these phenomena. It becomes evident that the optical surveys achieving lim mag better than 19 ... 23 for stars and/or 10 for 1 min exposures may detect OAs and OTs of GRBs. This opens the possibility of independent optical searches for GRBs. These searches must be of large field of view and high sensitivity i.e. CCD surveys and/or deep patrol plates are suitable. The prospects of optical surveys for GRBs may be summarized as follows. (1) The optical surveys may provide a larger sample (due to different beaming) and better localisation accuracy (1 arcmin or better) than provide gamma ray satellite detectors. (2) The larger sample of OAs and their host galaxies may be crucial for understanding the nature of GRBs. (3) The actual rate of OAs can place strong constraints on the afterglow appearance fraction and the initial beaming angle of GRB sources. (4) The UV flashes predicted by some theories such as Protheroe and Bednarek 1999 could be detected and studied. The corresponding delays regarding GRBs could serve to study the nature of the sources. This can be addressed only by surveys, not by follow-up devices since the flashes may precede the GRBs. (5) The optical surveys are cost-effective.

2 The Beaming

It is expected that sources emit jets from which the gamma ray emission is more beamed than the subsequent optical afterglow radiation due to the deceleration of the jet by the ambient gas and the corresponding decline in its relativistic beaming with time (Rhoads 1997). Because the shift to lower frequencies accompanies the shift to lower bulk Lorentz factor, the minimum solid angle into which the transient can radiate increases with time. A jet geometry hence implies a higher rate of OAs detection. If bursts are highly collimated, the gamma rays will radiate into a small solid angle, the optical light into a larger one, and radio into a still larger one. On the other hand, if bursts emit isotropically, we do not expect OAs unaccompanied by GRBs. The ratio of transients detected hence allows the ratio of the mean solid angle into which transients radiate to be estimated.

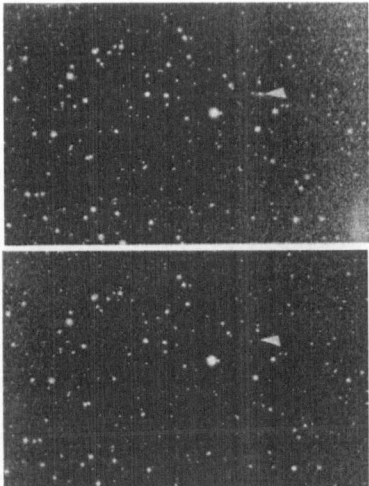

Fig. 1. The bright (6 magnitudes above the plate limit) OT image recorded on a Sonneberg astrograph plate (above, below is the comparison image)

3   The Estimated Event Rates

Theoretical Expectations. The number of optically selected OAs could be greater than the number of gamma-ray selected GRBs by a factor of $(\gamma_o/\gamma_a)^2$ where the γ_o is the initial Lorentz factor, the initial gamma-ray emission is beamed to an angle of about $1/\gamma_o$ and the afterglow emission is produced when the fireball has been decelerated to a modest Lorentz factor γ_a. The typical bursts have $\gamma \leq 10^{2-3}$ (e.g. Fenimore et al. 1993, Woods and Loeb 1995) while the optical afterglow emission occurs at γ_a nearly 10^{3-4}, hence this could boost the expected OAs rates by up to four orders of magnitude (boosting factor).

Optical Transients (Prompt Emission). The observed rate of GRBs (by BATSE) is about 1 GRB/day. The expected beaming factor is 1 ... 10 (no more since the time after GRB is small), i.e. the boosting factor is 1 ... 100. The actual GRB fraction with OTs (deduced from recent observational results) is 0.01–1 (brighter than mag 12). The estimated OT rate is accordingly 0.01–100 OTs/day for the whole sky sphere, this means 2.5×10^{-7} ... $2.5 \times 10^{-2} deg^{-2}$. Example: astrograph plate (100 deg^2) should include 2.5×10^{-5} ... 2.5×10^{-1} OTs per plate.

Optical Afterglows (Delayed Emission). For the delayed optical emission, the assumed beaming factor amounts to 1 ... 100, i.e. the boosting factor is 1 ... 10 000. The actual GRB fraction with OAs deduced from recent observational results is 0.5 (brighter than mag 23) and/or 0.05 (brighter than mag 18). The estimated OA rate is correspondingly 0.5 ... 5 000 OAs/day for the whole sky sphere (limiting magnitude 23), or 0.05 ... 500 OAs/day for the whole sphere (lim mag 18). This means 1×10^{-5} ... $1 \times 10^{-1} deg^{-2}$ for the lim mag 23 and 1×10^{-6} ... 1×10^{-2} for the lim mag 18, not in contrast with results obtained by plate searches. Example: UKSTU plate area of 41 deg^2 : 23 mag limit ... 4

x 10^{-4} ... 4 objects per plate, 18 mag limit ... 4 x 10^{-5} ... 4 x 10^{-1} objects per plate. Constraints from observations/plate analyses yield $\leq$ 0.15 events/square degree for the lim mag 23 i.e. boosting factor $\leq$ 10000.

Note however that the background by SNe, variable AGNs, variable stars, flare stars etc is higher. Depending on galactic latitude, their integrated rate may achieve $\sim$ 1 000 to 5 000 variable objects per UKSTU plate, lim mag 23.

4   Conclusions

It proves to be feasible that both flaring (OTs) as well as fading (OAs) optical emission related to GRBs may be detected by optical sky patrols. The the rate of OAs and OTs may be (significantly) higher than the rate of GRBs due to different beaming. The recent tools for distinguishing real OAs related to GRBs from background phenomena are as follows: (1) light curve, (2) peak luminosity (only for objects with known redshift), and (3) color information: most of OAs have R–I = 0.50 plus minus 0.25, V–R= 0.44 plus minus 0.25, B–V=0.44 plus minus 0.18 (Šimon et al., 2001). There is however a background of false triggers (not related to GRBs but with similar transient behavior, e.g. Hudec, 1993) with poorly known statistics (for faint magnitudes). This background is due to supernovae, AGN/QSOs, stellar flares, variable stars, optical transients of unknown origin and non-astrophysical triggers. The detected OAs and especially OTs (since they will be recorded only once due to their short duration in most cases) must be further studied in detail to eliminate them from background triggers. Among the suitable methods/systems, one can mention the OMC (Optical Monitoring Camera) on the ESA INTEGRAL satellite with FOV of 5 x 5 deg^2, lim mag 19, the CCD based devices and telescopes (ASPA, ROTSE, OTM, BOOTES,...), the digitised plate surveys, as well as the digitised deep archival plates (e.g. UKSTU plate collection, Siding Springs Schmidt, 17 000 plates with limiting magnitude 20-23, different filters/colours).

Acknowledgement. The investigations of gamma-ray bursts and optical transients are supported by the project KONTAKT ES002 provided by the Ministry of Education and Youth of the Czech Republic and by the grant 205/99/0145 provided by the Grant Agency of the Czech Republic. The investigation of plate defects has been supported by the Academic Link between the University of Westminster and Astronomical Institute Ondrejov provided by the British Council in Prague.

References

1. Šimon V. et al., A&A, submitted, 2001.
2. Fenimore E. F., Epstein R. I. and Ho C. 1993, AAS 97, 59
3. Protheroe R. J. and Bednarek W., 1999, astro-ph/9904279, Astropart. Phys., submitted.
4. Rhoads J. E. 1997, ApJ 487, L1.
5. Woods E. and Loeb A. 1995, ApJ 453, 583
6. Hudec R. 1993 Astroph. Letters and Communications 28, 359

High Precision Space Astrometry of Cosmic GRBs

S.M. Kopeikin[1,3], V.G. Kurt[1], O.S. Ougolnikov[1], and A.A. Sukhanov[2]

[1] Astro Space Center of the P.N. Lebedev Physical Institute Russian Academy of Sciences, Russia, Moskow 117810, Profsojuznaya str. 84/32

[2] Space Research Institute of the Russian Academy of Sciences, Russia, Moscow 117810, Profsojuznaya str. 84/32,

[3] Department of Physics and Astronomy, University of Missouri-Columbia, Physics Building 223, Columbia, MO 65211, USA
vkurt@asc.rssi.ru

1 Introduction

The problem of low angular resolution of detectors in gamma-ray part of spectrum is the main one in gamma-astronomy now. Impossibility of precise measuring of cosmic GRBs coordinates had stayed their nature unknown for a long time. Last years the step forward was made, and coordinates of several bursts was measured with high accuracy. But the measurement was made for optical afterglow of the burst, so it was only indirect. Low amount of GRBs with measured coordinates and inconveniency of their identification remains the problem of effective method of bursts astrometry opened.

One of the most efficient methods of direct coordinates measuring for gamma-ray bursts with accuracy up to 1" is the "space triangulation" method – the measuring of time delay between burst registration moments on different spacecrafts. We consider this method in detail – as from technical point of view so from theoretical one. We consider all technical requirements for the spacecrafts system and estimate the limit accuracy of this method at the case of ideal solution of the technical problems.

2 Configuration Models for Spacecrafts

Imagine two spacecrafts with distance L one from another detecting the signals from gamma-ray bursts (Fig. 1). If we have Δt interval between times of detection of one signal, we can calculate the α angle by the formula:

$$\cos \alpha = \frac{c\Delta t}{L} \qquad (1)$$

where c is the velocity of light. The amount of α uncertainty E_α will be determined by uncertainties of time and distance measuring and α itself:

$$E_\alpha = \frac{1}{L \sin \alpha}(E_t c + E_L \cos \alpha) \qquad (2)$$

Here we can see that accuracy of α measurements will be the best at $\alpha = 90°$. Thus, system of two spacecrafts can locate the burst position at the ring

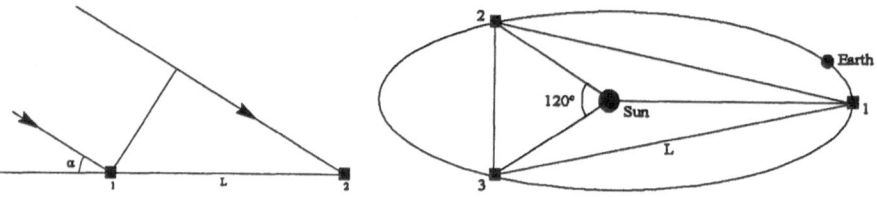

Fig. 1. Detection of Gamma-ray **Fig. 2.** The system of three spacecrafts at near-
Burst signal by two different space- solar orbits (model C)
crafts

centered at the line connecting them and with E_α thickness. And if we have
three spacecrafts not situated at one line, we can locate the burst in two small
regions at the sky, and to choose right one will not be difficult.

If we want to measure burst coordinates with the accuracy up to 1", we
have to know spatial coordinates of spacecrafts with accuracy of 10 km, and
craft's clock must have 100 μs exactness. The measured time amounts must
be transformed from spacecraft reference frame to the barycentric time scale of
Solar System, also taking into account General Relativity effects.

We suggest three different configurations of three spacecrafts – A, B and C.
In all cases all three crafts are moving at ecliptic plane, and one of them has
circular orbit with radius equal to 1 a.u. In model A two other spacecrafts have
elliptical orbits with perihelion and aphelion distances equal to 0.625 a.u. and 1
a.u. respectively for one craft and 1 a.u. and 1.33 a.u. for second one. In models
B and C all spacecrafts have circular orbits with radius of 1 a.u., two crafts are
moving at 60° before and after third in model B, and at 120° in model C. In C
three crafts make an equal-sided triangle with side of 1.72 a.u. (Fig. 2).

The model A is the best from technological point of view, but it brings a lot
of problems decreasing the coordinate measuring accuracy. From this point of
view the model C is the best since the distances between crafts are largest and
constant, and the amount of relativistic correction sufficiently smaller.

Since all crafts are situated at ecliptic plane, the best accuracy of measurements will be near ecliptic poles, and two gamma-ray detectors should have
semi-spherical diagrams directed to these poles.

3   Reference Frame and Time Transformations

Spacecraft will detect the signals from gamma-ray burst in its own reference
frame (τ, ξ). The Mission Control will work in Geocentric reference frame $(u, \mathbf{w})$.
All time moments have to be transformed to the Barycentric reference frame of
the Solar System (t, $\mathbf{x}$). Geocentric time u and spacecraft time τ are related
with barycentric time t by similar equations:

$$u = t - \frac{1}{c^2}\left\{\int_{t_0}^{t}\left(\frac{1}{2}v_E^2 + U(\mathbf{x}_E)\right)dt + [\mathbf{V}_E \cdot \mathbf{R}_E]\right\} + O(c^{-4}) \qquad (3)$$

$$\tau = t - \frac{1}{c^2} \left\{ \int_{t_0}^{t} \left(\frac{1}{2} v_S^2 + U(\mathbf{x}_S) \right) dt + [\mathbf{V}_S \cdot \mathbf{R}_S] \right\} + O(c^{-4}) \qquad (4)$$

t_0 is the start of observations, $\mathbf{v}_E$ and $\mathbf{v}_S$ are barycentric velocities of the geocenter and spacecraft respectively, U is gravitational potential (in (3) the gravitational field of the Earth is not taken into account), $\mathbf{R}_E$ and $\mathbf{R}_S$ are radius-vectors of observation point in geocentric and spacecrafts frame respectively.

Taking into account that spacecraft reference frame is used only at this space-craft and thus $\mathbf{R}_S = 0$, we can relate amounts of τ and u with each other, expressing the transformation of spacecraft time to the geocentric scale:

$$\tau = u - \frac{1}{c^2} \left\{ \int_{t_0}^{t} \left[\frac{1}{2} \left(v_E^2 - v_S^2 \right) + (U(\mathbf{x}_E) - U(\mathbf{x}_S)) \right] dt + [\mathbf{v}_E \cdot (\mathbf{x}_S - \mathbf{x}_S)] \right\} + Oc^{-4}$$
$$(5)$$

The transformations (4) and (5) of spacecraft time to barycentric and geocentric reference frames have centurial and periodic components. The numerical calculations of these values were conducted for A, B and C spacecraft configurations. For the case of circular orbits (B and C system) difference between spacecraft and barycentric times will not have periodic components, and the centurial shift (-0.4673 sec/yr) will be easy to reduce. And the difference between spacecraft and geocentric times will have no centurial components, and periodic one will be express by the formula:

$$\tau = u + \frac{4\pi a^2 e}{c^2 T_0} \sin \frac{\varphi_0}{2} \cos \left(\frac{2\pi t}{T_0} - \frac{\varphi_0}{2} \right) + Oc^{-4} \qquad (6)$$

where a is the radius of the orbit (1 a.u.), e is the excentricity of the Earth's orbit (0.01667), T_0 is the orbital period (1 year), φ_0 is the orbital phase difference between spacecraft and Earth. Here we can see that the period of this value is equal to one year, and the amplitude will not exceed $4\pi a^2 e/(c^2 T_0) = 1.656\ msec$, that is slightly more than limit accuracy of time delay measurements and can be easily reduced too. In the case of configuration A with elliptical spacecraft orbits the relativistic time differences will be more and complicated (Fig. 3). All of them will have big centurial and periodic component that must be taken into account. It is another advantage of configurations B and especially C over configuration A. Knowing time delays between signal detection at three spacecrafts Δ_{12}, Δ_{23} and Δ_{31} ($\Delta_{12} + \Delta_{23} + \Delta_{31} = 0$ in the case of absence of systematic and accidental errors), we can easily obtain the vector n directed to the gamma-burst in barycentric reference frame.

4   Estimation of the Limit Accuracy

The accuracy of "cosmic triangulation" method will be determined by technical parameters of spacecraft detectors, timescales synchronization and parameters of the burst. Here we will estimate the accuracy of this method depending on burst parameters and detector square assuming ideal work of all technique and

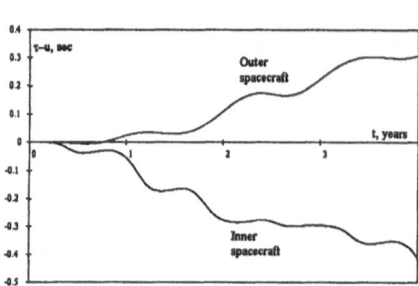

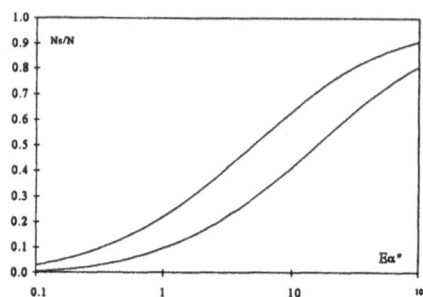

Fig. 3. The difference between spacecraft and geometric time for two spacecraft at elliptical orbits in model A

Fig. 4. The relative number of Gamma-ray Bursts which coordinates could be measured with given accuracy

Table 1. Critical N values

Burst Energy	J	$erg\ cm^{-2}$	$10^{-8}; 10^{-7}; 10^{-6}; 10^{-5}; 10^{-4}; 10^{-3}$
Burst Duration	T	sec	$1; 10.$
Burst Variability Scale	$\Delta T\ T^{-1}$		$0.01; 0.1; 1.$
Detector Square	S	cm^2	$100; 1000$

100 % quantum output of the detector. We imitated the time delay measuring process for different values of burst and detector parameters, given in Tab.1.

Here we have 72 combinations of parameters. For each combination we had calculated average value of time delay error for 100 conditional gamma-ray bursts. As the result, we have obtained the empirical formula expressing from all these parameters the average error in seconds:

$$\log E_t = -4.240 - 0.515 \log J + 0.947\ \log T + 0.522\ \log(\Delta T\ T^{-1}) - 0.497\ logS \tag{7}$$

Taking in account the isotropy and the known brightness and duration distribution, we can estimate the relative number of the bursts which coordinates can be measured with a given accuracy by detectors with given surface. The results are shown in fig.4. We can see that the coordinates for the half of all burst could me measured with the exactness of $10''$, and for 10 % of the bursts it could reach $1''$ and better. This way the method of "space triangulation" can provide exact coordinates of a large number of gamma-ray bursts, that would allow to make their identification with known object on the sky.

5   Analysis of the Transfer

The impulsive and low-thrust transfers in 60 and 120 degrees both ahead and behind the Earth is analyzed below. The spacecraft mass was estimated for

Rockot launch vehicle based on the SS19 missile. It was assumed that a solid rocket kick stage fit for a given launch C_3 is available.

Note that a detailed transfer analysis is not a goal of the paper and only basic results of the preliminary analysis are given.

We analyze first the case of Impulsive Transfer. In order to move the spacecraft along the Earth's orbit in a given angle, it should be launched in an orbit tangential to the Earth's one and having a period P different from one year. After one revolution around the sun the spacecraft will return to the Earth's orbit and the angle Earth-Sun-SC will be equal to $360\,|P-1|$ degrees (assuming that P is given in year). Whe n the given Earth-Sun-SC angle is reached, the spacecraft should be inserted in the earth orbit. The transfer trajectory in the angle 120° behind the Earth position for three spacecraft orbits is shown in Fig. (5). The frame rotating with the Earth is used in Fig. (5), i.e. the Earth is always positioned on the x-axis. The dots on the trajectory mark one-month time intervals.

Let us consider the transfer for one to four spacecraft orbits around the Sun. The transfer parameters are given in Table 1. The spacecraft dry mass includes the engine mass, which can be quite significant, if the delta-V at the end point (i.e. delta-V inserting the spacecraft into the Earth's orbit) is high. The dashes are put instead of the spacecraft mass if the transfer is impossible with a single kick stage and with one-stage spacecraft engine.

For the low-thrust transfer1 the launch C_3 and the transfer time can be changed continuously. Let us assume that the effective power of the solar electric propulsion (i.e. taking into account the propulsion efficiency) is equal to 3 W

Table 2. The transfer parameters

	Ahead				Behind			
Numb. of orbits	Transfer time, yr	Launch C_3, $(km/s)^2$	ΔV at the end point	SC dry mass, kg	Transfer time, yr	Launch C_3, $(km/s)^2$	ΔV at the end point	SC dry mass, kg

Earth-Sun-SC angle= 60°

1	0.83	13.2	3.63	92	1.17	9.5	3.07	120
2	1.83	3.0	1.73	222	2.17	2.5	1.59	232
3	2.83	1.2	1.14	280	3.17	1.2	1.07	284
4	3.83	0.8	0.85	312	4.17	0.8	0.81	316

Earth-Sun-SC angle= 120°

1	0.67	66.9	8.18	-	1.33	32.9	5.74	-
2	1.67	13.2	3.63	92	2.33	9.5	3.07	120
3	2.67	5.4	2.34	169	3.33	4.3	2.09	189
4	3.67	3.0	1.73	222	4.33	2.5	1.59	232

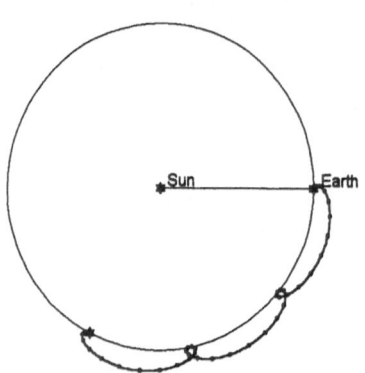

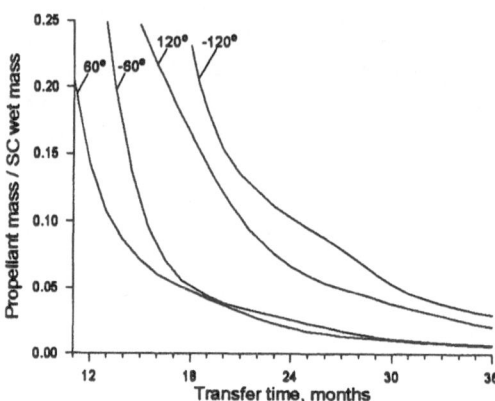

Fig. 5. Transfer in 120° for three orbits around the Sun

Fig. 6. Propellent mass as a fraction of the spacecraft wet mass versus transfer time for zero launch C_3

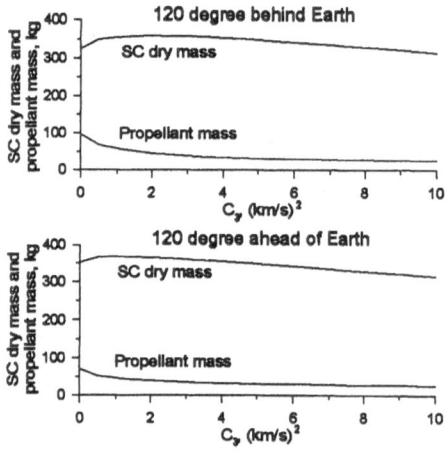

Fig. 7. Spacecraft dry mass and propellant mass versus C_3 value

per 1 Kg of the spacecraft wet mass. That means that, if the electric propulsion efficiency is equal to 0.6, then the total electric power should be not less than 5 W per 1 kg of the spacecraft wet mass. Let us consider the transfer with zero launch C_3 as an example. The propellant mass as a fraction of the spacecraft wet mass versus the transfer time for this case is shown in Fig. (6).

Let us assume a Rockot launch vehicle and a transfer duration of 18 months. The spacecraft dry mass and the propellant mass versus C_3 value are shown in Fig. (7) for the transfer in 120 degrees behind and ahead of the Earth. As is seen in Fig. (7), the maximum of the spacecraft dry mass is 358 kg for the transfer behind Earth ($C_3 = 2$) and 367 kg for the transfer ahead of Earth ($C_3 = 1$). However it may be reasonable to select higher C_3 value with slightly lower SC mass and significantly lower propellant consumption. The spacecraft dry mass for the 18-month transfer in 60 degrees both forward and backward is about 400 kg and the propellant consumption is 8 to 12 kg for $C_3 > 0.5(km/s)^2$.

6   Conclusions

The low thrust delivers in the operational point higher spacecraft dry mass for a shorter time. However the solar electric propulsion needs use of large solar panels. The optimal low thrust control is rather complicated, but it can be dramatically simplified with a small increase of the propellant consumption.

New Version of Optical Transient Monitor for BOOTES Project

Petr Páta[1], Martin Bernas[1], Alberto J. Castro-Tirado[2], and René Hudec[3]

[1] Czech Technical University in Prague, department of radioelectronics,
 Technická 2, Prague, 166 27, Czech Republic
[2] LAEFF-INTA Madrid, Spain
[3] Astronomical Institute Ondrejov, Czech Republic

Abstract. We present controlling software for two wide field cameras of BOOTES station in this paper. Optical Transient Monitor (OTM) system is prepared for automatically evaluating of data and monitoring of sky also. Our system is windows 95 or MSDOS based.

1 OTM Conception

Optical Transient Monitor (OTM) is a controlling software for one experiment of the BOOTES project [2]. The program controls the wide field cameras ST8 (Santa Barbara Instruments Group) for continuous monitoring of the same fields of view as BeppoSAX, HETE-2 or INTEGRAL. The scheme of the OTM conception is described in Fig. 1. There are two versions of OTM. The first is MSDOS based program and it requires the acceleration board Photomate 20 with two signal processors TMS320 C40 (Texas Instruments) [3]. This version has been in operation at BOOTES-1 station (El Aeronosillo, INTA, Spain) since July 1998. The second version, which is designed as a full 32 bit application for W95/W98, has been in trial operation for one year (since June 2000). The acceleration board Photomate 20 is recommended but not required. Detail hardware requirements of both versions of OTM programs are shown in Tables 1, 2.

2 OTM Description

2.1 Supported Modes of Work

The OTM system supports both the manual and full automatic regime (i.e. without interaction with operator). Each step of the image processing from data acquisition to its evaluation in the manual mode can be managed separately according to user's requirements. The second mode allows to set up the system so that the cameras would start and finish the data acquisition automatically every night. The real time image processing can be carried out in the MSDOS version only. The important information are available during observation in a form of table containing temperatures of CCDs and air, system state, log file, etc. The cameras can be controlled separately or synchronously. There are two algorithms of image capturing. The main difference between them is the fact, that the both cameras work parallel despite the second algorithm, where one camera takes exposures while the second is read out at each time.

Table 1. Hardware requirements of 16 bit MS DOS version.

MSDOS 5.0 or higher
PC 386DX/40 or better
32 MB RAM
Two paralell ports lpt1, lpt2
Two CCD cameras ST-8
Photomate 20 with processors TMS320C40
VGA card

Table 2. Hardware requirements of 32 bit W95/W98 version.

Microsoft Windows 95 or Windows 98
PC Pentium 70 MHz or better
64 MB RAM
Two paralell ports lpt1, lpt2
Two CCD cameras SBIG ST-8 or compatibile type
SVGA, optimalized for 800 × 600 × 16 bit

2.2 Image Processing Functions

There are implemented many useful functions for evaluation image data in OTM program. The Photomate 20 card is used in MSDOS version as an accelerated image processing board independence on main PC.
Supported functions are:
1. Automatic compensation of dark frame and flat fielding.

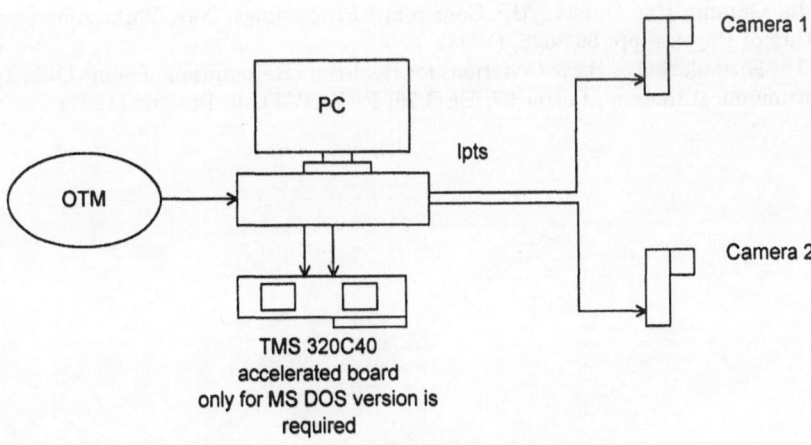

Fig. 1. Description of OTM system conception.

2. Polynom fit of the background value using histogram, iteration method or least square fit.

3. Searching for objects – elimination of optic's defects , CCD's and false objects.

4. Building of photometric table of found stars and others objects.

5. Comparison of the new table with previous tables from the same camera. Construction of light curves for each object in the image.

6. Pseudocorrelation method – finding of transformation matrix, translation vector.

7. Differential photometry – automatic or manual selection of standard stars.

8. Searching for new objects based on change of brightness.

9. Comparison of actual results from both cameras – elimination of suspicion and false objects.

Acknowledgement

This work has been conducted at the department of radioelectronics of the Faculty of Electrical Enegineering of the Czech Technical University in Prague and has been supported by grant No. ESO 0036 Kontakt. A part of this research work has been partially supported by the research program No. 4/98:212300014 "Research in the Area of Information Technologies and Communications" of the Czech Technical University in Prague.

References

1. Bernas M., Páta P., Hudec R., Rezek T., 'Lossless and Lossy Compression of Image from the OMC Experiment of Integral Project', Astrophysical Letters & Communications, 429/[897], vol. **39**, part II., (1999)

2. Castro-Tirado, A. J., Soldán J., Bernas M., Páta P., Rezek T., Hudec R., Sanguino T., M., De La Morena B., Berná B., Rodriguez J., Pena A., Gorosabel J., Más-Hesse J., Giménez A., 'The Burst Observer and Optical Transient Exploring System (Bootes)', A&AS, 138..583C, (1999)

3. Bernas M., Páta P., Hudec R., Rezek T., Castro-Tirado A., 'Optical Transient Monitor' In: Gamma Ray Bursts, AIP Conference Proceedings, New York: American Institute of Physics, pp. 864-868, (1998)

4. Páta P., 'Formulation of Basic Criterions for Real-time Recognition of Some Objects in Astronomical Images', Poster 97, EEC 25, FEE CVTU in Prague, (1997)

High Resolution Spectroscopy of the X-Ray Emission of GRBs by IMXS-BOSS on the ISS

L. Piro[1], L. Colasanti[1], E. Costa[1], G. Gandolfi[1], P. Soffitta[1], F. Gatti[2],
M. Razeti[2], D. Pergolesi[2], R. Vaccarone[2], G. Testera[2], M. Pallavicini[2],
A. Ferrari[3], E. Trussoni[3], M. Orio[3], D. Mc Cammon[4], T. Sanders[4],
M. Galeazzi[4], A. Szymkowiak[5], and S. Porter[5]

[1] Istituto di Astrofisica Spaziale, CNR, Rome, Italy
[2] INFN, Universita' di Genova, Italy
[3] Osservatorio Astronomico di Torino, Italy
[4] Wisconsin University, Madison, USA
[5] GSFC-NASA, USA

Abstract. The IMXS (Interstellar and Intergalactic Medium X-ray Survey)-BOSS (gamma-ray Burst Observatory and Spectroscopy Survey) is an experiment proposed to fly on the ISS, in order to perform an all-sky X-ray survey with high spectroscopy resolution and to meas ure the spectra of Gamma-Ray Burst (GRB) with an high resolution in the 0.1-10 keV energy band. In a 3 years lifetime mission, the experiment will detect 20/40 GRBs. In several events, we can make an high energy resolution spectroscopy of the iron emission lines and absorption edges. Recently, such a components have been observed by BeppoSAX and CHANDRA. The measurements of this features would provide a direct diagnostic of the physical and kinematical state of the medium surrounding a GRB. Furthermore, they would supply informations about the origin of the progenitors of the GRBs (probably a supernova explosion of massive stars) and their site of formation (possibly a star-forming region).

1 Technical Description

The 3 main parts (figure 1) of experiments are: the *collimators* and the *detectors*; the *dewar* and the *cryogenic insert*; the *external analogical* and *digital electronics*.

There will be two *detectors* with different fields of view. The first, by a *collimator* of 10°×10°, will be used to make a survey of the diffuse X-ray emission produced by Interstellar and Intergalactic Media; the second, by a *collimator* with a minimum field of view of 40°×40° and a final goal of 60°×60°, to observe and achieve an high resolution spectroscopy of X-ray emission of GRBs. Each one *detector* will be an array of 36 silicon bolometers, with HgTe absorber, a total area of 1.44 cm^2 and an energy resolution of 4 eV. The *dewar* will contain 100-120 L of superfluid helium (at 1.3 K), surrounding the magnet (a superconducting solenoid) with a maximum field of 4 T. The *cryogenic insert* will include the paramagnetic salt pill of ADR refrigerator (at operating temperature of 60 mK), the mechanical termal switch, 5 IR filters. The elements of *external analogical* and *digital electronics* are: J-Fet preamplifiers operating at 120 K; external low noise amplifiers with a gain of 20000; antialiasing filters; the analog to digital conversion (up to 100 kHz) of the signals, with a 14 bit resolution; the control

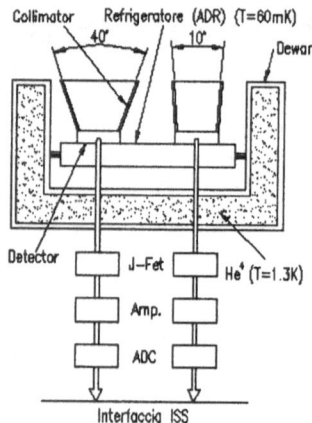

Fig. 1. IMXS-BOSS instrumental configuration.

logic for the refrigeration cycle; the interface to the ISS (cryostat power line, telemetry and telecommand interface).

2 How Many GRBs?

We have maked use of the data of BeppoSAX WFC (2-30 keV) to calculate the number of GRBs that we can detect with IMXS-BOSS. Taking into account the observed spectral shape (a power-law with photon index 1.5 and a tipical absorption column density, due to our Galaxy, of 2×10^{20} cm^{-2}), a count rate of the diffuse components of 180 cts (for t=10 s), assuming a signal to noise ratio of 5σ, we can detect, by IMXS-BOSS, GRBs with fluences of 4×10^{-7} erg/cm^2. It corresponds to a minimum fluence of about 3.4×10^{-7} erg/cm^2 for the BeppoSAX WFC. With a field of view of $40° \times 40°$, in a three years life mission we can detect 18 GRBs. With a field of view of $60° \times 60°$, we have about 250 count from diffuse component and the minimum fluence required for a 5σ signal is 4.7×10^{-7} erg/cm^2. In the BeppoSAX WFC energy range, this fluence corrisponds to 3.9×10^{-7} erg/cm^2. In that case, we can detect 42 GRBs.

3 Simulated Spectra

The observation of iron emission lines and absorption edges is an important tool to understand the origin of GRBs and the nature of their progenitors and to investigate the physical and kinematical state of the medium surrounding a GRB. If they are arised from a supernova explosion of a massive star, we expect to observe such a feature. The most significative detections of emission lines are those in GRB 970508 (Piro et al., ApJ 514, L73, 1999) and GRB 991216 (Piro et al., Science 290, 889, 2000). In the GRB 990705 (Amati et al., Science 290,

953, 2000) the BeppoSAX WFC have detected an iron absorption edge. On the basis of the above observed features, we have simulated some spectral emissions of GRBs, adding emission lines and absorption edge. The first simulation (figure 2) is for a field of view of $40° \times 40°$, assuming a fluence of 4.8×10^{-6} erg/cm^2 (5 GRBs in3 years). To the GRB emission (continuous line), we have added two emission lines (E=6.4 keV, E.W.= 0.3 keV, σ_L=0 eV and E=6.9 keV, E.W.=4 keV, σ_L = 0.4 keV in the restframe) and an absorption edge (E=7.1 keV and τ=1.5) at redshift z=4. In the second, we have taken a field of view of $60° \times 60°$, a fluence S=8.9×10^{-6} erg/cm^2 for the GRB emission, two emission lines ($E = 6.4$ keV, $E.W. = 0.3$ keV, $\sigma_L = 0$ eV and $E = 6.9$ keV, $E.W. = 1$ keV, $\sigma_L = 0.4$ keV in the restframe) and an absorption edge ($E = 7.1$ keV and $\tau = 1.5$ in the restframe) at redshift z=1.

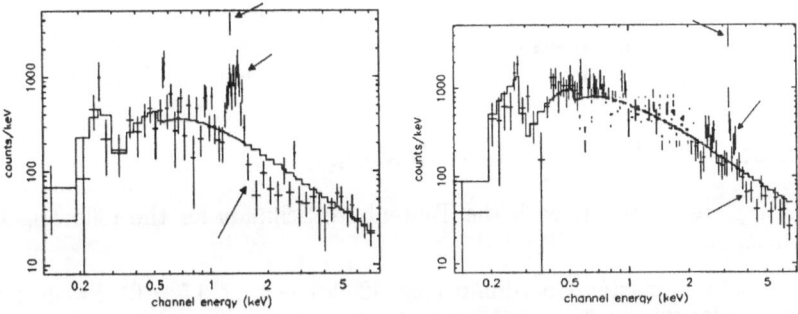

Fig. 2. *Left*: Simulated spectrum for a field of view of $40° \times 40°$. *Right*: Simulated spectrum for a field of view of $60° \times 60°$. The deviations from the best-fit power-law (continuous histogram line), indicated by arrows, represent the three added components (see text).

4 Conclusions

The above simulated spectra show that we can perform an high resolution spectroscopy of iron emission lines and absorption edges. We can so discriminate the nature of emission lines by the determination of their width σ_L. A narrow line is generated from a medium at rest in the space. Therefore it should be associated to stellar forming regions. Broad lines are index of moving medium, probably pre-ejected from the supernova explosion of a massive star. The line sensitivity of IMXS-BOSS will allow us to observe features in 15 GRBs (at least) and broad lines in about 7 GRBs, assuming a field of view of $60° \times 60°$. By an high resolution spectroscopy of X-ray emission of GRBs, we can investigate the state and the nature of medium that surrounds them. In this way, it is possible to draw important informations about the origin of GRBs and the nature of their progenitors.

Statistics of Faint Variable Sources for GRB OA Analyses

Jiri Polcar[1], Rene Hudec[2], and Helmut Meusinger[2]

[1] Masaryk University, Kotlarska 2, 611 37, Czech Republic
[2] Astronomical Institute Ondrejov, Ondrejov 251 65, Czech Republic
[3] Thueringer Landessternwarte Tautenburg, Germany

Abstract. In GRB Optical Afterglows analyses, good knowledge and statistics of variable background sources is required. For statistical analyses of faint variable objects, many large archives of archival plates from several observatories can be used. Recent technologies allow digitizing of photographies and subsequent computer processingi. Archival Schmidt plates from the Tautenburg Observatory have been used for method testing and for automated searches for variable objects. The results obtained are presented and discussed with the focus on applications in GRBs OAs analyses.

1 Data Used for Method Testing

Schmidt plates taken at with the Tautenburg Schmidt for the following fields have been used:

M92 – globular cluster; Coordinates (eq. J2000): $\alpha = 17^{\mathrm{h}}17^{\mathrm{m}}7.0^{\mathrm{s}}$, $\delta = +43°08'10''$ ($l_{\mathrm{II}} = 44°14'31.2''$, $b_{\mathrm{II}} = 46°45'50.4''$). Size: 14.0'.

NGC1275 – galaxy; Coordinates (eq. J2000): $\alpha = 03^{\mathrm{h}}19^{\mathrm{m}}48.2^{\mathrm{s}}$, $\delta = +41°30'42''$ ($l_{\mathrm{II}} = 151°5'36.6''$, $b_{\mathrm{II}} = 13°28'58.8''$). Size: 2.2'×1.8'. Also known as **Perseus A** and **3C84**.

2 Method Itself

- Photograph acquisition (Tautenburg Schmidt telescope)
- Photograph digitizing (Tautenburg Plate Scanner)
- Photograph reduction (Tautenburg)
- Finding corresponding objects on different photographs
- Finding corresponding object magnitudes on different photographs
- Variable objects detection
- Objects classification

3 Finding Corresponding Objects on Different Plates

From a certain set of photographies one has been chosen as referential. Coordinates of objects from the others photographs were transformed with respect to the referential one. The transformation was found using the `munimatch` program (`ftp://integral.muni.cz/pub/munipack/`). Only common area of all the photographs was processed. In case of extraordinary big deviation some photographs were discarded.

4 Finding Corresponding Object Magnitudes on Different Plates

One photograph is chosen as referential again. Objects displayed on it are searched on other photographs. Linear transformation was determined between such object couples. The x-axis represents the referential object magnitude while the y-axis means means magnitude of object from the transformed photograph.

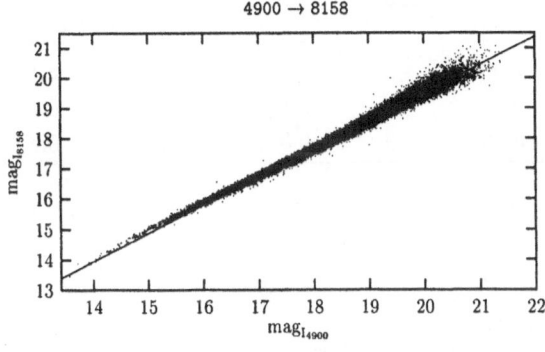

Fig. 1.

5 Variable Objects Detection

Before the main search, the objects are sorted by one coordinate (e.g. X). This allows use much faster searching algorithm. Vicinity of the searched object can be found within $\log_2 N$ steps. Selecting of suitable vicinity size is very important with respect to desired results. Number of the object found is shown in the table on the following page.

6 Variable Objects Statistics

Figure shows fraction of SO (stellar-like objects) and NSO (non-stellar objects) per square degree against the amplitude of variability. Clasification of SO and NSO was made using effective radius.

7 Discussion

This paper suggests method of variable object detection followed by subsequent statistical processing. This method was tested on a set of digitized photographic plates from Tautenburg Observatory. The routine described can be applied without big changes to the digitized photographic plates from many other plate archives. Images of Optical Afterglows of GRBs as well as other types of transient phenomena may be detected and studied this way.

Table 1.

Δm_I	N_{M92}	N_{M92} $(\deg^{-2})$	$N_{NGC1275}$	$N_{NGC1275}$ $(\deg^{-2})$
0.00	43 418	4499.7	68 005	8266.1
0.50	39 504	4094.1	63 737	7747.3
1.00	8 442	874.9	41 241	5012.9
1.50	1 206	125.0	4 763	578.9
2.00	228	23.6	634	77.1
2.50	42	4.4	104	12.6
3.00	9	0.9	25	3.1
3.50	1	0.1	6	0.7
4.00	0	0.0	2	0.2

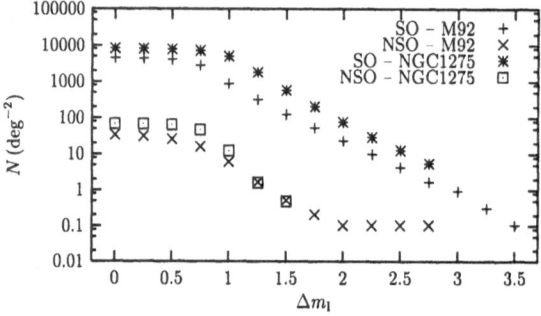

Fig. 2.

Acknowledgements

We acknowledge the support provided by the Grant Agency of the CR, grant 205/99/1546, and by the Ministry of Education of the CR, Project KONTAKT ES002.

References

1. Polcar, J: Statistics of Faint Variable Galactical and Extragalactical Objects, Diploma Thesis, Masaryk University Brno, 2000

Event Rates in SuperAGILE and HETE

Barbara Preger[1], Edward E. Fenimore[1], and Enrico Costa[2]

[1] Los Alamos National Laboratory, Los Alamos NM 87545, USA
[2] IAS/CNR, via Fosso del Cavaliere, 00133 Roma, Italy

1   Introduction

Among the present and near future missions aimed at detecting and studying Gamma Ray Bursts, the Wide Field X-Ray Monitor (WXM, [1]) on board HETE-2 and SuperAGILE ([2]) on board AGILE have very similar characteristics. We applied the same simulation tools to these two experiments and compared their expected performances, with a particular focus on the expected event rates and the instruments' localizing capabilities. Both experiments use the one-dimensional coded aperture technique to localize sources, and both have several detectors equipped with masks that encode the field of view in perpendicular directions. The two-dimensional position of a detected source can be determined by crossing the two one-dimensional positions.

SuperAGILE has four separate units (two for X and two for Y) and the detector is composed of Silicon microstrips. The total geometric area is 1444 cm^2 and the operational energy range is 10 to 40 keV. AGILE is scheduled for launch in 2003. WXM is made up of two units (one X and one Y), with two detectors each. The detectors are one-dimensional PSPCs; WXM's total geometric area is 350 cm^2 and it is sensitive from 2 to 25 keV. HETE-2 has been succesfully launched in October 2000. WXM is aided by the FREGATE Gamma-Ray Telescope (6 to 400 keV): WXM can perform a position localization when the trigger is provided by data from FREGATE.

2   Simulations

We have performed our simulations using a Monte Carlo technique, which tracks individual photons through the various components of the experiments and simulates the possible interactions of radiation with matter. We use actual GRB time profiles and spectral characteristics in our foreground sources, and a diffuse radiation whose count rate and overall spectrum is the one expected for the instruments as background. We choose simulated GRB intensities with a distribution based on the BATSE LogN-LogP relationship, and we choose their location in the field of view from a uniform distribution, so that our estimate of the event rate is as reliable as possible. Once the simulation is run, we apply a trigger-like algorithm to decide whether each GRB was detected by the instrument and by the different units of each experiment. During these simulations and the following analysis we consider an event "detected" if the signal to noise ratio in the light curves registered by the instrument is greater than 5.5.

For WXM we also apply an image reconstruction algorithm to the "detected" GRBs, find the position of the source, and compare it with the simulated source (the equivalent algorithm for SuperAGILE is not yet ready because the mission is still in its early phases). This way we also evaluate the localization capability of the experiment. We consider a localization "good" when the discrepancy between the simulated and the reconstructed position is less than 34 arcmins. In this work we ran two sets of simulations for HETE-2: in the first we distinguish detected from non-detected GRBs and perform the imaging and the position reconstruction on the basis of WXM data, in the second we run the detection algorithm on the FREGATE data and the imaging and position reconstruction on the WXM data, as in the first case.

Table 1. Results summary on WXM and SuperAGILE simulations. All values express number of GRBs per year. "Det." and "d." correspond to number of detected GRBs, while "l." correspond to number of GRBs localized with good accuracy. "X and Y" means that the detection or the good localization appears *both* in the X and in the Y detectors.

	Det.	X d.	Y d.	X and Y d.	X l.	Y l.	X and Y l.
SuperAGILE	46	35	33	26	n/a	n/a	n/a
WXM	33	n/a	n/a	n/a	15	19	12
WXM + FREGATE	58	n/a	n/a	n/a	16	25	13

3   Results and Discussion

The results on the expected performances of the WXM and SuperAGILE experiments are shown in Fig. 1 and summarized in Tab. 1. Note that SuperAGILE sees fewer events in both detectors than in each of the one-dimensional detectors. This is because the overall field of view is non-symmetric: the field of view of each detector is rectangular, so the overlapping field of view is smaller than the sum of the areas. In HETE-2 we see that the number of detected GRBs increases if we use FREGATE data for triggering. This is due both to FREGATE's slightly larger field of view and to the greater energy dynamic range achievable when using also the gamma-ray data. In this case, moreover, the number of events that yield to a good location increases with respect to the WXM triggering. This means that WXM can produce a good localization even if the signal-to-noise ratio in the light curves is not high enough to induce a trigger. We also see that there is a difference in the expected performance of the two units: the Y-direction encoding unit gives better localization results. This is an interesting issue because no obvious explanation has been found, although various members of the WXM team are investigating it.

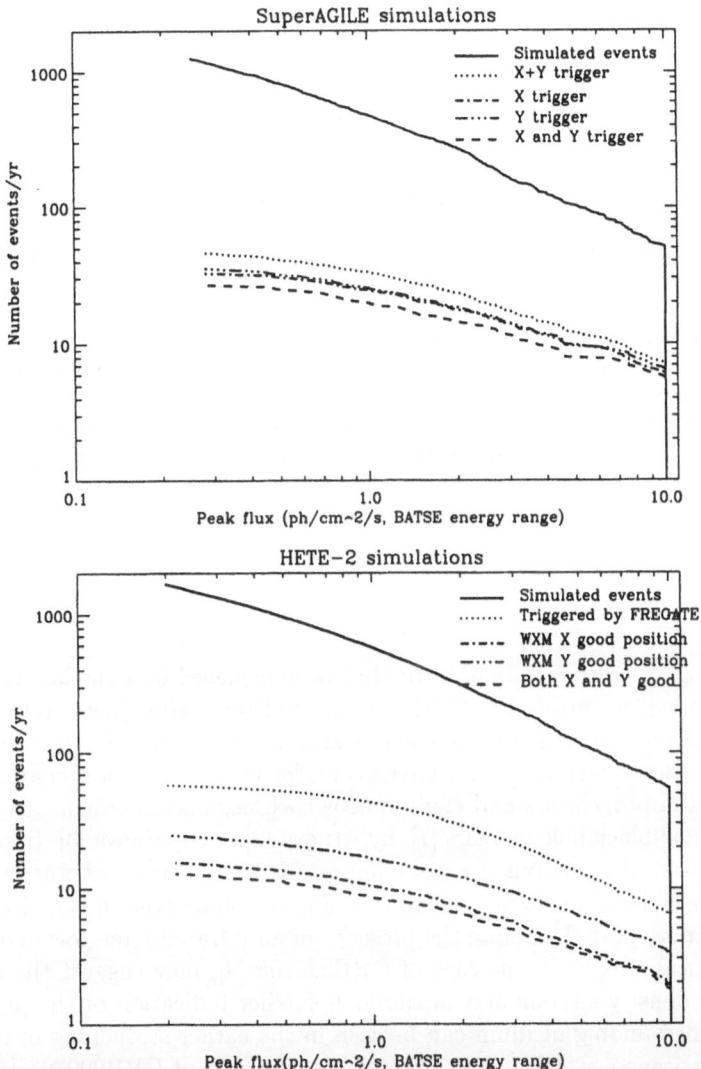

Fig. 1. *Top* The cumulative distribution of the simulated and detected GRBs for SuperAGILE. The different curves show the detection with the X-decoding detectors, the Y-decoding ones and their sum (X+Y) and intersection (X and Y) *Bottom* The same study for WXM, with FREGATE trigger. The different curves show the number of events detected with FREGATE data and the well localized GRBs using WXM data, by the X-decoding detectors the Y-decoding ones and the intersection of the two separately.

References

1. N. Kawai, M. Matsuoka, A. Yoshida et al., A&A Supp. Ser. **138**, 563 (1999)
2. Soffitta P., Costa E., Del Monte E. et al., *Proc. SPIE Conf.* 4140, 283 (2000)

IR and Optical Observations of GRB from Campo Imperatore

R. Speziali, F. D'Alessio, L. A. Antonelli, A. Di Paola, L. Burderi, F. Fiore, G. Israel, D. Lorenzetti, F. Pedichini, L. Stella, and F. Vitali

Osservatorio Astronomico di Roma, Monte Porzio Catone, Rome, ITALY

Abstract. In this poster we present a description of the Campo Imperatore Observatory facilities, suitable for very fast reaction to GRB triggers and multiwavelength observations. Two complementary instruments are available in the Observatory: the AZT-24 (1.1m Ritchey Cretien equipped with the NIR camera SWIRCAM), one of the few IR telescope working in the northern emisphere, and a Schmidt telescope (60/90/180 cm, equipped with a 2Kx2K back illuminated CCD). The IR detection of the GRB 000926, obtained with the AZT-24, is also reported.

1 Introduction

Much progress in the study of GRBs has been achieved over the last three years from detailed multi-wavelength observations of their afterglows. Yet the origin and the physics of GRB phenomenon is still actively debated. The most widely discussed theoretical models for GRBs consider vastly different scenarios both in terms of the progenitors and environment such as,neutron star-neutron star or neutron star-black hole mergers [1], hypernova [2] or supranova [3]. In particular near-infrared observations are very important to ascertain whether the burst goes off in a dense medium of a star-forming region or even in the ejecta of the pre-supernova star. Therefore the presence of an infrared afterglow and the lack of an optical one, as in the case of GRB990705 [4], may suggest the existence of a high density circumburst medium. A further indication of the presence of a dense sorrounding medium can be seen in the earlier steepening of the decay law as envisaged by [5] and as observed in the case of GRB990705 [4], in the case of GRB990123 [6], GRB990510 [7] and, GRB000315c [8].

The prompt triggers from BeppoSAX (1-3 hours) and the Wide Field Cameras error-circles ($\sim 3'$) provide at the moment the best opportunity for successful follow-up observations. In the near future (October 2000) the quality of the trigger will be significantly improved with the launch of HETE2. This satellite will detect GRBs in few seconds with an accuracy of 3 arcmin, giving small telescopes a chance to substantially contribute to GRBs science.

2 The Schmidt Telescope and ROSI

The Schimdt telescope, installed in 1958, is now placed under the second renewing phase. The old mechanics was completely overhauled and the same control system of the AZT-24 allow remote operations now [9]. The very fast optics (F# = 3) and the large FOV make this telescope a unique instrument among the 1m class telescopes, both for fast photometry and the search of optical GRB counterparts with large error boxes. ROSI [10] is the new camera of the Schmidt telescope. Based on the 2Kx2K thinned EEV chip cooled down to 180K, has a FOV of 55x55 arcmin. The high Q.E. of the array allow to reach m_v >22 in a few minutes.

The first detection of a GRB from Campo Imperatore was the famous optical afterglow of GRB 970228. The field was imaged 16 hours after the burst with the old camera [11] and five hours before the well-known observation of this GRB [12].

Fig. 1. The Schmidt and the AZT-24 Telescopes

3 The AZT-24 Telescope and SWIRCAM

The 1.1 m AZT-24 [13] is the new telescope of the station. Placed in the East dome was opened in the end of 1997 and has been fully operative for two years. Highly automatic it's the only telescope in Italy dedicated to the near infrared observations.

SWIRCAM [14], [15] is based on a the 256x256 PICNIC array. Equipped with a set of standard NIR filters (J,H,K,K') has a FOV of 4.5'x4.5'. The camera is equipped with two grism [16] that will also allow to work with a low resolution (R=300) spectroscopic mode.

4   IR Detection of GRB 000926

This is the first detection of a GRB with the AZT-24. In the image, taken on September 28th, is clear the IR counterpart of the GRB 000926 [17]. This image was obtained in the J band with a 30 min. exposure [18] measuring $m_J = 18.6 \pm 0.3$ with a S/N=5. The object is not present in the K band were it was only possible to determine an upper limit of $m_K \simeq 17.0$ with a 30 min. exposure.

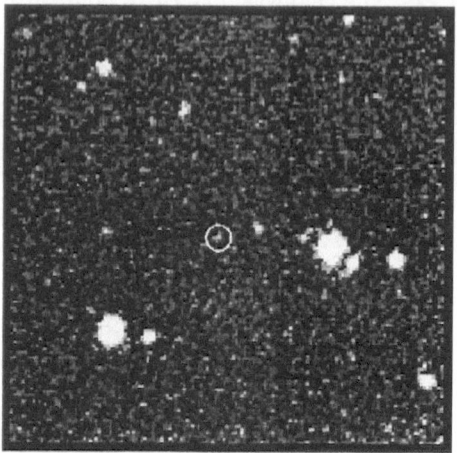

Fig. 2. The IR counterpart of GRB 000926 detected with the AZT-24

References

1. Meszaros & Rees: ApJ **482**, L29 (1997)
2. Paczynski: ApJ **494**, L95 (1998)
3. Vietri & Stella: ApJ, **507**, L45 (1998)
4. Masetti et al.: A&A, **354**, 473 (2000)
5. Dai & Lu (1999)
6. Castro-Tirado et al.: Sci.,283,2069 (1999)
7. Stanek et al.: ApJL, **522**, 39 (1999)
8. Jensen et al.: astro-ph/0005609 (2000)
9. Di Paola et al.: SPIE, **4009**, 317-326 (2000)
10. Speziali et al.: SPIE, **4008**, 389-395 (2000)
11. Pedichini et al.: A&A, **327**, L36-38 (1997)
12. Paradijs et al.: Nat., **386**, 686 (1997).
13. Abalakin et al.: SAIT (1998)
14. Vitali et al.: OAR/IR5 (1997)
15. D'Alessio et al.:SPIE, **4008**, 748-758 (2000)
16. Speziali, Vitali: OAR/IR6 (1997)
17. Gorosabel et al.: GCN 803 (2000)
18. Di Paola et al.: GCN 816 (2000)

A GRB Detection System Using the BGO-Shield of the INTEGRAL-Sectrometer SPI

Andreas von Kienlin, Nikolas Arend, and Giselher G. Lichti

Max-Planck-Institut für extraterrestrische Physik, 85741 Garching, Germany

Abstract. The anticoincidence shield (ACS) of the INTEGRAL-spectrometer SPI consists of 512 kg of BGO crystals. This massive scintillator allows the measurement of gamma-ray bursts (GRBs) with a very high sensitivity. Estimations have shown that with the ACS some hundred gamma-ray bursts per year on the 5 σ level can be detected, having an equivalent sensitivity to BATSE. The GRB detection will be part of the real-time INTEGRAL burst-alert system (IBAS). The ACS branch of IBAS will produce burst alerts and light curves with 50 ms resolution. It is planned to use ACS burst alerts in the 4th interplanetary network (IPN) [1].

1 The Anticoincidence Shield of SPI

The spectrometer SPI [2] is one of the two main instruments on INTEGRAL, one of ESA's next missions devoted to γ-ray research. Fig. 1 shows a drawing of the spectrometer SPI. The camera of SPI, which consists of 19 cooled high-purity germanium detectors, is shielded on the side walls and rear side by a large anticoincidence shield (ACS). The field of view of the camera is defined by the upper opening of the ACS. The imaging capability of the instrument is attained by a passive-coded mask on the top. Below the mask a plasticscintillator anticoincidence (PSAC) takes care for the reduction of the 511 keV background, which is mainly generated by particle interactions in the mask.

The ACS consists of 91 BGO crystals which are arranged in 4 subunits. The units of the upper veto shield (UVS) consists of the upper collimator ring (UCR), the lower collimator ring (LCR) and the side shield assembly (SSA), each containing 18 crystals which are arranged hexagonally around the cylindrical axis of SPI. The lower veto shield (LVS), consisting of 36 crystals, is assembled as a hexagonal shell. The thickness of the crystals increases from 16 mm at the top (UCR) to 50 mm at the bottom (LVS). The total mass of BGO used for the ACS is 512 kg resulting in the obvious use of the ACS as a burst monitor.

Each BGO crystal of the ACS (with one exception) is viewed by two photo-multipliers (PMTs). Due to the redundancy concept used for the ACS, each of the 91 front-end electronic boxes (FEEs) sums the anode signals of two PMTs, which are viewing different BGO crystals (in most cases neighbouring crystals). This cross strapping of FEEs and BGO units leads in a failure case of one single PMT or FEE not to the loss of a complete BGO crystal. It emerges that a disadvantage of this method is an uncertainty in the energy-threshold value of individual FEEs, caused by a different light yield of neighbouring BGO-crystals and different PMT properties like quantum efficiency and amplification. A result

428   A. von Kienlin, N. Arend, and G.G. Lichti

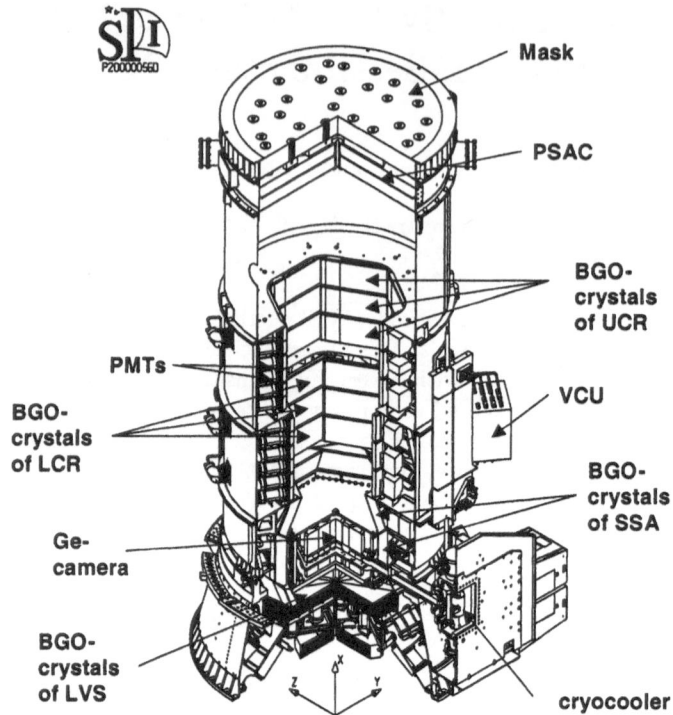

Fig. 1. INTEGRAL Spectrometer SPI

of this is that the threshold extends over a wide energy range and is not at all sharp. The energy-threshold settings of the ACS depend on a tradeoff between background reduction and deadtime for the SPI camera.

2   The ACS as GRB Monitor

The main task of the ACS as a detector is the veto generation for charged particles and γ-rays coming from outside the FoV. But there are also data which can be used for scientific purposes. The ACS housekeeping (HK) data include the values of the overall veto counter of the veto control unit (VCU) and the individual ratemeter values of each FEE. Both HK data are suitable for burst detection. The count rate of the overall veto counter (ORed veto signals of all 91 FEEs) is sampled every 50 ms. A packet, containing 160 consecutive count rates, will be transmitted every 8 sec to ground. If no gap in the telemetry stream occurs one could have a continuous ACS veto-rate light curve with 50 ms binning. The measurement time of the individual FEE ratemeter can be adjusted between 0.1 and 2 sec. All 91 FEEs are read out successively in groups of 8 FEEs every 8 sec. The read out of all 91 ratemeter values thus needs 96 sec. In distinction to the VCU overall veto counter the individual ratemeter values do not yield a

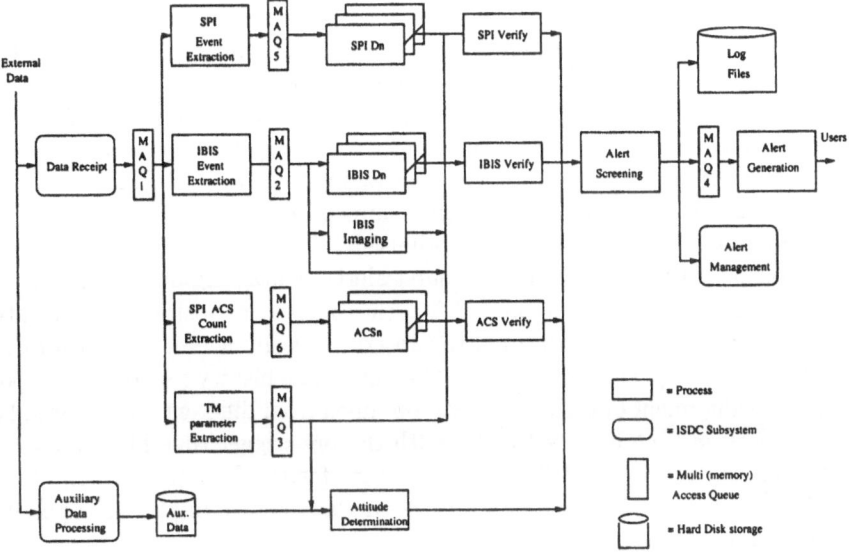

Fig. 2. IBAS structure

continuous stream. Additionally the values of different FEE groups are shifted by a time interval of 8 sec. So it is very difficult to derive the burst-arrival direction from these individual counting rates.

3 SPI/ACS Burst-Alert System

The search for GRBs in the ACS veto-rate light curve and the generation of alerts will be performed automatically on ground at the INTEGRAL Science Data Center (ISDC). The ISDC system responsible for the monitoring of the data for burst occurence of all INTEGRAL instruments is the INTEGRAL Burst Alert System (IBAS) [3]. The structure of IBAS is shown in Fig.2. The SPI/ACS Burst Alert System (SACS-BAS) is one branch of IBAS. The telemetry (TM) files of the INTEGRAL satellite are transmitted via the Mission Operation Center (MOC) to the ISDC and are then directly fed into the real-time telemetry (RTTM) receiver of the IBAS system. After distribution and extraction of the relevant data, each IBAS branch is processing a burst-search algorithm with a subsequent verify procedure. The trigger algorithm used for SACS-BAS is looking for a significant excess with respect to a running average, comparable to the trigger algorithm used for other spacecrafts (e.g. ULYSSES). Due to the sufficient computing power available on ground it is possible to run several burst-search and burst-verify processes in parallel. SACS-BAS has also implemented this option: several trigger processes with different time bin durations will run in parallel in order to be able to trigger on bursts with different temporal behaviour; several verify processes with different criteria, gain for the supression of fault triggers generated by backgound variations, could be started in SACS-BAS after a burst alert. Up to now SACS-BAS is only reading the overall-veto counter values. But it is planned also to include the read out of rate meter-values of individual FEEs into the SACS-BAS routine. This will allow a rough estimation of the GRB arrival direction. An accuracy of about $10°$–$20°$ will be enough to distinguish between the two arrival-cone intersections of the interplanetary network (IPN). The SACS-BAS trigger algorithm will be tested with generated TM data of simulated burst data plus background. After launch all parameters for the trigger algorithm and verify criteria will be optimised.

The output of SACS-BAS will be burst alerts, containing information about the trigger time in universal time (UT), the spacecraft position and its attitude. The alerts will be transmitted to subscribed users by e-mail and/or direct TCP/IP socket. Especially for the IPN the burst time history (~ 100 s) is important for the alignment of the light curves obtained from different spacecrafts. For this purpose the time history togther with the pre-trigger time-history (~ 5 s) will be transmitted to the IPN. It is important for the IPN to know the burst arrival time with a millisecond accuracy. As already shown in [4] this is possible for the ACS overall counter values.

4   Sensitivity Estimation

An estimation of the expected rate of GRBs detected by ACS has already been given in [4]. For an effective area of about 3000 cm^2, an ACS background rate between ~ 80000 cts/s and ~ 160000 cts/s, an ACS threshold of 80 keV and a time binning of 50 ms the minimal detectable energy flux lies between $2 \leftrightarrow 2.8 \times 10^{-6} \frac{erg}{cm^2 sec}$. For a time binning of 1 sec the sensitivity increases to $5 \times 10^{-7} \frac{erg}{cm^2 sec}$. Using the logN-logP distribution, measured by BATSE and PVO [5] one can derive the number of bursts which will be observed with the ACS in one year. The resulting values are ~ 50 bursts/year for 50 ms integration time and ~ 280 bursts/year for 1 sec integration time. The response of ACS depends on the infalling direction of a burst due to projection effects of other ACS crystals and due to shielding of neighbouring instruments and spacecraft structure. The burst intensity could be determined once the infalling direction is known. This is possible only via Monte-Carlo simulations using the INTEGRAL mass model. Similar simulations have been performed by P. Jean et al. [6] in order to determine the sensitivity of the ACS for detection of novae.

References

1. K. Hurley, ESA SP-**382**, 491–493 (1997)
2. G. Vedrenne et al., Astrophys. Letters and Communications **39**, 325–329 (1999)
3. S. Mereghetti et al., Proc. 4th INTEGRAL Workshop, Alicante, 2000
4. G. G. Lichti et al., AIP Conf. Proc. **510**, 722–726 (2000)
5. E. E. Fenimore et al., Nature **366**, 40 (1993)
6. P. Jean et al., Astrophys. Letters and Communications **38**, 421–424 (1999)

A Preview of the Swift UVOT Capabilities: Imaging and Lightcurves from *XMM-Newton* OM

S. Zane[1], K.O. Mason[1], M.S. Cropper[1], T.E. Kennedy[1], J. Nousek[2],
P. Roming[2], and M. McLelland[3]

[1] Mullard Space Science Laboratory, UCL,
    Holmbury St Mary, Dorking, Surrey, RH5 6NT, UK
[2] Astronomy Department, Davey Laboratory, Pennsylvania State University,
    University Park, PA 16802, US
[3] South-West Research Institute,
    6220 Culebra Road, P.O. Drawer 28510, San Antonio, Texas 78228-0510, US

Abstract. The Swift MIDEX mission, scheduled for launch on 2003, is expected to capture $\sim$ 1000 new γ-Ray Bursts (GRBs), making a comprehensive study of their afterglow characteristics, their origin, evolution and interaction with the surroundings. Swift will include a wide-field γ-ray Burst Alert Telescope, plus narrow-field X-ray and UV/optical telescopes. We will build the Swift UVOT as a clone of the Optical Monitor flying on board XMM-Newton, and the first stage of calibrating XMM-OM has now been completed. Here we present imaging, spectra and light curves from selected XMM-OM fields, as a preview of the UVOT capabilities and time-resolution.

1 Introduction

The Swift Observatory is a Midex mission, selected for flight by NASA in November 1999 and scheduled for launch in the second half of 2003. It is particularly

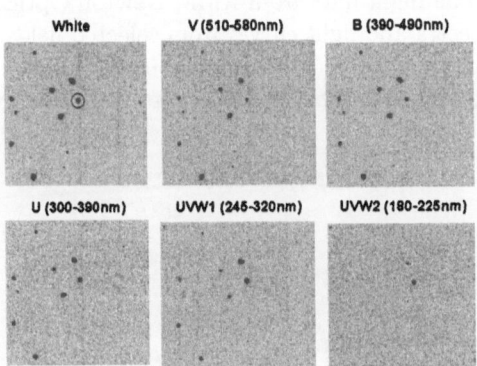

Fig. 1. An image of the Lockman Hole field taken in various XMM-OM filters. The object circled in the White Light image is an AGN, referred to as R32 by [2]. The size of the image is about 3.5 arcmin and the exposure times were 1500 s (white), 1000 s (V, B and U), 2200 s (UVW1) and 4400 s (UVW2).

designed to identify GRBs and to make multiwavelength observations. It will be equipped with a wide-field γ-ray Burst Alert Telescope (BAT), a narrow-field X-ray Telescope (XRT) and a narrow-field UV/Optical telescope (UVOT). Swift will be a rapid response instrument and after the first detection of the event it will manouvre rapidly to bring the burst into the field of view of narrow-field X-ray and UV/Optical telescopes.

We will build the UVOT instrument as a clone of the Optical/UV Monitor telescope currently flying on XMM-Newton. Both instruments are sensitive to photons between 170nm and 650nm over a 17 arcmin square field of view. Good color discrimination is provided by a white light filter and a series of 6 broad-band filters, while for brighter targets 2 Grisms provide low resolution spectra. Both telescopes are also equipped with a field magnifier, to ensure high spatial resolution in the 380nm–650nm band.

In addition, the UVOT will have improved data handling capability to accommodate the demands of rapidly varying burst sources. Due to the impressive imaging sensitivity, already demonstrated by XMM-OM, the UVOT will be able to detect counterparts and afterglow of bursts as bright as B∼ 8 and as faint as B=24. It will be designed to operate simultaneously in Image Mode or Timing Mode, with emphasis on the spatial coverage or on timing information respectively, and it will be able to deliver a finding chart of the burst counterpart within a few hundred seconds of the γ-ray flash. This will allow a rapid follow-up of the burst afterglow with ground-based telescopes. Furthermore, by locating the observed energy of the Lyman edge, UVOT can provide direct information on the redshift of the burst.

2   Results From XMM-Newton OM

In order to illustrate the capabilities of UVOT, we have presented at this meeting a preview based on flight data from XMM-Newton Optical Monitor. Results include imaging, spectra and light curves from selected fields, and some of them are reported here. For all technical details about XMM-OM we refer the interested reader to [1], while a number of further results from XMM-OM have been

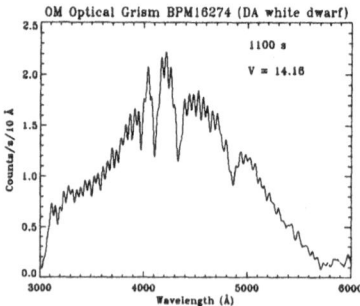

Fig. 2. Optical Grism spectrum of the DA WD BPM16274.The Balmer absorption lines are clearly visible.

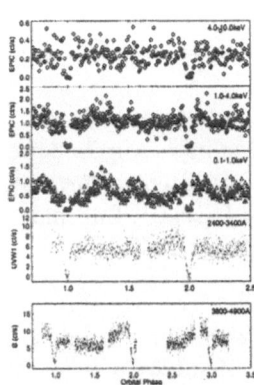
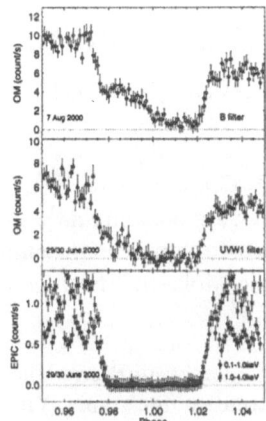

Fig. 3. Left: simultaneous EPIC and OM lightcurves of OY Car. Right: the mean eclipse profile in X-rays, near UV and B. There is a rapid ingress/egress in X-rays: 20–30 s, close to the time to eclipse the WD. The eclipse start earlier in B and UVW1 because the source of the emission, probably the accretion disk, is larger than the WD [3].

published in the A&A special issue, vol. 365. To illustrate the capabilities of the instrument, we show in Fig. 1 images of the Lockman Hole field taken in White Light and in five of the six color filters. The images contain an $R = 18.1$ AGN identified by ROSAT (R32, [2]). The AGN is clearly UV bright: it gives ~ 12 counts/s in White Light, while the count rate in the color filters ranges from 2.9 counts/s in U down to ~ 0.25 counts/s in UVW2. The limiting magnitudes after 1000 s have been calculated from analysis of the Lockman Hole field, and turn out to be 21.0 in V, 22.0 in B, 21.5 in U (6 sigma). The value for the White Light Filter depends on the spectral type, and is ~ 23.5 for an A0 star.

Spectral capabilities of XMM-OM are illustrate in Fig. 2, that shows the spectrum of the standard DA white dwarf (WD) BPM 16264, while an example of high quality timing studies is given in Fig. 3, in which we report the simultaneous X-ray and optical lightcurves from the cataclysmic variable OY Car.

3 Conclusion

XMM-OM successfully extends the spectral coverage of XMM-Newton to the UV/Optical range. Selected results have been presented, to demonstrate the ability of UVOT to make Swift a fast-response, multi-wavelength observatory.

References

1. K.O. Mason, et al., A&A, 365, L36 (2001)
2. M. Schmidt, et al., A&A, 329, 495 (1998)
3. G. Ramsay, et al., A&A, 365, L288 (2001)

REM – Rapid Eye Mount. A Fast Slewing Robotized Telescope to Monitor the Prompt Infra-red Afterglow of GRBs

Filippo M. Zerbi[1], Guido Chincarini[1], Marcello Rodonó[2], Angelo Antonelli[3], Luciano Burderi[3], Sergio Campana[1], Paolo Conconi[1], Stefano Covino[1], Giuseppe Cutispoto[2], Gabriele Ghisellini[1], Davide Lazzati[1], Eugenio Martinetti[2], Emilio Molinari[1], Stefano Sardone[2], Luigi Stella[3], and Fabrizio Vitali[3]

[1] Osservatorio Astronomico di Brera, Via Bianchi 46 – I-23807 – Merate (Lc), Italy
[2] Osservatorio Astrofisico di Catania, Via S.Sofia, 78 – I-95123 Catania – Italy
[3] Osservatorio Astronomico di Roma, Via di Frascati, 33 – I-00044 – Rome – Italy

Abstract. We present REM, a fully robotized fast slewing telescope equipped with a NIR (Z', J, H, K') camera dedicated to monitoring of the prompt IR afterglow of GRBs. REM can discover objects at extremely high red-shift and to trigger large telescopes to observe them.

1 Science Case

The science case which dictates the REM characteristics is the study of gamma-ray bursts simultaneously with SWIFT or other similar space-borne observatories. The SWIFT satellite will detect ∼300 bursts per year, locating them accurately through three instruments: a coded mask gamma-ray detector, with a precision of a few arcmin and a 60×60 degree field of view, an X-ray telescope, with the precision of a few arcsec, and a 30 cm UV-optical monitor, with sub-arcsec precision. The latter instrument is sensitive in the wavelength range 1700–6500 Å: therefore it is blind in the infrared.

Although nearly all bursts have an X-ray afterglow, for about half of them we could not detect an optical afterglow. This is likely due to absorption by dust, or to absorption by Lyα clouds if the bursts is at high redshift. Important information about both types of absorption can be obtained observing the Infrared Afterglow. Indeed, in the NIR band, light is much less affected by reddening, and free from Lyα absorption for redshifts $z < 10$–15.

The infrared-optical prompt is expected to be extremely bright, at least according to the 8.9 V-magnitude detected in GRB 990123 50 seconds after the gamma-ray trigger. Despite its modest aperture REM can record high quality measurements of infra-red afterglows.

In the case of a high redshift burst the Optical Monitor on-board SWIFT will not detect any signal, but REM will instead record NIR flux. Since both data will be available a few minutes after the trigger (when the magnitude is still expected to be around 14–15) REM can alert large telescopes such as KECK or

VLT to record spectra of the most distant objects in the universe. Indeed even if the monochromatic flux is decaying fast ($\propto t^{-1}$–$t^{-1.5}$), it will still be possible to obtain high quality infrared spectra up to redshifts as high as 15.

The case of dust reddening will be recognized by examination of color–color diagrams since dust absorption is more gradual than Ly-α absorption. In this case REM can help to measure the amount of reddening, and hence to establish if gamma-ray bursts explode in dense, probably dusty, star forming regions, giving information on the nature of their progenitor population. The simultaneous infrared and optical light curves will instead tell about the energy spent by the burst to destroy, at least partially, the circumburst dust.

· About 3/4 of the bursts detected by SWIFT will be unobservable because of sun constraint, giving ~75 bursts per year observable by REM. This implies that a considerable amount of time can be spent on additional science. Relatively bright, variable objects are obvious candidates. Among the research fields where REM can contribute there are blazars, compact galactic objects, and variable stars.

2   REM Characteristics

REM has a classical Ritchey-Chretien optical Scheme mounted in an alt-azimuthal configuration with 2 Nasmyth focal stations. The primary mirror, with a diameter of 60 cm and total focal ratio (F/5.3) is tuned to give a focal length of f=3200 mm. Such optical figures allow a compact structure that provides the needed stiffness for fast motion and windy conditions.

REM is designed to be located in a major Observatory in the Southern Hemisphere. The most favorable location would be Cerro Paranal since it is probably the best IR site in the Southern Emisphere to which European Astronomers have access. Furthermore the number of clear nights per year is extremely important while observing in coordination with a satellite. The Telescope is foreseen to be lodged in a fully-deployable dome and has been designed accordingly to face wind and environmental conditions.

One of the Nasmyth focal station will be equipped with a fully cryogenic NIR (0.9–2.3 μm) camera. The camera design had been developed with high throughput as major goal. The camera is afocal (1:1) and made of two identical detached doublets of Silica and CaF_2, the second reversed in the optical path to reform the telescope focal plane on the array. The pupil of the system is reformed between the doublets allowing to locate either imaging filters or prisms to perform low resolution slit-less spectroscopy. The camera has quasi diffraction-limited Optical Quality. A representation of the camera layout and related spot diagrams are reported in figure 1.

The camera will be equipped with a 512x512 Rockwell HAWAII LPE HgCdTe chip (18 μm pitch) with a peak efficiency of 64% and values never lower than 56% between 1 and 2.5 μm. The possibility to upgrade to an MBE HgCdTe chip (80% peak efficiency) is currently under examination. The camera (telescope and filter excluded) is expected to have a transmission of $T >53\%$ ($T >68\%$ with

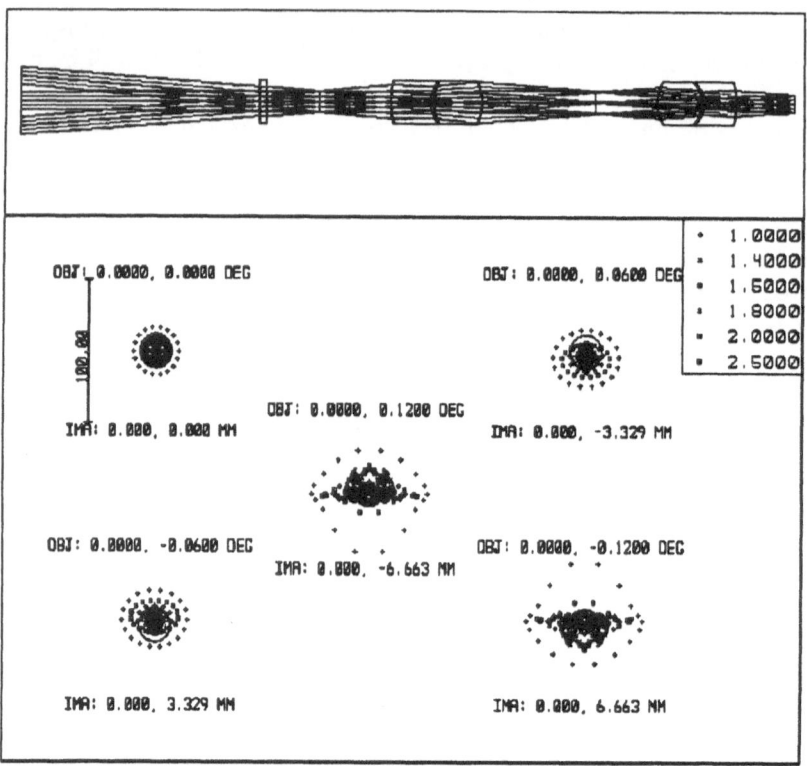

Fig. 1. REM camera layout and spot diagrams.

MBE). With a scale of 1.16 arcsec/pixel the camera covers a FOV of 9.8x9.8 arcmin² well beyond the 4 arcmin diameter 3-σ error-box of BAT, from which REM will receive the trigger. This ensures the presence of the transient in the field. At the same time the scale allows to provide coordinates of the transient with a precision of a few tenth of an arcsecond.

Given the intrinsic rapidity of the phenomena a rapid photometry is as essential as the rapid mount drive. Based on a flux model which evolves as $f \sim t^{-\delta}$ with $0.8 < \delta < 2.0$ we reckon that a measurement of each filter J, H and K every 5 seconds in the first observations after targeting is the minimum acceptable frequency. Slower data flow will be onset after a 1.5 minutes and a measurement every 30 seconds will be delivered. One hour after the BAT signal the frequency will be definitely slowed down to one measure every 10 minutes. This is to ensure a reasonable S/N which follows the initial foreseen time evolution of the IR burst counterpart, and it is open to policy change after the hitherto unknown parameters will be measured (colour index, typical IR magnitude).

List of Participants

Amati Lorenzo
Istituto TESRE / CNR
Via Gobetti, 101
I-40126 Bologna, Italy
amati@tesre.bo.cnr.it

Andersen Michael
Division of Astronomy, Univ. of Oulu
P.O. Box 3000
FIN-90014 Oulu, Finland
manderse@sun3.oulu.fi

Anfimov Dmitrij
Space Research Institute, RAS
Profsojuznaya 84/32
117810 Moscow, Russia
dima@cgrsmx.iki.rssi.ru

Antonelli Angelo
Osservatorio Astronomico di Roma
Via Frascati, 33
I-00040 Monteporzio, Italy
angelo@coma.mporzio.astro.it

Aptekar Rafail
Ioffe Physico-Technical Inst.
Polytekhnicheskaya 26
194021 St. Petersburg, Russia
aptekar@pop.ioffe.rssi.ru

Arend Nikolas
MPI für Extraterr. Physik
Postfach 1312
D-85741 Garching, Germany
nka@mpe.mpg.de

Argan Andrea
IFCTR / CNR
Via Bassini, 15
I-20133 Milan, Italy
argan@ifctr.mi.cnr.it

Asano Katsuaki
University of Osaka
Toyonaka 560-0043
Osaka, Japan
asano@vega.ess.sci.osaka-u.ac.jp

Auricchio Natalia
Istituto TESRE / CNR
Via Gobetti, 101
I-40126 Bologna, Italy
natalia@tesre.bo.cnr.it

Bagoly Zsolt
Lab. for Information Technology
Eötvös University
H-1518 Budapest, Hungary
bagoly@ludens.elte.hu

Balázs Lajos
Konkoly Observatory
Box 67,
H-1505 Budapest, Hungary
balazs@konkoly.hu

Band David
Los Alamos National Laboratory
Los Alamos, NM
87505, USA
dband@lanl.gov

Barat Claude
CESR
Toulouse Cedex
4346–31029 France
Claude.Barat@cesr.fr

Barbiellini Guido
Universitaà di Trieste
Via Valerio, 2
I-34127 Trieste, Italy
guido.barbiellini@ts.infn.it

Barrena Rafael
Instituto de Astrofísica de Canarias
E-38200 La Laguna, Tenerife
Canary Islands, Spain
rbarrena@ll.iac.es

Barthelmy Scott
Goddard Space Flight Center, NASA
Mail Code 660, Greenbelt, MD
20771, USA
scott@lheamail.gsfc.nasa.gov

Bartolini Corrado
Dip. Astronomia, Univ. Bologna
Via Ranzani, 1
I-40129 Bologna, Italy
bartolini@astbo3.bo.astro.it

Belli Bianca Maria
IAS / CNR
Via Fosso del Cavaliere, 100
I-00133 Rome, Italy
bianca@ias.rm.cnr.it

Beloborodov Andrei
Stockholm Observatory
Saltsjöbaden
SE-133 36, Sweden
andrei@astro.su.se

Bennett Kevin
Space Science Department of ESA
P.O. Box 299
NL-2200 Noordwijk, The Netherlands
kbennett@astro.estec.esa.nl

Berger Edo
California Institute of Technology
Pasadena, CA
91125, USA
ejb@astro.caltech.edu

Berná Josè Ángel
FISTS, UA
03690 San Vicente del Raspeig
Alicante, Spain
jberna@disc.ua.es

Bernas Martin
Technical University in Prague
Technická 2, Prague
CZ-16627, Czech Republic
bernas@fel.cvut.cz

Beskin Gregory
SAO / RAS
Nizhnij Arkhyz
369167 Russia
beskin@sao.ru

Biryukov Sergej
Physics Dept., Ben-Gurion Univ.
P.O. Box 653, Beer-Sheva
84105, Israel
beryukov@bgumail.bgu.ac.il

Björnsson Gunnlaugur
Science Inst., Univ. of Iceland
Dunhagi 3
IS-107 Reykjavik, Iceland
gulli@raunvis.hi.is

Bloom Joshua
California Institute of Technology
Pasadena, CA
91125, USA
jsb@astro.caltech.edu

Bonnell Jerry
Goddard Space Flight Center, NASA
Mail Code 660, Greenbelt, MD
20771, USA
jerry@milkyway.gsfc.nasa.gov

Borkowski Jurek
INTEGRAL Science Data Center
Chemin d'Ecogia 16
CH-1290 Versoix, Switzerland
Jerzy.Borkowski@obs.unige.ch

Böttcher Markus
Rice University
Houston, TX
77005 USA
mboett@spacsun.rice.edu

Bradt Hale
Massachusetts Institute of Technology
Cambridge, MA
02139 USA
bradt@mit.edu

Briggs Michael
Nat. Space Sci. and Technol. Center
320 Sparkman Drive
Huntsville, AL 35805, USA
michael.briggs@msfc.nasa.gov

Buckley Jason
Washington University
St. Louis, MO
63130, USA
buckley@cosray2.wustl.edu

Burderi Luciano
Osservatorio Astronomico di Roma
Via Frascati, 33
I-00040 Monteporzio, Italy
burderi@coma.mporzio.astro.it

Butterworth Paul
Goddard Space Flight Center, NASA
Mail Code 660, Greenbelt, MD
20771, USA
butterworth@lheavx.gsfc.nasa.gov

Čadež Andrej
Univ. of Ljubljana
Pot na Golovec 25
1000 Ljubljana, Slovenia
andrej.cadez@uni-lj.si

Calura Francesco
Dip. Fisica, Univ. of Ferrara
Via Paradiso, 12
I-44100 Ferrara, Italy
fcalura@fe.infn.it

Campana Sergio
Osservatorio Astronomico di Brera
Via Bianchi, 46
I-23807 Merate, Italy
campana@merate.mi.astro.it

Capalbi Milvia
BeppoSAX SDC
Via Galileo Galilei
I-00044 Frascati, Italy
capalbi@asi.it

Caraveo Patrizia
IFCTR / CNR
Via Bassini, 15
I-20133 Milan, Italy
pat@ifctr.mi.cnr.it

Castro Sandra
California Institute of Technology
Pasadena, CA
91125, USA
smc@astro.caltech.edu

Castro Cerón Josè Maria
Real Inst. y Obs. de la Armada
11110 San Fernando–Naval
Cádiz, Spain
josemari@roa.es

Castro-Tirado Alberto
IAA-CSIC
P.O. Box 03004
18080 Granada, Spain
ajct@iaa.es

Cei Fabrizio
Dip. Fisica, Univ. of Pisa
Via Livornese, 1291/a
I-56010, S. Piero a Grado, Pisa, Italy
fabrizio.cei@pi.infn.it

Chaffee Frederic
W.M. Keck Observatory
Kamuela, HI
96743, USA
fchaffee@keck.hawaii.edu

Chen Andrew
IFCTR / CNR
Via Bassini, 15
I-20133 Milan, Italy
chen@ifctr.mi.cnr.it

Cherepashchuk Anatol
Sternberg Astronomical Institute
Universitetsky Pr., 13
119899 Moscow, Russia
cher@sai.msu.ru

Cherry Michael
Louisiana State University
Baton Rouge, LA
70803, USA
cherry@phunds.phys.lsu.edu

Chevalier Roger
Dept. of Astronomy, Univ. of Virginia
Charlottesville, VA
22903, USA
rac5x@virginia.edu

Chincarini Guido
Osservatorio Astronomico di Brera
Via Bianchi, 46
I-23807 Merate, Italy
guido@merate.mi.astro.it

Christensen Lise
Astron. Obs., Univ. of Copenhagen
Juliane Maries Vej 30
DK-2100 Copenhagen Ø, Denmark
lise@astro.ku.dk

Cline David
Dept. of Phys. and Astron., UCLA
Box 951457 Los Angeles, CA
90095 USA
dcline@physics.ucla.edu

Cline Thomas
Goddard Space Flight Center, NASA
Mail Code 660, Greenbelt, MD
20771, USA
cline@apache.gsfc.nasa.gov

Cocco Veronica
Università di Roma "Tor Vergata"
Via della Ricerca Scientifica, 1
I-00133 Rome, Italy
cocco@roma2.infn.it

Colasanti Luca
IAS / CNR
Via Fosso del Cavaliere, 100
I-00133 Rome, Italy
colasant@ias.rm.cnr.it

Conconi Paolo
Osservatorio Astronomico di Brera
Via Bianchi, 46
I-23807 Merate, Italy
conconi@merate.mi.astro.it

Cosentino Giuseppe
Dip. Astronomia, Univ. Bologna
Via Ranzani, 1
I-40129 Bologna, Italy
cosentino@astbo3.bo.astro.it

Costa Enrico
IAS / CNR
Via Fosso del Cavaliere, 100
I-00133 Rome, Italy
costa@ias.rm.cnr.it

Covino Stefano
Osservatorio Astronomico di Brera
Via Bianchi, 46
I-23807 Merate, Italy
covino@merate.mi.astro.it

Cremonesi Davide
IFCTR / CNR
Via Bassini, 15
I-20133 Milan, Italy
davide@ifctr.mi.cnr.it

Cropper Mark
MSSL, UCL
Holmbury St. Mary
Dorking, Surrey, RH5 6NT, UK
msc@mssl.ucl.ac.uk

Csabai István
Dept. of Physics for Complex Systems
Eötvös University
H-1518 Budapest, Hungary
csabai@ludens.elte.hu

Cusumano Giancarlo
IFCAI / CNR
Via U. La Malfa, 153
I-90146 Palermo, Italy
cusumano@ifcai.pa.cnr.it

Cutispoto Giuseppe
Oss. Astrofisico di Catania
Via S. Sofia, 78
I-95123 Catania, Italy
gcutispoto@alpha4.ct.astro.it

D'Alessio Francesco
Osservatorio Astronomico di Roma
Via Frascati, 33
I-00040 Monteporzio, Italy
dalessio@coma.mporzio.astro.it

Daigne Frédéric
Max Planck Institut für Astrophysik
Karl-Schwarzschild-Str. 1
D-85741 Garching, Germany
daigne@mpa-garching.mpg.de

de la Morena Benito
CEA-INTA
21130 Mazagón
Huelva, Spain
morenacb@inta.es

de Ugarte Postigo Antonio
Facultad de CC.FF.
UCM
28040 Madrid Spain
ugarte@retemail.es

Del Monte Ettore
IAS / CNR
Via Fosso del Cavaliere, 100
I-00133 Rome, Italy
delmonte@ias.rm.cnr.it

Derishev Evgeny
Institute of Applied Physics, RAS
46 Ulyanov Street
603600 Nizhny Novgorod, Russia
derishev@appl.sci-nnov.ru

Dermer Charles
Naval Research Laboratory
Washington, DC
20375 USA
dermer@gamma.nrl.navy.mil

Di Cocco Guido
Istituto TESRE / CNR
Via Gobetti, 101
I-40126 Bologna, Italy
dicocco@tesre.bo.cnr.it

Di Paola Andrea
Osservatorio Astronomico di Roma
Via Frascati, 33
I-00040 Monteporzio, Italy
dipaola@coma.mporzio.astro.it

Diehl Ronald
MPI für Extraterr. Physik
Postfach 1312
D-85741 Garching, Germany
rod@mpe.mpg.de

Diercks Alan
California Institute of Technology
Pasadena, CA
91125, USA
ad@astro.caltech.edu

Djorgovski George
California Institute of Technology
Pasadena, CA
91125, USA
george@astro.caltech.edu

Dray Lynnette
Institute of Astronomy
Madingley Road, Cambridge
CB3 0HA, United Kingdom
ldray@ast.cam.ac.uk

Duggan Paul
University College Dublin
Stillorgan Road
Dublin 4, Ireland
pduggan@bermuda.ucd.ie

Efremov Yuri
Sternberg Astronomical Institute
Universitetsky Pr., 13
119899 Moscow, Russia
efremov@sai.msu.ru

Eichler David
Physics Dept., Ben-Gurion Univ.
P.O. Box 653, Beer-Sheva
84105, Israel
eichler@bgumail.bgu.ac.il

Faiman David
Physics Dept., Ben-Gurion Univ.
P.O. Box 653, Beer-Sheva
84105, Israel
faiman@bgumail.bgu.ac.il

Fatkhullin Timur
SAO / RAS
Nizhnij Arkhyz
369167 Russia
timur@sao.ru

Fargion Daniele
Univ. di Roma "La Sapienza"
Piazzale Aldo Moro, 5
I-00185, Rome, Italy
daniele.fargion@roma1.infn.it

Fedel Giulio
INFN, Sez. di Trieste
Padriciano, 99
I-34012 Trieste, Italy
giulio.fedel@ts.infn.it

Fenimore Edward
Los Alamos National Laboratory
Los Alamos, NM
87505, USA
efenimore@lanl.gov

Feroci Marco
IAS / CNR
Via Fosso del Cavaliere, 100
I-00133 Rome, Italy
feroci@ias.rm.cnr.it

Fernández-Soto Alberto
Osservatorio Astronomico di Brera
Via Bianchi, 46
I-23807 Merate, Italy
fsoto@merate.mi.astro.it

Ferrari Attilio
Oss. Astron. di Torino
Strada Osservatorio, 20
I-10025 Pino Torinese, Italy
ferrari@to.astro.it

Fiore Fabrizio
Osservatorio Astronomico di Roma
Via Frascati, 33
I-00040 Monteporzio, Italy
fiore@quasar.mporzio.astro.it

Fiorini Mauro
IFCTR / CNR
Via Bassini, 15
I-20133 Milan, Italy
fiorini@ifctr.mi.cnr.it

Fischer Olaf
Friedrich-Schiller-Universität Jena
Max-Wien-Platz 1
D-07743 Jena, Germany
fischer@astro.uni-jena.de

Fishman Gerald
Marshall Space Flight Center, NASA
Huntsville, AL
35812 USA
Gerald.Fishman@msfc.nasa.gov

Florian Jan
ASÚ, AVČR
CZ-25165 Ondřejov
Czech Republic
florian@asu.cas.cz

Fox Derek
California Institute of Technology
Pasadena, CA
91125, USA
derekfox@astro.caltech.edu

Frail Dale
NRAO
P.O. Box O, Socorro, NM
87801, USA
dfrail@nrao.edu

Frederiks Dmitry
Ioffe Physico-Technical Inst.
Polytekhnicheskaya 26
194021 St. Petersburg, Russia
fred@pop.ioffe.rssi.ru

Frontera Filippo
Dip. Fisica, Univ. of Ferrara
Via Paradiso, 12
I-44100 Ferrara, Italy
frontera@fe.infn.it

Fryer Chris
Los Alamos National Laboratory
Los Alamos, NM
87505, USA
clf@lanl.gov

Fynbo Johan
European Southern Observatory
Karl-Schwarzschild-Str. 2
D-85748 Garching, Germany
jfynbo@eso.org

Galama Titus
California Institute of Technology
Pasadena, CA
91125, USA
tjg@astro.caltech.edu

Galassi Mark
Los Alamos National Laboratory
Los Alamos, NM
87505, USA
rosalia@nis.lanl.gov

Galeazzi Massimiliano
University of Wisconsin
1150 University Ave., Madison, WI
53706 USA
galeazzi@wisp.physics.wisc.edu

Galli Marcello
ENEA, Sez. di Bologna
Via Don Fiammelli, 2
I-40129 Bologna, Italy
galli@bologna.enea.it

Gandolfi Giangiacomo
IAS / CNR
Via Fosso del Cavaliere, 100
I-00133 Rome, Italy
gandolfi@ias.rm.cnr.it

Gatti Flavio
INFN, Sez. di Genova
Via Dodecaneso, 33
I-16146 Genoa, Italy
gatti@ge.infn.it

Gehrels Neil
Goddard Space Flight Center, NASA
Mail Code 660, Greenbelt, MD
20771, USA
gehrels@gsfc.nasa.gov

Georgii Robert
MPI für Extraterr. Physik
Postfach 1312
D-85741 Garching, Germany
rog@mpe.mpg.de

Ghisellini Gabriele
Osservatorio Astronomico di Brera
Via Bianchi, 46
I-23807 Merate, Italy
gabriele@merate.mi.astro.it

Giménez Álvaro
LAEFF-INTA
28692 Villafranca del Castillo
Madrid, Spain
ag@laeff.esa.es

Giuliani Andrea
IFCTR / CNR
Via Bassini, 15
I-20133 Milan, Italy
giuliani@ifctr.mi.cnr.it

Golenetskii Sergey
Ioffe Physico-Technical Inst.
Polytekhnicheskaya 26
194021 St. Petersburg, Russia
golen@pop.ioffe.rssi.ru

Gomboc Andreja
Univ. of Ljubljana
Pot na Golovec 25
1000 Ljubljana, Slovenia
andreja@fiz.uni-lj.si

Goodrich Robert
W.M. Keck Observatory
Kamuela, HI
96743, USA
goodrich@keck.hawaii.edu

Gorosabel Javier
Danish Space Research Institute
Juliane Maries Vej 30
DK-2100 Copenhagen Ø, Denmark
jgu@dsri.dk

Granot Jonathan
Racah Institute of Physics
Hebrew University
91904 Jerusalem, Israel
jgranot@fireball.fiz.huji.ac.il

Graziani Carlo
Univ. of Chicago
Chicago, IL
60637 USA
carlo@kazoo.uchicago.edu

Greiner Jochen
Astrophysikalisches Institut
An der Sternwarte 16
D-14482 Potsdam, Germany
jgreiner@aip.de

Guarnieri Adriano
Dip. Astronomia, Univ. Bologna
Via Ranzani, 1
I-40129 Bologna, Italy
adriano@astbo3.bo.astro.it

Guetta Dafne
Oss. Astrofisico di Arcetri
Largo E. Fermi, 5
I-50125 Florence, Italy
dafne@arcetri.astro.it

Guidorzi Cristiano
Dip. Fisica, Univ. of Ferrara
Via Paradiso, 12
I-44100 Ferrara, Italy
guidorzi@fe.infn.it

Guzik Gregory
Louisiana State University
Baton Rouge, LA
70803, USA
guzik@phunds.phys.lsu.edu

Haardt Francesco
Università dell'Insubria
Via Valleggio 11
I-22100 Como, Italy
francesco.haardt@mib.infn.it

Haglin David
Minnesota State University
Mankato, MN
56001 USA
David.Haglin@mnsu.edu

Hakkila Jon
College of Charleston
Charleston, SC
29424 USA
hakkilaj@cofc.edu

Hanlon Lorraine
University College Dublin
Stillorgan Road
Dublin 4, Ireland
lhanlon@bermuda.ucd.ie

Harrison Fiona
California Institute of Technology
Pasadena, CA
91125, USA
fiona@srl.caltech.edu

Heise John
SRON
Sorbonnelaan 2
NL-3584 CA Utrecht, The Netherlands
j.heise@sron.nl

Hink Paul
Washington University
St. Louis, MO
63130, USA
hink@wuphys.wustl.edu

Hjorth Jens
Astron. Obs., Univ. of Copenhagen
Juliane Maries Vej 30
DK-2100 Copenhagen Ø, Denmark
jens@astro.ku.dk

Holland Stephen
Dep. of Physics, Univ. of Notre Dame
Notre Dame, IN
46556 USA
sholland@nd.edu

Horváth István
Physics Dept., Bolyai Military Univ.
Box 12
H-1456 Budapest, Hungary
hoi@bjkmf.hu

Hroch Filip
Faculty of Science, Masaryk Univ.
Kotlarska 2
CZ-61137 Brno, Czech Republic
hroch@monoceros.physics.muni.cz

Hudcová Věra
ASÚ, AVČR
CZ-25165 Ondřejov
Czech Republic
vhudcova@asu.cas.cz

Hudec Renè
ASÚ, AVČR
CZ-25165 Ondřejov
Czech Republic
rhudec@asu.cas.cz

Hughes Philip
Dept. of Astronomy, Univ. of Michigan
Ann Arbor, MI
48109, USA
hughes@astro.lsa.umich.edu

Hunt Leslie
CAISMI / CNR
Largo E. Fermi 5
I-50125 Firenze, Italy
hunt@arcetri.astro.it

Hurley Kevin C.
SSL, Univ. of California
Berkeley, CA
94720, USA
khurley@sunspot.ssl.berkeley.edu

Hurley Kevin J.
University College Dublin
Stillorgan Road
Dublin 4, Ireland
khurley@bermuda.ucd.ie

Il'inskii Vadim
Ioffe Physico-Technical Inst.
Polytekhnicheskaya 26
194021 St. Petersburg, Russia
ilyin@pop.ioffe.rssi.ru

in 't Zand Jean
SRON
Sorbonnelaan 2
NL-3584 CA Utrecht, The Netherlands
jeanz@sron.nl

Inneman Adolf
Centre for Advanced X-ray Technol.
Reflex, Novodvorská 994
CZ-14000 Praha 4, Czech Republic
adolf.inneman@reflex-co.cz

Israel Gian Luca
Osservatorio Astronomico di Roma
Via Frascati, 33
I-00040 Monteporzio, Italy
gianluca@coma.mporzio.astro.it

Jakobsson Pall
Astron. Obs., Univ. of Copenhagen
Juliane Maries Vej 30
DK-2100 Copenhagen Ø, Denmark
pallja@astro.ku.dk

Jaunsen Andreas
European Southern Observatory
Casilla 19001
Santiago 19, Chile
ajaunsen@eso.org

Jensen Brian Lindgren
Astron. Obs., Univ. of Copenhagen
Juliane Maries Vej 30
DK-2100 Copenhagen Ø, Denmark
brian_j@astro.ku.dk

Jimenez Raul
Rutgers University
Piscataway, NJ
08854, USA
raulj@physics.rutgers.edu

Kappadath Cheenu
Louisiana State University
Baton Rouge, LA
70803, USA
cheenu@rouge.phys.lsu.edu

Kasimova Ekaterina
Institute of Physics, Rostov Univ.
Stachki 194, Rostov-on-Don
344090 Russia
katrin@ip.rsu.ru

Kennedy Tom
MSSL, UCL
Holmbury St. Mary
Dorking, Surrey, RH5 6NT, UK
tek@mssl.ucl.ac.uk

Kinsella Derek
University College Dublin
Stillorgan Road
Dublin 4, Ireland
kinsella@bermuda.ucd.ie

Kippen Marc
Nat. Space Sci. and Technol. Center
320 Sparkman Drive, Huntsville, AL
35805, USA
marc.kippen@msfc.nasa.gov

Klose Sylvio
Thüringer Landessternw. Tautenburg
Sternwarte 5
D-07778 Tautenburg, Germany
klose@tls-tautenburg.de

Kobayashi Shiho
University of Osaka
Toyonaka 560-0043
Osaka, Japan
shiho@vega.ess.sci.osaka-u.ac.jp

Kocharovsky Vitaly
Institute of Applied Physics, RAS
46 Ulyanov Street
603600 Nizhny Novgorod, Russia
kochar@appl.sci-nnov.ru

Kocharovsky Vladimir
Institute of Applied Physics, RAS
46 Ulyanov Street
603600 Nizhny Novgorod, Russia
kochar@appl.sci-nnov.ru

Komarova Victoria
SAO / RAS
Nizhnij Arkhyz
369167 Russia
vkom@sao.ru

Kompaneets Dmitrii
Lebedev Physical Institute
Profsoyuznaya 84/32
117810 Moscow, Russia
dkompan@lukash.asc.rssi.ru

Kopeikin Sergei
Dept. of Physics and Astronomy
Univ. of Missouri-Columbia, MO
65211 USA
KopeikinS@missouri.edu

Korchagin Vladimir
Institute of Physics, Rostov Univ.
Stachki 194, Rostov-on-Don
344090 Russia
vik@rsuss1.rnd.runnet.ru

Kouveliotou Chryssa
Marshall Space Flight Center, NASA
Huntsville, AL
35812 USA
Chryssa.Kouveliotou@msfc.nasa.gov

Kozyrev Alexandr
Space Research Institute, RAS
Profsojuznaya 84/32
117810 Moscow, Russia
kozyrev@mx.iki.rssi.ru

Kubánek Petr
Faculty of Math., Charles Univ.
Malostranske nam. 1
CZ-11000 Praha 1, Czech Republic
pkubanek@email.cz

Kulkarni Shrinivas
California Institute of Technology
Pasadena, CA
91125, USA
srk@astro.caltech.edu

Kurt Vladimir
Lebedev Physical Institute
Profsoyuznaya 84/32
117810 Moscow, Russia
vkurt@asc.rssi.ru

Kvick Ake
ESRF
BP 220, Grenoble Cedex
F-38043 France
kvick@esrf.fr

Labanti Claudio
Istituto TESRE / CNR
Via Gobetti, 101
I-40126 Bologna, Italy
labanti@tesre.bo.cnr.it

Lamb Donald
Univ. of Chicago
Chicago, IL
60637 USA
lamb@oddjob.uchicago.edu

Lapshov Igor
IAS / CNR
Via Fosso del Cavaliere, 100
I-00133 Rome, Italy
lapa@ias.rm.cnr.it

Lazzati Davide
Osservatorio Astronomico di Brera
Via Bianchi, 46
I-23807 Merate, Italy
lazzati@merate.mi.astro.it

Levine Alan
Massachusetts Institute of Technology
Cambridge, MA
02139 USA
aml@space.mit.edu

Liang Edison
Rice University
Houston, TX
77005 USA
liang@spacsun.rice.edu

Lichti Giselher
MPI für Extraterr. Physik
Postfach 1312
D-85741 Garching, Germany
grl@mpe.mpg.de

Lindfors Elina
Science Inst., Univ. of Iceland
Dunhagi 3
IS-107 Reykjavik, Iceland
ejl@rhi.hi.is

Lipari Paolo
Univ. di Roma "La Sapienza"
Piazzale Aldo Moro, 5
I-00185, Rome, Italy
paolo.lipari@roma1.infn.it

Litvak Maxim
Space Research Institute, RAS
Profsojuznaya 84/32
117810 Moscow, Russia
litvak@mx.iki.rssi.ru

Longo Francesco
INFN, Sez. di Ferrara
Via Paradiso, 12
I-44100 Ferrara, Italy
longo@fe.infn.it

Lorenzetti Dario
Osservatorio Astronomico di Roma
Via Frascati, 33
I-00040 Monteporzio, Italy
dloren@coma.mporzio.astro.it

Lund Niels
Danish Space Research Institute
Juliane Maries Vej 30
DK-2100 Copenhagen Ø, Denmark
nl@dsri.dk

Macri John
University of New Hampshire
Durham, NH
03824, USA
jmacri@unh.edu

Madau Piero
Institute of Astronomy
Madingley Road, Cambridge
CB3 0HA, United Kingdom
pmadau@ast.cam.ac.uk

Maeda Keiichi
Department of Astronomy
University of Tokyo
Tokyo, Japan
maeda@astron.s.u-tokyo.ac.jp

Mallozzi Robert*
Science Communications, Inc.
Huntsville, AL
35899 USA
*Deceased

Martinetti Eugenio
Oss. Astrofisico di Catania
Via S. Sofia, 78
I-95123 Catania, Italy
emartin@alpha4.ct.astro.it

Marshall Francis
Goddard Space Flight Center, NASA
Mail Code 660, Greenbelt, MD
20771 USA
marshall@lheamail.gsfc.nasa.gov

Masetti Nicola
Istituto TESRE / CNR
Via Gobetti, 101
I-40126 Bologna, Italy
masetti@tesre.bo.cnr.it

Más-Hesse Miguel
LAEFF-INTA
28692 Villafranca del Castillo
Madrid, Spain
mm@laeff.esa.es

Mason Keith
MSSL, UCL
Holmbury St. Mary
Dorking, Surrey, RH5 6NT, UK
kom@mssl.ucl.ac.uk

Mateo Sanguino Tomàs de Jesùs
CEA-INTA
21130 Mazagón
Huelva, Spain
mateost@inta.es

Matteson James
Univ. of California at San Diego
La Jolla, CA
92093, USA
jmatteson@ucsd.edu

Matthey Christina
Dept. of Phys. and Astron., UCLA
Box 951457 Los Angeles, CA
90095 USA
Christina.Matthey@cern.ch

Mazets Evgueni
Ioffe Physico-Technical Inst.
Polytekhnicheskaya 26
194021 St. Petersburg, Russia
mazets@pop.ioffe.rssi.ru

Mazzali Paolo
Oss. Astron. di Trieste
Via Tiepolo, 11
34131 Trieste, Italy
mazzali@ts.astro.it

McBreen Brian
University College Dublin
Stillorgan Road
Dublin 4, Ireland
bmcbreen@ucd.ie

McBreen Sheila
University College Dublin
Stillorgan Road
Dublin 4, Ireland
smcbreen@bermuda.ucd.ie

McCammon Dan
University of Wisconsin
1150 University Ave., Madison, WI
53706 USA
mccammon@wisp.physics.wisc.edu

McConnell Mark
University of New Hampshire
Durham, NH
03824, USA
mark.mcconnell@unh.edu

McLelland Michael
South-West Research Institute
6220 Culebra Road, San Antonio, TX
78228, USA
mmclelland@swri.org

Mediavilla Evencio
Instituto de Astrofísica de Canarias
E-38200 La Laguna, Tenerife
Canary Islands, Spain
emg@ll.iac.es

Meegan Charles
Marshall Space Flight Center, NASA
Huntsville, AL
35812 USA
Charles.Meegan@msfc.nasa.gov

Mereghetti Sandro
IFCTR / CNR
Via Bassini, 15
I-20133 Milan, Italy
sandro@ifctr.mi.cnr.it

Merloni Andrea
Institute of Astronomy
Madingley Road, Cambridge
CB3 0HA, United Kingdom
am@ast.cam.ac.uk

Mészáros Attila
Astron. Inst., Charles Univ.
V Holešovičkách 2
CZ-18000 Prague 8, Czech Republic
meszaros@mbox.cesnet.cz

Mészáros Peter
Pennsylvania State Univ.
University Park, PA
16802 USA
nnp@astro.psu.edu

Metcalfe Leo
Astrophysics Division of ESA
Vilspa, PO Box 50727
E-28080 Madrid, Spain
lmetcalf@iso.vilspa.esa.es

Meusinger Helmut
Thüringer Landessternw. Tautenburg
Sternwarte 5
D-07778 Tautenburg, Germany
meus@tls-tautenburg.de

Miller Mark
Washington University
St. Louis, MO
63130, USA
mamiller@void.wustl.edu

Mitrofanov Igor
Space Research Institute, RAS
Profsojuznaya 84/32
117810 Moscow, Russia
imitrofa@space.ru

Mochkovitch Robert
Institut d'Astrophysique
98bis Boulevard Arago
F-75014 Paris, France
mochko@iap.fr

Møller Palle
European Southern Observatory
Karl-Schwarzschild-Str. 2
D-85748 Garching, Germany
pmoller@eso.org

Molinari Emilio
Osservatorio Astronomico di Brera
Via Bianchi, 46
I-23807 Merate, Italy
molinari@merate.mi.astro.it

Montanari Enrico
Dip. Fisica, Univ. of Ferrara
Via Paradiso, 12
I-44100 Ferrara, Italy
montana@fe.infn.it

Montaruli Teresa
Dip. Fisica, Univ. of Bari
Via Amendola, 173
I-70126 Bari, Italy
teresa.montaruli@ba.infn.it

Morelli Ennio
Istituto TESRE / CNR
Via Gobetti, 101
I-40126 Bologna, Italy
morelli@tesre.bo.cnr.it

Morselli Aldo
Università di Roma "Tor Vergata"
Via della Ricerca Scientifica, 1
I-00133 Rome, Italy
aldo.morselli@roma2.infn.it

Murakami Toshio
ISAS
3-1-1, Yoshinodai, Sagamihara
229-8510 Kanagawa, Japan
murakami@astro.isas.ac.jp

Murphy Noel
University College Dublin
Stillorgan Road
Dublin 4, Ireland
nmurphy@bermuda.ucd.ie

Nakamura Takayoshi
Department of Astronomy
University of Tokyo
Tokyo, Japan
nakamura@astron.s.u-tokyo.ac.jp

Nakar Ehud
Racah Institute of Physics
Hebrew University
91904 Jerusalem, Israel
udini@merger.phys.huji.ac.il

Natarajan Priyamvada
Dept. of Astronomy, Yale Univ.
New Haven, CT
06520, USA
priya@astro.yale.edu

Nekola Martin
ASÚ, AVČR
CZ-25165 Ondřejov
Czech Republic
ford@asu.cas.cz

Nicastro Luciano
IFCAI / CNR
Via U. La Malfa, 153
I-90146 Palermo, Italy
nicastro@ifcai.pa.cnr.it

Nomoto Ken'ichi
Department of Astronomy
University of Tokyo
Tokyo, Japan
nomoto@astron.s.u-tokyo.ac.jp

Norris Jay
Goddard Space Flight Center, NASA
Mail Code 660, Greenbelt, MD
20771, USA
jnorris@lheapop.gsfc.nasa.gov

Nousek John
Pennsylvania State Univ.
University Park, PA
16802 USA
jnousek@astro.psu.edu

Olsen Lisbeth
Astron. Obs., Univ. of Copenhagen
Juliane Maries Vej 30
DK-2100 Copenhagen Ø, Denmark
lisbeth@astro.ku.dk

Orio Marina
Oss. Astron. di Torino
Strada Osservatorio, 20
I-10025 Pino Torinese, Italy
orio@to.astro.it

Otwinowski Stanislaw
Dept. of Phys. and Astron., UCLA
Box 951457 Los Angeles, CA
90095 USA
s.otwinowski@cern.ch

Ougolnikov Oleg
Lebedev Physical Institute
Profsoyuznaya 84/32
117810 Moscow, Russia
ugol@tanatos.asc.rssi.ru

Paciesas William
University of Alabama
Huntsville AL
35899 USA
William.Paciesas@msfc.nasa.gov

Palazzi Eliana
Istituto TESRE / CNR
Via Gobetti, 101
I-40126 Bologna, Italy
eliana@tesre.bo.cnr.it

Pálek Jiři
ASÚ, AVČR
CZ-25165 Ondřejov
Czech Republic
rhudec@asu.cas.cz

Pallavicini Marco
INFN, Sez. di Genova
Via Dodecaneso, 33
I-16146 Genoa, Italy
pallavicini@ge.infn.it

Pal'shin Valentin
Ioffe Physico-Technical Inst.
Polytekhnicheskaya 26
194021 St. Petersburg, Russia
val@pop.ioffe.rssi.ru

Páta Petr
Technical University in Prague
Technická 2, Prague
CZ-16627, Czech Republic
pata@fel.cvut.cz

Pazzi Roberto
Dip. Fisica, Univ. of Pisa
Via Livornese, 1291/a
I-56010, S. Piero a Grado, Pisa, Italy
roberto.pazzi@pi.infn.it

Pedersen Holger
Astron. Obs., Univ. of Copenhagen
Juliane Maries Vej 30
DK-2100 Copenhagen Ø, Denmark
holger@astro.ku.dk

Pedichini Fernando
Osservatorio Astronomico di Roma
Via Frascati, 33
I-00040 Monteporzio, Italy
pedik@coma.mporzio.astro.it

Pellizzoni Alberto
IFCTR / CNR
Via Bassini, 15
I-20133 Milan, Italy
alberto@ifctr.mi.cnr.it

Pendleton Geoffrey
University of Alabama
Huntsville AL
35899 USA
geoff.pendleton@msfc.nasa.gov

Pergolesi Daniele
INFN, Sez. di Genova
Via Dodecaneso, 33
I-16146 Genoa, Italy
pergoles@ge.infn.it

Perna Rosalba
Harvard-Smithsonian CfA
60 Garden Str., Cambridge, MA
01238, USA
rperna@cfa.harvard.edu

Perola Cesare
Universitá "Roma Tre"
Via della Vasca Navale, 84
I-00146 Rome, Italy
perola@fis.uniroma3.it

Perotti Francesco
IFCTR / CNR
Via Bassini, 15
I-20133 Milan, Italy
perotti@ifctr.mi.cnr.it

Pian Elena
Oss. Astron. di Trieste
Via Tiepolo, 11
34131 Trieste, Italy
pian@ts.astro.it

Piccioni Adalberto
Dip. Astronomia, Univ. Bologna
Via Ranzani, 1
I-40129 Bologna, Italy
piccioni@ermione.bo.astro.it

Picozza Pier Giorgio
Università di Roma "Tor Vergata"
Via della Ricerca Scientifica, 1
I-00133 Rome, Italy
picozza@roma2.infn.it

Pina Ladislav
Technical University in Prague
V Holešovičkách 2
CZ-18000 Prague 8, Czech Republic
pina@troja.fjfi.cvut.cz

Piran Tsvi
Racah Institute of Physics
Hebrew University
91904 Jerusalem, Israel
tsvi@nikki.phys.huji.ac.il

Piro Luigi
IAS / CNR
Via Fosso del Cavaliere, 100
I-00133 Rome, Italy
piro@ias.rm.cnr.it

Pittori Carlotta
Università di Roma "Tor Vergata"
Via della Ricerca Scientifica, 1
I-00133 Rome, Italy
Carlotta.Pittori@roma2.infn.it

Pizzichini Graziella
Istituto TESRE / CNR
Via Gobetti, 101
I-40126 Bologna, Italy
graziell@tesre.bo.cnr.it

Polašek Cyril
ASÚ, AVČR
CZ-25165 Ondřejov
Czech Republic
polasek@asu.cas.cz

Polcar Jiri
Faculty of Science, Masaryk Univ.
Kotlarska 2
CZ-61137 Brno, Czech Republic
polcar@monoceros.physics.muni.cz

Pontoni Cristian
INFN, Sez. di Trieste
Padriciano, 99
I-34012 Trieste, Italy
c.pontoni@trieste.infn.it

Porter Frederick Scott
Goddard Space Flight Center, NASA
Mail Code 660, Greenbelt, MD
20771, USA
Frederick.S.Porter.1@gsfc.nasa.gov

Postnov Konstantin
Sternberg Astronomical Institute
Universitetsky Pr., 13
119899 Moscow, Russia
pk@sai.msu.ru

Pozanenko Aleksej
Space Research Institute, RAS
Profsojuznaya 84/32
117810 Moscow, Russia
apozanen@vm1.iki.rssi.ru

Preece Robert
Nat. Space Sci. and Technol. Center
320 Sparkman Drive, Huntsville, AL
35805, USA
robert.preece@msfc.nasa.gov

Preger Barbara
Los Alamos National Laboratory
Los Alamos, NM
87505, USA
bpreger@nis.lanl.gov

Prest Michela
INFN, Sez. di Trieste
Padriciano, 99
I-34012 Trieste, Italy
michela.prest@ts.infn.it

Price Paul
RSAA, ANU
Canberra
ACT 2601, Australia
pap@mso.anu.edu.au

Quilligan Fergus
University College Dublin
Stillorgan Road
Dublin 4, Ireland
fquillig@bermuda.ucd.ie

Ramirez-Ruiz Enrico
Institute of Astronomy
Madingley Road, Cambridge
CB3 0HA, United Kingdom
enrico@ast.cam.ac.uk

Rapisarda Massimo
ENEA, Sez. di Roma
Via Enrico Fermi, 45
I-00044 Frascati, Italy
rapisarda@frascati.enea.it

Razeti Marco
INFN, Sez. di Genova
Via Dodecaneso, 33
I-16146 Genoa, Italy
razeti@ge.infn.it

Reichart Daniel
California Institute of Technology
Pasadena, CA
91125, USA
der@astro.caltech.edu

Remillard Ronald
Massachusetts Institute of Technology
Cambridge, MA
02139 USA
rr@space.mit.edu

Rees Martin
Institute of Astronomy
Madingley Road, Cambridge
CB3 0HA, United Kingdom
mjr@ast.cam.ac.uk

Rodonó Marcello
Oss. Astrofisico di Catania
Via S. Sofia, 78
I-95123 Catania, Italy
mrodono@alpha4.ct.astro.it

Roiger Richard
Minnesota State University
Mankato, MN
56001 USA
Richard.Roiger@mnsu.edu

Roming Peter
Pennsylvania State Univ.
University Park, PA
16802 USA
roming@astro.psu.edu

Ronga Francesco
INFN
Laboratori Nazionali di Frascati
I-00044 Frascati, Italy
ronga@lnf.infn.it

Rossi Elio
Istituto TESRE / CNR
Via Gobetti, 101
I-40126 Bologna, Italy
rossi@tesre.bo.cnr.it

Rubini Alda
IAS / CNR
Via Fosso del Cavaliere, 100
I-00133 Rome, Italy
alda@ias.rm.cnr.it

Ryan James
University of New Hampshire
Durham, NH
03824, USA
jryan@unh.edu

Ryde Felix
Stanford University
Stanford, CA
94305, USA
felix@ahoor.stanford.edu

Sanders Wilt
University of Wisconsin
1150 University Ave., Madison, WI
53706 USA
sanders@wisp.physics.wisc.edu

Sanin Anton
Space Research Institute, RAS
Profsojuznaya 84/32
117810 Moscow, Russia
sanin@mx.iki.rssi.ru

Sardone Stefano
Oss. Astrofisico di Catania
Via S. Sofia, 78
I-95123 Catania, Italy
ssardone@alpha4.ct.astro.it

Sari Re'em
California Institute of Technology
Pasadena, CA
91125, USA
sari@tapir.caltech.edu

Scargle Jeffrey
NASA Ames Research Center
Moffett Field, CA
94035, USA
jeffrey@cosmic.arc.nasa.gov

Schönfelder Volker
MPI für Extraterr. Physik
Postfach 1312
D-85741 Garching, Germany
vos@mpe.mpg.de

Šimon Vojtěch
ASÚ, AVČR
CZ-25165 Ondřejov
Czech Republic
simon@asu.cas.cz

Smida Radomir
ASÚ, AVČR
CZ-25165 Ondřejov
Czech Republic
radomir_smida@yahoo.com

Smith Donald
Massachusetts Institute of Technology
Cambridge, MA
02139 USA
dasmith@space.mit.edu

Smith Ian
Rice University
Houston, TX
77005 USA
ian@spacsun.rice.edu

Soderbergh Alicia
DAMTP
Silver Street, Cambridge
CB3 9EW, United Kingdom
ams76@hermes.cam.ac.uk

Soffitta Paolo
IAS / CNR
Via Fosso del Cavaliere, 100
I-00133 Rome, Italy
soffitta@ias.rm.cnr.it

Sokolov Vladimir
SAO / RAS
Nizhnij Arkhyz
369167 Russia
sokolov@sao.ru

Soldán Jan
ASÚ, AVČR
CZ-25165 Ondřejov
Czech Republic
jsoldan@asu.cas.cz

Spada Maddalena
Oss. Astrofisico di Arcetri
Largo E. Fermi, 5
I-50125 Florence, Italy
spada@arcetri.astro.it

Speziali Roberto
Osservatorio Astronomico di Roma
Via Frascati, 33
I-00040 Monteporzio, Italy
speziali@coma.mporzio.astro.it

Stacy Gregory
Louisiana State University
Baton Rouge, LA
70803, USA
gstacy@phunds.phys.lsu.edu

Stecklum Bringfried
Thüringer Landessternw. Tautenburg
Sternwarte 5
D-07778 Tautenburg, Germany
stecklum@tls-tautenburg.de

Stella Luigi
Osservatorio Astronomico di Roma
Via Frascati 33
I-00040 Monteporzio, Italy
stella@ulysses.mporzio.astro.it

Stern Boris
Institute for Nuclear Research
Russian Academy of Sciences
117312 Moscow, Russia
stern@lukash.asc.rssi.ru

Stoklasova Ivana
ASÚ, AVČR
CZ-25165 Ondřejov
Czech Republic
ivana@sirrah.troja.mff.cuni.cz

Stratta Giulia
IAS / CNR
Via Fosso del Cavaliere, 100
I-00133 Rome, Italy
stratta@ias.rm.cnr.it

Suen Wai-Mo
Washington University
St. Louis, MO
63130, USA
wms@wugrav.wustl.edu

Sukhanov Alexander
Space Research Institute, RAS
Profsojuznaya 84/32
117810 Moscow, Russia
asuhanov@vm1.iki.rssi.ru

Svensson Roland
Stockholm Observatory
Saltsjöbaden
SE-133 36, Sweden
svensson@astro.su.se

Szymkowiak Andrew
Goddard Space Flight Center, NASA
Mail Code 660, Greenbelt, MD
20771, USA
andrew.szymkowiak@gsfc.nasa.gov

Takeshima Toshiaki
Goddard Space Flight Center, NASA
Mail Code 660, Greenbelt, MD
20771 USA
takeshim@ginpo.gsfc.nasa.gov

Tagliaferri Gianpiero
Osservatorio Astronomico di Brera
Via Bianchi, 46
I-23807 Merate, Italy
gtagliaf@merate.mi.astro.it

Tanvir Nial
Dept. of Physical Sciences
Univ. of Hertfordshire, Hatfield
AL10 9AB, United Kingdom
nrt@star.herts.ac.uk

Tavani Marco
IFCTR / CNR
Via Bassini, 15
I-20133 Milan, Italy
tavani@ifctr.mi.cnr.it

Testera Gemma
INFN, Sez. di Genova
Via Dodecaneso, 33
I-16146 Genoa, Italy
testera@ge.infn.it

Thompson Christopher
CITA
60 St. George St., Toronto
M5S 3H8, Canada
thompson@cita.utoronto.ca

Thomsen Bjarne
Institute of Physics and Astronomy
University of Århus
DK-8000 Århus C, Denmark
bt@ifa.au.dk

Tikhomirova Yana
Lebedev Physical Institute
Profsoyuznaya 84/32
117810 Moscow, Russia
jana@anubis.asc.rssi.ru

Torres Riera Josè
DCE-INTA
28850 Torrejón de Ardoz
Madrid, Spain
torresrj@inta.es

Tout Cristopher
Institute of Astronomy
Madingley Road, Cambridge
CB3 0HA, United Kingdom
cat@ast.cam.ac.uk

Trifoglio Massimo
Istituto TESRE / CNR
Via Gobetti, 101
I-40126 Bologna, Italy
trifoglio@tesre.bo.cnr.it

Trombka Jacob
Goddard Space Flight Center, NASA
Mail Code 660, Greenbelt, MD
20771, USA
Jacob.I.Trombka.1@gsfc.nasa.gov

Trussoni Edoardo
Oss. Astron. di Torino
Strada Osservatorio, 20
I-10025 Pino Torinese, Italy
trussoni@to.astro.it

Vaccarone Renzo
INFN, Sez. di Genova
Via Dodecaneso, 33
I-16146 Genoa, Italy
vaccarone@ge.infn.it

Vallazza Erik
INFN, Sez. di Trieste
Padriciano, 99
I-34012 Trieste, Italy
erik.vallazza@ts.infn.it

Vaughan Gavin
ESRF
BP 220, Grenoble Cedex
F-38043 France
vaughan@esrf.fr

Vavrek Roland
Konkoly Observatory
Box 67
H-1505 Budapest, Hungary
vavrek@konkoly.hu

Vercellone Stefano
IFCTR / CNR
Via Bassini, 15
I-20133 Milan, Italy
stefano@ifctr.mi.cnr.it

Vernetto Silvia
Istituto di Cosmogeofisica, CNR
Corso Fiume, 4
I-10133 Turin, Italy
vernetto@to.infn.it

Vietri Mario
Universitá "Roma Tre"
Via della Vasca Navale, 84
I-00146 Rome, Italy
vietri@fis.uniroma3.it

Vitali Fabrizio
Osservatorio Astronomico di Roma
Via Frascati 33
I-00040 Monteporzio, Italy
vitali@coma.mporzio.astro.it

von Kienlin Andreas
MPI für Extraterr. Physik
Postfach 1312
D-85741 Garching, Germany
azk@mpe.mpg.de

Watson Darach
University College Dublin
Stillorgan Road
Dublin 4, Ireland
dwatson@bermuda.ucd.ie

Waxman Eli
Weizmann Institute of Science
Rehovot 76100
Israel
waxman@wicc.weizmann.ac.il

Weinberg Nevin
Univ. of Chicago
Chicago, IL
60637 USA
nweinber@midway.uchicago.edu

Wenzel Wolfgang
Sternwarte Sonneberg
Sternwartestr. 32
D-96515 Sonneberg, Germany
office@stw.tu-ilmenau.de

Wijers Ralph
Dept. of Phys. and Astron., SUNY
Stony Brook, NY
11794, USA
rwijers@mail.astro.sunysb.edu

Williams Rees
Space Science Department of ESA
P.O. Box 299
NL-2200 Noordwijk, The Netherlands
owilliam@astro.estec.esa.nl

Winkler Chris
Space Science Department of ESA
P.O. Box 299
NL-2200 Noordwijk, The Netherlands
cwinkler@astro.estec.esa.nl

Woods Peter
Nat. Space Sci. and Technol. Center
320 Sparkman Drive, Huntsville, AL
35805, USA
peter.woods@msfc.nasa.gov

Woosley Stanford
University of California, Santa Cruz
1156 High Street, Santa Cruz, CA
95064 USA
woosley@ucolick.org

Yonetoku Daisuke
ISAS
3-1-1, Yoshinodai, Sagamihara
229-8510 Kanagawa, Japan
yonetoku@astro.isas.ac.jp

Yoshida Atsumasa
Inst. of Phys. and Chem. Research
2-1 Hirosawa Wako Saitama
351-0198 Japan
ayoshida@crab.riken.go.jp

Yost Sarah
California Institute of Technology
Pasadena, CA
91125, USA
yost@srl.caltech.edu

Zane Silvia
MSSL, UCL
Holmbury St. Mary
Dorking, Surrey, RH5 6NT, UK
sz@mssl.ucl.ac.uk

Zanello Dino
Univ. di Roma "La Sapienza"
Piazzale Aldo Moro, 5
I-00185, Rome, Italy
dino.zanello@roma1.infn.it

Zerbi Filippo
Osservatorio Astronomico di Brera
Via Bianchi, 46
I-23807 Merate, Italy
zerbi@merate.mi.astro.it

Zhang Bing
Pennsylvania State Univ.
University Park, PA
16802 USA
bzhang@astro.psu.edu